AF869660

# Springer Transactions in Civil and Environmental Engineering

**Editor-in-Chief**

T. G. Sitharam, Indian Institute of Technology Guwahati, Guwahati, Assam, India

Springer Transactions in Civil and Environmental Engineering (STICEE) publishes the latest developments in Civil and Environmental Engineering. The intent is to cover all the main branches of Civil and Environmental Engineering, both theoretical and applied, including, but not limited to: Structural Mechanics, Steel Structures, Concrete Structures, Reinforced Cement Concrete, Civil Engineering Materials, Soil Mechanics, Ground Improvement, Geotechnical Engineering, Foundation Engineering, Earthquake Engineering, Structural Health and Monitoring, Water Resources Engineering, Engineering Hydrology, Solid Waste Engineering, Environmental Engineering, Wastewater Management, Transportation Engineering, Sustainable Civil Infrastructure, Fluid Mechanics, Pavement Engineering, Soil Dynamics, Rock Mechanics, Timber Engineering, Hazardous Waste Disposal Instrumentation and Monitoring, Construction Management, Civil Engineering Construction, Surveying and GIS Strength of Materials (Mechanics of Materials), Environmental Geotechnics, Concrete Engineering, Timber Structures.

Within the scopes of the series are monographs, professional books, graduate and undergraduate textbooks, edited volumes and handbooks devoted to the above subject areas.

More information about this series at http://www.springer.com/series/13593

Ravinesh C. Deo · Pijush Samui ·
Ozgur Kisi · Zaher Mundher Yaseen
Editors

# Intelligent Data Analytics for Decision-Support Systems in Hazard Mitigation

## Theory and Practice of Hazard Mitigation

Springer

*Editors*
Ravinesh C. Deo
School of Sciences
University of Southern Queensland
Springfield Central, QLD, Australia

Pijush Samui
Department of Civil Engineering
National Institute of Technology Patna
Patna, Bihar, India

Ozgur Kisi
Department of Civil Engineering
Ilia State University
Tbilisi, Georgia

Zaher Mundher Yaseen
Faculty of Civil Engineering
Ton Duc Thang University
Ho Chi Minh City, Vietnam

ISSN 2363-7633 ISSN 2363-7641 (electronic)
Springer Transactions in Civil and Environmental Engineering
ISBN 978-981-15-5771-2 ISBN 978-981-15-5772-9 (eBook)
https://doi.org/10.1007/978-981-15-5772-9

© Springer Nature Singapore Pte Ltd. 2021
This work is subject to copyright. All rights are reserved by the Publisher, whether the whole or part of the material is concerned, specifically the rights of translation, reprinting, reuse of illustrations, recitation, broadcasting, reproduction on microfilms or in any other physical way, and transmission or information storage and retrieval, electronic adaptation, computer software, or by similar or dissimilar methodology now known or hereafter developed.
The use of general descriptive names, registered names, trademarks, service marks, etc. in this publication does not imply, even in the absence of a specific statement, that such names are exempt from the relevant protective laws and regulations and therefore free for general use.
The publisher, the authors and the editors are safe to assume that the advice and information in this book are believed to be true and accurate at the date of publication. Neither the publisher nor the authors or the editors give a warranty, expressed or implied, with respect to the material contained herein or for any errors or omissions that may have been made. The publisher remains neutral with regard to jurisdictional claims in published maps and institutional affiliations.

This Springer imprint is published by the registered company Springer Nature Singapore Pte Ltd.
The registered company address is: 152 Beach Road, #21-01/04 Gateway East, Singapore 189721, Singapore

*This work is dedicated to my father Mr. Bisun Deo (1944–2010) of Labasa, Fiji Islands*

*A farmer, a community worker and a source of inspiration for the leading Editor*

*—Ravinesh C. Deo*

# Foreword

Global environmental change driven by hotter temperatures, uneven distribution of precipitation, and elevated greenhouse gases bring significant challenges to humanity as a whole—with rising incidences and severity of extreme weather events, including drought, heatwaves, floods, tsunamis, tidal waves, earthquakes and other forms of natural disasters.

The socio economic development and human well-being depends on how people draw on and manage the natural resources available to them with the current phase of a highly fragile and changing climate system. The mitigation of disaster impacts is a way to save humanity, their survival or well-being. Disaster risk reduction through mitigation and adaptation is an integral part of social and economic development, and is essential if development is to be sustainable for the future.

This edited book is a consolidated account of the theory and practice of disaster mitigation techniques using artificial intelligence and data analytics methods.

The work presented therein supports the initiatives of United Nations sustainable Development Goal Target 11.5: "By 2030, significantly reduce the number of deaths and the number of people affected and substantially decrease the direct economic losses relative to global gross domestic product caused by disasters, including water-related disasters, with a focus on protecting the poor and people in vulnerable situations".

The studies presented in this edited book, through expert writers of various chapters, are aimed at developing measures for adopting and implementing integrated policies and plans towards mitigation and adaptation to climate change, resilience to disasters, and developing and implementing in line with the Sendai Framework for Disaster Risk Reduction 2015–2030, a science-based disaster risk management.

The chapters presented provide very relevant methodologies used by researchers and policy makers that data science offers to support positive economic, social and environmental decisions and strengthening national and regional development planning for disaster risk mitigation.

It is my hope that this book will serve a wide range of audience from graduate students, novice researchers, academics, and people working in areas of meteorology, hydrology, climate change and policy development.

April 2020

Dr. Ashok K. Mishra
Dean's Professor in Civil Engineering
Clemson University
Clemson, SC, USA

# Preface

Data describes many concealed facets of humanity, environment and universe. Intelligent analytics (big data techniques) explore such data to unveil real truths about patterns and laws of nature. Natural hazards can be investigated and their effect mitigated through big data technique. Hazards trigger catastrophic social and hydrological imbalance, exacerbate climate extremes such as bushfires, drought and heatwaves and lead to aberration in water supply and its quality. They also hinder domestic, industrial and agricultural needs that humans require for survival. Hazards cause detrimental impact on urban and rural infrastructures, flora, fauna and biodiversity, bringing health and economic consequences to developing and the first world nations through their impact on food and water security.

The book describes some of the latest artificial intelligence machine learning and intelligent analytics used to design disaster mitigation systems. Such systems advance disaster policy and practice and provide end-users prior knowledge on hazards, their monitoring and forecasting in a natural environment. These techniques use an array of data, on-site measurements, satellite geographic positioning systems and reanalysis products. Intelligent analytics can thus be used in forewarning and monitoring of extremes that support practical decision systems.

The book will increase a reader's curiosity, offering unending knowledge and opportunities to learn more about artificial intelligence on which intelligent analytic methods are based. It will help readers to successfully learn how to develop methods that monitor, analyse and forecast hydro-meteorological variables. The tools can be practically implemented in construction of models for hazard risk-mitigation designed as a framework for in expert predictive systems. Generally, predictive modelling is a consolidated discipline used to forewarn possibility of a hazard. By fostering strategic decisions, expert system models can be a cost-effective way to forewarn abnormal events with overarching aim to develop and improve decision-support system in disaster mitigation.

The book has welcomed contributions from diverse authors including original research, reviews and discussions and debates in light of latest intelligent analytics. The chapters are written by experts, dwelling on applications of primitive and modern day soft computing strategies for disaster forecasting, ideas on decision-support systems for natural hazard mitigation. It illustrates data-intelligent approaches that can advance knowledge in hydro-meteorological sciences. The works augments machine learning with pre-processing algorithms that can enhance decision making, understanding and predictions, and explore interrelationships between hydro-metrological and natural hazard mitigation systems via data-intelligent approach. Extensions, applications and case studies advancing evolutionary computing in hazard risk mitigating have also been welcomed.

The book provides description of relevant theory and practical applications. It has published some of the finest cutting-edge application. Chapters are drawn from a consortium of experts in mathematics, computing, weather forecasting, meteorology, hydrology, engineering, agriculture, economics, environmental science, disaster management and policy-makers and climate advocacies.

It is the our hope that all readers will benefit significantly in learning about the state-of-the-art machine learning models, decision support systems, including disaster management science and policy perspectives.

Happy reading and learning!

Springfield Central, Australia
January 2020

Dr. Ravinesh C. Deo

# Contents

# Editors and Contributors

## About the Editors

**Dr. Ravinesh C. Deo** is currently an Associate Professor, and Associate Editor for *Stochastic Environmental Research & Risk Assessment*, *ASCE Journal Hydrologic Engineering* journal (USA), Director for Postgraduate Data Science Programs and a Research Leader in Artificial Intelligence at the University of Southern Queensland, Australia. As Applied Scientist with proven leadership in artificial intelligence, his research aims to develop decision-systems with machine learning, heuristic and metaheuristic algorithms to improve real-life predictive systems, especially using deep learning, convolutional neural network and long- short-term memory network. He was awarded many internationally competitive fellowships including Queensland Government U.S. Smithsonian Fellowship, Australia-India Strategic Fellowship, Australia-China Young Scientist Exchange Award, Japan Society for Promotion of Science Fellowship, Chinese Academy of Science Presidential International Fellowship and the Endeavour Executive Fellowship. He is a member of numerous scientific bodies and has won multiple Publication Excellence Awards in both School and University Categories, Head of Department Research Award, Dean's Commendation for Postgraduate Research Supervision, BSc Gold Medal for Academic Excellence and he was also the Dux of Fiji in Year 13 examinations. Professor Deo held visiting positions at the United Stations Tropical Research Institute (Panama), Chinese Academy of Sciences, Peking University, Northwest Normal University, University of Tokyo, Kyoto and Kyushu University, University of Alcala Spain, McGill University and National University of Singapore. He has undertaken knowledge exchange research programs in Singapore, Japan, Europe, China, USA and Canada and also secured international standing by researching innovative problems with global researchers. He has published several Books with Springer Nature, Elsevier and IGI Global and achieved more than 185 publications, more than 140 *Q*1 journals, including refereed conferences, Edited Books and book chapters. Professor Deo's research papers have been cited over 4,000 times that provides a Google Scholar H-Index as 36 and a Field Weighted Citation Index exceeding 3.5.

**Dr. Pijush Samui** is currently an Associate Professor at National Institute of Technology, Patna, India. He is an established researcher in the application of Artificial Intelligence (AI) for solving different problems in engineering. He developed a new method for prediction of response of soil during an earthquake. He has given charts for prediction of response of soil during an earthquake and developed equations for prediction of lateral spreading of soil due to earthquake. He developed equations for determination of seismic liquefaction potential of soil based on strain energy and prediction of seismic attenuation. He developed efficient models for prediction of magnitude of reservoir induced earthquake. He has developed models for determination of medical waste generation in hospitals with equations used for practical purpose. The developed models can be used for clean India project. He determined frequency effects on liquefaction by using Shake Table. He has applied AI techniques for determination of bearing capacity and settlement of foundation and equations for determination of bearing capacity and settlement of shallow foundation. He also developed equations for determination of compression index and angle of shearing resistance of soil. he developed equations for prediction of uplift capacity of suction caisson. He also developed equations for determination of fracture parameters of concrete. His active research activity is evident from his extensive citation of publications in google scholar (total frequency of 1280) with H-Index of 22. Dr Samui has published journal papers, books/book chapters and peer reviewed conference papers that involved co-authors from Australia, India, Korea and several other nations. He is also holding the position of Visiting Professor at Far East Federal University (Russia).

**Prof. Ozgur Kisi** is currently a Professor of Engineering at the Ilia Sate University, Georgia. He received his B.Sc. in Civil Engineering from the Cukurova University, Turkey (1997), his M.Sc. in Hydraulics from the Erciyes University, Turkey (1999), and his Ph.D. from Istanbul Technical University, Turkey (2003). His research fields include developing novel algorithms and methods towards the innovative solution of hydrologic forecasting and modeling, suspended sediment modeling, forecasting and estimating hydrological variables such as precipitation, streamflow, evaporation, evapotranspiration, groundwater, lake level, hydroinformatics and trend analysis. He is an active participant in numerous national research projects and supervisor of several MSc and PhD works. He is serving as an Editorial Board Member of several reputed journals (e.g. Journal of Hydrologic Engineering (ASCE), Arabian Journal of Geosciences, ISH Journal of Hydraulic Engineering, Irrigation & Drainage Systems Engineering and Austin Journal of Irrigation). He has also served as a Guest Editor for Special Issues published in Hydrological Hazards in a Changing Environment-Early Warning, Forecasting, and Impact Assessment (Advances in Meteorology) in 2015. He is also serving as a

Reviewer for more than 80 journals indexed in Science Citation Index (SCI) in the fields of hydrology, irrigation, water resources and hydro-informatics (e.g. ASCE Journal of Hydrologic Engineering, ASCE Journal of Irrigation and Drainage Engineering, Water Resources Research, Journal of Hydrology, Hydrological Processes, Hydrology Research, Water Resources Management, Hydrological Sciences Journal, Journal of Hydroinformatics, Water Science and Technology). He has authored more than 200 research articles, 2 chapters and 24 discussions. He is the recipient of the 2006 International Tison Award, given by the International Association of Hydrological Sciences (IAHS). He is a member of Turkish Academy of Science (selected in 2012).

**Dr. Zaher Mundher Yaseen** is a Senior Lecturer in Civil Engineering at Ton Duc Thang University, Vietnam with an expertise in machine learning, advanced data analytics and environmental sciences. He has expertise in Hydrology, Water Resources Engineering, Hydro-logical processes modeling, Environmental Engineering and Civil Engineering. He has recently published in top hydrology and water resources journals such as Journal of Hydrology, Water Resources Management and Stochastic Environmental research and Risk Assessment.

## Contributors

**Rana Muhammad Adnan** State Key Laboratory of Hydrology-Water Resources and Hydraulic Engineering, Hohai University, Nanjing, China

**Oluwole Agbede** Department of Civil Engineering, University of Ibadan, Ibadan, Nigeria

**Oluwatobi Aiyelokun** Department of Civil Engineering, University of Ibadan, Ibadan, Nigeria

**Meysam Alizamir** Department of Civil Engineering, Hamedan Branch, Islamic Azad University, Hamedan, Iran

**Armando A. Apan** School of Civil Engineering, University of Southern Queensland, Toowoomba, QLD, Australia

**Mohammad Danish Azim** Department of Civil Engineering, National Institute of Technology Patna, Patna, India

**Bhuwan Chand** School of Sciences, University of Southern Queensland, Toowoomba, QLD, Australia

**Dhrishna Charan** Department of Science, School of Science and Technology, The University of Fiji, Saweni, Lautoka, Fiji

**Eun-Sung Chung** Department of Civil Engineering, Seoul National University of Science and Technology, Seoul, Republic of Korea

**Kavina S. Dayal** Commonwealth Scientific and Industrial Research Organisation (CSIRO), Hobart, TAS, Australia

**Ravinesh C. Deo** School of Sciences, University of Southern Queensland, Springfield Central, QLD, Australia

**Mohammad Ali Ghorbani** Department of Water Engineering, Faculty of Agriculture, University of Tabriz, Tabriz, Iran;
Sustainable Management of Natural Resources and Environment Research Group, Faculty of Environment and Labour Safety, Ton Duc Thang University, Ho Chi Minh City, Vietnam

**Salim Heddam** University 20 Août 1955 Skikda, Faculty of Science, Agronomy Department, Hydraulics Division, Laboratory of Research in Biodiversity Interaction Ecosystem and Biotechnology, Skikda, Algeria;
Agronomy Department, Hydraulics Division, Laboratory of Research in Biodiversity Interaction Ecosystem and Biotechnology, University 20 Août 1955 Skikda, Skikda, Algeria

**Dar Himayoun** Department of Civil Engineering, National Institute of Technology Patna, Patna, India

**Abolfazl Jaafari** Research Institute of Forests and Rangelands, Agricultural Research, Education and Extension Organization (AREEO), Tehran, Iran

**Lionel Joseph** Department of Science, School of Science and Technology, The University of Fiji, Saweni, Lautoka, Fiji

**Ercan Kahya** Department of Civil Engineering, Istanbul Technical University, Istanbul, Turkey

**Vahid Karimi** Department of Water Engineering, Faculty of Agriculture, University of Tabriz, Tabriz, Iran

**Sungwon Kim** Department of Railroad Construction and Safety Engineering, Dongyang University, Yeongju, Republic of Korea

**Ozgur Kisi** Department of Civil Engineering, Ilia State University, Tbilisi, Georgia;
School of Technology, Ilia State University, Tbilisi, Georgia

**Aiding Kornejady** Department of Watershed Management, Gorgan University of Agricultural Sciences and Natural Resources, Gorgan, Iran

**Anil Kumar** Department of Soil and Water Conservation Engineering, College of Technology, G.B. Pant University of Agriculture & Technology, Pantnagar, Uttarakhand, India

**Alban Kuriqi** CERIS, Instituto Superior Técnico, Universidade de Lisboa, Lisbon, Portugal

**Anurag Malik** Department of Soil and Water Conservation Engineering, College of Technology, G.B. Pant, University of Agriculture and Technology, Pantnagar, Uttarakhand, India;
Punjab Agricultural University, Regional Research Station, Bathinda, Punjab, India

**Shahbaz Mushtaq** Centre for Applied Climate Sciences, University of Southern Queensland, Toowoomba, Australia

**M. Z. Naser** Glenn Department of Civil Engineering, Clemson University, Clemson, SC, USA

**Thong Nguyen-Huy** Centre for Applied Climate Sciences, University of Southern Queensland, Toowoomba, QLD, Australia;
Vietnam National Space Center, Vietnam Academy of Science and Technology, Hanoi, Vietnam

**Gbenga Ogunsanwo** Department of Computer and Information Science, Tai Solarin University of Education, Ijebu-Ode, Ogun, Nigeria

**Akintunde Ojelabi** Department of Civil Engineering, University of Ibadan, Ibadan, Nigeria

**Kulwinder Singh Parmar** Department of Mathematics, IKG Punjab Technical University, Jalandhar, Kapurthala, India

**Ramendra Prasad** Department of Science, School of Science and Technology, The University of Fiji, Saweni, Lautoka, Fiji

**Omid Rahmati** Soil Conservation and Watershed Management Research Department, Kurdistan Agricultural and Natural Resources Research and Education Center, AREEO, Sanandaj, Iran

**Thendiyath Roshni** Department of Civil Engineering, National Institute of Technology Patna, Patna, India

**Sinan Q. Salih** Computer Science Department, College of Computer Science and Information Technology, University of Anbar, Ramadi, Iraq

**Farzin Salmasi** Department of Water Engineering, Faculty of Agriculture, University of Tabriz, Tabriz, Iran

**Shamsuddin Shahid** Department of Water and Environmental Engineering, School of Civil Engineering, Faculty of Engineering, Universiti Teknologi Malaysia (UTM), Skudai, Johor Bahru, Malaysia

**Jalal Shiri** Water Engineering Department, Faculty of Agriculture, University of Tabriz, Tabriz, Iran

**Mohammed Sanusi Shiru** Department of Civil Engineering, Seoul National University of Science and Technology, Seoul, Republic of Korea;
Department of Environmental Sciences, Faculty of Science, Federal University Dutse, Dutse, Nigeria

**Sanjay Singh** Department of Science, School of Science and Technology, The University of Fiji, Saweni, Lautoka, Fiji

**Sarbjit Singh** Guru Nanak Dev University College, Narot Jaimal Singh, Pathankot, Punjab, India;
Guru Nanak Dev University, Amritsar, Punjab, India

**Kirti Soni** CSIR-National Physical Laboratory, New Delhi, India

**Bin Wang** Department of Atmospheric Science, School of Ocean and Earth Science and Technology, University of Hawaii at Manoa, Honolulu, HI, USA

**Zaher Mundher Yaseen** Sustainable Developments in Civil Engineering Research Group, Faculty of Civil Engineering, Ton Duc Thang University, Ho Chi Minh City, Vietnam

**Huanting Zhou** School of Civil Engineering and Architecture, Wuhan University of Technology, Wuhan, China

**Mohammad Zounemat-Kermani** Department of Water Engineering, Shahid Bahonar University of Kerman, Kerman, Iran

# Chapter 1
# Drought Index Prediction Using Data Intelligent Analytic Models: A Review

**Zaher Mundher Yaseen and Shamsuddin Shahid**

## Abbreviation

| | |
|---|---|
| ANN | Artificial neural network |
| ARIMA | Autoregressive integrated moving average |
| BPNN | Backpropagation neural network |
| BRT | Boosted regression tree |
| DT | Decision tree |
| DBN | Deep belief network |
| EDI | Effective drought index |
| NINO 3.4 index | El Niño–Southern Oscillation indicator |
| ELM | Extreme learning machine |
| ERT | Extremely randomized trees |
| WFL | Fuzzy-wavelet |
| GRNN | Generalized regression neural network |
| HANN | Hybrid artificial neural network |
| LSSVR | Least square support vector machine |
| MC | Markov chain |
| MPMR | Minimum probability machine regression |
| MLP | Multilayer perceptron |
| MLP | Multilayer perceptron neural network |
| MLR | Multiple linear regression |

Z. M. Yaseen (✉)
Sustainable Developments in Civil Engineering Research Group, Faculty of Civil Engineering, Ton Duc Thang University, Ho Chi Minh City, Vietnam
e-mail: yaseen@tdtu.edu.vn

S. Shahid
School of Civil Engineering, Faculty of Engineering, Universiti Teknologi Malaysia (UTM), Skudai, Johor 81310, Malaysia

© Springer Nature Singapore Pte Ltd. 2021

R. C. Deo et al. (eds.), *Intelligent Data Analytics for Decision-Support Systems in Hazard Mitigation*, Springer Transactions in Civil and Environmental Engineering,
https://doi.org/10.1007/978-981-15-5772-9_1

| | |
|---|---|
| MARS | Multivariate adaptive regression splines |
| ANFIS | Neuro-fuzzy inference system |
| PDSI | Palmer Drought Severity Index |
| PMDI | Palmer's modified drought index |
| RBF | Radial basis function |
| RF | Rainfall |
| SSI | Standardized streamflow index |
| SVR | Support vector regression |
| WANN | Wavelet neuro-wavelet |
| WA-ANFIS | Wavelet-adaptive-neuro-fuzzy inference system |
| WA-ANN | Wavelet-ANN |
| WA-ARIMA | Wavelet-ARIMA |
| WLGP | Wavelet-linear genetic programing |

## 1.1 Introduction

One of the natural phenomena that have an extreme influence on the climate is drought (Mishra and Singh 2011, 2010). It has a significant influence on the sustainability of water resources, as well as on agricultural production and the environment (Dai 2011; Samarah 2005). There is no definite way of defining drought because it is not straightforward to determine the exact beginning and duration of a drought event. Drought can slowly build over time, leave a prolonged influence over a large geographical space without necessarily causing any significant infrastructural damage (Choat et al. 2012; Passioura 1996; Reddy et al. 2004). Therefore, forecasting of when a drought is likely to begin or to come to an end is extremely difficult or impossible (Alamgir et al. 2019; Cordery and McCall 2000). Characterisation and forecasting of droughts through drought indices are generally used to support drought monitoring and mitigation (Alamgir et al. 2015).

Numerous indices have been developed to identify the severity of drought conditions (Wilhite and Glantz 1985). For example, *Z*-index (Palmer 1965), Palmer Modified Drought Index (PMDI) (Palmer 1965), Rainfall Anomaly Index (RAI) (Rooy 1965), Quartiles and Deciles (Gibbs 1967), Bhalme and Mooly Drought Index (BMDI) (Bhalme and Mooley 1980), Keetch–Byram Drought Index (KBDI) (Byram and Keetch 1988), Standardized Precipitation Index (SPI) (Mckee et al. 1993), percent of normal (Willeke et al. 1994), Effective Drought Index (EDI) (Byun and Wilhite 1999), Drought Frequency Index (DFI) (González and Valdés 2006), Reconnaissance Drought Index (RDI) (Tsakiris et al. 2007), Resiliency-Reliability-Vulnerability (RRV) Drought Index (Panaou 2018), Standardized Precipitation Evapotranspiration Index (SPEI) (Vicente-Serrano et al. 2010). Precipitation has been widely used as a drought variable in drought indices. However, temperature and evaporation were also used for conducting meteorological drought analysis.

In general, statistical and dynamical models are used for the prediction of drought indices for early warning (Beyaztas and Yaseen 2019; Mishra and Singh 2011; Pozzi et al. 2013; Shahid 2010). Statistical relationships between climate variables or drought index estimated for preceding months and drought index of following months are developed for statistical forecasting of droughts (Fung et al. 2019). In contrast, the physics of climate circulation due to interaction of land–ocean–atmosphere is used for the development of dynamic drought forecasting model. Besides that, a combination of statistical and dynamic models also known as physical-empirical model or hybrid model has been developed for drought forecasting (Strazzo et al. 2019). Statistical models are often preferred among the other types of models considering less complexity and computational requirements.

Linear or nonlinear statistical models are generally used for the development of relationship between climate variable and drought index for the forecasting of droughts. However, drought is a natural hazard having a highly stochastic and nonlinear characteristic. It is often difficult to capture the highly nonlinear relationship of climate variables of preceding months with drought index using conventional statistical methods. Data intelligent analytics or machine learning (ML) methods have an automatic learning capability of relationship between inputs and output from data and therefore found highly efficient in modeling nonlinear and complex relationship (Lantz 2013). Different types of ML models have been developed for simulation of nonlinear hydrological phenomena in last three decades. It has also been used for prediction of drought index (Danandeh Mehr et al. 2018; Fung et al. 2019; Ganguli and Reddy 2014; Morid et al. 2007; Yaseen et al. 2018).

Although there have been couple review researches on the drought indices using the feasibility of ML models. The research area is still in the stage of development where more insightful point of views is required to be discussed and evaluated. In addition, establishing a new survey for the implementation of the ML models can attribute to the base knowledge of climate drought understanding and the potential to overcome the existed drawbacks over the literature. The literature review studies have been summarized in Table 1.1 where indicating the type of the models developed, the investigated region, the employed historical climate data, the number of the studies stations, the utilized input variables, the targeted drought index, and a summary remark of the main finding of the researches.

## 1.2 Literature Evaluation

Drought is characterized by the changes of several hydrological and meteorological processes (e.g., air temperature, precipitation, soil moisture, streamflow, etc.), and thus, it is considered a complex natural hazard. The development of effective model for the measurement of droughts requires an accurate understanding of the actual relationships between such variables. Therefore, the relationship between drought events and the related hydrological and meteorological variables must be understood for better mitigation of the effects of drought in a proactive manner. This also can be

**Table 1.1** The surveyed research works on drought indices using data intelligent analytic models over the last two decades

| Scholars | Region | Models | Data period/no. stations | Input/output variables | Remarks |
|---|---|---|---|---|---|
| (Kim and Valdés 2003) | Conchos River basin, Mexico | ANN, WA-ANN | (1955–2000)/1 | RF/PDSI | Complementary wavelet-ANN model is developed to forecast drought events represented in the form of PDSI. Different lead times including 1, 3, 6, and 12 months are investigated. Results evidenced the potential of the preprocessing approach prior the ANN predictive model |
| (Paulo and Pereira 2007) | Altentejo region, Portugal | Markov chains | 67 years/7 | RF/SPI | Markov chains are utilized to characterize the drought pattern that is allowing to predict SPI up to three months ahead using homogenous and non-homogenous formulations. The results indicated the possibility to predict the drought successfully |

(continued)

**Table 1.1** (continued)

| Scholars | Region | Models | Data period/no. stations | Input/output variables | Remarks |
|---|---|---|---|---|---|
| (Moreira et al. 2008) | Altentejo and Algarve region, Southern Portugal | Loglinear model | 1932–2006/(14) | RF/SPI | SPI was predicted by loglinear model, the results indicate that loglinear prediction of drought class transitions is useful for short-term drought warning over the validation phase (2004–2006). Only a few sites did not perform well. However, two months are mainly the correlated leads times to the drought patter detection |
| (Keskin et al. 2011) | Lakes district, Turkey | ANN | 1964–2006/(5) | RF/SPI | ANN model is applied to predict SPI pattern based on four different data spans including 3, 6, 9, and 12 months. The results showed good agreement between the developed ANN model and the calculated SPI value with 12 months span dataset |

(continued)

**Table 1.1** (continued)

| Scholars | Region | Models | Data period/no. stations | Input/output variables | Remarks |
|---|---|---|---|---|---|
| (Rezaeian-Zadeh and Tabari 2012) | Abadan, Bandar Anzali, Kermanshah, Mashhad, Iran Shahr | MLP | 39–55 years/5 | RF/SPI | Different climate conditions are investigated for SPI forecasting over Iran including arid, very humid, semi-arid and arid. The developed MPL model indicated a good performance |
| (Belayneh and Adamowski 2012) | Awash river basin in Ethiopia | ANN, SVR, WA-ANN | 1970–2005/3 | RF/SPI | SPI is forecast using three different data intelligent analytic models ANN, SVM, WA-ANN. The modeling performed based on different lead time dataset. The results indicated that the coupled WA-ANN as a prior data preprocessing of time series to the predictive model enhance the model predictability |

(continued)

**Table 1.1** (continued)

| Scholars | Region | Models | Data period/no. stations | Input/output variables | Remarks |
|---|---|---|---|---|---|
| (Chen et al. 2012a) | Huaihe river basin, china | BPNN, DBN | 1958–2006/(4) | RF/SPI | DBN is developed as new deep learning data intelligent analytic model for SPI prediction. The feasibility of the developed model validated against classical BPNN. The results showed that DBN has better performance in term of predictability potential |
| (Chen et al. 2012b) | Huaihe river basin, China | RF, ARIMA | 1966–2004/(4) | RF/SPI | Newly data-mining predictive model is investigated for SPI modeling over four stations located over Huaihe river basin in China. ARIMA model is employed for accuracy verification. RF model exhibited good performance with limitation of conservative extreme patterns |

(continued)

**Table 1.1** (continued)

| Scholars | Region | Models | Data period/no. stations | Input/output variables | Remarks |
|---|---|---|---|---|---|
| (Abarghouei et al. 2013) | Ardakan region, Iran | ANN | 1969–2008 (40)/4 | RF/SPI | SPI is computed using classical ANN model. The modeling is constructed based on various spans of dataset including 3, 6, 9, 12, and 24 months in hyper-arid region of Iran. ANN displayed good results for the region |
| (Shirmohammadi et al. 2013) | Azerbaijan province, Iran | ANFIS, ANN, wavelet-ANN, wavelet-ANFIS | 1952–2011/(1) | RF/SPI | The authors integrated wavelet-transform approach with ANFIS as a robust predictive model for solving the uncertainty problem of SPI simulation. The results showed that the wavelet transform can improve the metrological drought modeling. WA-ANFIS also demonstrated good potential in solving the non-stationarity problem |

(continued)

**Table 1.1** (continued)

| Scholars | Region | Models | Data period/no. stations | Input/output variables | Remarks |
|---|---|---|---|---|---|
| (Danandeh Mehr et al. 2014) | State of Texas | W-LGP, WANN, WFL | 1951–2007/(1) | NINO 3.4 index, PMDI/PMDI | The modeling performed on 3-, 6-, and 12-month lead times. The wavelet transformed features applied affectively in parallel with the LGP for draught forecasting |
| (Belayneh et al. 2014) | Awash river basin, Ethiopia | WA-ANN, WA-SVR | 1970–2005/(3) | RF/SPI | The results of the applied complementary predictive models (data preprocessing integrated with AI models) for SPI prediction successfully accomplished. The WA-ANN was found superior to WA-SVR model |
| (Hosseini-Moghari and Araghinejad 2015) | Gorganround basin, Iran | Several ANN algorithms | (1976-2006)/14 | RF/SPI | SPI is computed based on 3-, 6-, 9-, 12-, and 24-month time scales. SPI forecasted reliably using two ANN algorithms including GRNN and RBF |

(continued)

**Table 1.1** (continued)

| Scholars | Region | Models | Data period/no. stations | Input/output variables | Remarks |
|---|---|---|---|---|---|
| (Deo and Şahin 2015a) | Eastern Australia | ELM, ANN | 1957–2011/(4) | RF, Temperature, Southern Oscillation Index, Pacific Decadel Oscillation, Southern Annular Mode, Indian Ocean Dipole moment/EDI | As new version of artificial neural network called extreme learning machine is used. The learning process convergence exhibited an excellent intelligence model for EDI detection over the studied region |
| (Jalalkamali et al. 2015) | Yazd province, Iran | MLP ANN, ANFIS, SVM, ARIMA | 1961–2012/(9) | RF, Temperature/SPI | The authors evidenced an interesting conclusion where ARIMA model could gave best results in prediction of nine-month timescale SPI. A noticeable superiority is observed in comparison with the other AI models |

(continued)

**Table 1.1** (continued)

| Scholars | Region | Models | Data period/no. stations | Input/output variables | Remarks |
|---|---|---|---|---|---|
| (Deo and Şahin 2015b) | Eastern Australia | ANN | 1915–2012/8 | RF, Temperature, Evapotranspiration/SPEI | The authors attempted to establish an intelligence scheme for SPEI using the classical ANN model and a century of climatological data for the Eastern Australia. The finding of the study evidenced the potential of the ANN model |
| (Belayneh et al. 2016) | Awash river basin, Ethiopia | ANN, SVR, WA-ANN, WA-SVR | 1970–2005/3 | RF/SPI | SPI was computed using five AI models including two classical (e.g., ANN, SVR) and complementary (e.g., WA-ANN, WA-SVR). The results showed that WA-ANN gives the best performance for forecasting SPI with 3- and 6-month lead time |
| (Le et al. 2016) | Khanhhoa province, Vietnam | ANN | 1977–2014/(3) | RF, Temperature/SPEI | SPEI was predicted using ANN model. The results showed that adding climate signals can potentially enhance the prediction capability |

(continued)

**Table 1.1** (continued)

| Scholars | Region | Models | Data period/no. stations | Input/output variables | Remarks |
|---|---|---|---|---|---|
| (Maca and Pech 2016) | US catchment | ANN, HANN | 1948–2002/(2) | RF/SPI, SPEI | SPI and SPEI was predicted using two versions of ML models, the classical ANN and the hybrid model compromising several primary learning procedures. The authors acknowledged the efficiency of the HANN over the classical ANN for predicting drought indicators |

(continued)

**Table 1.1** (continued)

| Scholars | Region | Models | Data period/no. stations | Input/output variables | Remarks |
|---|---|---|---|---|---|
| (Park et al. 2016) | Arizona and New Mexico, USA | Random forest, BRT, cubist | 1975–2012/(54) | Drought factors based remote sensing, RF/SPI, crop yield | SPI and crop yield computed using three version of soft computing models. The approaches applied over vegetated region where remote sensing data are available. Random forest produced the best prediction over the other models. The authors found that land surface and evapotranspiration are the most correlated factor with the SPI |

(continued)

**Table 1.1** (continued)

| Scholars | Region | Models | Data period/no. stations | Input/output variables | Remarks |
|---|---|---|---|---|---|
| (Rezaeianzadeh et al. 2016) | Doroodzan Reservoir within Fars province, Iran | ANN, MC | 1976–2009/(5) | RF/SSI | Stochastic and machine models were used to predict drought conditions one-month ahead and the inflow volume. Rainfall information was incorporated to calculate SSI. The results evidenced the potential of the integration of the MC with ANN to predict SSI |
| (Choubin et al. 2016) | Maharlu—Bakhtaran catchment, Iran | MLP, M5P, ANFIS | 1967–2009/3 | Large-scale climate signal/SPI | SPI for 12-month scale was predicted using large-scale synoptic climate information. The results of the prediction process approved the capability of the MLP model over the others applied models |

(continued)

**Table 1.1** (continued)

| Scholars | Region | Models | Data period/no. stations | Input/output variables | Remarks |
|---|---|---|---|---|---|
| (Deo et al. 2016) | Merredin, Bathurst Agricultural station, Wilsons Lighthouse Station, Australia | WA-ELM, ELM, WA-ANN, ANN, WA-LSSVR, LSSVR | 1916–2012/(3) | RF/EDI | EDI was forecasted using different standalone and complementary AI models. The results revealed the importance of using WA-ELM over the other applied AI models |
| (Djerbouai and Souag-Gamane 2016) | Algeria | ANN, WA-ANN, ARIMA, SARIMA | 1936–2003/(17) | RF/SPI | SPI was forecasted for 3-, 6- and 12-month leads time using an integrative model based on wavelet-artificial neural network. Results indicated good capacity of the developed model over the traditional predictive models |

(continued)

**Table 1.1** (continued)

| Scholars | Region | Models | Data period/no. stations | Input/output variables | Remarks |
|---|---|---|---|---|---|
| (Rhee and Im 2017) | South Korea region | DT, Random forest, ERT | 1973–2015/(61) | RF, Temperature/SPI, SPEI | SPI and SPEI with 3-, 6-, 9-, and 12-month scales were predicted. The developed models exhibited an optimistic intelligence system that can be used for drought-related decision making, particularly for the ungauged watershed |
| (Deo et al. 2017) | Bathurst Agricultural, Peak Hill, Collarenebri, Barraba, Yamba, Eastern Australia | MARS, LSSVR, M5Tree | 1915–2012/(5) | RF/SPI | SPI predicted using MARS, LSSVR, and M5Tree models. The results highlighted the importance of periodicity as a predictor variable of SPI. However, the performance of the models displayed a variance results among the investigated cases studies |

(continued)

**Table 1.1** (continued)

| Scholars | Region | Models | Data period/no. stations | Input/output variables | Remarks |
|---|---|---|---|---|---|
| (Ali et al. 2018) | Pakistan | Ensemble-ANFIS, M5 tree, MPMR | (1981–2015)/3 | RF/SPI | SPI predicted at three major locations in Pakistan using novel data intelligent analytic predictive model called ensemble-ANFIS. The proposed model showed a notable prediction for the 6- and 12-month time scales compared to the three-month forecasts. The model also displayed good performance due to the capacity of the fuzzy set to handle the uncertainty phenomena |

(continued)

Table 1.1 (continued)

| Scholars | Region | Models | Data period/no. stations | Input/output variables | Remarks |
|---|---|---|---|---|---|
| (Mouatadid et al. 2018) | Drought–prone region, Australia | ELM, MLR, ANN, LSSVR | 1915–2012/(6) | RF, Temperature/SPEI | The authors adopted new methodology where the most correlated information toward the SPEI investigated. Different data intelligent analytic models are used for the modeling. Overall, the results indicated good agreement between observed and predicted models |
| (Soh et al. 2018) | Langat river basin, Malaysia | WA-ANN, WANFIS | 1976–2015(6) | RF, Temperature/SPEI | SPEI over tropics predicted using WA-ANN and WA-ANFIS. The results showed that the WA-ANFIS model has satisfactory prediction of the mid-term drought for all stations. the WA-ANN model gives better accuracy for both the short-term and mid-term droughts |

(continued)

**Table 1.1** (continued)

| Scholars | Region | Models | Data period/no. stations | Input/output variables | Remarks |
|---|---|---|---|---|---|
| (Agana and Homaifar 2018) | Lower Colorado river basin | DBN, EMD-DBF, MLP, EMD-MLP, SVR, EMD-SVR | 1906–2014/(10) | SF/SSI | Authors integrated empirical mode decomposition with new ML model called deep belief network. The integrative model is examined in predicting streamflow drought events. Result provided promising prediction outcome |

used to address the vulnerabilities via a risk management approach. Drought events have been reported in almost every part of the world in the last few decades due to spatiotemporal changes in climatic patterns (Ahmed et al. 2017), increasing water demands (Wang et al. 2016), and deficient water sources (Ahmed et al. 2018). It is necessary that water resources managers (considering the impacts of frequent drought episodes) to consider the influence of droughts based on historical, current, and future (potential) scenarios. The knowledge gathered from this review is as follows.

Significant progress toward the derivation of efficient drought indices which can provide better monitoring of drought events for useful early warning and derive better drought variables. The current drought indices are generally performed based on climatic and hydrology variables to reflect drought conditions; however, the quantification of the economic deficiencies is not really possible, thereby, creating the room for further improvement of drought indices to acquire better information based on the users' demand. Drought indices can further be explored toward drought classification based on their different characteristics.

The derivation of drought indices must be cautiously done for any region; for instance, the problem of the most used drought index, SPI often arises in the dry climates when the precipitation contains several significant zero values (Shiru et al. 2019, 2018). This usually constitutes a problem due to inapplicability of gamma distribution, which is usually fitted in most cases, as well as the issue of highly left-skewed distribution. There might be errors during the simulation of precipitation based on a probability distribution to derive drought indices due to data limitations. This is the case when there is an improperly chosen probability distribution. Similarly, the occasional changes in precipitation over time make it necessary to always check parameters' distribution probability over different time scales to ensure there is no significant effect on the drought indices (Nasrollahi et al. 2018).

Prior to the execution of any new water engineering project, it is important to properly analyze the historical drought events and review the existing projects on water resources in order to assist in the provision of valuable information related to water deficits/demands during drought events. With this information, water resources structures can address the likely challenges which may occur during droughts. Data shortage can be addressed by exploring paleoclimatic data through the extension of hydrometeorological time series in order to have a better understanding of historical droughts. Based on the reported studies in Table 1.1, current researches are focused on regions severely affected by increasing drought patterns. It is required to initiate studies to understand whether drought in such regions is periodic, climate change-related, or human-made due to increasing demand for water.

Globally, drought is estimated using large-scale climate indices; as such, most research on drought has concentrated on the regional or national scale. Meanwhile, the local scale understanding is still a problem owing to the heterogeneity in spatiotemporal hydrology and climatology. The understanding of the linkage between climate indices and the streamflow pattern is important in achieving proper information regarding the patterns of climate changes since several factors which are under human intervention can affect abstracted streamflow.

Being that drought events are multivariate, the determination of a better modeling approach for the description of drought characteristics requires for the derivation of the joint drought distribution based on its characteristics (Nabaei et al. 2019). This is necessary because the conventional models for the derivation of the joint probability distribution are mainly based on the marginal distribution of each drought parameter; hence, the recent computer-aid advancements have proven flexibility for the simulation of multivariate drought characterization. Although different ML models give different outcomes based on the actual trend of hydrology and climate variables, there is still a need to identify suitable ML models for multivariate drought characterization. Despite the number of reports on the different aspects of droughts, there is still no proper for the channeling of the results of these studies to the decision makers, thereby creating the need for the development of a decision support systems that will consider different climate change scenarios and also able to quantify uncertainties for risks assessment, issuing warnings, and taking precautionary measures (Wang and Davies 2015).

The prospect of the advanced ML models for real-time forecasting revealed an optimistic progression on the simulation of drought indices (Fung et al. 2019). Being that remote sensing data can provide real-time information, it can be an excellent input attribute for drought model. Meanwhile, such data may contain biased information; hence, such bias is required to correct by performing Bayesian merging with observed climatology to improve the forecasts.

The modern methods of data collection and retrieval have made water resources data available from both traditional recorded sources and from different real-time data, thereby allowing the analysis of such data using a combination of several datasets simultaneously from different sources. Thus, future drought modeling approaches would depend on the use of big data integrated system to produce real-time and robust forecasts. A big data approach, for instance, is a combination of remotely sensed and meteorological data streams with other types of datasets, such as wildfire occurrence, vegetation type, and pest activity. This approach can help in determining the direct effects of drought.

The applicability of the SC models such as ANN, FL, and SVR has been approved by the existing literature in recent drought prediction studies. These approaches can predict drought events that do not have a proper mathematical solution. They have been demonstrated to have the ability to capture the inherent features (such as white noise, nonlinearity, and non-stationary) in the time series. The outcome of this review showed that different drought indices, such as the NADI, SPI, SPEI, EDI, and PDSI, can be reliably predicted using the AI models.

Land surface model predictions can be improved by developing data assimilation models based on the combination of data from different sources with different resolutions. It is certain that many countries are still unable to achieve real-time drought prediction, and therefore, care must be taken during land data assimilation model development since drought prediction can be affected by changes in climatic patterns.

Both short-term and long-term water management systems demand spatiotemporal drought analysis based on the integration of the duration, severity, area, and inter-arrival time of drought. Although numerous works have been performed in this regard, the missing values in the gauged data used on the spatial scale have made it impossible to achieve accurate results; this is also complicated by the large distances between the gauging stations (Qutbudin et al. 2019). The problem can only be solved if there are available remote sensing data. Drought regionalization based on remote sensing data must, therefore, be examined. The space-time variability of droughts can also be explored from local to regional scale based on the connection between large-scale atmospheric patterns and regional droughts; hence, this demands to be investigated. Until now, drought regionalization is mainly dependent on climatic and hydrological variables; however, the growing water demand and the associated declining water sources are the major factors affecting droughts. As such, drought can be regionalized based on temporal and spatial water demands.

## 1.3 Conclusion and Possible Future Research Directions

A brief review of the models is provided in this chapter to comprehend the recent advances in forecasting droughts indices using ML models. The study revealed that a wide range of AI models including ANN, SVM, SVR, ANFIS, RF, ELM, M5, DT, MARS and their modified standalone and hybrid versions have been used for the forecasting of different drought indices including SPI, SEPI, SSI, PMDI, and PDSI. The studies revealed higher performance of AI models in forecasting drought indices. The literature review identified some of the research gaps in drought forecasting using ML methods, which are outlined below:

i. The non-stationary in climate due to climate change alters the relationship between physical predictors and drought indices dynamically over time. None of the ML-based models developed so for prediction of drought indices is able to address the non-stationary in climate or the dynamic changes in the relationship between the predictors and the drought index. The frequency and severity of droughts are increasing all over the globe due to climate change. Research toward development of climate resilient drought forecasting model is extremely important for adaptation to climate change.
ii. The ML models drought indices prediction developed so far used monthly rainfall data. Such models can be used for forecasting droughts having a duration of a month or multiple months. Droughts in tropical and humid regions occur for short periods, such as for few days. The indices used for measuring droughts using daily data show highly variability of drought index time series over time. Forecasting of such highly nonlinear and stochastic time series of drought indices is still remain a challenge.
iii. Droughts are more devastating when occur during crop growing seasons (Ahmed et al. 2016; Mohsenipour et al. 2018). The conventional drought indices can be used with some modification for estimation of droughts for different seasons. No study has been conducted so far for forecasting drought index estimated for

a climatic or crop growing season. However, forecasting of seasonal droughts is very important for reduction of economic damages of droughts.

iv. The devastation by a drought event depends on different characteristics of the event, such as intensity, duration, and affected area. Severity-area-duration (SAF) curves often estimated from conventional drought index to show devastating nature of droughts (Ahmed et al. 2019). ML models can be developed for the forecasting of SAF for better management of drought risk.

v. ML models developed so far are used for forecasting of drought index estimated at a station or averaged over a region. Such forecasting model does not provide information of spatial distribution of drought characteristics. However, region-specific knowledge on possible occurrence of droughts is very important. Models can be developed for forecasting of spatial variability of drought characteristics.

## References

Abarghouei H, Kousari MR, Asadi Zarch MA (2013) Prediction of drought in dry lands through feed forward artificial neural network abilities. Arab J Geosci 6:1417–1433. https://doi.org/10.1007/s12517-011-0445-x

Agana NA, Homaifar A (2018) EMD-based predictive deep belief network for time series prediction: an application to drought forecasting. Hydrology 5:18. https://doi.org/10.3390/hydrology5010018

Ahmed K, Chung E-S, Song J-Y, Shahid S (2017) Effective design and planning specification of low impact development practices using Water Management Analysis Module (WMAM): case of Malaysia. Water 9:173. https://doi.org/10.3390/w9030173

Ahmed K, Shahid S, Bin Harun S, Wang XJ (2016) Characterization of seasonal droughts in Balochistan Province, Pakistan. Stoch Env Res Risk Assess 30(2):747–762. https://doi.org/10.1007/s00477-015-1117-2

Ahmed K, Shahid S, Nawaz N (2018) Impacts of climate variability and change on seasonal drought characteristics of Pakistan. Atmos Res 214:364–374. https://doi.org/10.1016/j.atmosres.2018.08.020

Ahmed K, Shahid S, Sachindra DA, Nawaz N, Chung E-S (2019) Fidelity assessment of general circulation model simulated precipitation and temperature over Pakistan using a feature selection method. J Hydrol 573:281–298. https://doi.org/10.1016/j.jhydrol.2019.03.092

Alamgir M, Mohsenipour M, Homsi R, Wang X, Shahid S, Shiru M, Alias N, Yuzir A (2019) Parametric assessment of seasonal drought risk to crop production in Bangladesh. Sustainability 11:1442. https://doi.org/10.3390/su11051442

Alamgir M, Shahid S, Hazarika MK, Nashrrullah S, Harun S Bin, Shamsudin S (2015) Analysis of meteorological drought pattern during different climatic and cropping seasons in Bangladesh. JAWRA J Am Water Resour Assoc. 51:794–806

Ali M, Deo RC, Downs NJ, Maraseni T (2018) An ensemble-ANFIS based uncertainty assessment model for forecasting multi-scalar standardized precipitation index. Atmos Res 207:155–180. https://doi.org/10.1016/j.atmosres.2018.02.024

Belayneh A, Adamowski J (2012) Standard Precipitation Index Drought Forecasting Using Neural Networks, Wavelet Neural Networks, and Support Vector Regression. Appl Comput Intell Soft Comput 2012:1–13. https://doi.org/10.1155/2012/794061

Belayneh A, Adamowski J, Khalil B (2016) Short-term SPI drought forecasting in the Awash River Basin in Ethiopia using wavelet transforms and machine learning methods. Sustain Water Resour Manag 2:87–101. https://doi.org/10.1007/s40899-015-0040-5

Belayneh A, Adamowski J, Khalil B, Ozga-Zielinski B (2014) Long-term SPI drought forecasting in the Awash River Basin in Ethiopia using wavelet neural network and wavelet support vector regression models. J Hydrol 508:418–429. https://doi.org/10.1016/j.jhydrol.2013.10.052

Beyaztas U, Yaseen ZM (2019) Drought interval simulation using functional data analysis. J Hydrol 124141

Bhalme HN, Mooley DA (1980) Large-scale droughts/floods and monsoon circulation. Mon Weather Rev 108:1197–1211

Byram G, Keetch J (1988) A drought index for forest fire control. United States Dep. Agric. Serv. Res. Pap. SE-38 1–33

Byun HR, Wilhite DA (1999) Objective quantification of drought severity and duration. J Clim. https://doi.org/10.1175/1520-0442(1999)012%3c2747:OQODSA%3e2.0.CO;2

Chen J, Jin Q, Chao J (2012a) Design of deep belief networks for short-term prediction of drought index using data in the Huaihe river Basin. Math Probl Eng. https://doi.org/10.1155/2012/235929

Chen J, Li M, Wang W (2012b) Statistical uncertainty estimation using random forests and its application to drought forecast. Math Probl Eng 1–13. https://doi.org/10.1155/2012/915053

Choat B, Jansen S, Brodribb TJ, Cochard H, Delzon S, Bhaskar R, Bucci SJ, Feild TS, Gleason SM, Hacke UG, Jacobsen AL, Lens F, Maherali H, Martínez-Vilalta J, Mayr S, Mencuccini M, Mitchell PJ, Nardini A, Pittermann J, Pratt RB, Sperry JS, Westoby M, Wright IJ, Zanne AE (2012) Global convergence in the vulnerability of forests to drought. Nature 491(7426):752–755. https://doi.org/10.1038/nature11688

Choubin B, Malekian A, Golshan M (2016) Application of several data-driven techniques to predict a standardized precipitation index. Atmosfera 29:121–128. https://doi.org/10.20937/ATM.2016.29.02.02

Cordery I, McCall M (2000) A model for forecasting drought from teleconnections. Water Resour Res. https://doi.org/10.1029/1999WR900318

Dai A (2011) Drought under global warming: a review. Wiley Interdiscip Revi Clim Chang 2(1):45–65

Danandeh Mehr A, Kahya E, O¨zger M (2014) A gene-wavelet model for long lead time drought forecasting. J Hydrol 517:691–699. https://doi.org/10.1016/j.jhydrol.2014.06.012

Danandeh Mehr A, Nourani V, Kahya E, Hrnjica B, Sattar AMA, Yaseen ZM (2018) Genetic programming in water resources engineering: a state-of-the-art review. J Hydrol. https://doi.org/10.1016/j.jhydrol.2018.09.043

Deo RC, Kisi O, Singh VP (2017a) Drought forecasting in eastern Australia using multivariate adaptive regression spline, least square support vector machine and M5Tree model. Atmos Res 184:149–175. https://doi.org/10.1016/j.atmosres.2016.10.004

Deo RC, Şahin M (2015a) Application of the extreme learning machine algorithm for the prediction of monthly effective drought index in Eastern Australia. Atmos Res 153:512–525. https://doi.org/10.1016/j.atmosres.2014.10.016

Deo RC, Şahin M (2015b) Application of the artificial neural network model for prediction of monthly standardized precipitation and evapotranspiration index using hydrometeorological parameters and climate indices in Eastern Australia. Atmos Res 161–162:65–81. https://doi.org/10.1016/j.atmosres.2015.03.018

Deo RC, Tiwari MK, Adamowski JF, Quilty JM (2017b) Forecasting effective drought index using a wavelet extreme learning machine (W-ELM) model. Stoch Env Res Risk Assess 31(5):1211–1240. https://doi.org/10.1007/s00477-016-1265-z

Djerbouai S, Souag-Gamane D (2016) Drought forecasting using neural networks, wavelet neural networks, and stochastic models: case of the Algerois Basin in North Algeria. Water Resour Manag. https://doi.org/10.1007/s11269-016-1298-6

Fung KF, Huang YF, Koo CH, Soh YW (2019) Drought forecasting: a review of modelling approaches 2007–2017. J Water Clim Chang. https://doi.org/10.2166/wcc.2019.236

Ganguli P, Reddy MJ (2014) Evaluation of trends and multivariate frequency analysis of droughts in three meteorological subdivisions of Western India. Int J Climatol 34(3):911–928. https://doi.org/10.1002/joc.3742

Gibbs WJ (1967) Rainfall deciles as drought indicators

González J, Valdés JB (2006) New drought frequency index: definition and comparative performance analysis. Water Resour Res. https://doi.org/10.1029/2005WR004308

Hosseini-Moghari SM, Araghinejad S (2015) Monthly and seasonal drought forecasting using statistical neural networks. Environ Earth Sci 74:397–412. https://doi.org/10.1007/s12665-015-4047-x

Jalalkamali A, Moradi M, Moradi N (2015) Application of several artificial intelligence models and ARIMAX model for forecasting drought using the Standardized Precipitation Index. Int J Environ Sci Technol 12:1201–1210. https://doi.org/10.1007/s13762-014-0717-6

Kim T-W, Valdés JB (2003) Nonlinear model for drought forecasting based on a conjunction of wavelet transforms and neural networks. J Hydrol Eng 8:319–328. https://doi.org/10.1061/(asce)1084-0699(2003)8:6(319)

Lantz B (2013) Machine learning with R. Packt Publishing Ltd

Le MH, Perez GC, Solomatine D, Nguyen LB (2016) Meteorological drought forecasting based on climate signals using artificial neural network—a case study in Khanhhoa Province Vietnam. Procedia Eng. 154:1169–1175. https://doi.org/10.1016/j.proeng.2016.07.528

Keskin ME, Terzi O, Taylan ED, Küçükyaman D (2011) Meteorological drought analysis using artificial neural networks. Sci Res Essays 6(21):4469–4477. https://doi.org/10.5897/SRE10.1022

Maca P, Pech P (2016) Forecasting SPEI and SPI drought indices using the integrated artificial neural networks. Comput Intell Neurosci. https://doi.org/10.1155/2016/3868519

Mckee TB, Doesken NJ, Kleist J (1993) The relationship of drought frequency and duration to time scales. In: AMS 8th conference applied climatology, pp 179–184. https://doi.org/10.1175/citeulike-article-id:10490403

Mishra AK, Singh VP (2010) A review of drought concepts. J Hydrol. https://doi.org/10.1016/j.jhydrol.2010.07.012

Mishra AK, Singh VP (2011) Drought modeling—a review. J Hydrol. https://doi.org/10.1016/j.jhydrol.2011.03.049

Mohsenipour M, Shahid S, Chung ES, Wang X (2018) Changing pattern of droughts during cropping seasons of Bangladesh. Water Resour Manag. https://doi.org/10.1007/s11269-017-1890-4

Moreira EE, Coelho CA, Paulo AA, Pereira LS, Mexia JT (2008) SPI-based drought category prediction using loglinear models. J Hydrol 354:116–130. https://doi.org/10.1016/j.jhydrol.2008.03.002

Morid S, Smakhtin V, Bagherzadeh K (2007) Drought forecasting using artificial neural networks and time series of drought indices. Int J Climatol 27:2103–2111. https://doi.org/10.1002/joc.1498

Mouatadid S, Raj N, Deo RC, Adamowski JF (2018) Input selection and data-driven model performance optimization to predict the standardized precipitation and evaporation index in a drought-prone region. Atmos Res 212:130–149. https://doi.org/10.1016/j.atmosres.2018.05.012

Nabaei S, Sharafati A, Yaseen ZM, Shahid S (2019) Copula based assessment of meteorological drought characteristics: regional investigation of Iran. Agric For Meteorol 276:107611

Nasrollahi M, Khosravi H, Moghaddamnia A, Malekian A, Shahid S (2018) Assessment of drought risk index using drought hazard and vulnerability indices. Arab J Geosci 11:606

Palmer WC (1965) Meteorological Drought. US Weather Bureau, Res. Pap. No. 45

Panaou T (2018) Assessing the impacts of climate change on streamflow and reservoir operation in central Florida

Park S, Im J, Jang E, Rhee J (2016) Drought assessment and monitoring through blending of multi-sensor indices using machine learning approaches for different climate regions. Agric For Meteorol 216:157–169. https://doi.org/10.1016/j.agrformet.2015.10.011

Passioura JB (1996) Drought and drought tolerance. Plant Growth Regul. https://doi.org/10.1007/BF00024003

Paulo AA, Pereira LS (2007) Prediction of SPI drought class transitions using Markov chains. Water Resour Manag 21:1813–1827. https://doi.org/10.1007/s11269-006-9129-9

Pozzi W, Sheffield J, Stefanski R, Cripe D, Pulwarty R, Vogt JV, Heim RR, Brewer MJ, Svoboda M, Westerhoff R, Van Dijk AIJM, Lloyd-Hughes B, Pappenberger F, Werner M, Dutra E, Wetterhall F, Wagner W, Schubert S, Mo K, Nicholson M, Bettio L, Nunez L, Van Beek R, Bierkens M, De Goncalves LGG, De Mattos JGZ, Lawford R (2013) Toward global drought early warning capability: Expanding international cooperation for the development of a framework for monitoring and forecasting. Am Meteorol Soc, Bull. https://doi.org/10.1175/BAMS-D-11-00176.1

Qutbudin I, Shiru MS, Sharafati A, Ahmed K, Al-Ansari N, Yaseen ZM, Shahid S, Wang X (2019) Seasonal drought pattern changes due to climate variability: case study in Afghanistan. Water 11:1096. https://doi.org/10.3390/w11051096

Reddy AR, Chaitanya KV, Vivekanandan M (2004) Drought-induced responses of photosynthesis and antioxidant metabolism in higher plants. J Plant Physiol. https://doi.org/10.1016/j.jplph.2004.01.013

Rezaeian-Zadeh M, Tabari H (2012) MLP-based drought forecasting in different climatic regions. Theor Appl Climatol 109:407–414. https://doi.org/10.1007/s00704-012-0592-3

Rezaeian-Zadeh M, Stein A, Cox JP (2016) drought forecasting using markov chain model and artificial neural networks. Water Resour Manag 30:2245–2259. https://doi.org/10.1007/s11269-016-1283-0

Rhee J, Im J (2017) Meteorological drought forecasting for ungauged areas based on machine learning: Using long-range climate forecast and remote sensing data. Agric For Meteorol 237–238:105–122. https://doi.org/10.1016/j.agrformet.2017.02.011

Rooy V (1965) A rainfall anomaly index independent of time and space. Notos

Samarah, N.H., 2005. Effects of drought stress on growth and yield of barley. Agron. Sustain. Dev. https://doi.org/10.1051/agro

Shahid S (2010) Recent trends in the climate of Bangladesh. Clim Res. https://doi.org/10.3354/cr00889

Shirmohammadi B, Moradi H, Moosavi V, Semiromi MT, Zeinali A (2013) Forecasting of meteorological drought using Wavelet-ANFIS hybrid model for different time steps (case study: Southeastern part of east Azerbaijan province, Iran). Nat Hazards 69:389–402. https://doi.org/10.1007/s11069-013-0716-9

Shiru MS, Shahid S, Alias N, Chung ES (2018) Trend analysis of droughts during crop growing seasons of Nigeria. Sustain 10:1–13. https://doi.org/10.3390/su10030871

Shiru MS, Shahid S, Chung ES, Alias N (2019) Changing characteristics of meteorological droughts in Nigeria during 1901–2010. Atmos Res. https://doi.org/10.1016/j.atmosres.2019.03.010

Soh YW, Koo CH, Huang YF, Fung KF (2018) Application of artificial intelligence models for the prediction of standardized precipitation evapotranspiration index (SPEI) at Langat River Basin Malaysia. Comput Electron Agric 144:164–173. https://doi.org/10.1016/j.compag.2017.12.002

Strazzo S, Collins DC, Schepen A, Wang QJ, Becker E, Jia L (2019) Application of a hybrid statistical-dynamical system to seasonal prediction of North American temperature and precipitation. Mon Weather Rev 147:607–625

Tsakiris G, Pangalou D, Vangelis H (2007) Regional drought assessment based on the Reconnaissance Drought Index (RDI). Water Resour. Manag. https://doi.org/10.1007/s11269-006-9105-4

Vicente-Serrano SM, Beguería S, López-Moreno JI (2010) A multiscalar drought index sensitive to global warming: the standardized precipitation evapotranspiration index. J Clim 23:1696–1718. https://doi.org/10.1175/2009JCLI2909.1

Wang K, Davies EGR (2015) A water resources simulation gaming model for the Invitational Drought Tournament. J Environ Manage. https://doi.org/10.1016/j.jenvman.2015.06.007

Wang XJ, Zhang JY, Ali M, Shahid S, He RM, Xia XH, Jiang Z (2016) Impact of climate change on regional irrigation water demand in Baojixia irrigation district of China. Mitig Adapt Strat Glob Change 21(2):233–247

Wilhite DA, Glantz MH, (1985) Understanding: the drought phenomenon: the role of definitions. Water Int. 10

Willeke G, Hosking JRM, Wallis JR, Guttman NB (1994) The national drought atlas. Inst water Resour, Rep, p 94

Yaseen ZM, Sulaiman SO, Deo RC, Chau K-W (2018) An enhanced extreme learning machine model for river flow forecasting: state-of-the-art, practical applications in water resource engineering area and future research direction. J Hydrol 569:387–408. https://doi.org/10.1016/j.jhydrol.2018.11.069

# Chapter 2
# Bayesian Markov Chain Monte Carlo-Based Copulas: Factoring the Role of Large-Scale Climate Indices in Monthly Flood Prediction

**Thong Nguyen-Huy, Ravinesh C. Deo, Zaher Mundher Yaseen, Ramendra Prasad, and Shahbaz Mushtaq**

## 2.1 Introduction

Floods are the most destructive and dangerous natural hazards, causing enormous damage to human life, infrastructure, and agriculture all over the world (Posthumus et al. 2009; Johnson et al. 2016). In 2011–2012, a La Niña year, approximately 200 million people were affected by floods with a total damage of nearly US$95 billion worldwide (Ceola et al. 2014). Extreme rainfall in the summer of 2007 caused extensive flooding in parts of England resulting in an estimated 42,000 ha of farmland with significant effects on yields and farm incomes (Posthumus et al. 2009). Under the future outlook of climate conditions, damages due to flood events can be more serious. For example, the 1993 US Midwest floods caused damages to farmers valued at about US$6–8 billion while the 1997 North Dakota Red River floods caused a total damage of roughly USD$1 billion to agricultural production (Rosenzweig et al. 2002).

T. Nguyen-Huy (✉) · S. Mushtaq
Centre for Applied Climate Sciences, University of Southern Queensland, Toowoomba, Australia
e-mail: thonghuy.nguyen@usq.edu.au

T. Nguyen-Huy
Vietnam National Space Center, Vietnam Academy of Science and Technology, Hanoi, Vietnam

R. C. Deo
School of Sciences, University of Southern Queensland, Springfield Central, QLD 4300, Australia

Z. M. Yaseen
Sustainable Developments in Civil Engineering Research Group, Faculty of Civil Engineering, Ton Duc Thang University, Ho Chi Minh City, Vietnam

R. Prasad
Department of Science, School of Science and Technology, The University of Fiji, Saweni, Lautoka, Fiji

© Springer Nature Singapore Pte Ltd. 2021

R. C. Deo et al. (eds.), *Intelligent Data Analytics for Decision-Support Systems in Hazard Mitigation*, Springer Transactions in Civil and Environmental Engineering,
https://doi.org/10.1007/978-981-15-5772-9_2

In Australia, floods are the most costly of all disaster types, contributing 29% of the total cost of the nation's economy and the built environment (Hasanzadeh Nafari et al. 2016). Floods left a damage bill of over A$900 million (US$540 million) in 1998, which was more than twice the estimated average annual cost of floods, which is the country's most expensive natural (Yeo 2002). A total of 253 major floods over the period 1860–2013 occurred in the coastal catchments in the east from Brisbane in Queensland to Eden in New South Wales (Callaghan and Power 2014). In particular, the 2010–2011 Queensland floods caused over A$2 billion infrastructure damage and even larger indirect costs to the economy (Johnson et al. 2016). Recently, the 2019 Townsville flood in the city of Townsville and the surrounding areas caused damage bills in the billions (Adekunle et al. 2019).

Fluvial floods in Australia are caused by a diversity of weather systems (Yeo 2002), in which heavy rainfall is the primary driver. A suite of large-scale drivers of rainfall variability across Australian regions and seasons has been identified and documented in the work of Risbey et al. (2009). Among such climate mode indices, El Niño–Southern Oscillation (ENSO) is the key driver in terms of broad influence and impact on rainfall. ENSO is associated with rainfall over much of the continent at different times, particularly in the north and east, with the regions of influence shifting with the seasons. For example, the intense rainfall that generated floods at Townsville and Katherine in January 1998 was related to a shift from a highly negative Southern Oscillation Index (SOI), an ENSO indicator, to a positive value pointed to the waning of a severe El Niño event in May–June (Yeo 2002). Floods caused by a series of cold fronts bringing steady rainfall to the Namoi Valley over the eastern states are likely linked to the SOI value higher than +10.0, a La Niña conditions (Yeo 2002). These events evidently established a relationship between large-scale drivers such as ENSO and floods.

Flood prediction is the most important component in every early flood warning system that allows the much-needed preparation time for flood protection and reduction of flood impacts. Yet, flood prediction is one of the most challenging tasks in hydrology mostly due to the complex dynamic processes. Generally, there are three different approaches for flood prediction including numerical and physical, data-driven and statistical models. Physical models (Zeinivand and De Smedt 2010; Pappenberger et al. 2012) describe the physical processes based on the mathematical equations of mass, momentum, and energy conservation (Khac-Tien Nguyen and Hock-Chye Chua 2012). Such physical-based models showed great capabilities for predicting different flooding scenarios, and however, they often require a large amount of hydro-geomorphological monitoring datasets, intensive computation, in-depth knowledge and expertise regarding hydrological parameters, and have a gap in short-term prediction capability (Mosavi et al. 2018). Similarly, while numerical prediction models (Horritt and Bates 2002; Lin et al. 2006) represented an advancement in deterministic calculations, they required a complex and meticulous stimulation of physical equations and were not reliable due to systematic errors (Shrestha et al. 2013; Xingjian et al. 2015).

Data-driven machine learning (ML) models offer high prediction performance for floods due to their capability to handle complicated relationships between input variables and extract their significant features (Bui et al. 2019). A wide range of data-driven models has been developed for flood prediction such as artificial neural networks (ANN) (Wei et al. 2002; Do Hoai et al. 2011), support vector machines (SVM) (Liong and Sivapragasam 2002; Han et al. 2007), support vector regression (SVR) (Yu et al. 2006), fuzzy inference systems (FISs) (Lohani et al. 2014), and decision trees (DT) (Solomatine and Xue 2004). Other studies reported that a high level of predictive accuracy for flood prediction can be achieved through hybrid ML models, e.g., wavelet–bootstrap–ANN (WBANN) (Tiwari and Chatterjee 2010), and neuro-fuzzy (Nayak et al. 2005). Some limitations of ML models involve the interpretability, the considerable visual deformations (Tarsha-Kurdi et al. 2007), the need for large training data, and the computational cost.

The statistical approach provides simple models for flood prediction based on the historical relationship between input variables. The most common statistical algorithm is linear regression (Chau et al. 2005), which assumes a linear relationship between response and predictors. The statistical methods for time series analysis such as autoregressive moving average (ARMA) or ARMA with exogenous inputs, or autoregressive integrated moving average (ARIMA) have been used for real-time flood and river-level forecasting (Toth et al. 1999, 2000; Galavi et al. 2013). However, these linear regressive models do not describe the highly nonlinear dynamics inherent in flooding processes well and hence may not always perform adequately in practices.

Advanced statistical copula models are robust alternatives for describing dependence structure among random variables without any assumption on individual distributions. Sklar (1959) has introduced the copula theorem since 1959 but a recent decade has witnessed a sturdy revival of this approach. An extensive application of copula models has been carried out across different fields such as insurance and financial (Pfeifer and Nešlehová 2003; Fang and Madsen 2013), rainfall (Nguyen-Huy et al. 2020), drought (Dodangeh et al. 2017), flood (Durocher et al. 2016), streamflow (Chen et al. 2015), climate risk, and agricultural system (Nguyen-Huy et al. 2018, 2019).

This chapter evaluates the potential utility of large-scale climate information for flood prediction at Lockyer Valley station in Queensland, Australia. Lag relationships between monthly SOI and Flood Index (FI) were first explored. Copula functions were used to model such lag relationships in a joint distribution that then served the return period analysis. Copula parameters were numerically derived from under a hybrid-evolution Markov Chain Monte Carlo (MCMC) approach within a Bayesian framework (Ali et al. 2018a, b), developed by Sadegh et al. (2017). This global optimization method can handle the local minima problem in the parameter estimation, improve the description of the dependence structure, and evaluate the underlying uncertainties associated with fitting copulas to observations with limited length of the record. The commonly used local optimization was also performed for comparison. The performance of copula models was evaluated by different statistical tests and criteria.

## 2.2 Materials and Methods

### *2.2.1 Study Region and Data Preparation*

Lockyer Valley (152.34° E, 27.55° S) has rich farmlands making it become one of Queensland's most important regions of diversified agriculture (Fig. 2.1). The region has experienced severe floods that have caused massive damage to human life, agriculture, and infrastructure. The 2011 Brisbane flood resulted in a damage bill of over A$176 million with the loss of 24 lives and over 120 homes (van den Honert and McAneney 2011). The Lockyer Valley was also severely affected by the January 2013 flood in Queensland (Setunge et al. 2014). Therefore, flood prediction is essential to manage and reduce the impacts of future floods.

This chapter applied the method presented in the study of Deo et al. (2015) to compute daily FI. Specifically, FI is derived from daily effective precipitation $(P_E)$ mathematically expressed as:

$$\mathrm{FI} = \frac{P_E - {}_{2012}^{1915}\overline{P_E^{\max}}}{\sigma\left({}_{2012}^{1915}\overline{P_E^{\max}}\right)}, \tag{2.1}$$

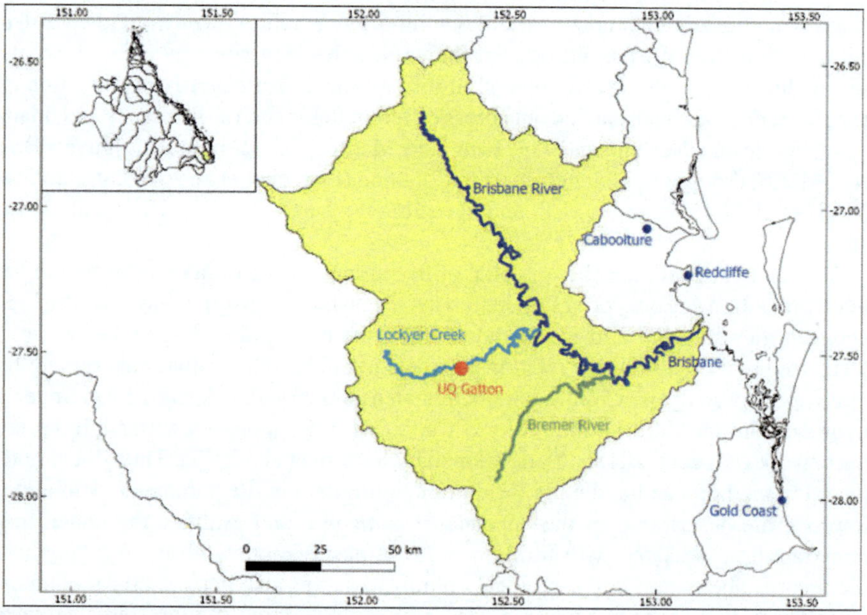


**Fig. 2.1** Study region with location of UQ Gatton weather station in Brisbane Basin and major tributaries of Brisbane River system

where ${}^{1915}_{2012}\overline{P_E^{\max}}$ and $\sigma\left({}^{1915}_{2012}\overline{P_E^{\max}}\right)$ denote means and standard deviations, respectively, of yearly maximum daily across the recorded hydrological period 1915–2012. Flood is identified using the criterion whether $FI > 0$. $P_E$ is the summed value of rainfall for current and antecedent day determined by a time-dependent reduction function and so $P_E$ for $i$th day is:

$$
\begin{aligned}
P_{E_i} &= \sum_{N=1}^{D}\left[\frac{\sum_{m=1}^{N} P_m}{N}\right] \\
&= P_1 + \frac{P_1 + P_2}{2} + \frac{P_1 + P_2 + P_3}{3} + \cdots + \frac{P_1 + P_2 + P_3 + \cdots + P_{365}}{365} \\
&= P_1\left(1 + \frac{1}{2} + \frac{1}{3} + \cdots + \frac{1}{365}\right) + P_2\left(\frac{1}{2} + \frac{1}{3} + \cdots + \frac{1}{365}\right) \\
&+ P_{365}\left(\frac{1}{365}\right) \approx P_1 + 0.85P_2 + 0.77P_3 + \cdots + 4.23 \times 10^{-4} P_{365},
\end{aligned} \quad (2.2)
$$

where $P_m$ is the rainfall recorded on any day, $m \in [1, 365]$ and $N$ is the duration of summation of the preceding period. Equation (2.2) defines the degree by which $P_m$ is converted into $P_E$ for $i$th day but more importantly, the formula considers antecedent $P$ with reduced weights. The data from January 01, 1950, till December 31, 2012, were acquired from the Australian Bureau of Meteorology (BoM).

Monthly SOI data for the same period with precipitation were acquired from BoM's Web site (http://www.bom.gov.au/climate/current/soihtm1.shtml). SOI is an indication of the development and intensity of ENSO events (i.e., El Niño or La Niña) in the Pacific Ocean. The SOI is calculated using the pressure differences between Tahiti and Darwin station. According to BoM, SOI is usually computed on a monthly basis, with values over longer periods such a year being sometimes used. Daily or weekly values of the SOI do not convey much in the way of useful information about the current state of the climate since daily values, in particular, can fluctuate markedly because of daily weather patterns. Therefore, it is suggested that daily SOI should not be used for climate purposes. To make the data consistent for modeling purposes, daily FI was converted into monthly data by averaging all values in each month.

An El Niño episode is often defined when there are persistent negative values of SOI below −7. These negative values are usually associated with a reduction in winter and spring rainfall over much of eastern Australia. By contrast, a La Niña episode is associated with persistent positive values of the SOI above +7 increasing the probability of being wetter than normal in eastern and northern Australia. Figure 2.2 shows maximum FI values in each year together with five-month moving average SOI series over the period 1915–2012. It can be seen that FI values are greater than zero implying flood occurrence associated with La Niña events, for example, the 2011 flood.

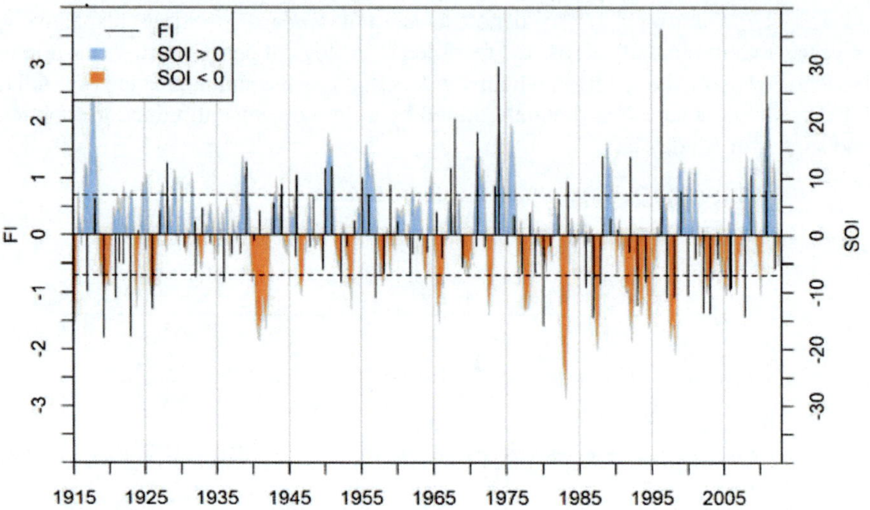


**Fig. 2.2** Annual maximum FI values plotted with five-month moving average SOI series over the period 1915–2012. FI values are greater than zero implying flood occurrence. Consecutive SOI values above +7 or below -7 (dashed lines) indicates La Niña and El Niño events, respectively

### 2.2.2 *Copulas*

Copulas are mathematical functions that "join" or "couple" two or more random variables irrespective of their marginal univariate distributions (Nelsen 2006). They are an efficient way to investigate the underlying dependence structure and provide a descriptive basis for constructing families of bivariate or multivariate distributions. According to Sklar's theorem (Sklar 1959), if $X$ and $Y$ are continuous random variables with their marginal distribution functions $F(x) = P(X \leq x)$ and $G(y) = P(Y \leq y)$, and $H(x, y) = P(X \leq x, Y \leq y)$ is their joint cumulative distribution function, then there exists a copula $C$ uniquely defined as:

$$H(x, y) = C[F(x), G(y)]. \tag{2.3}$$

Conversely, if there exists a joint distribution with continuous marginals $F$ and $G$, it can be always established associated copulas as $C(u, v) = H\left[F^{-1}(u), G^{-1}(v)\right]$ with $u = F(x)$ and $v = G(y)$. $F^{-1}$ and $G^{-1}$ are the inverse cumulative distribution functions. For example, the Gaussian copula is defined as:

$$C_\theta(u, v) = \Phi_\Sigma\left[\phi^{-1}(u), \phi^{-1}(v)\right], \tag{2.4}$$

where $\Phi_\Sigma$ signifies the joint cumulative distribution function of a bivariate normal vector with zero means and covariance matrix $\Sigma$, and $\phi$ is the cumulative distribution function of a standard normal.

### 2.2.3 Copula Model Development

In short, under a Bayesian framework, the posterior distribution of copula parameters is numerically estimated using a hybrid-evolution MCMC approach. The MCMC approach has several advantages compared to the commonly used local optimization. First, it can address the local minima problem in the parameter estimation and improve the description of the dependence structure. Second, it allows the evaluation of underlying uncertainties associated with fitting copulas to datasets with limited length of the record. This chapter applies the Multivariate Copula Analysis Toolbox (MvCAT) developed by Sadegh et al. (2017) for copula selection. Interested readers can refer to Sadegh et al. (2017) for more detail.

Many parametric bivariate copulas have been developed in the literature (see Nelsen (2006) and Joe (2014) for more detail about their properties). The toolbox provides 26 parametric copula families (see Sadegh et al. (2017), Table 2.1). In a copula model inference, one wishes to estimate the parameters $\theta$ by tuning them so that the copula predicted probability values $Y$ fit the joint probability of observed variables $\widetilde{Y}$ through a model hypothesis $M$ given a forcing $\widetilde{\mathrm{I}}$ as:

$$Y = M\left(\theta, \widetilde{\mathrm{I}}\right). \tag{2.5}$$

A vector $\mathrm{e} = \widetilde{Y} - Y$ that represents error residuals of $n$ observations as $\mathrm{e} = \{e_1, e_2, \ldots, e_n\}$ involves the effects of model structural errors arising from different sources, detailed in Sadegh et al. (2017).

**Table 2.1** Lag correlation coefficients between monthly SOI and monthly FI over Jan 1915–Dec 2012. For example, the value of Sep–lag2 intersection is the correlation coefficient ($r$) of 0.50 between Sep SOI–Nov FI

| | Jan | Feb | Mar | Apr | May | Jun | Jul | Aug | Sep | Oct | Nov | Dec |
|---|---|---|---|---|---|---|---|---|---|---|---|---|
| lag0 | **0.30** | **0.31** | **0.38** | **0.31** | **0.28** | **0.27** | **0.22** | **0.32** | **0.40** | **0.39** | **0.41** | **0.35** |
| lag1 | 0.17 | **0.34** | **0.25** | **0.21** | **0.32** | **0.28** | **0.28** | **0.47** | **0.36** | **0.37** | **0.35** | **0.35** |
| lag2 | 0.18 | **0.23** | 0.12 | 0.15 | **0.34** | **0.35** | **0.39** | **0.44** | **0.48** | **0.37** | **0.38** | **0.26** |
| lag3 | 0.12 | 0.11 | 0.13 | 0.11 | **0.35** | **0.37** | **0.29** | **0.44** | **0.46** | **0.37** | **0.32** | **0.24** |
| lag4 | 0.10 | 0.11 | 0.13 | 0.15 | **0.40** | **0.23** | **0.35** | **0.41** | **0.42** | **0.26** | **0.22** | **0.21** |
| lag5 | 0.05 | 0.03 | 0.11 | 0.19 | 0.19 | **0.26** | **0.36** | **0.40** | 0.19 | **0.21** | 0.16 | 0.07 |
| lag6 | 0.03 | 0.03 | 0.18 | 0.14 | **0.26** | **0.30** | **0.35** | **0.21** | 0.16 | 0.16 | 0.12 | 0.05 |
| lag7 | 0.01 | −0.01 | **0.21** | **0.23** | **0.30** | **0.29** | **0.20** | **0.22** | 0.15 | 0.14 | 0.01 | −0.07 |
| lag8 | −0.04 | −0.03 | **0.30** | **0.25** | **0.23** | 0.16 | **0.22** | 0.18 | 0.13 | 0.02 | −0.05 | −0.06 |
| lag9 | −0.01 | 0.04 | **0.27** | 0.08 | 0.01 | 0.16 | 0.11 | 0.16 | 0.04 | 0.04 | −0.10 | −0.03 |
| lag10 | 0.06 | −0.01 | 0.06 | −0.09 | 0.06 | 0.02 | 0.13 | 0.07 | −0.01 | 0.02 | −0.13 | −0.02 |
| lag11 | 0.01 | −0.14 | 0.03 | −0.01 | 0.03 | 0.05 | 0.03 | 0.05 | −0.01 | −0.06 | **−0.23** | 0.06 |

The bold values are significant at the level of 0.05

Bayesian analysis is carried for model inference and uncertainty quantification purposes since Bayes' theorem updates the prior probability (belief) of a certain hypothesis when new information is acquired. Bayes' law assigns all modeling uncertainties to the parameters and estimates the posterior distribution of model parameters by the following equations:

$$p(\theta|\widetilde{Y}) = \frac{p(\theta)p(\widetilde{Y}|\theta)}{p(\widetilde{Y})}, \tag{2.6}$$

where $p(\theta)$ and $p(\theta|\widetilde{Y})$ denote prior and posterior distribution of parameters, respectively, and $p(\widetilde{Y}) = \int_\theta p(\widetilde{Y}|\theta)d\theta$ is coined evidence, i.e., a constant value in modeling practices that can be removed from the analysis and thus the posterior distribution of parameters can be estimated as:

$$p(\theta|\widetilde{Y}) \propto p(\theta)p(\widetilde{Y}|\theta) \tag{2.7}$$

Also, $p(\widetilde{Y}|\theta) \cong L(\theta|\widetilde{Y})$ signifies the likelihood function. For simplicity and numerical stability, this likelihood function can be presented as:

$$\ell(\theta|\widetilde{Y}) \simeq -\frac{n}{2}\ln\left\{\frac{\sum_{i=1}^{n}[\widetilde{y}_i - y_i(\theta)]^2}{n}\right\}. \tag{2.8}$$

To solve Eq. (2.5) analytically and numerically, an MCMC simulation technique will be adopted to sample from the posterior distribution. For more details, readers are referred to Sadegh et al. (2017).

### 2.2.4 Goodness-of-Fit Criteria

Several goodness-of-fit criteria are used to rank the performance of copula models. Given the number of parameters of the statistical model $D$ and a constant $CS$, the mathematical formulations of all these measures are expressed as follows (Sadegh et al. 2017):

I. Akaike information criterion (AIC):

$$\text{AIC} = 2D + n\ln\left\{\frac{\sum_{i=1}^{n}[\widetilde{y}_i - y_i(\theta)]^2}{n}\right\} - 2\text{CS}. \tag{2.9}$$

II. Bayesian information criterion (BIC):

$$\mathrm{BIC} = D\ln n + n\ln\left\{\frac{\sum_{i=1}^{n}[\widetilde{y}_i - y_i(\theta)]^2}{n}\right\} - 2\mathrm{CS}. \tag{2.10}$$

III. Nash-Sutcliffe coefficient (NSE):

$$\mathrm{NSE} = 1 - \frac{\sum_{i=1}^{n}[\widetilde{y}_i - y_i(\theta)]^2}{\sum_{i=1}^{n}\left[\widetilde{y}_i - \overline{\widetilde{y}}_i\right]^2}. \tag{2.11}$$

IV. Root mean square error (RMSE):

$$\mathrm{RMSE} = \sqrt{\frac{\sum_{i=1}^{n}[\widetilde{y}_i - y_i(\theta)]^2}{n}}. \tag{2.12}$$

All these assessment metrics evaluate the copula performance in terms of how close modeled bivariate probabilities $Y$ are to their empirical observed counterparts $\widetilde{Y}$. A lower AIC value associates with a better model fit. BIC is similar to AIC, however, the penalty for two-parameter families is stronger than when using the AIC. A perfect model fit is associated with $\mathrm{NSE} = 1,\ \mathrm{NSE} \in (-\infty, 1]$ and $\mathrm{RMSE} = 0,\ \mathrm{RMSE} \in [0, \infty)$.

## 2.3 Results and Discussion

### *2.3.1 Influence of ENSO on Flood Index*

Figure 2.3a describes the number of times FI peak the highest value corresponding with each month during the period 1915–2012. It can be observed that the maximum FI values mostly occur from Dec to Mar. Figure 2.3b shows the frequency of FI values that are greater than zero, i.e., flood occurrence. There are 95 times out of 98 years that the maximum FI values of each month are greater than zero, which mostly occur from Jan to Apr.

The cross-correlation coefficients between monthly SOI and monthly FI, from one to eleven months ahead, are shown in Table 2.1. The significant values range from 0.20 to 0.50 depending on months and lags. The empirical results show that the lag correlation coefficients are even higher than the concurrent ones, i.e., at lag0. The cross-correlation coefficients are statistically significant (at the level of 0.05) from lag1-4 (up to lag8), indicating monthly FI can be predicted at least four months in advance using SOI information. The correlation coefficients are positive, indicating

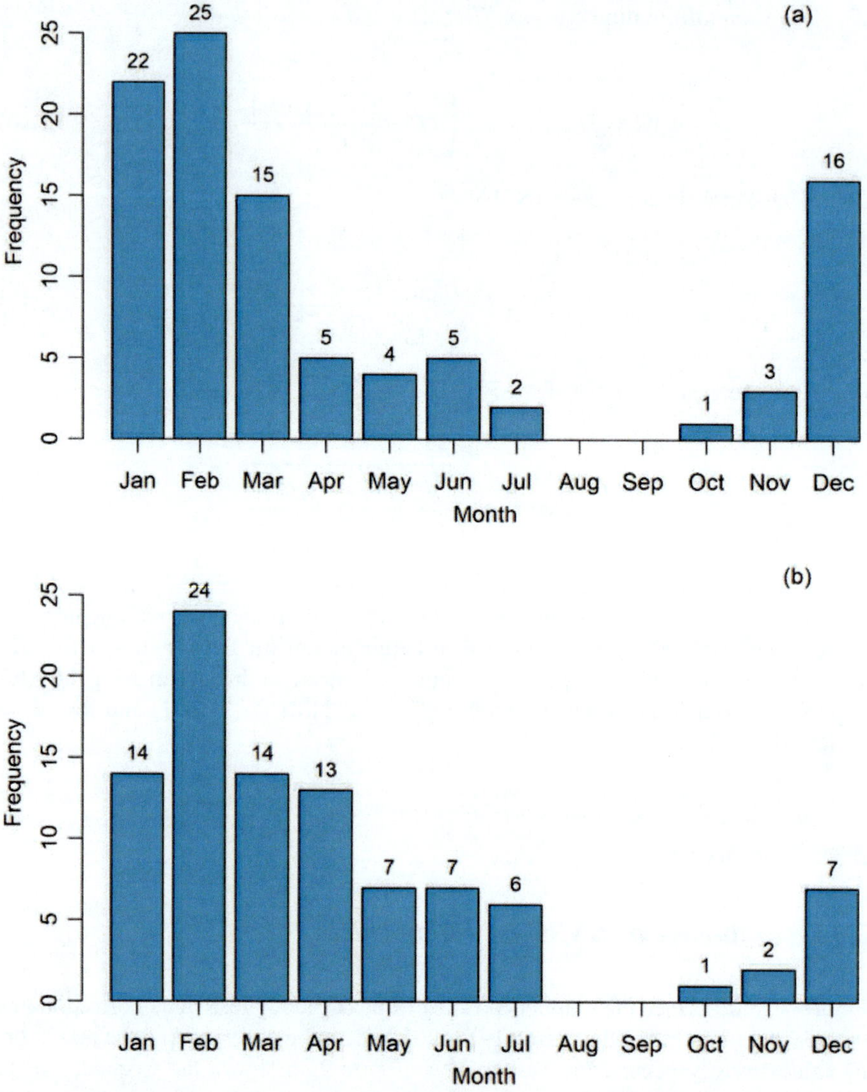


**Fig. 2.3** Frequency of FI values that are **a** highest and **b** greater than zero in each year during the period 1915–2012

that an above-average value of SOI is likely to lead to an above-average value of FI about four months later and conversely. These results are expected since ENSO events strongly influence rainfall across eastern and northern Australia (Risbey et al. 2009). In particular, the correlation coefficients are strongest in between Aug and Dec that are in agreement with the development of ENSO events. A typical ENSO event may show its first signs of development during the southern hemisphere autumn

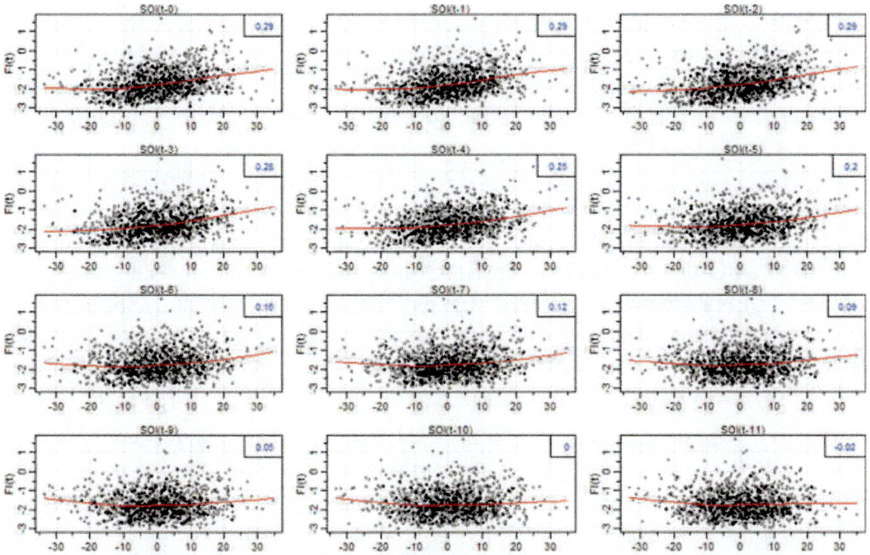


**Fig. 2.4** Scatter plots of the lag relationship between monthly SOI and FI time series over Jan 1915–Dec 2012 with a lowess fit (red lines) emphasizing nonlinearity

(Mar to May) and strengthens over winter (Jun to Aug) and spring (Sep to Nov). It will often start to decay in the mid to late southern summer (Dec to Feb) and finally dissipate in the subsequent autumn.

The dependence structure between large-scale climate index and floods was found to be complex and applying the traditional linear regression method to model such dependence may not be adequate. Figure 2.4 shows the lag plots with lowess fits superimposed clearly indicating nonlinear behavior between monthly SOI and FI time series. These findings justify the use of copula function to model the dependence among these two variables. The results of marginal and copula models are represented in the Sect. 2.3.2.

### *2.3.2 Copula Selection*

Since the highest frequency of flood occurrence is in Feb (Fig. 2.3b), Nov SOI that exhibits the strongest correlation coefficient with Feb FI (0.32) was selected to perform the copula-based model using Bayesian inference with the MCMC simulation technique. The marginal distributions of Nov SOI and Feb FI data can be estimated parametrically or nonparametrically. The MvCAT used the parametric method, i.e., the data were fitted to a number of theoretical distributions and the most appropriate distribution was selected based on chi-square statistics, a goodness-of-fit test, at 5% significant. The results indicate that monthly SOI and FI can be fitted to the logistic and generalized extreme value distribution, respectively. The estimated parameters are shown in Fig. 2.5, which are graphical tools for a comparison of fitted and empirical plots using the probability and quantile functions. It is clear that both marginal distributions were modeled appropriately by these distributions, in particular at the tails highlighted by the quantile-quantile plot.

Figure 2.6 shows the histogram of the posterior distribution of the first parameters for several commonly used copulas derived by the MCMC simulation within a Bayesian framework. Frank copula was ranked in the first place based on maximum likelihood, AIC, and BIC. Frank copula is a symmetric Archimedean copula that implies a symmetric dependence, i.e. the extreme high and low values of Nov SOI are associated with extreme high and low values of Feb FI. Also, differences between the copula parameters derived from the global MCMC (green square) and local (red

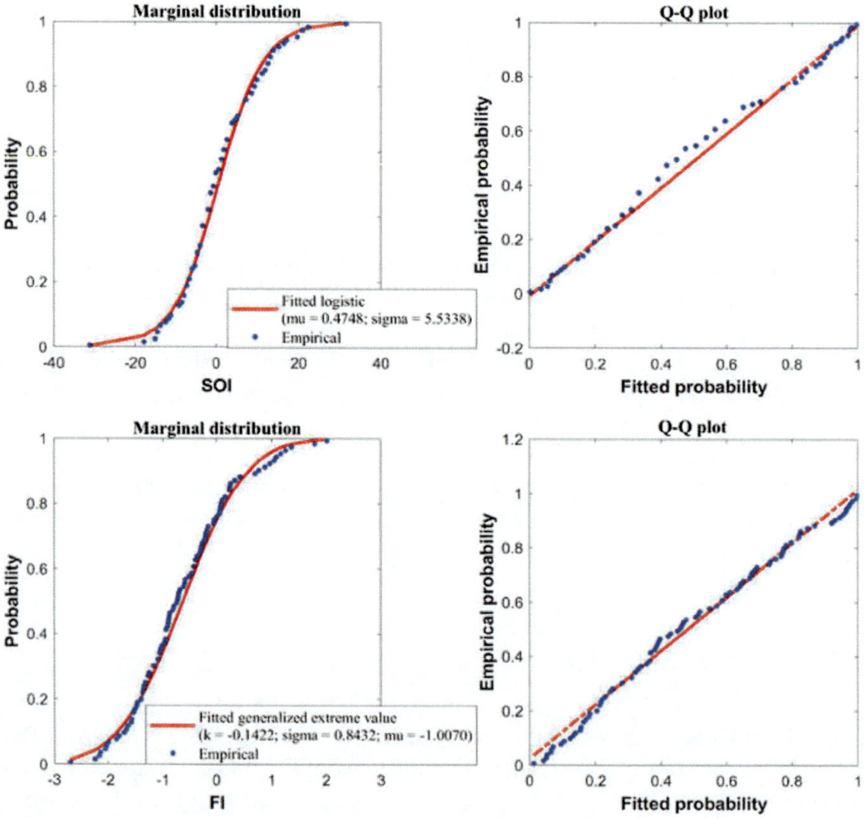


**Fig. 2.5** Goodness-of-fit plots indicate Nov SOI and Feb FI can be approximately fitted to a logistic and generalized distribution. The left panels compare the empirical and fitted probability distribution. The right panels illustrate the quantiles of the fitted distribution versus the empirical quantiles

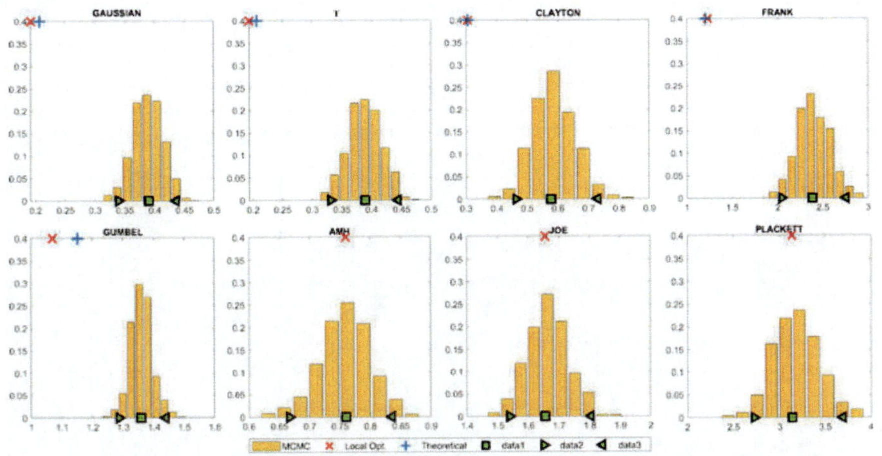


**Fig. 2.6** Posterior distribution of first parameters for different copulas derived by the MCMC simulation within a Bayesian framework. Red crosses show the copula parameter value derived by the local optimization while the blue plus signs are the theoretical parameter. The orange bins are the MCMC-derived parameters and green squares show the maximum likelihood parameters (Par. 1) with its lower and upper limit of the MCMC

cross) optimization technique can be observed. The estimated parameters using local optimization are much closer to the theoretical parameters for Gaussian, t, Clayton, Frank, and Gumbel copulas. However, the estimated parameters using the global optimization are closer to the theoretical parameters for the AMH, Joe, and Plackett copulas.

Table 2.2 shows the estimated copula parameters using local and global (MCMC within a Bayesian framework) optimization with RMSE, NSE, and the uncertainty at 95% confidence. It is visual that the Frank copula has the lowest and highest values of RMSE = 0.2294 and NSE = 0.9905, respectively, which coincide with the results of maximum likelihood, AIC, and BIC as the "best"-fit model. The estimated local copula parameter $\theta_{l,1} = 1.2316$ and global $\theta_{g,1} = 2.3838$ can be converted to the respective Kendall's tau coefficient $\tau_l = 0.1329$ and $\tau_g = 0.2460$ using the formula $\tau = 1 - \frac{4}{\theta}[1 - D_1(\theta)]$, where $D_1$ denotes the Debye function defined as $D_n(x) = \frac{n}{x^n}\int_0^x \frac{t^n}{e^t-1}dt$ for $n$ a positive interger. Comparing to the empirical $\widetilde{\tau} = 0.1324$, the local value is closer than the global one (see also Fig. 2.5). However, as mentioned above, the global optimization approach allows retrieving the estimation uncertainties as one of the superior attributes to the local technique as mentioned above.

Figure 2.7 explains how Nov SOI and Feb FI observations were transformed into copula data in [0, 1] using their marginal cumulative distributions of FI and SOI and then their joint distribution is modeled by Frank copula. Color lines present joint probability isolines derived from Frank copula that are color-coded with joint density levels normalized to the highest density value. For example, the extreme values of

**Table 2.2** Estimated copula parameters ($\theta$) using local and global (MCMC within a Bayesian framework) optimization with RMSE, NSE, and the uncertainty [lower, upper] at 95% confidence

| Copula Name | RMSE | NSE | Local | | | MCMC | | |
|---|---|---|---|---|---|---|---|---|
| | | | $\theta_{l,1}$ | $\theta_{l,2}$ | $\theta_{l,3}$ | $\theta_{g,1}$ | $\theta_{g,2}$ | $\theta_{g,3}$ |
| Gaussian | 0.2374 | 0.9899 | 0.1917 | | | 0.3897 [0.3391, 0.4359] | | |
| t | 0.2388 | 0.9898 | 0.1932 | 15700041.8581 | | 0.3896 [0.3311, 0.4442] | 34.8307 [8.3211, 34.7358] | |
| Clayton | 0.2786 | 0.9861 | 0.3049 | | | 0.5785 [0.4633, 0.7308] | | |
| Frank | 0.2294 | 0.9905 | 1.2316 | | | 2.3838 [2.0412, 2.7544] | | |
| Gumbel | 0.2354 | 0.9900 | 1.0703 | | | 1.3577 [1.2865, 1.4388] | | |
| Independence | 0.4328 | 0.9663 | | | | | | |
| AMH | 0.2545 | 0.9884 | 0.7599 | | | 0.7601 [0.6668, 0.8358] | | |
| Joe | 0.2397 | 0.9897 | 1.6559 | | | 1.6559 [1.5394, 1.8043] | | |
| FGM | 0.2297 | 0.9905 | 1.0000 | | | 1.0000 [0.8872, 0.9993] | | |
| Plackett | 0.2323 | 0.9903 | 3.1387 | | | 3.1384 [2.7226, 3.6932] | | |
| Cuadras-Auge | 0.2492 | 0.9888 | 0.4231 | | | 0.4232 [0.3697, 0.4737] | | |
| Raftery | 0.2962 | 0.9842 | 0.2918 | | | 0.2918 [0.2413, 0.3410] | | |
| Shih-Louis | 0.2608 | 0.9878 | 0.3427 | | | 0.3428 [0.2953, 0.3954] | | |
| Linear-Spearman | 0.2608 | 0.9878 | 0.3427 | | | 0.3427 [0.2961, 0.3947] | | |
| Cubic | 0.4325 | 0.9664 | -0.3781 | | | -0.3781 [ -0.9434, 1.5566] | | |
| Burr | 0.2334 | 0.9902 | 1.3970 | | | 1.3969 [1.1794, 1.6951] | | |
| Nelsen | 0.2294 | 0.9905 | 29.2529 | | | 2.3837 [2.0716, 2.7914] | | |
| Galambos | 0.2346 | 0.9901 | 0.6246 | | | 0.6246 [0.5675, 0.7031] | | |
| Marshal-Olkin | 0.2474 | 0.9890 | 32.0885 | 25.6876 | | 0.4636 [0.3819, 0.6240] | 0.3863 [0.2895, 0.4689] | |
| Fischer-Hinzmann | 0.2446 | 0.9893 | 0.5177 | -1.2419 | | 0.5164 [0.4096, 0.6383] | -1.2330 [-3.1809, -0.2157] | |
| Roch-Alegre | 0.2352 | 0.9901 | 0.8803 | 1.4067 | | 0.8897 [0.3155, 1.7145] | 1.4048 [1.1241, 1.6722] | |
| Fischer-Kock | 0.2299 | 0.9905 | 1.0000 | 1.0000 | | 1.0009 [1.0016, 1.1481] | 0.9978 [0.9078, 0.9989] | |
| BB1 | 0.2355 | 0.9900 | 0.0001 | 1.3574 | | 0.0009 [0.0021, 0.1700] | 1.3538 [1.2330, 1.4004] | |
| BB5 | 0.2346 | 0.9901 | 1.0020 | 0.6217 | | 1.0328 [1.0158, 1.4082] | 0.5796 [0.0058, 0.6546] | |
| Tawn | 0.2356 | 0.9900 | 1.0000 | 0.9299 | 1.3782 | 0.9736 [0.3809, 0.9701] | 0.8327 [0.3648, 0.9985] | 1.4151 [1.34963, 4.4318] |

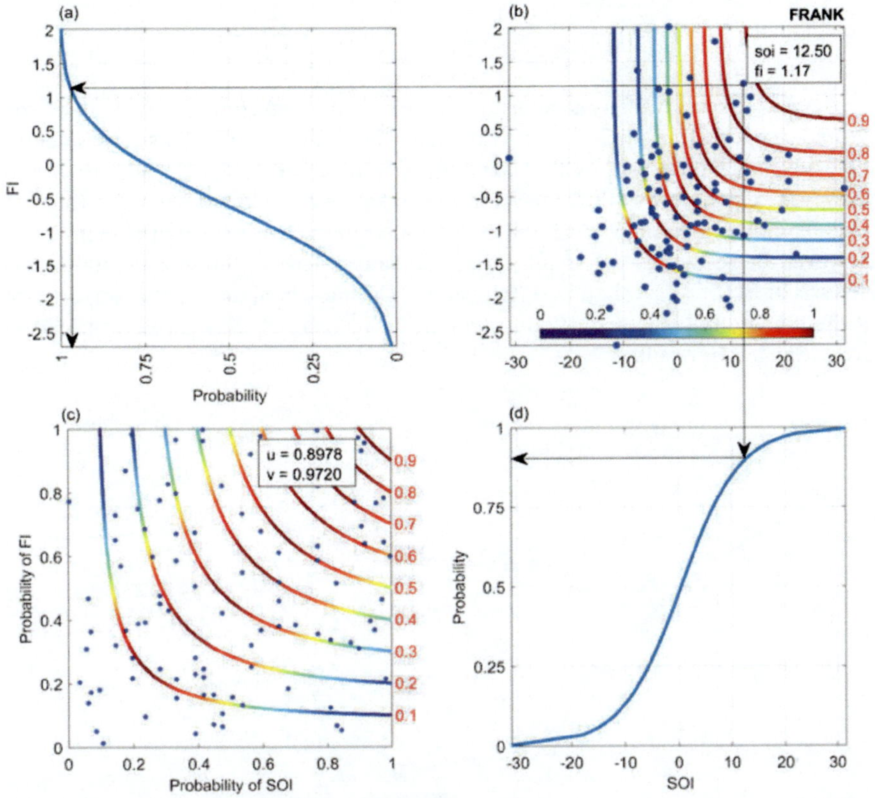


**Fig. 2.7** Plots illustrate how Nov SOI and Feb FI observations are transformed into pseudo data using their marginal cumulative distributions of FI (**a**) and SOI (**c**) and their joint distribution is modeled by Frank copula. Color lines present joint probability isolines derived from Frank copula (**b**) that are color-coded with joint density levels normalized to the highest density value. Symmetric dependence structure of SOI and FI observations presented in probability space (**d**). Blue dots show observed data and the black line is the empirical probability isoline

soi $= 12.50$ and $fi = 1.17$ were first converted into the unit interval $u = 0.8978$ and $v = 0.9720$ using their cumulative distribution functions (logistic and generalized extreme value). The joint probability value $P(\text{SOI} < \text{soi}, \text{FI} < fi) = 0.8649$ was derived from Eq. (2.3). Figure 2.7d shows the symmetric dependence structure of Nov SOI and Feb FI observations presented in probability space. The probability isolines derived with the Frank copula are visibly centered to the diagonal direction, i.e., high probability of SOI associated with a high probability of FI, and conversely.

The return period $T$ (years) is commonly calculated as $T = \mu/P$, where $\mu$ (year) denotes the average inter-arrival time between two consecutive realizations and $P$ is the probability of observing realizations exceeding reference threshold values. Let $\mu$ is equal to 1 year, and the return period (RP), in which SOI or FI, or both exceed a prespecified threshold value (soi, $fi$), is then estimated as (Aghakouchak 2014):

$$T = \frac{1}{P(\text{SOI} \geq \text{soi}, \text{FI} \geq fi)} = \frac{1}{1 - C(u, v)}, \tag{2.13}$$

Figure 2.8 shows different return period levels derived from Eq. (2.13) and the dependence structure of SOI and FI modeled by Frank copula. This figure shows relatively large observations that have the return period level of fewer than two years. One example was illustrated for an extreme event with an Nov SOI value of 12.50 and Feb FI value of 1.12, where the joint return period is less than ten years. However, these figures are smaller than the values derived through the univariate analysis of FI (25 years) and SOI (ten years). This clearly highlights the importance of the climate influences on flood prediction and ignoring this driver can lead to overestimated FI results.

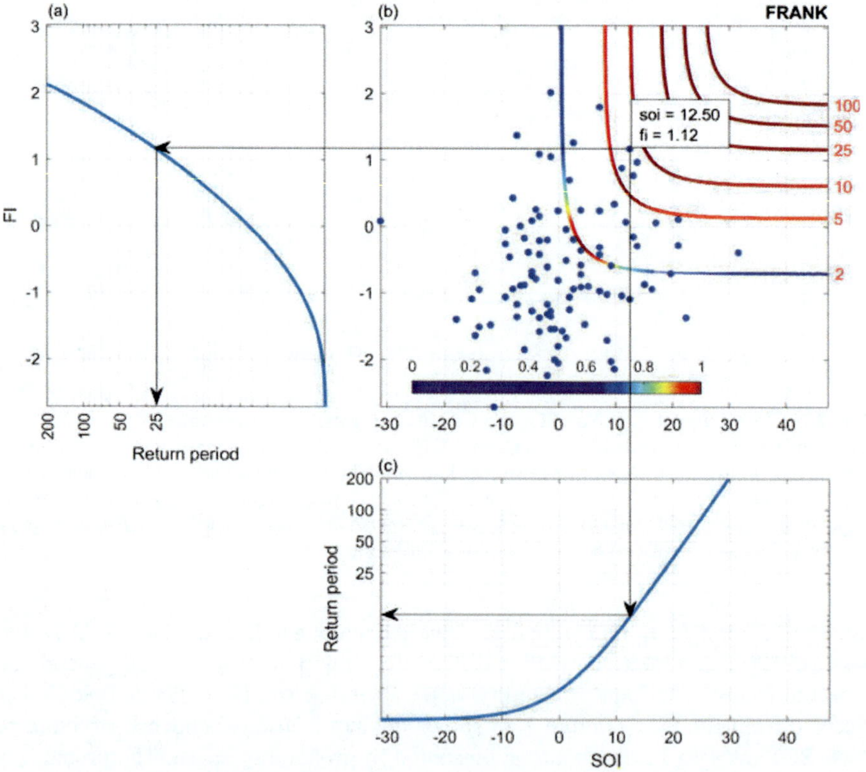


**Fig. 2.8** Different return period levels of Aug SOI and Oct FI modeled by Frank copula, in which SOI or FI, or both exceed a prespecified threshold value. Bivariate return period isolines are color-coded with joint density levels normalized to the highest density value, and univariate return periods of FI (**a**) and SOI (**c**) (*y*-axis presented in log scale). Blue dots show observed data of SOI and FI

## 2.4 Conclusions

Floods in Australia are predominately caused by heavy rainfall as a result of a diversity of weather systems. Large-scale climate mode indices such as ENSO have been identified as main drivers of rainfall variability across Australian regions and seasons. The results indicate that monthly SOI data from Aug to Dec have a significant correlation with monthly FI that can be predicted at least four months ahead using ENSO information. Flood mostly occurs from Jan to Apr, particularly in Feb. Frank copula was chosen based on maximum likelihood, AIC, and BIC for modeling and understanding the lag relationship between Nov SOI and Feb FI data after they were fitted to the respective logistic and generalized extreme value distribution. The local optimization is a comparative performance in estimating copula parameters compared to the global technique implemented by MCMC within a Bayesian framework. However, the global method provided 95% confidence bounds of copula parameters that allowed the evaluation of estimated uncertainty. The return period values derived from the joint probability distribution is smaller than those based on the univariate analysis. This implied that ignoring the influence of climate information may lead to an overestimation of FI. These advanced flood prediction models are indeed imperative for civil protection and important to early warning and risk reduction systems.

**Acknowledgements** The authors are grateful to the Australian Bureau of Meteorology for providing the relevant meteorological data for the study region.

## References

Adekunle AI, Adegboye OA, Rahman KM (2019) Flooding in Townsville, North Queensland, Australia, in February 2019 and its effects on mosquito-borne diseases. Int J Environ Res Public Health 16(8):1393

Aghakouchak A (2014) Entropy–copula in hydrology and climatology. J Hydrometeorol 15(6):2176–2189

Ali M, Deo RC, Downs NJ, Maraseni T (2018a) Multi-stage hybridized online sequential extreme learning machine integrated with Markov Chain Monte Carlo copula-Bat algorithm for rainfall forecasting. Atmos Res 213:450–464

Ali M, Deo RC, Downs NJ, Maraseni T (2018b) Cotton yield prediction with Markov Chain Monte Carlo-based simulation model integrated with genetic programing algorithm: a new hybrid copula-driven approach. Agric For Meteorol 263:428–448

Bui DT, Ngo P-TT, Pham TD, Jaafari A, Minh NQ, Hoa PV, Samui P (2019) A novel hybrid approach based on a swarm intelligence optimized extreme learning machine for flash flood susceptibility mapping. CATENA 179:184–196

Callaghan J, Power SB (2014) Major coastal flooding in southeastern Australia 1860–2012, associated deaths and weather systems. Australian Meteorol Oceanographic J 64(3):183–213

Ceola S, Laio F, Montanari A (2014) Satellite nighttime lights reveal increasing human exposure to floods worldwide. Geophys Res Lett 41(20):7184–7190

Chau K, Wu C, Li Y (2005) Comparison of several flood forecasting models in Yangtze River. J Hydrol Eng 10(6):485–491
Chen L, Singh VP, Guo S, Zhou J, Zhang J (2015) Copula-based method for multisite monthly and daily streamflow simulation. J Hydrol 528:369–384
Deo RC, Byun H-R, Adamowski JF, Kim D-W (2015) A real-time flood monitoring index based on daily effective precipitation and its application to Brisbane and Lockyer Valley flood events. Water Resour Manage 29(11):4075–4093
Do Hoai N, Udo K, Mano A. (2011) Downscaling global weather forecast outputs using ANN for flood prediction. J Appl Mathe
Dodangeh E, Shahedi K, Shiau J-T, MirAkbari M (2017) Spatial hydrological drought characteristics in Karkheh River basin, southwest Iran using copulas. J Earth Syst Sci 126(6):80
Durocher M, Chebana F, Ouarda TB (2016) On the prediction of extreme flood quantiles at ungauged locations with spatial copula. J Hydrol 533:523–532
Fang Y, Madsen L (2013) Modified Gaussian pseudo-copula: applications in insurance and finance. Insurance Mathe Econo 53(1):292–301
Galavi H, Mirzaei M, Shul LT, Valizadeh N (2013) Klang River–level forecasting using ARIMA and ANFIS models. J Am Water Works Assoc 105(9):E496–E506
Han D, Chan L, Zhu N (2007) Flood forecasting using support vector machines. J Hydroinformatics 9(4):267–276
Hasanzadeh Nafari R, Ngo T, Mendis P (2016) An assessment of the effectiveness of tree-based models for multi-variate flood damage assessment in Australia. Water 8(7):282
Horritt M, Bates P (2002) Evaluation of 1D and 2D numerical models for predicting river flood inundation. J Hydrol 268(1–4):87–99
Joe H (2014) Dependence modeling with copulas. Chapman and Hall/CRC
Johnson F, White CJ, van Dijk A, Ekstrom M, Evans JP, Jakob D, Kiem AS, Leonard M, Rouillard A, Westra S (2016) Natural hazards in Australia: floods. Clim Change 139(1):21–35
Khac-Tien Nguyen P, Hock-Chye Chua L (2012) The data-driven approach as an operational real-time flood forecasting model. Hydrol Process 26(19):2878–2893
Lin B, Wicks JM, Falconer RA, Adams K (2006) Integrating 1D and 2D hydrodynamic models for flood simulation. In: Proceedings of the institution of civil engineers-water management. Citeseer, pp 19–25
Liong SY, Sivapragasam C (2002) Flood stage forecasting with support vector machines 1. JAWRA J Am Water Res Association 38(1):173–186
Lohani AK, Goel N, Bhatia K (2014) Improving real time flood forecasting using fuzzy inference system. J Hydrol 509:25–41
Mosavi A, Ozturk P, Chau K-w (2018) Flood prediction using machine learning models: Literature review. Water 10(11):1536
Nayak P, Sudheer K, Rangan D Ramasastri K (2005) Short-term flood forecasting with a neurofuzzy model. Water Resour Res 41(4)
Nelsen RB (2006) An introduction to copulas, 2 edn. Springer
Nguyen-Huy T, Deo RC, Mushtaq S, Khan S (2020) Probabilistic seasonal rainfall forecasts using semiparametric d-vine copula-based quantile regression. In: Handbook of probabilistic models. Elsevier, pp 203–27
Nguyen-Huy T, Deo RC, Mushtaq S, Kath J, Khan S (2018) Copula-based agricultural conditional value-at-risk modelling for geographical diversifications in wheat farming portfolio management. Weather Clim Extremes 21:76–89
Nguyen-Huy T, Deo RC, Mushtaq S, Kath J, Khan S (2019) Copula statistical models for analyzing stochastic dependencies of systemic drought risk and potential adaptation strategies. Stochastic Environ Res Risk Assessment
Pappenberger F, Dutra E, Wetterhall F, Cloke HL (2012) Deriving global flood hazard maps of fluvial floods through a physical model cascade. Hydrol Earth Syst Sci 16(11):4143–4156
Pfeifer D, Nešlehová J (2003) Modeling dependence in finance and insurance: the copula approach. Blätter der DGVFM 26(2):177–191

Posthumus H, Morris J, Hess T, Neville D, Phillips E, Baylis A (2009) Impacts of the summer 2007 floods on agriculture in England. J Flood Risk Manag 2(3):182–189
Risbey JS, Pook MJ, McIntosh PC, Wheeler MC, Hendon HH (2009) On the remote drivers of rainfall variability in Australia. Mon Weather Rev 137(10):3233–3253
Rosenzweig C, Tubiello FN, Goldberg R, Mills E, Bloomfield J (2002) Increased crop damage in the US from excess precipitation under climate change. Glob Environ Change 12(3):197–202
Sadegh M, Ragno E, AghaKouchak A (2017) Multivariate Copula Analysis Toolbox (MvCAT): describing dependence and underlying uncertainty using a Bayesian framework. Water Resour Res 53(6):5166–5183
Setunge S, Lokuge W, Mohseni H, Karunasena W (2014) Vulnerability of road bridge infrastructure under extreme flood events. In: AFAC & Bushfire & Natural Hazards CRC Conference 2014. University of Southern Queensland
Shrestha D, Robertson D, Wang Q, Pagano T, Hapuarachchi H (2013) Evaluation of numerical weather prediction model precipitation forecasts for short-term streamflow forecasting purpose. Hydrol Earth Syst Sci 17(5):1913–1931
Sklar M (1959) Fonctions de répartition à n dimensions et leurs marges. Université Paris 8
Solomatine DP, Xue Y (2004) M5 model trees and neural networks: application to flood forecasting in the upper reach of the Huai River in China. J Hydrol Eng 9(6):491–501
Tarsha-Kurdi F, Landes T, Grussenmeyer P Koehl M (2007) Model-driven and data-driven approaches using LIDAR data: analysis and comparison. In: ISPRS workshop, photogrammetric image analysis (PIA07), pp. 87–92
Tiwari MK, Chatterjee C (2010) Development of an accurate and reliable hourly flood forecasting model using wavelet–bootstrap–ANN (WBANN) hybrid approach. J Hydrol 394(3–4):458–470
Toth E, Montanari A, Brath A (1999) Real-time flood forecasting via combined use of conceptual and stochastic models. Phys Chem Earth Part B 24(7):793–798
Toth E, Brath A, Montanari A (2000) Comparison of short-term rainfall prediction models for real-time flood forecasting. J Hydrol 239(1–4):132–147
van den Honert RC, McAneney J (2011) The 2011 Brisbane floods: causes, impacts and implications. Water 3(4):1149–1173
Wei Y, Xu W, Fan Y, Tasi H-T (2002) Artificial neural network based predictive method for flood disaster. Comput Ind Eng 42(2–4):383–390
Xingjian S, Chen Z, Wang H, Yeung D-Y, Wong W-K, Woo W-C (2015). Convolutional LSTM network: A machine learning approach for precipitation nowcasting. In: Advances in neural information processing systems, pp 802–10
Yeo SW (2002) Flooding in Australia: a review of events in 1998. Nat Hazards 25(2):177–191
Yu P-S, Chen S-T, Chang I-F (2006) Support vector regression for real-time flood stage forecasting. J Hydrol 328(3–4):704–716
Zeinivand H, De Smedt F (2010) Prediction of snowmelt floods with a distributed hydrological model using a physical snow mass and energy balance approach. Nat Hazards 54(2):451–468

# Chapter 3
# Gaussian Naïve Bayes Classification Algorithm for Drought and Flood Risk Reduction

**Oluwatobi Aiyelokun, Gbenga Ogunsanwo, Akintunde Ojelabi, and Oluwole Agbede**

## 3.1 Introduction

Natural disasters such as flood and drought have been declared as sources of setback to sustainable development efforts in countries where their risk is high (Mochizuki and Naqvi 2019). Rainfall patterns usually have spatial and temporal variability which affect the agricultural production, water supply, transportation, the entire economy of a region, and the existence of its people (Oduro-Afriyie and Adukpo 2006). In regions where the year-to-year variability is high, people often suffer great calamities due to floods or droughts. Whereas damage due to extremes of rainfall cannot be avoided completely, a forewarning could certainly be useful (Oduro-Afriyie and Adukpo 2006).

Vargas and Paneque (2019) distinguished the difference between drought as a natural phenomenon and as a risk. As a natural phenomenon, drought is perceived as the availability of rainfall too less than normal at a place for a given period (Pita 2007; Aiyelokun et al. 2017); whereas, drought risk is the impact of decreased rainfall on the available water resources, while trying to maintain availability and demand (Vargas and Paneque 2017). Flood is risk is the probability of flood occurrence and its potential effects. Flood risk has been projected to amplify in many regions of the world as a result of climate change, urbanization, and population growth (Bradford et al. 2012; Liu et al. 2018). Flooding is globally perceived as a huge problem that impacts millions of people annually (Henstra et al. 2019) and has grown over the years in frequency and magnitude (Berghuijs et al. 2017). The World Meteorological

O. Aiyelokun (✉) · A. Ojelabi · O. Agbede
Department of Civil Engineering, University of Ibadan, Ibadan, Nigeria
e-mail: aiyelokuntobi@gmail.com

G. Ogunsanwo
Department of Computer and Information Science, Tai Solarin University of Education, Ijebu-Ode, Ogun, Nigeria

© Springer Nature Singapore Pte Ltd. 2021

R. C. Deo et al. (eds.), *Intelligent Data Analytics for Decision-Support Systems in Hazard Mitigation*, Springer Transactions in Civil and Environmental Engineering,
https://doi.org/10.1007/978-981-15-5772-9_3

Organization (WMO 2012) encourages the adoption of the Standardized Precipitation Index (SPI) to monitor drought events; however, extant literature (Bordi et al. 2004; Kumar et al. 2009; Juan et al. 2013; Kumari et al. 2018) also commended the use of SPI for flood related studies. This is because negative SPI symbolizes rainfall deficit, while positive SPI values represent rainfall surplus leading to flooding (Wu 2013).

Due to its robustness and convenience to use, SPI has already been widely used to characterize drought and flooding conditions in many countries and regions, such as the USA (Wu et al. 2007), Argentina (Seiler et al. 2002), Canada (Quiring and Papakryiakou 2003), Italy (Piccarreta et al. 2004; Vergni and Todisco 2010), Iran (Moradi et al. 2011; Nafarzadegana et al. 2012), Korea (Min et al. 2003; Kim et al. 2009), Ghana (Nyatuame and Agodzo 2017), China (Liu et al. 2018), and in Nigeria (Aiyelokun et al. 2017; Omonijo and Okogbue 2014). It is based on the validity of the adoption of SPI in drought and flood risk investigations that this chapter focuses on the use of SPI for risk reduction of these natural disasters in North Central Nigeria.

## 3.2 Intelligent Algorithms for Drought and Flood Risk Reduction

Models are decision support tools for disaster risk reduction, and stakeholders usually rely on models to support and aid in the decision-making process (Dolcine et al. 2010). Intelligent data analytic, machine learning (ML), or data-driven models, as they are known, are alternative to physical models and are being popularly experimented in a variety of hydrologic and climatic studies (Deo and Şahin 2015).

ML models employ, incorporate, and learn from past observational datasets to predict future events (Deo and Şahin 2015). A lot of ML algorithms have lately been proposed in the literature for hydro-meteorological investigations, including the co-integration methods (Kaufmann and Stern 2002; Kaufmann et al. 2011), support vector machine (Bray and Han 2004; Botsis et al. 2011) regression approaches (Douglass et al. 2004; Stone and Allen 2005), artificial neural networks (Abbot and Marohasy 2012; Abbot and Marohasy 2014), wavelet or vector regression (Belayneh and Adamowski 2012), rule-based fuzzy inference systems (Asklany et al. 2011), Naïve Bayes (Sriram and Suresh 2016), and random forest (Park et al. 2019). The use ML algorithm has been wildly acknowledged because of their ability to explain outwardly driven climate not requiring to deploy complex physical models, simplicity of experimentation, validation and evaluation, near to the ground computational trouble, trouble-free and prompt training and the testing stages, the pertinence to data and viable performance in comparison with physical models (Deo and Şahin 2015). ML models have the competence to recognize multifaceted nonlinear relationships between input and output data sets without the necessity of understanding the nature of the phenomena and without making any underlying assumptions regarding linearity or normality (Abudu et al. 2011). Because of all these advantages, ML model has been used to study the natural behavior of hydrological processes

(El-shafie et al. 2011), most especially those in relation to rainfall, whose surplus or deficit determines flood or drought. ANN, for example, was used by Sahai et al. (2000) to predict the seasonal and monthly mean summer monsoon rainfall over the whole of India, using only rainfall time series as inputs. Toth et al. (2000) developed an autoregressive moving average (ARMA) model, ANN, and k-nearest neighbor (K-NN) method to forecast rainfall. Kihoro et al. (2004) comparatively evaluated the performance of ANN to the univariate time series forecasting model ARIMA in forecasting various monthly time series data and showed that the ANN is relatively better than ARIMA models in forecasting ability. Somvanshi et al. (2006) made a comparative study of the complexity of the nature and behavior of annual rainfall record as obtained by ARIMA and ANN techniques and revealed that ANN model outperformed ARIMA model. Iseri et al. (2005) computed the partial mutual information between August rainfall in Fukuoka, Japan, and hydro-climatic variables in order to identify the predictors and forecasted August rainfall with the identified predictors using ANN. Sarkar et al. (2006) developed backpropagation ANN runoff models to simulate and forecast daily runoff for a part of the Satluj River basin of India. Kumar et al. (2007) adopted ANN for individual months and for seasonal rainfall prediction using climate indices as predictor variables. Karamouz et al. (2009) compared ANN with a statistical downscaling model (SDSM) for rainfall prediction and concluded that the SDSM performance is better than the ANN model, even though, in comparison, it is a more data-intensive model than ANN. Dastorani et al. (2010) applied ANN as well as ANFIS models to predict future precipitation in the hyper-arid region of Yazd in Iran and found comparable prediction performance of these tools. Khalili et al. (2011) used ANNs to obtain a forecasting model for the daily rainfall of Mashhad, Iran, using only the past information of the system and got satisfactory prediction performance. But Geetha and Selvaraj (2011) used ANN to predict monthly rainfall in Chennai and concluded that ANN could not predict the sharp peak values.

In addition, ML has been used in a number of studies as a drought forecasting tool (Belayneh et al. 2014; Deo and Şahin 2015; Aiyelokun et al. 2017; Liu et al. 2018; Park et al. 2019); as well as for flood-related investigations (Song et al. 2012; Kumar and Rajpoot 2013; Aiyelokun et al. 2018). Having instituted the wild applicability of ML in flood and drought related investigations, the present chapter seeks to apply the Naïve Bayes classification algorithm for the prediction of monthly SPI in North Central Nigeria, which will serve as a decision support tool for the flood and drought risk reduction in the study area.

## 3.3 Materials and Method

### 3.3.1 *Study Area*

The study area lies between latitudes 9°37′–9°79′ N and longitude 6°16′–6°65′ E. Minna is the capital of Niger state, and one of the major growing states of North Central Nigeria (63). Minna is located at about 150 km from Abuja which serves as the capital of the Federal Republic of Nigeria and has a total population of approximately 506,113 with an average population density of about 3448 persons per km$^2$ (UNDP/NISEPA 2009). The population growth in the city is higher than the average of the whole country because of its proximity to Abuja, the new administrative capital of the country (Musa 2012).

The geologic formation of the study area is based on the undifferentiated basement complex of mainly gneiss and magnetite (Ishaku 2011). The climate of Minna lies within a region described as a tropical climate (Ishaku 2011). The region has a tropical dry and wet climate characterized by double rainfall maxima with a mean annual precipitation of 1300 mm (Ishaku 2011). The rainy season commences most of the time in April and lasts till October, with fluctuations in quantity received annually. The highest mean monthly rainfall occurs September with almost 300 mm. Temperature is uniformly high throughout the year reaching the peaks of 40 °C (Feb./March) and 30 °C (Nov./Dec.) (Ishaku 2011) (Fig. 3.1).

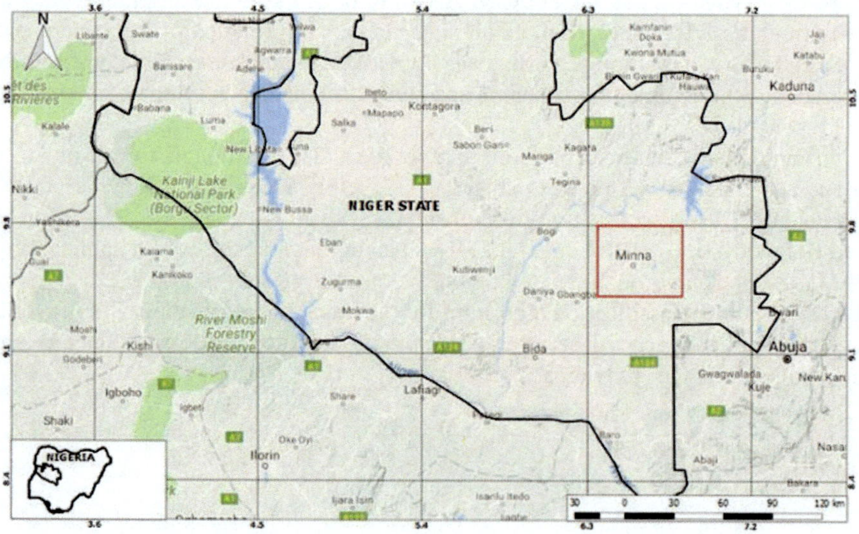


**Fig. 3.1** Map of the study area

### 3.3.2 Materials

Hydro-meteorological data for Minna rainfall station was obtained from the Nigerian Meteorological Agency (NIMET), which includes rainfall, maximum temperature, minimum temperature, sunshine hour, relative humidity, pan evaporation, and wind speed. The time series of the data covered a period of 660 months (1961–2015).

### 3.3.3 Modified Penman–Monteith's Method

The Penman–Monteith method, which was modified by the Food and Agricultural Organization (FAO) in 1963, was adopted for estimating reference evapotranspiration of the study area. The modified Penman–Monteith method according to Doorenbos and Pruitt (1977) and Jennifer (2001) can be mathematically expressed as follows:

$$ET_r = \frac{C}{\rho_w}\left[W\frac{(R_n - G)}{\lambda} + (1 - W)f(u)(e_a - e_d)\right] \quad (3.1)$$

where $ET_r$ is the reference evapotranspiration, mm/day; $W$ is a temperature and the altitude of the area related weighting factor; $R_n$ is the net solar radiation, $MJm^{-2}d^{-1}$; $G$ is the soil heat flux in $MJm^{-2}d^{-1}$; $\lambda$ is the latent heat of evaporation, MJ $kg^{-1}$; $\rho_w = 1000$ $kgm^{-3}$ is the density of water; $f(u)$ is the wind-related function, kg $hPa^{-1}m^{-2}d^{-1}$; $e_a$ is the saturation vapor pressure at mean air temperature, hPa; $e_d$ is the mean actual vapor pressure of the air, hPa; and $C$ is an adjustment factor to account for day and night weather conditions.

### 3.3.4 Standardized Precipitation Index (SPI)

The most commonly used distribution for SPI calculation is the two-parameter gamma distribution with a shape and scale parameter, which is defined by its probability density function:

$$G(x) = \frac{1}{\beta^{\alpha}\tau(\alpha)}\int_0^x x^{\alpha-1}e^{-\alpha/\beta}dx \; for\, x > 0 \quad (3.2)$$

where $\alpha$ is the shape parameter, $\beta$ is the scale parameter, $x$ is the precipitation value, and $\tau(\alpha)$ is the gamma function. The gamma distribution is undefined for $x = 0$, but the precipitation may have zero value, so the cumulative probability distribution given a zero value is derived as follows:

$$H(x) = q + (1 - q)\,G(x) \quad (3)$$

**Table 3.1** Dryness/wetness categories according to SPI values

| Classification | Class | SPI values |
|---|---|---|
| Extremely wet | EW | ≥ 2.0 |
| Severely wet | SW | 1.50–1.99 |
| Moderate wet | MW | 1.00–1.49 |
| Near normal | NN | 0.99 to −0.99 |
| Moderately dry | MD | −1.00 to −1.49 |
| Severely dry | SD | −1.50 to −1.99 |
| Extremely dry | ED | ≤ − 2.0 |

where q is the probability of the zero-precipitation value. The cumulative probability distribution is then transformed into the standard normal distribution to calculate SPI. The value of SPI indicates the strength of the anomaly. Mckee et al. (1993) suggested a classification system to define the intensity of dry/wet phases (Table 3.1). In this study, one-month timescale SPI was adopted, while drought and flood conditions were adopted to represent dry and wet phases, respectively.

### 3.3.5 Naïve Bayes Classification Algorithm

Naïve Bayes classification originates from the Bayes theorem, and it is both a supervised learning and statistical algorithm for classification. Naïve Bayes is an extremely straightforward Bayesian network that consists of directed acyclic graphs (DAG) (Kotsiantis 2007). Although Naïve Bayes is simple and straightforward, it has been confirmed to perform even better than known complicated classification methods (Sriram and Suresh 2016). According to Sriram and Suresh (2016), Naïve Bayes classification in relation to Bayes theorem endeavors to estimate the posterior probability, $P(c|x_i,...,x_n)$, from $P(c)$, $P(x)$ and $P(x|c)$; for each of c possible outcomes or class (Narasimha and Susheela 2011). By decomposing the conditional probability, $P(c|x_i,...,x_n)$ can be represented by Eq. 3.1 and simplified in Eq. 3.2.

$$P(c|x_i\,...,x_n) = \frac{P(C)P(X|C)}{P(X)} \tag{3.4}$$

$$\text{Posterior} = \frac{\text{Prior} * \text{likelihood}}{\text{evidence}} \tag{3.5}$$

Because this study employed the use of meteorological data which are continuous, the Gaussian Naïve Bayes technique was adopted, based on the assumption that the weather parameters are normally distributed (Gaussian distribution). In order to develop the Naïve Bayes models used in predicting drought and flood, the monthly data were randomly divided into two groups with a ratio from 70 to 30%. The 70% of the data was used for training of model development, while the 30% was used for the testing. In general, two models were developed, the first model was constructed using SPI classification as the response variable, and meteorological parameters such as rainfall, maximum temperature, minimum temperature, sunshine hour, relative humidity, pan evaporation, wind speed and potential evapotranspiration as predictors; while the computed SPI values for each month were added to the predictors of the second model. The Waikato Environment for Knowledge Analysis (WEKA) was used for Naïve Bayes model development.

### 3.3.6 *Performance Evaluation*

The performance of the models developed was assessed by seven criteria which include correct classification instance, incorrect classification instance, Kappa statistics, mean absolute error, root-mean-square error, relative absolute error, and root relative squared error.

## 3.4 Results

The probability of occurrence of the different SPI classes based on the period of study is shown in Fig. 3.2. The figure shows that the months of the near-normal class have the highest probability of 0.72, followed by moderately wet with 0.09, severely wet and moderately dry with 0.05, respectively, severely dry had a probability of occurrence of 0.04, while extremely wet and extremely dry has a probability of

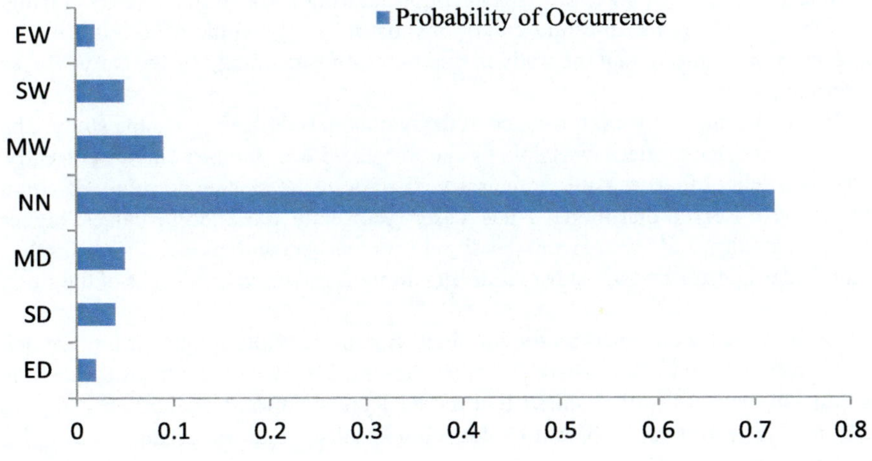


**Fig. 3.2** Probability of occurrence of SPI classes

**Table 3.2** Classification of dry and wet conditions based on meteorological parameters

| | | Intensity of dry and wet | | | | | | |
|---|---|---|---|---|---|---|---|---|
| | | ED | SD | MD | NN | MW | SW | EW |
| Rainfall (mm) | Mean | 45.21 | 48.52 | 72.62 | 89.94 | 162.53 | 213.68 | 236.44 |
| | Std | 25.88 | 38.26 | 31.37 | 85.30 | 112.96 | 119.15 | 149.22 |
| Max temperature (°C) | Mean | 31.50 | 31.12 | 30.91 | 32.46 | 32.44 | 32.30 | 33.15 |
| | Std | 1.84 | 2.88 | 2.08 | 2.50 | 2.31 | 2.60 | 1.90 |
| Min temperature (°C) | Mean | 22.00 | 21.66 | 21.80 | 21.16 | 21.13 | 21.39 | 21.47 |
| | Std | 1.10 | 1.17 | 1.34 | 1.75 | 1.99 | 1.24 | 1.05 |
| Relative humidity (%) | Mean | 79.51 | 77.56 | 79.00 | 73.30 | 76.12 | 78.04 | 77.64 |
| | Std | 4.45 | 9.48 | 8.95 | 13.32 | 10.70 | 9.70 | 8.53 |
| Evaporation (mm) | Mean | 3.99 | 4.34 | 3.75 | 5.16 | 4.43 | 4.38 | 4.60 |
| | Std | 1.51 | 2.96 | 2.13 | 2.87 | 2.47 | 2.75 | 2.48 |
| Sunshine duration (h) | Mean | 6.39 | 5.18 | 5.40 | 6.35 | 6.47 | 5.87 | 6.77 |
| | Std | 1.16 | 1.66 | 1.76 | 1.51 | 1.36 | 1.41 | 0.86 |
| Wind speed ($m^2s$) | Mean | 4.16 | 4.43 | 4.61 | 4.24 | 4.33 | 4.17 | 4.00 |
| | Std | 1.04 | 1.18 | 1.35 | 1.45 | 1.56 | 1.53 | 0.88 |
| Potential Evapotranspiration (mm) | Mean | 151.58 | 152.02 | 150.71 | 160.27 | 161.50 | 157.36 | 164.32 |
| | Std | 21.37 | 34.29 | 22.29 | 25.80 | 25.70 | 27.55 | 15.37 |
| SPI | Mean | −2.35 | −1.71 | −1.19 | 0.12 | 1.20 | 1.70 | 2.35 |
| | Std | 0.36 | 0.15 | 0.14 | 0.48 | 0.15 | 0.14 | 0.28 |

occurrences of 0.02 each. Implying that based on a one-month timescale, extreme weather condition leading to drought and flood has a high probability of not occurring.

Table 3.2 shows the different conditions that necessitated the occurrence of dry and wet phases throughout the study period based on data mining by the Naïve Bayes algorithm.

Figure 3.3 shows the performance of the two models developed for the study. The figure shows that model 1 was able to classify 92.93% of the SPI classes correctly, against model 2 that was able to classify 65.65% of the classes correctly. It could be further observed that model 1 has lesser incorrectly classified instances, higher Kappa statistics, and lower error statistics, in comparison with model 2. This implies that model 1 performed better in classifying drought and flood conditions of the study area.

Both correctly and incorrectly classified instances of the models are presented in the form of confusion matrix in Tables 3.3 and 3.4. A visual comparison of the confusion matrices further shows that model 1 outperforms model 2. This implies that the Naïve Bayes classification algorithm is sensitive to the predictors used for its development.

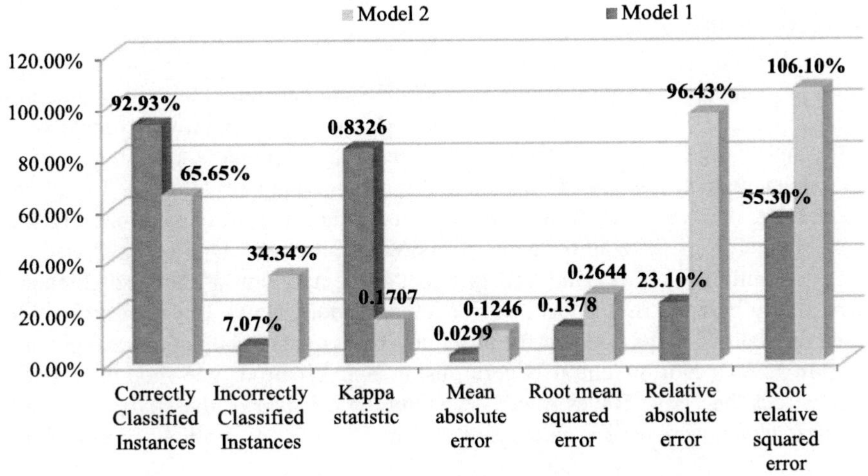


**Fig. 3.3** Performance evaluation of Naïve Bayes models

**Table 3.3** Confusion matrix for model 1

| NN | SD | MD | MW | SW | EW | ED | |
|---|---|---|---|---|---|---|---|
| 144 | 0 | 2 | 1 | 0 | 0 | 0 | NN |
| 1 | 3 | 0 | 0 | 0 | 0 | 1 | SD |
| 1 | 0 | 9 | 0 | 0 | 0 | 0 | MD |
| 3 | 0 | 0 | 17 | 0 | 0 | 0 | MW |
| 0 | 0 | 0 | 1 | 6 | 0 | 0 | SW |
| 0 | 0 | 0 | 0 | 2 | 2 | 0 | EW |
| 1 | 1 | 0 | 0 | 0 | 0 | 3 | ED |

**Table 3.4** Confusion matrix 2

| NN | SD | MD | MW | SW | EW | ED | |
|---|---|---|---|---|---|---|---|
| 122 | 2 | 7 | 1 | 12 | 3 | 0 | NN |
| 3 | 2 | 0 | 0 | 0 | 0 | 0 | SD |
| 7 | 0 | 3 | 0 | 0 | 0 | 0 | MD |
| 11 | 0 | 0 | 0 | 6 | 2 | 1 | MW |
| 5 | 0 | 0 | 0 | 2 | 0 | 0 | SW |
| 4 | 0 | 0 | 0 | 0 | 0 | 0 | EW |
| 2 | 0 | 2 | 0 | 0 | 0 | 1 | ED |

## 3.5 Discussion

There are general speculations and scientifically proven indications that climate-induced hazards will increase due to climate change and variability. Climate variability and change have been confirmed to affect rainfall patterns, runoff, as well as the frequency and intensity of floods and droughts (Traore et al. 2018). Therefore, stakeholders need to assess different options in the quest to proffer solutions targeted at the reduction of risks associated with flood and drought. One of those options in risk reduction of climate-induced extremes is the adoption of models, which are increasingly being harnessed as decision support tools. Naïve Bayes classification based on Gaussian was adopted for this study by assessing short-term drought and flood in Minna, North Central Nigeria, using a timescale of one month. In order to construct the Naïve Bayes models, meteorological data of the study area were assessed, which were used in calculating potential evapotranspiration and SPI for six hundred and sixty (660) months.

Based on the Naïve Bayes models constructed, it was observed that those months with near-normal SPI had a higher probability of occurrence (0.72) than the combination of dry phases (0.11) and wet phases (0.16). Implying that based on past records of the study area, wet conditions are more likely to occur than dry ones. This corroborates the findings of Omonijo and Okogbue (2014) which unraveled that Minna experienced drought irrespective of timescales, in three decades out of the ten decades that were investigated; many other investigations have been performed to ascertain that flood is a major problem in Minna (Doorenbos and Pruitt 1977; Dalil et al. 2015; Adeleye and Ayangbile 2016).

The study also presents the meteorological conditions that necessitate the classification of the data into different dry and wet phases in Table 3.1, from the table that rainfall of 45.2 mm $\pm$ 25.9, maximum temperature of 31.5 °C $\pm$ 1.8, minimum temperature of 22.0 °C $\pm$ 1.1, relative humidity of 79.5 $\pm$ 4.4, evaporation of 4.0 $\pm$ 2, sunshine hour of 6.4 h $\pm$ 1.2, wind speed of 4.1 ms$^{-1}$ $\pm$ 1.0, potential evapotranspiration of 151.6 mm $\pm$ 21.4, and SPI of -2.4 $\pm$ 0.4, necessitate the occurrence of extreme droughts; while rainfall of 236.4 mm $\pm$ 149.2, maximum temperature of 33.2 °C $\pm$ 1.9, minimum temperature of 21. °C $\pm$ 1.1, relative humidity of 77.6 $\pm$ 8.5, evaporation of 4.6 + 2.5, sunshine hour of 6.8 h $\pm$ 0.8 wind speed of 4.0 ms$^{-1}$ $\pm$ 1.0, potential evapotranspiration of 164.3 mm $\pm$ 15.4, and SPI of 2.4 $\pm$ 0.3 necessitate extreme floods.

Furthermore, ML algorithms have high a dependency on data than physical based models; the performance of the models constructed was tested by seven statistics and confusion matrices, which confirmed that Naïve Bayes models with SPI values as part of its predictors performed better than the second model, whose predictors did not include SPI values. This could have been as a result of the variability of climate data in form of noise. Extant literature has revealed that characteristics of data used in building ML models can affect classification accuracy and performance, for instance, Maksoud et al. (2019) posited that noise in data reduces ML system performance by increasing time taken to build classifiers, while Gill et al. (2007)

showed that the performance accuracy of ANN and SVM reduced with increase in the percentage of missing data.

Finally, application of ML in real-time extreme event monitoring systems has increased globally. Therefore, the finding of this study will serve as guidelines for hydrologists, climatologists, water resources engineers, drainage and irrigation engineers, and other stakeholders whose major interest is reducing risk of extreme natural hazard in the study area.

## 3.6 Concluding Remarks

Majority of the months in the study area were in the near-normal class, while drought and flood had low probability of occurrence. The meteorological conditions that necessitate flood and drought have been mined and presented in the study, while it was instituted that predictive performance of Gaussian Naïve Bayes classification algorithm used in predicting climate-induced hazards is sensitive to predictor variables employed.

## References

Abbot J, Marohasy J (2012) Application of artificial neural networks to rainfall forecasting in Queensland. Australia Adv Atmos Sci 29(4):717–730

Abbot J, Marohasy J (2014) Input selection and optimisation for monthly rainfall forecasting in Queensland, Australia, using artificial neural networks. Atmos Res 138:166–178

Abudu S, King JP, Bawazir AS (2011) Forecasting monthly streamflow of spring-summer runoff season in Rio Grande headwaters basin using stochastic hybrid modeling approach. J Hydrol Eng 16(4):384–390

Adeleye B, Ayangbile O (2016) Flood vulnerability: impending danger in Sabon-Gari Minna, Niger State, Nigeria. Ethiopian J Environ Stud Manage 9(1):35–44

Aiyelokun O, Ogunsanwo G, Fabiyi O (2017) Artificial intelligence based drought predictions in part of the tropics. J Urban Environ Eng 11(2):165–173

Aiyelokun O, Ogunsanwo G, Adelere J, Agbede O (2018) Modeling and simulation of river discharge using artificial neural networks. Ife J Sci 20(2):207–214. https://doi.org/10.4314/ijs.v20i2.17

Asklany SA, Elhelow K, Youssef I, Abd El-wahab M (2011) Rainfall events prediction using rule-based fuzzy inference system. Atmos Res 101(1):228–236

Belayneh A, Adamowski J (2012) Standard precipitation index drought forecasting using neural networks, wavelet neural networks, and support vector regression. Appl Comput Intell Soft Comput 2012:6

Belayneh AJ, Adamowski B Khalil, Ozga-Zielinski B (2014) Long-term SPI drought forecasting in the Awash River Basin in Ethiopia using wavelet neural network and wavelet support vector regression models. J Hydrol 508:418–429

Berghuijs WR, Aalbers EE, Larsen JR, Trancoso R, Woods RA (2017) Recent changes in extreme floods across multiple continents, Environ Res Lett 12: 1–8. https://doi.org/10.1088/1748-9326/aa8847

Bordi I, Fraedrich K, Jiang M, Sutera A (2004) Spatio-temporal variability of dry and wet periods in eastern China. Theoret Appl Climatology 79:81–91. https://doi.org/10.1007/s00704-004-0053-8

Botsis D, Latinopulos P, Diamantaras K (2011) Rainfall runoff modeling using support vector regression and artificial neural networks, CEST2011- Rhodes, Greece

Bradford RA, O'Sullivan JJ, van der Craats IM, Krywkow J, Rotko P, Aaltonen J, Bonaiuto M, De Dominicis S, Waylen K, Schelfaut K (2012) Risk perception—issues for flood management in Europe. Nat Hazards Earth Syst Sci 12:2299–2309

Bray M, Han D (2004) Identification of support vector machines for runoff modeling. J Hydro 06.4, IWA Publishing

Dalil M, Mohammad N, Husaini A, Mohammed S (2015) An assessment of flood vulnerability on physical development along drainage channels in Minna, Niger State, Nigeria. Afr J Environ Sci Technol 9(1):38–46

Dastorani MT, Afkhami H, Sharifidarani H, Dastorani M (2010) Application of ANN and ANFIS models on dry land precipitation prediction (Case study: Yazd in Central IRAN). J Appl Sci 10(20):2387–2394

Deo R, Şahin M (2015) Application of the extreme learning machine algorithm for the prediction of monthly effective drought index in eastern Australia. Atmos Res 153:512–525

Dolcine L, Prévil C, Brham A, Ahluwalia H, El-Menaoui (2010) Inflow Modelling and reservoir management in Souss-Massa (Morocco). (Eds) UNESCO (2010). Technical Documents in Hydrology. France: International Hydrological Programme (IHP) of the United Nations Educational, Scientific and Cultural Organization (UNESCO)

Doorenbos J, Pruitt WO (1977) Crop water requirements. Irrigation and drainage paper (24): 144, (rev.) FAO, Rome, Italy

Douglass DH, Blackman EG, Knox RS (2004) Temperature response of Earth to the annual solar irradiance cycle. Phys Lett A 323(3):315–322

El-shafie A, Mukhlisin M, Najah AA, Taha MR (2011) Performance of artificial neural network and regression techniques for rainfall-runoff prediction. International Journal of the Physical Sciences. 6(8):1997–2003

Geetha G, Selvaraj RS (2011) Prediction of monthly rainfall in Chennai using back propagation neural network. Int J Eng Sci Technol 3(1):211–213

Gill MK, Asefa T, Kaheil Y, McKee M (2007) Effect of missing data on performance of learning algorithms for hydrologic predictions: implications to an imputation technique. Water Resour Res 43:W07416. https://doi.org/10.1029/2006WR005298

Henstra D, Minano A, Thistlethwaite J (2019) Communicating disaster risk? an evaluation of the availability and quality of flood maps. Nat Hazards Earth Syst Sci 19:313–323

Iseri Y, Dandy G, Maier H, Kawamura A, Jinno K (2005) Medium term forecasting of rainfall using artificial neural networks. Int Congress on Model Simul 1834–1840

Ishaku JM (2011) Assessment of groundwater quality index for JimetaYola area, north eastern Nigeria. J Geol and Min Res 3(9): 219–23 I

Jennifer MJ (2001) Faculty investigator evaluation of reference evapotranspiration methodologies and afsirs crop water use simulation model (Final Report). Department of Civil and Coastal Engineering University of Florida, Gainesville, Florida

Juan D, Jian F, Wei X, Peijun S (2013) Analysis of dry/wet conditions using the standardized precipitation index and its potential usefulness for drought/flood monitoring in Hunan Province. China, Stochastic Environ Res Risk Assesment 27:377–387. https://doi.org/10.1002/joc.1371

Karamouz M, Fallahi M, Nazif S, Farahani MR (2009) Long lead rainfall prediction using statistical downscaling and artificial neural network modeling. Trans A: Civil Eng 16(2):165–172

Kaufmann RK, Stern DI (2002) Cointegration analysis of hemispheric temperature relations. J Geophys Res Atmos (1984–2012) 107 (D2), 8–10 (ACL 8- 1-ACL)

Kaufmann RK, Kauppi H, Mann ML, Stock JH (2011) Reconciling anthropogenic climate change with observed temperature 1998–2008. Proc Natl Acad Sci 108(29):11790–11793

Khalili N, Khodashenas SR, Davary K, Karimaldini F (2011) Daily rainfall forecasting for Mashhad synoptic station using artificial neural networks. Int Conference on Environ Comput Sci 19:118–123

Kihoro JM, Otieno RO, Wafula C (2004) Seasonal time series forecasting: a comparative study of ARIMA and ANN models. African J ScimTechnol, Sci and Eng Series 5(2):41–49

Kim DW, Byun HR, Choi KS (2009) Evaluation, modification, and application of the effective drought index to 200-year drought climatology of Seoul, Korea. J Hydrol 378:1–12

Kotsiantis SB (2007) Supervised machine learning: a review of classification techniques. Informatica 31(3):249–268

Kumar A, Rajpoot PS (2013) Assessment of hydro-environmental loss as surface runoff using CN method of Pahuj River Basin Datia, India. Proceedings of the international academy of ecology and environmental sciences 3(4):324–329

Kumari A, Mayoor M, Mahapatra S, Singh H, Parhi P (2018) Flood risk monitoring of koshi river basin in north plains of bihar state of india, using standardized precipitation index. Int J Adv Innovative Res 5(3):21–30

Kumar DN, Reddy MJ, Maity R (2007) Regional rainfall forecasting using large scale climate teleconnections and artificial intelligence techniques. J Intell Syst 16(4):307–322

Kumar MN, Murthy CS, Saiand MVR, Roy SPS (2009) On the use of standardized precipitation index (SPI) for drought intensity assessment. R Meteorological Soc. https://doi.org/10.1002/joc.799

Liu DL, Li Y, Shen X, Xie YL, Zhang YL (2018a) Flood risk perception of rural households in western mountainous regions of Henan Province. China Int J Disaster Risk Reduct 27:155–160

Liu DH, You JF, Xie QJ, Huang YY, Tong HJ (2018b) Spatial and temporal characteristics of drought and flood in quanzhou based on standardized precipitation index (SPI) in recent 55 years. J Geosci Environ Prot 6:25–37. https://doi.org/10.4236/gep.2018.68003

Maksoud EAA, Ramadan M, Elmogy M (2019) A computer-aided diagnoses system for detecting multiple ocular diseases using color retinal fundus images. In: Nilanjan D, Surekha B, Amira A, Fuqian S Machine Learning in Bio-Signal Analysis and Diagnostic Imaging. Science Direct. 19:52

McKee, T.B. Doesken, N.J. Kleist, J. (1993). The relationship of drought frequency and duration to time scales. In: proceedings of the 8th conference on applied climatology. American Meteorology Society, Boston. 1993:179–184

Min SK, Kwon WT, Park EH, Choi Y (2003) Spatial and temporal comparisons of droughts over Korea with East Asia. Int J Climatol 23:223–233

Mishra AK, Desai VR (2006) Drought forecasting using feed-forward recursive neural network. Ecol Model 198(1–2):127–138

Mochizuki J, Naqvi A (2019) Reflecting disaster risk in development indicators. Sustainability 11(4):2–14

Moradi HR, Rajabi M, Faragzadeh M (2011) Investigation of meteorological drought characteristics in Fars Province. Iran. Catena. 84:35–46

Musa JJ (2012) Environmental appraisal of drainage system in Minna, Niger State-Nigeria. Int J Appl Biological Res 4(1&2):85–93

Nafarzadegana AR, Zadeha MR, Kherada M, Ahania H, Gharehkhania A, Karampoora MA, Kousari MR (2012) Drought area monitoring during the past three decades in Fars Province. Iran. QuatInt 250:27–36

Narasimha M, Susheela D (2011) Pattern recognition: an algorithmic approach. ISBN 978-0857294944

Nyatuame M, Agodzo S (2017) Analysis of extreme rainfall events (drought and flood) over tordzie watershed in the volta region of ghana. J Geosci Environ Prot 5:275–295. https://doi.org/10.4236/gep.2017.59019

Oduro-Afriyie K, Adukpo DC (2006) Spectral characteristics of the annual mean rainfall series in ghana. West Afr J Appl Ecology 19:1–9

Omonijo T, Okogbue E (2014) Trend analysis of drought in the guinea and sudano-sahelian climatic zones of Northern Nigeria (1907–2006). Atmos Climate Sci 4:483–507. https://doi.org/10.4236/acs.2014.44045

Pita MF (2007) Recomendaciones para el establecimiento de un sistema de Indicadores para la previsión, el seguimiento y la gestión de la sequía. In La Sequía en España. Directrices Para Minimizar su Impacto. Comité de Expertos en Sequía del Ministerio de Medio Ambiente; Ministerio de Medio Ambiente: Madrid. Spain 2007:69–85

Park H, Kim K, Lee D (2019) Prediction of severe drought area based on random forest: using satellite image and topography data. Water 11(705):2–15

Piccarreta M, Capolongo D, Boenzi F (2004) Trend analysis of precipitation and drought in Basilicata from 1923 to 2000 within a southern Italy context. Int J Climatol 24:907–922

Quiring SM, Papakryiakou TN (2003) An evaluation of agricultural drought indices for the Canadian prairies. Agric For Meteorol 118:49–62

Sahai AK, Soman MK, Satyan V (2000) All India summer monsoon rainfall prediction using an artificial neural network. Clim Dyn 16:291–302

Sarkar A, Agarwal R, Singh D (2006) Artificial neural network models for rainfall-runoff forecasting in a hilly catchment. J Indian Water Res Soc 26(3–4):1–4

Seiler RA, Hayes M, Bressan L (2002) Using the standardized precipitation index for flood riskmonitoring. Int J Climatology 22: 1365–1376. https://doi.org/10.1002/met.136

Somvanshi VK, Pandey OP, Agrawal PK, Kalanker NV, RaviPrakash M, Chand R (2006) Modeling and prediction of rainfall using artificial neural network and ARIMA techniques. J Indian geophys Union 10(2):141–151

Song Y, Gong J, Gao S, Wang D, Cui T, Li Y, Wei B (2012) Susceptibility assessment of earthquake-induced landslides using Bayesian network: a case study in Beichuan, China. Comput Geosci 42:189–199. https://doi.org/10.1016/j.cageo.2011.09.011

Sriram K, Suresh K (2016) Machine learning perspective for predicting agricultural droughts using Naïve Bayes algorithm. Middle-East J Sci Res 24:178–184

Stone DA, Allen M (2005) Attribution of global surface warming without dynamical models. Geophys Res Lett 32(18)

WMO (2012) Standardized precipitation index user guide. In: WMO-No. 1090. World meteorological organization, Geneva 2, Switzerland

Toth E, Brath A, Montanari A (2000) Comparison of short-term rainfall prediction models for real-time flood forecasting. J Hydrol 239:132–147

Traore V, Ndiaye M, Diouf R, Malomar G, bakhoum P, Faye M, Abderaman M, Mbow C, Sarr J, Beye A, Diaw A (2018) Variability and change analysis in temperature time series at Kolda Region, Senegal. J Wat Env Sci 2(2): 337–358

UNDP/NISEPA (2009) Niger state framework for integrated sustainable waste management. Niger state strategic waste management framework (Unpublished)

Vargas J, Paneque P (2017) Methodology for the analysis of causes of drought vulnerability on river basin scale. Nat Hazards 89:609–621

Vargas J, Paneque P (2019) Challenges for the integration ofwater resource and drought-risk management in Spain. Sustainability 11(308): 2–16

Vergni L, Todisco F (2010) Spatio-temporal variability of precipitation, temperature and agricultural drought indices in central Italy. Agric For Meteorol 151(3):301–313

Wu M (2013) A brief introduction to standardised precipitation index (SPI). Hong Kong Observatory

Wu H, Svobod MD, Hayes MJ, Wilhite DA, Wen F (2007) Appropriate application of the standardized precipitation index in arid locations and dry seasons. Int J Climatol 27:65–79

# Chapter 4
# Hydrological Drought Investigation Using Streamflow Drought Index

**Anurag Malik, Anil Kumar, Sinan Q. Salih, and Zaher Mundher Yaseen**

## 4.1 Introduction

The negative effects of drought are noticed in almost every aspect of the ecosystem (Nabaei et al. 2019; Qutbudin et al. 2019). As a natural disaster, drought affects various sectors including water supply, hydropower generation, agriculture, and industry (Dobrovolski 2015). The proper water resources planning and management land area of about 3.28 million $km^2$, but out of this land area, nearly 1.07 million $km^2$ is prone to various levels of water unavailability and drought episodes (Subramanya 2005). A study by Wilhite and Glantz (1985) explained the definition of drought using conceptual and operational terms (Wilhite and Glantz 1985). Conceptually, drought is defined in general terms, such as lack of precipitation which causes crop damages and low crop yield. The conceptual definition of drought is significant for the establishment of drought regulations. On the other hand, the operational definition of drought can be helpful in identifying the beginning, severity, and end of

A. Malik · A. Kumar
Department of Soil and Water Conservation Engineering, College of Technology, G.B. Pant University of Agriculture & Technology, Pantnagar, Uttarakhand 263145, India
e-mail: anuragmalik_swce2014@rediffmail.com

A. Malik
Punjab Agricultural University, Regional Research Station, Bathinda, Punjab 151001, India

S. Q. Salih
Computer Science Department, College of Computer Science and Information Technology, University of Anbar, Ramadi, Iraq

Z. M. Yaseen (✉)
Sustainable Developments in Civil Engineering Research Group, Faculty of Civil Engineering, Ton Duc Thang University, Ho Chi Minh City, Vietnam
e-mail: yaseen@tdtu.edu.vn

© Springer Nature Singapore Pte Ltd. 2021

R. C. Deo et al. (eds.), *Intelligent Data Analytics for Decision-Support Systems in Hazard Mitigation*, Springer Transactions in Civil and Environmental Engineering,
https://doi.org/10.1007/978-981-15-5772-9_4

drought episodes based on the assessment of the average of a 30-year record with the current situation.

Drought has affected both groundwater and surface water that includes rivers, reservoirs, ponds, lakes, aquifers, wells, and others (Beyaztas and Yaseen 2019). This means general consequence reduction in water supply, poor water quality, reduced crop productivity, low hydropower generation, and suspended water activities that affect both social and economic activities (Riebsame et al. 1991). An appropriate water resource planning and management must be preceded by a proper drought assessment, and this concept demands a correct understanding of history of droughts episodes, as well as the impact of drought in the given area (Sayl et al. 2016). It is, therefore, necessary to have a good understanding of different concepts of droughts when striving to develop models for investigating drought episodes and their consequences. According to the National Drought Mitigation Center (NDMC), the USA, the appropriate way to mitigate drought is to perform an overall risk analysis based on the previous drought experiences during drought planning (Wilhite 2000). As per (Hayes et al. 2004), it is important to understand drought risk to be in a better position to know the natural hazards and set up a practical model that will help mitigate drought on geographical and political scales.

Globally, the past decade witnessed several studies for hydrological droughts analysis based on several indices (Nalbantis and Tsakiris 2009; Nikbakht et al. 2013; Pathak and Channaveerappa 2016; Razmkhah 2017). For instance, the SDI was developed by (Nalbantis and Tsakiris 2009) to study the characteristics of the hydrological drought of Evinos River basin in Greece. This SDI is similar to SPI and calculates for 3-, 6-, 9-, and 12-month timescale in each hydrological year. The outcome of study reported the effectiveness of SDI in hydrological drought characterization and suggested to apply at a global scale. In another study, Tabari et al. (2013) used SDI to assess the hydrological drought in Urmia Lake and the Kloy River basins in Iran (Tabari et al. 2013). Here, log-normal distribution for the streamflow data was used to calculate the SDI, and the results showed that almost of all the investigated stations experienced extreme drought. They also reported that extreme drought mostly occurred from 1997 to 1998 and 2008 to 2009, respectively. A study by (Sardou and Bahremand 2014) used SDI to examine hydrological drought in Halil Rud basin in Iran. The output of the study showed the variation of drought magnitudes all over the region from upstream to downstream. Another outcome of the study is that the study region had a high correlation between SPI and SDI. Pathak and Channaveerappa (2016) examined the standardized runoff index (SRI) and SDI to evaluate the multiple scale hydrological drought for a 36-year period (1972–2007) in Ghataprabha River basin, India (Pathak and Channaveerappa 2016). From the analysis, there was a moderate drought in the region from 1986 to 1988 and 2001 to 2005, respectively. A good correlation was also established between SRI and SDI for a 9-month period which later increased to a 12-month period in the region. The severity-duration-frequency (SDF) curves are derived by the threshold level method for streamflow drought of Roudzard River basin in Iran which was compared by (Razmkhah 2017). The study also assessed the runoff data of Mashin station (from 1970 to 2012) using 70 and 90% of mean daily runoff, and 70% of monthly average runoff as threshold levels. From the results, the SDF curves revealed an increase in the deficit-volume with a nonlinear trend as the duration increases. Regarding the duration and severities, the threshold levels were observed to vary significantly.

Studies have recently been conducted on the relationship inspection between meteorological and hydrological droughts. For instance, David and Davidová (2017) reported the SPI and SDI approaches in Blanice River catchment located in Bohemia. During the assessment, the monthly precipitation and streamflow data of 50 years record were used to calculate the SPI and SDI. From the results, the hydrological and meteorological drought indices showed a strong correlation from 3 to 6 months. A study by (Gumus and Algin 2017) used the SPI and SDI approaches to analyze meteorological and hydrological droughts in Seyhan–Ceyhan River basins in Turkey. From the observed correlation between SPI and SDI, a specific lag time of one year existed for drought such that in the following year, the meteorological drought occurred as hydrological drought. Choi et al. (2018) used the EDI and threshold-level (TL) approaches to examine meteorological and streamflow droughts in the Milwaukee River basin in Wisconsin for the period of 1951–2006. From the results, a decrease was found in the intensity and duration of both meteorological and streamflow droughts over time. As such, they suggested that the EDI and TL approaches were effective in diagnosing the streamflow and meteorological droughts events of all durations in the study area.

Following these literatures, the current research is conducted on the SDI approach to analyze the hydrological drought and wet characteristics at Naula and Kedar stations situated at the upper Ramganga River catchment, Uttarakhand, India. From the results of this multi-scalar SDI-based analysis, hydrologist, water managers, and policymakers can have a better understanding of the risks and danger associated with human activities and climate changes in the study area.

## 4.2 Materials and Method

### *4.2.1 Study Area and Data Collection*

Among several tributaries on Ganga River, Ramganga River is the foremost. It originates in the outer Himalayas, Chamoli District of Uttarakhand and travels through the districts of Chamoli, Almora, Bageshwar, Nainital, and Pauri Garhwal before entering the plains near Kalagarh dam after coursing about 168 km in the hilly terrain. The Ramganga River catchment area to Kalagarh dam site is 3134 $km^2$ in the shoe-shape and lying between 78° 35′ and 79° 34′ E longitudes and 29° 30′ and 30° 06′ N latitudes, with the altitude varying from 260 to 2950 m above mean sea level (MSL). For hydrological drought prediction, two streamflow gauging stations (i.e., Naula and Kedar) were selected in the upper Ramganga River catchment (Fig. 1). The location details of these stations are given in Table 1. The monthly streamflow data at the selected stations (i.e., Naula and Kedar) were collected from the Divisional

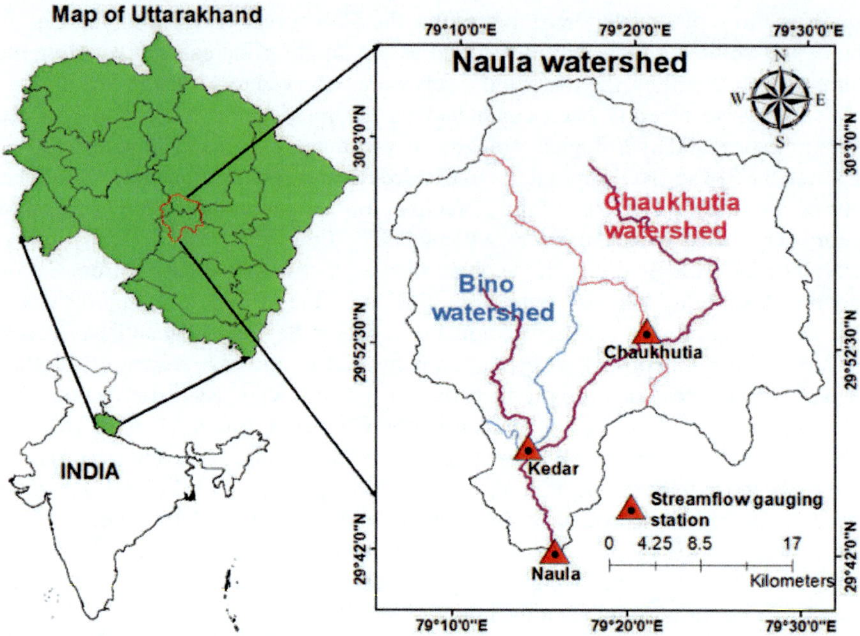


**Fig. 1** Location map of Naula and Kedar stream gauging stations in Ramganga river catchment

**Table 4.1** Details of study stations and streamflow data availability

| Hydrological station | Latitude (N) | Longitude (E) | Altitude (m) | Streamflow data (year) |
|---|---|---|---|---|
| Naula | 29° 04' 20" | 70° 15' 20" | 724 | 1975 to 2007 |
| Kedar | 29° 47' 36" | 79° 14' 12" | 929 | 1975 to 2007 |

Office, Forest and Soil Conservation Department, Ranikhet, Uttarakhand (Table 1). The streamflow at Kedar station is collected from the Bino watershed (295 $km^2$) alone, whereas the streamflow at Naula station includes the combinational flows from Bino and Chaukhutia (570 $km^2$) watersheds and some portions of ungauged watershed (206 $km^2$). In fact, the Chaukhutia watershed contributed to the streamflow at Naula station compared to Bino watershed.

### *4.2.2 Streamflow Drought Index*

The main concept of the SDI was proposed by (Nalbantis 2008) for the analysis of hydrological drought characteristics (i.e., duration, severity, and intensity) at multi-timescales. The computational procedure of the SDI is analogous to the SPI as (Mckee et al. 1993) with the difference that the rainfall values are replaced by streamflow values. The detailed information and application of the SDI can be found in hydrological drought monitoring by (Nalbantis 2008; Borji et al. 2016; Myronidis et al. 2018). In order to compute SDI, firstly, the streamflow data are fitted to the probability distribution functions based on Kolmogorov–Smirnov (K-S). In this study, three probability distributions such as normal, log-normal, and gamma were applied to the streamflow series (i.e., 1-, 3-, 6-, 9-, 12-, and 24-months) and the best one was selected based on K-S test statistic as described below.

#### 4.2.2.1 Normal Distribution Function

The concept of normal distribution was first given by English mathematician Moivre (1738). The probability density function (PDF) of normal distribution is written as (Angelidis et al. 2012; Mandal et al. 2015):

$$f_N(x) = \frac{1}{\sigma\sqrt{2\pi}} \exp\left[-\frac{1}{2}\left(\frac{x-\mu}{\sigma}\right)^2\right], \text{ for } -\infty < \mathrm{x},\ \mu < \infty;\ \sigma > 0 \quad (4.1)$$

where $x$ = the random variable (streamflow); $\mu$ = the mean or location parameter of random variable; and $\sigma$ = the standard deviation or scale parameter of random variable.

#### 4.2.2.2 Log-Normal Distribution Function

Based on the conceptual theory of the probability, the positive random variable (e.g., streamflow) is followed by the log-normal distribution. This is in the case where the logarithm is distributed normally. The PDF of log-normal distribution is given as (Bhattacharjya 2004; Angelidis et al. 2012; Mandal et al. 2015):

$$f_{LN}(x) = \frac{1}{x\sigma_{\ln(x)}\sqrt{2\pi}} \exp\left[-\frac{1}{2}\left(\frac{\ln(x)-\mu_{\ln(x)}}{\sigma_{\ln(x)}}\right)^2\right], \text{ for } \mathrm{x},\ \sigma > 0$$
$$-\infty < \mu < \infty \quad (4.2)$$

where $\mu_{\ln(x)}$ = the scale and $\sigma_{\ln(x)}$ = the shape parameters of the distribution.

#### 4.2.2.3 Gamma Distribution Function

The gamma distribution can be defined using the shape and scale parameters. The PDF of the random variable (e.g., rainfall) is given as (Bhattacharjya 2004; Angelidis et al. 2012; Mandal et al. 2015):

$$f_G(x; \alpha, \beta) = \frac{\beta^\alpha x^{\alpha-1} e^{-\beta x}}{\Gamma(\alpha)} \quad \text{for } x \geq 0;\ \alpha, \beta > 0 \tag{4.3}$$

where $x \geq 0$ = the amount of streamflow; $\alpha, \beta > 0$ = shape and scale parameter; and $\Gamma$ = the gamma function.

#### 4.2.2.4 Kolmogorov–Smirnov Test

After fitting the normal, log-normal, or gamma distributions on streamflow data of multi-timescales of SDI, their goodness-of-fit was evaluated using the Kolmogorov–Smirnov (K-S) test which is a nonparametric approach. The K-S test is not valid for discrete variables. However, it is a statistic test for one sample in two-sided test, which is the maximum absolute difference between the empirical and theoretical cumulative distribution functions (CDFs) (Stephens 1974; Lloyd-Hughes and Saunders 2002; Olea and Pawlowsky-Glahn 2009; Hassani and Silva 2015) such that:

$$D_{\text{cal}} = \max_x |F_n(x) - F(x)| \tag{4.4}$$

where $D_{\text{cal}}$ = the calculated value of the K-S test; $n$ = the number of observations in the population $x$; $\max_x$ = the maximum of the set of distances; $F_n(x)$ and $F(x)$ = the empirical and theoretical CDFs, respectively. The empirical and theoretical CDFs for a continuous variate $X$ ($x_{(1)}, x_{(2)}, \ldots x_{(n)}$ represent the order statistics) of a sample of size n are defined as:

$$F_n(x) = P_n(X \leq x) = \frac{1}{n}\sum_{i=1}^{n} I(X_i - x) \tag{4.5}$$

and

$$F(x) = P(X \leq x) \tag{4.6}$$

where $P$ = the probability of $X$ taking on a value less than or equal to $x$; $I$ = the indicator function, and $F_n(x) = 0$, for $x < x_{(1)}$, $F_n(x) = 1$, for $x \geq x_{(n)}$. when sample size $n > 30$, then the critical value or tabulated value ($D_{\text{tab}}$) at 1% and 5% significance levels are calculated using (Lindgren 1968):

$$\text{D}_{\text{tab}} = \frac{1.628}{\sqrt{n}} \quad \text{(for 1\% significance level)} \tag{4.7}$$

$$\text{D}_{\text{tab}} = \frac{1.358}{\sqrt{n}} \quad \text{(for 5\% significance level)} \tag{4.8}$$

where $D_{\text{tab}}$ = the tabulated value of the K-S test. The critical value ($D_{\text{tab}}$) at 1% and 5% significance levels is 0.2768 and 0.2308 for 33 years streamflow data. Based on these critical values, null ($H_0$) and alternate ($H_1$) hypotheses are proposed as:

$$H_0\text{: data are drawn from the theoretical distribution} \quad (4.9)$$

$$H_1\text{: data are not drawn from the theoretical distribution} \quad (4.10)$$

If the calculated value of K-S test statistic ($D_{\text{cal}}$) smaller than the tabulated value ($D_{\text{tab}}$), then $H_0$ is accepted and $H_1$ is rejected.

### 4.2.3 Classification of Drought and Wet Conditions

The run theory (RT) for the definition of hydrologic drought characteristics (e.g., duration, severity, and intensity) was proposed by (Yevjevich 1967) and was used for drought identification (Fig. 2). It is useful in describing the duration, severity, and intensity of drought event. The truncation or threshold level remains the most important parameter for orienting these characteristics as it may be constant or time-dependent. Therefore, it is defined as a timeseries of drought variable ($X_t$) wherein all the values are either above or below a given truncation level ($\pm 1$); thus, it is either referred to as a negative run (depicting drought) or a positive run (depicting wet condition). The variable ($X_t$), in Fig. 2, intersected the truncation level at several positions, suggesting that the variable may either be a deterministic variable, a stochastic variable, or both. The five major components of drought or wet events based on the run theory are as follows (Yevjevich 1967; Dracup et al. 1980):

Drought initiation time ($t_{di}$) is the starting time of water shortage (SDI $\leq -1$), indicating the beginning of the drought condition; while wet initiation time ($t_{wi}$) is the starting time of water surplus (SDI $\geq$ 1), indicating the beginning of the wet condition (Fig. 2).

Drought termination time ($t_{dt}$) is the end of water shortage (SDI $\geq$ 1), indicating the beginning of the normal or wet condition; while wet termination time ($t_{wt}$) is the end of water surplus (SDI $\leq -1$), indicating the beginning of the normal or drought condition (Fig. 2).

Drought duration ($D_d$) is the time period during which the drought condition prevails while wet duration ($W_d$) is the time period during which the wet condition prevails (Fig. 2). It is expressed in terms of week, month, and year, etc., and mathematically expressed as:

$$D_d = t_{dt} - t_{di} \quad (4.11)$$

$$W_d = t_{wt} - t_{wi} \quad (4.12)$$

Drought severity ($S_d$) is the cumulative value of a drought condition (sum of SDI values during $D_d$), while wet severity ($S_w$) is the cumulative value of a wet

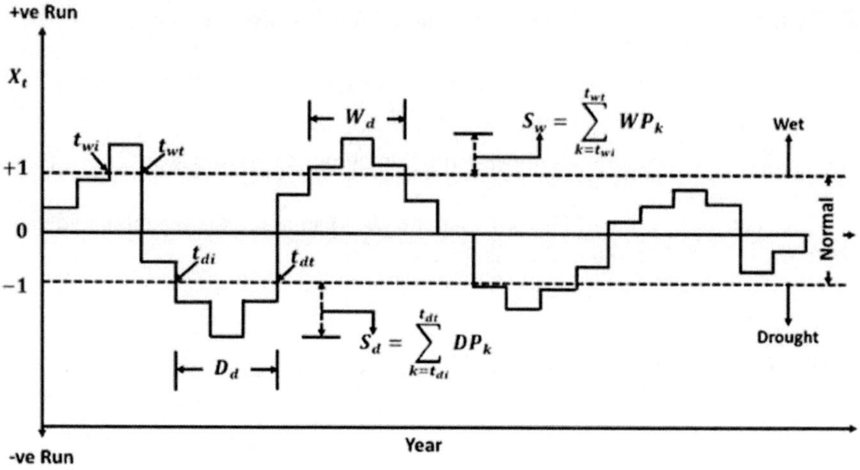


**Fig. 2** Characteristics of drought and wet patterns based on Run Theory

condition (sum of SDI values during $W_d$) (Fig. 2). The drought or wet severity is mathematically written as:

$$S_d = \sum_{k=t_{di}}^{t_{dt}} DP_k \tag{4.13}$$

$$S_w = \sum_{k=t_{wi}}^{t_{wt}} WP_k \tag{4.14}$$

where $DP_k$ and $WP_k$ = the drought or wet parameter (i.e., SDI with timescale $k$), respectively.

Drought intensity ($I_d$) is the average value of a drought condition, while wet intensity ($I_w$) is the average value of a wet condition (Fig. 2). The drought or wet intensity is mathematically written as:

$$I_d = \frac{S_d}{D_d} \tag{4.15}$$

$$I_w = \frac{S_w}{D_w} \tag{4.16}$$

Based on the SDI values, the drought and wet conditions existing at a place could be identified and categorized with specific symbol as presented in Table 2. The table reports that the classification of the SDI categories in which more than 2 present the extreme wet events, between 1.5 and 2 indicate severe wet events, between 1 and 1.5 indicate moderate wet events, between 1 and −1 present the normal events, between

**Table 4.2** Classification of drought and wet conditions based on SDI values

| SDI values | Category | Symbol |
|---|---|---|
| SDI ≥ 2.0 | Extremely wet | C |
| 1.50 ≤ SDI < 2.0 | Severely wet | B |
| 1.00 ≤ SDI < 1.5 | Moderately wet | A |
| -1.0 < SDI < 1.0 | Normal | N |
| -1.5 < SDI ≤ -1.0 | Moderately drought | 1 |
| -2.0 < SDI ≤ -1.5 | Severely drought | 2 |
| SDI ≤ -2.0 | Extremely drought | 3 |

$-1$ and $-1.5$ specify the moderate drought events, between $-1.5$ and $-2$ present severe drought events, and less than $-2$ defines the extreme drought events (Tabari et al. 2013). The drought and wet patterns are also classified based on SDI values by setting a threshold or truncation level. The SDI $\leq -1.0$ represents drought pattern, while SDI $\geq 1.0$ represents wet pattern in the study region.

## 4.3 Results and Discussion

The calculated value of the K-S test for normal, log-normal, and gamma probability distributions is given in Tables 3 and 4 in both stations. The critical values at 1% and 5% significance levels were calculated as 0.2768 and 0.2308 for 33-year streamflow data. As observed from Tables 3 and 4, the normal, log-normal, and gamma distributions fitted very well to the streamflow data series at all the investigated scales of the SDI and for all the months (January to December) at 1 and 5% significant levels. Since the gamma distribution was fitting well for most of months and all-timescales, it was used for further analysis of hydrological drought based on all the inspected SDI scales.

The hydrological drought and wet conditions were analyzed using all investigated SDI in both stations, Ramganga River catchment in Uttarakhand State. The characteristics of drought (SDI $\leq -1.0$) and wet (SDI $\geq 1.0$) conditions on multi-temporal scales of 1-, 3-, 6-, 9-, 12-, and 24-month are presented in Tables 5(a and b) to 7(a and b) in both stations, respectively. These tables provided the information on total number of drought and wet months, number of consecutive drought and wet incidents, drought and wet percentage, average duration, and the longest duration of drought and wet with total severity, average intensity and category, respectively.

**Table 4.3** K-S test statistic of three probability distributions for monthly streamflow data at different time scales at Naula station

| Month/ distribution | | K-S test statistic (D) for monthly streamflow data at different time scales | | | | | |
|---|---|---|---|---|---|---|---|
| | | 1-m | 3-m | 6-m | 9-m | 12-m | 24-m |
| Jan | Normal | 0.1957* | 0.1812* | 0.1475* | 0.1640* | 0.1404* | 0.1164* |
| | Log-normal | 0.1056* | 0.1515* | 0.1385* | 0.1204* | 0.1206* | 0.1178* |
| | Gamma | 0.1297* | 0.1375* | 0.1241* | 0.1196* | 0.1005* | 0.1091* |
| Feb | Normal | 0.1932* | 0.2248* | 0.1392* | 0.1707* | 0.1404* | 0.1181* |
| | Log-normal | 0.1058* | 0.1557* | 0.0711* | 0.1167* | 0.1269* | 0.1093* |
| | Gamma | 0.1777* | 0.1649* | 0.0842* | 0.1272* | 0.1068* | 0.1112* |
| Mar | Normal | 0.2458* | 0.1975* | 0.2112* | 0.2023* | 0.1568* | 0.1164* |
| | Log-normal | 0.1164* | 0.1041* | 0.1458* | 0.1498* | 0.1074* | 0.1035* |
| | Gamma | 0.1514* | 0.1361* | 0.1624* | 0.1666* | 0.1124* | 0.1090* |
| Apr | Normal | 0.1599* | 0.1761* | 0.1568* | 0.1396* | 0.1741* | 0.1130* |
| | Log-normal | 0.0584* | 0.0891* | 0.1176* | 0.1163* | 0.1150* | 0.1272* |
| | Gamma | 0.1025* | 0.1383* | 0.1324* | 0.1034* | 0.1319* | 0.1046* |
| May | Normal | 0.1405* | 0.1966* | 0.1642* | 0.1386* | 0.1867* | 0.1031* |
| | Log-normal | 0.0732* | 0.1116* | 0.0954* | 0.0981* | 0.1315* | 0.1333* |
| | Gamma | 0.0862* | 0.1209* | 0.1243* | 0.0988* | 0.1471* | 0.1090* |
| Jun | Normal | 0.2678* | 0.1654* | 0.2075* | 0.1958* | 0.1996* | 0.1021* |
| | Log-normal | 0.1595* | 0.0946* | 0.1080* | 0.1170* | 0.1471* | 0.1359* |
| | Gamma | 0.1646* | 0.1101* | 0.1425* | 0.1384* | 0.1636* | 0.1114* |
| July | Normal | 0.1809* | 0.1481* | 0.1696* | 0.1540* | 0.1128* | 0.0933* |
| | Log-normal | 0.0917* | 0.0761* | 0.1044* | 0.0855* | 0.1028* | 0.1060* |
| | Gamma | 0.0998* | 0.0949* | 0.1097* | 0.0917* | 0.0909* | 0.0985* |
| Aug | Normal | 0.1534* | 0.1342* | 0.1623* | 0.1499* | 0.1163* | 0.1069* |
| | Log-normal | 0.1350* | 0.0959* | 0.1041* | 0.1109* | 0.0689* | 0.1232* |
| | Gamma | 0.1205* | 0.1014* | 0.1083* | 0.0994* | 0.0674* | 0.1029* |
| Sep | Normal | 0.1759* | 0.1591* | 0.1310* | 0.1245* | 0.1127* | 0.1393* |
| | Log-normal | 0.1003* | 0.1045* | 0.0939* | 0.0933* | 0.0914* | 0.1289* |
| | Gamma | 0.0973* | 0.1084* | 0.0931* | 0.0792* | 0.0871* | 0.1292* |
| Oct | Normal | 0.1969* | 0.1501* | 0.1574* | 0.1447* | 0.1457* | 0.1236* |
| | Log-normal | 0.1063* | 0.1485* | 0.1218* | 0.1191* | 0.1105* | 0.1117* |
| | Gamma | 0.1248* | 0.1317* | 0.1067* | 0.0994* | 0.0939* | 0.1184* |
| Nov | Normal | 0.1398* | 0.2015* | 0.1572* | 0.1524* | 0.1539* | 0.1237* |
| | Log-normal | 0.0776* | 0.1113* | 0.1280* | 0.1164* | 0.1169* | 0.1121* |
| | Gamma | 0.0885* | 0.1376* | 0.1082* | 0.1053* | 0.1027* | 0.1185* |
| Dec | Normal | 0.1933* | 0.1555* | 0.2045* | 0.1575* | 0.1370* | 0.1247* |
| | Log-normal | 0.1390* | 0.0836* | 0.1438* | 0.1116* | 0.1164* | 0.1111* |

(continued)

**Table 4.3** (continued)

| Month/ distribution | | K-S test statistic (D) for monthly streamflow data at different time scales | | | | | |
|---|---|---|---|---|---|---|---|
| | | 1-m | 3-m | 6-m | 9-m | 12-m | 24-m |
| | Gamma | 0.2016* | 0.0983* | 0.1648* | 0.1148* | 0.0958* | 0.1139* |

**Note:** *Indicates significant probability distribution at $\alpha = 1\%$ (critical D = 0.2768)
**Indicates significant probability distribution at $\alpha = 5\%$ (critical D = 0.2308)

### *4.3.1 Assessment of Hydrological Drought and Wet Conditions at Naula Station*

The general pattern of drought events for Naula station is given in Table 5(a) which indicates that out of total available months, the number of drought months was 43 (SDI-1), 46 (SDI-3), 58 (SDI-6), 54 (SDI-9), 60 (SDI-12), and 72 (SDI-24). The percentage of drought occurrence was found to be 10.86, 11.68, 14.83, 13.92, 15.58, and 19.30% for 1-, 3-, 6-, 9-, 12-, and 24-SDI, respectively. The number of drought incidents (e.g., single or consecutive) was identified as 33, 37, 51, 50, 55, and 68 for 1-, 3-, 6-, 9-, 12-, and 24-SDI, respectively. The average duration of drought was found to be 1.30, 1.24, 1.14, 1.08, 1.09, and 1.06 months for 1-, 3-, 6-, 9-, 12-, and 24-SDI, respectively. The period of the longest duration of drought was also identified with its duration, period, severity, average intensity, and corresponding drought category. The longest drought duration was found to be of 8, 12, 28, 26, 28, and 38 months for 1-, 3-, 6-, 9-, 12-, and 24-SDI, respectively. Drought severity and average intensity were found to be −10.55 and −1.32, −16.95 and −1.41, −40.83 and −1.46, −40.24 and −1.55, −42.47 and −1.52, and −59.30 and −1.56 for 1-, 3-, 6-, 9-, 12-, and 24-SDI, respectively. Accordingly, the drought category varied from moderate drought (−1.32 to −1.46) for SDI-1, SDI-3 and SDI-6 to severe drought (−1.52 to −1.56) for SDI-9, SDI-12, and SDI-24. Figure 3(a–f) shows the distribution of drought events for 1-, 3-, 6-, 9-, 12-, and 24-SDI scales at Naula station for the study period (1975–2007).

The general pattern of wet events for Naula station is given in Table 5(b) which indicates that out of total available months, the number of wet months was 52 (SDI-1), 61 (SDI-3), 64 (SDI-6), 64 (SDI-9), 61 (SDI-12), and 60 (SDI-24). The percentage of wet occurrence was found to be 13.13, 15.48, 16.37, 16.49, 15.84, and 16.09% for 1-, 3-, 6-, 9-, 12-, and 24-SDI scales, respectively. The number of wet incidents (e.g., single or consecutive) was identified as 42, 48, 56, 57, 58, and 58 for 1-, 3-, 6-, 9-, 12-, and 24-SDI scales, respectively. The average duration of wet events was found to be 1.24, 1.27, 1.14, 1.12, 1.05, and 1.03 months for 1-, 3-, 6-, 9-, 12-, and 24-SDI scales, respectively. The period of the longest duration of wet events was also identified with its duration, period, severity, average intensity, and corresponding drought category. The longest wet duration was found to be of 10, 14, 24, 25, 24, and 36 months for 1-, 3-, 6-, 9-, 12-, and 24-SDI scales, respectively. Wet severity and average intensity were found to be 15.27 and 1.53, 23.83 and 1.70, 38.71 and 1.61, 39.66 and 1.59,

**Table 4.4** K-S test statistic of three probability distributions for monthly streamflow data at different time scales at Kedar station

| Month/ distribution | | K-S test statistic (D) for monthly streamflow data at different time scales | | | | | |
|---|---|---|---|---|---|---|---|
| | | 1-m | 3-m | 6-m | 9-m | 12-m | 24-m |
| Jan | Normal | 0.1015* | 0.1319* | 0.1343* | 0.1219* | 0.0960* | 0.0804* |
| | Log-normal | 0.1396* | 0.1184* | 0.0941* | 0.1009* | 0.0667* | 0.1365* |
| | Gamma | 0.1138* | 0.1006* | 0.0753* | 0.0747* | 0.0560* | 0.1043* |
| Feb | Normal | 0.2258* | 0.1791* | 0.1462* | 0.1542* | 0.0959* | 0.0923* |
| | Log-normal | 0.1246* | 0.1071* | 0.0711* | 0.1091* | 0.1163* | 0.1518* |
| | Gamma | 0.1688* | 0.1203* | 0.0801* | 0.1011* | 0.0931* | 0.1171* |
| Mar | Normal | 0.2327* | 0.1805* | 0.1197* | 0.1103* | 0.0852* | 0.1021* |
| | Log-normal | 0.1089* | 0.0976* | 0.1229* | 0.1507* | 0.1089* | 0.1672* |
| | Gamma | 0.1449* | 0.1214* | 0.0969* | 0.1191* | 0.0806* | 0.1375* |
| Apr | Normal | 0.1601* | 0.1574* | 0.1182* | 0.1330* | 0.0973* | 0.1144* |
| | Log-normal | 0.1165* | 0.0919* | 0.0782* | 0.1338* | 0.1271* | 0.1692* |
| | Gamma | 0.1092* | 0.0984* | 0.0812* | 0.1096* | 0.0986* | 0.1386* |
| May | Normal | 0.1528* | 0.1880* | 0.1615* | 0.1496* | 0.1028* | 0.1227* |
| | Log-normal | 0.0969* | 0.1352* | 0.0905* | 0.0952* | 0.1170* | 0.1842* |
| | Gamma | 0.1108* | 0.1446* | 0.1148* | 0.0872* | 0.0911* | 0.1523* |
| Jun | Normal | 0.1932* | 0.1436* | 0.1028* | 0.0972* | 0.0874* | 0.1241* |
| | Log-normal | 0.0882* | 0.1021* | 0.1010* | 0.1680* | 0.1085* | 0.1821* |
| | Gamma | 0.1166* | 0.0887* | 0.0812* | 0.1446* | 0.0779* | 0.1603* |
| July | Normal | 0.1395* | 0.1383* | 0.1300* | 0.1175* | 0.1266* | 0.1106* |
| | Log-normal | 0.1169* | 0.1524* | 0.1153* | 0.0793* | 0.1257* | 0.1721* |
| | Gamma | 0.0943* | 0.1372* | 0.1227* | 0.0838* | 0.0986* | 0.1437* |
| Aug | Normal | 0.1788* | 0.0867* | 0.1117* | 0.1256* | 0.1168* | 0.1000* |
| | Log-normal | 0.1553* | 0.1206* | 0.1582* | 0.1355* | 0.0843* | 0.1606* |
| | Gamma | 0.1194* | 0.1179* | 0.1377* | 0.1380* | 0.0878* | 0.1313* |
| Sep | Normal | 0.1870* | 0.1279* | 0.1287* | 0.1181* | 0.1083* | 0.0885* |
| | Log-normal | 0.1115* | 0.1092* | 0.0879* | 0.1215* | 0.1088* | 0.1468* |
| | Gamma | 0.1279* | 0.0943* | 0.0980* | 0.1052* | 0.1209* | 0.1196* |
| Oct | Normal | 0.1631* | 0.1134* | 0.1326* | 0.1076* | 0.0856* | 0.1205* |
| | Log-normal | 0.1130* | 0.0812* | 0.0890* | 0.0837* | 0.0850* | 0.1772* |
| | Gamma | 0.1139* | 0.0930* | 0.0869* | 0.0693* | 0.0784* | 0.1488* |
| Nov | Normal | 0.1084* | 0.1567* | 0.1335* | 0.0975* | 0.1019* | 0.0925* |
| | Log-normal | 0.1626* | 0.0801* | 0.1190* | 0.0740* | 0.0762* | 0.1480* |
| | Gamma | 0.1358* | 0.0978* | 0.0849* | 0.0724* | 0.0648* | 0.1176* |
| Dec | Normal | 0.2369* | 0.1558* | 0.1432* | 0.0984* | 0.0862* | 0.0899* |
| | Log-normal | 0.0852* | 0.1250* | 0.1255* | 0.0945* | 0.0871* | 0.1447* |

(continued)

**Table 4.4** (continued)

| Month/ distribution | | K-S test statistic (D) for monthly streamflow data at different time scales | | | | | |
|---|---|---|---|---|---|---|---|
| | | 1-m | 3-m | 6-m | 9-m | 12-m | 24-m |
| | Gamma | 0.2749* | 0.1075* | 0.0927* | 0.0670* | 0.0630* | 0.1138* |

**Note:** *Indicates significant probability distribution at $\alpha = 1\%$ (critical D = 0.2768)
**Indicates significant probability distribution at $\alpha = 5\%$ (critical D = 0.2308)

40.81, 38.81 and 1.70, 1.59, and 52.30 and 1.45 for 1-, 3-, 6-, 9-, 12- and 24-SDI scales, respectively. Accordingly, the wet category varied from moderate wet (1.15) for SDI-24 to severe wet (1.53–1.70) for 1-, 3-, 6-, 9-, and 12-SDI scales. Figure 3(a to f) shows the distribution of wet events for 1-, 3-, 6-, 9-, 12-, and 24-SDI scales at Naula station for the study period (1975–2007).

### *4.3.2 Probability of Occurrence of Drought and Wet Events at Naula Station*

The probability of occurrence of drought and wet (e.g., moderate, severe, and extreme) events at Naula station is summarized in Table 6 from SDI-1 to SDI-24, respectively. Table 6 reveals that out of total drought months, the probability of occurrence of moderate, severe, and extreme drought events was provided as 0.79, 0.16, and 0.05 for SDI-1; 0.74, 0.22, and 0.04 for SDI-3; 0.69, 0.31, and 0.00 for SDI-6; 0.67, 0.33, and 0.00 for SDI-9; 0.73, 0.27, and 0.00 for SDI-12; and 0.68, 0.32, and 0.00 for SDI-24, respectively. Similarly, out of total wet months, the probability of occurrence of moderate, severe, and extreme wet events was calculated as 0.58, 0.12, and 0.31 for SDI-1; 0.56, 0.28, and 0.16 for SDI-3; 0.61, 0.19, and 0.20 for SDI-6; 0.61, 0.23, and 0.16 for SDI-9; 0.54, 0.31, and 0.15 for SDI-12; and 0.58, 0.35, and 0.07 for SDI-24, respectively.

Figure 4(a and b) demonstrates the probability of occurrence of drought and wet events using 1-, 3-, 6-, 9-, 12-, and 24-SDI scales at Naula station. It was evident from Fig. 4(a) that the probability of occurrences of the moderate drought was higher than severe and extreme droughts, whereas the occurrence of severe drought for SDI-6, SDI-9 and SDI-24 has probabilities 0.31, 0.33, and 0.32, respectively. Similarly, Fig. 4(b) also shows that the probability of occurrences of moderate wet condition was higher than severe and extreme wet conditions. It is also observed from Fig. 4(b) that the chances of occurrence of wet conditions (e.g., inundation or flood) for SDI-1, SDI-12, and SDI-24 were marked as nearly 0.30.

**Table 4.5** (a) SDI-based drought pattern of streamflow series for Naula station

| SDI Category | Total months | No. of drought months | No. of drought Incidents | Drought percentage (%) | Average duration (month) | Longest drought pattern | | | | |
|---|---|---|---|---|---|---|---|---|---|---|
| | | | | | | Longest duration (month) | Period of longest drought | Drought severity | Average intensity | Drought category |
| SDI-1 | 396 | 43 | 33 | 10.86 | 1.30 | 8 | Oct, 2003 to May, 2004 | -10.55 | -1.32 | 1 |
| SDI-3 | 394 | 46 | 37 | 11.68 | 1.24 | 12 | Aug, 2003 to Jul, 2004 | -16.95 | -1.41 | 1 |
| SDI-6 | 391 | 58 | 51 | 14.83 | 1.14 | 28 | Apr, 2002 to Jul, 2004 | -40.83 | -1.46 | 1 |
| SDI-9 | 388 | 54 | 50 | 13.92 | 1.08 | 26 | Jul, 2002 to Aug, 2004 | -40.24 | -1.55 | 2 |
| SDI-12 | 385 | 60 | 55 | 15.58 | 1.09 | 28 | Jul, 2002 to Oct, 2004 | -42.47 | -1.52 | 2 |
| SDI-24 | 373 | 72 | 68 | 19.30 | 1.06 | 38 | Jul, 2002 to Aug, 2005 | -59.30 | -1.56 | 2 |

**Table 4.5** (b) SDI-based wet pattern of streamflow series for Naula station

| SDI Category | Total months | No. of wet months | No. of wet Incidents | Wet percentage (%) | Average duration (month) | Longest wet pattern | | | | |
|---|---|---|---|---|---|---|---|---|---|---|
| | | | | | | Longest duration (month) | Period of longest wet | Wet severity | Average intensity | Wet category |
| SDI-1 | 396 | 52 | 42 | 13.13 | 1.24 | 10 | Jun, 1978 to Mar, 1979 | 15.27 | 1.53 | B |
| SDI-3 | 394 | 61 | 48 | 15.48 | 1.27 | 14 | Mar, 1978 to Apr, 1979 | 23.83 | 1.70 | B |
| SDI-6 | 391 | 64 | 56 | 16.37 | 1.14 | 24 | Jul, 1977 to Jun, 1979 | 38.71 | 1.61 | B |
| SDI-9 | 388 | 64 | 57 | 16.49 | 1.12 | 25 | May, 1994 to May, 1996 | 39.66 | 1.59 | B |
| SDI-12 | 385 | 61 | 58 | 15.84 | 1.05 | 24 | Sep, 1977 to Aug, 1979<br>Jul, 1994 to Jun, 1996 | 40.81<br>38.13 | 1.70<br>1.59 | B<br>B |
| SDI-24 | 373 | 60 | 58 | 16.09 | 1.03 | 36 | Jul, 1977 to Jun, 1980 | 52.30 | 1.45 | A |

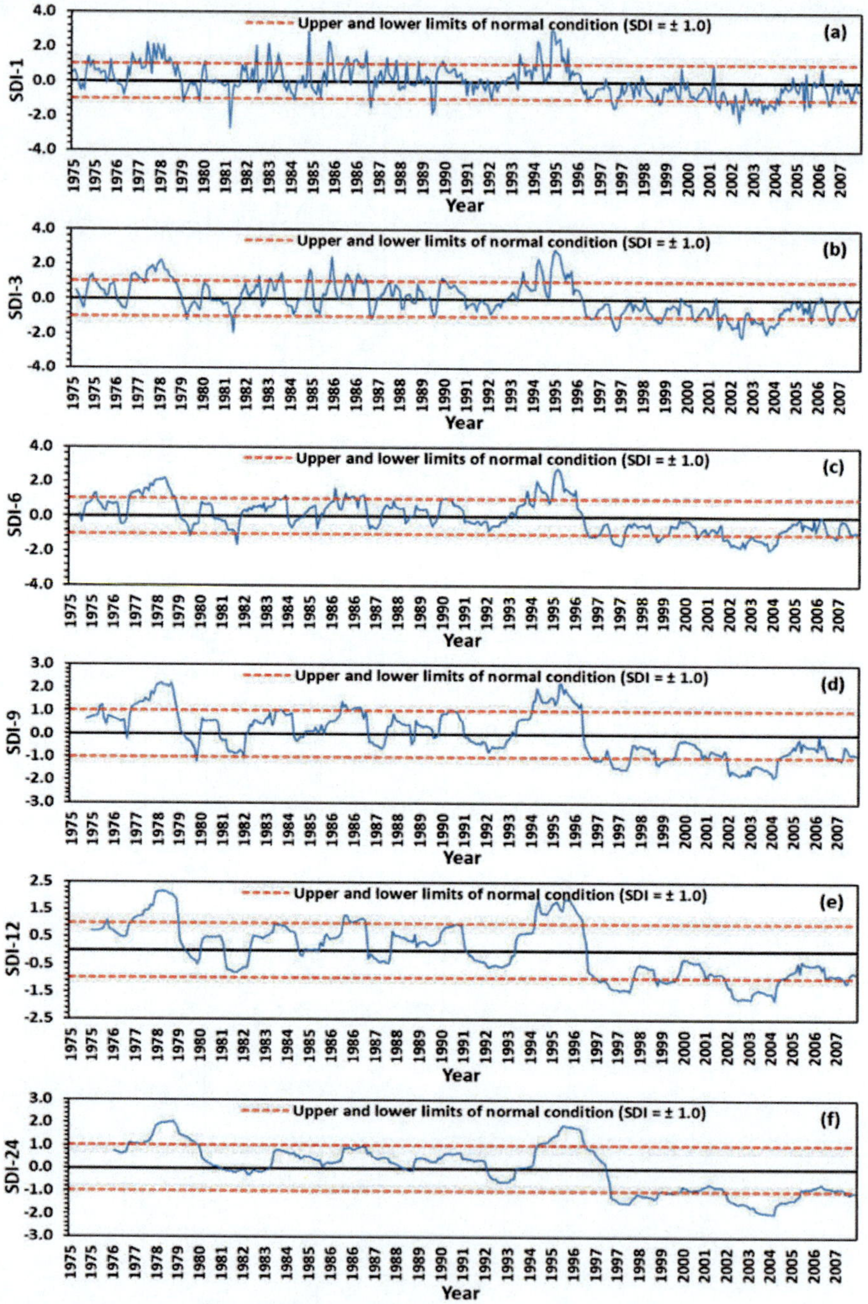


**Fig. 3** **(a–f)** Drought and wet patterns for SDI-1, SDI-3, SDI-6, SDI-9, SDI-12 and SDI-24 at Naula station

**Table 4.6** Probability of occurrence of drought and wet events at different SDI time scales for Naula station

| SDI time scale | Drought events | | | Wet events | | |
|---|---|---|---|---|---|---|
| | Moderate | Severe | Extreme | Moderate | Severe | Extreme |
| | 1 | 2 | 3 | A | B | C |
| **SDI-1** | | | | | | |
| Event months | 34 | 7 | 2 | 30 | 6 | 16 |
| Total months | 43 | 43 | 43 | 52 | 52 | 52 |
| Probability | 0.79 | 0.16 | 0.05 | 0.58 | 0.12 | 0.31 |
| **SDI-3** | | | | | | |
| Event months | 34 | 10 | 2 | 34 | 17 | 10 |
| Total months | 46 | 46 | 46 | 61 | 61 | 61 |
| Probability | 0.74 | 0.22 | 0.04 | 0.56 | 0.28 | 0.16 |
| **SDI-6** | | | | | | |
| Event months | 40 | 18 | 0 | 39 | 12 | 13 |
| Total months | 58 | 58 | 58 | 64 | 64 | 64 |
| Probability | 0.69 | 0.31 | 0.00 | 0.61 | 0.19 | 0.20 |
| **SDI-9** | | | | | | |
| Event months | 36 | 18 | 0 | 39 | 15 | 10 |
| Total months | 54 | 54 | 54 | 64 | 64 | 64 |
| Probability | 0.67 | 0.33 | 0.00 | 0.61 | 0.23 | 0.16 |
| **SDI-12** | | | | | | |
| Event months | 44 | 16 | 0 | 33 | 19 | 9 |
| Total months | 60 | 60 | 60 | 61 | 61 | 61 |
| Probability | 0.73 | 0.27 | 0.00 | 0.54 | 0.31 | 0.15 |
| **SDI-24** | | | | | | |
| Event months | 49 | 23 | 0 | 35 | 21 | 4 |
| Total months | 72 | 72 | 72 | 60 | 60 | 60 |
| Probability | 0.68 | 0.32 | 0.00 | 0.58 | 0.35 | 0.07 |

### *4.3.3 Assessment of Hydrological Drought and Wet Conditions at Kedar Station*

The general pattern of drought events for Kedar station is given in Table 7(a) which indicated that out of total available months, the number of drought months was 51, 61, 67, 66, 58, and 55 for 1-, 3-, 6-, 9-, 12-, and 24-SDI scales, respectively. The percentage of drought occurrence was found to be 12.88, 15.48, 17.17, 17.01, 15.06, and 14.75% for 1-, 3-, 6-, 9-, 12-, and 24-SDI scales, respectively. The number of drought incidents (e.g., single or consecutive) was identified as 40, 46, 56, 60, 56, and 53 for 1-, 3-, 6-, 9-, 12-, and 24-SDI scales, respectively. The average duration

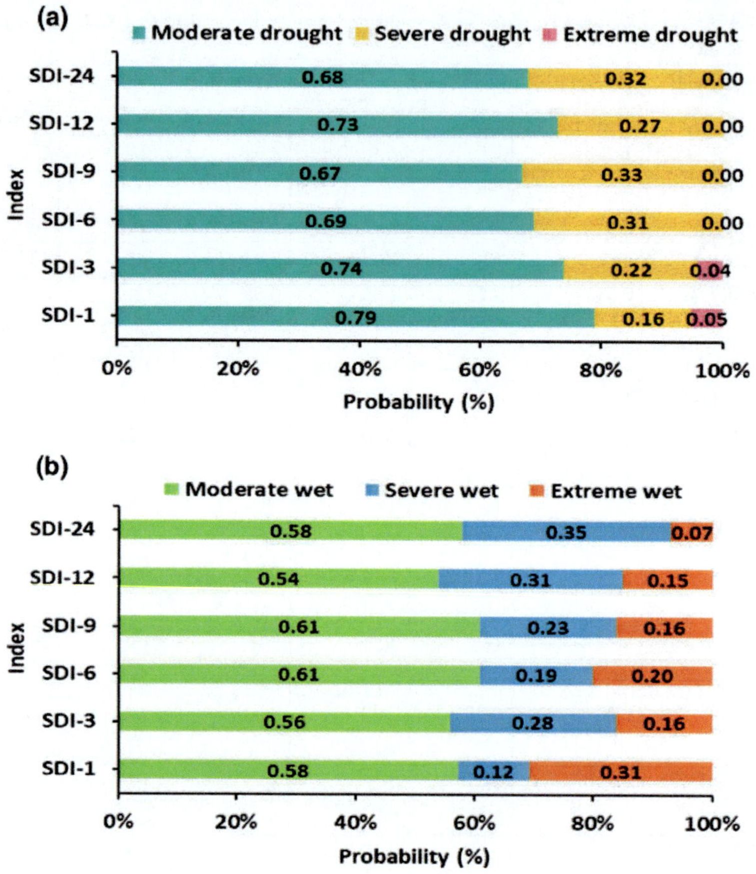


**Fig. 4** **(a & b)** Comparison of probability of occurrence of drought and wet categories demarcated using SDI-1, SDI-3, SDI-6, SDI-9, SDI-12 and SDI-24 for Naula station

of drought was found to be 1.28, 1.33, 1.20, 1.10, 1.04, and 1.04 months for 1-, 3-, 6-, 9-, 12-, and 24-SDI scales, respectively. The period of the longest duration of drought was also identified with its duration, period, severity, average intensity, and corresponding drought category. The longest drought duration was found to be of 6, 8, 14, 23, 35, and 36 months for 1-, 3-, 6-, 9-, 12-, and 24-SDI scales, respectively. Drought severity and average intensity were found to be −8.41 and −1.40, −11.15 and −1.39, −17.78 and −1.27, −44.88 and −1.95, −49.93 and −1.43, and −67.32 and −1.87 for 1-, 3-, 6-, 9-, 12-, and 24-SDI scales, respectively. Accordingly, the drought category varied from moderate drought (−1.39 to −1.43) for SDI-1, SDI-3, SDI-6, and SDI-12 to severe drought (−1.95 and −1.87) for SDI-9 and SDI-24. Figure 5(a to f) shows the distribution of drought events for 1-, 3-, 6-, 9-, 12-, and 24-SDI scales at Kedar station for the study period (1975–2007).

**Table 4.7** (a) SDI-based drought pattern of streamflow series for Kedar station

| SDI Category | Total months | No. of drought months | No. of drought Incidents | Drought percentage (%) | Average duration (month) | Longest drought pattern | | | | |
|---|---|---|---|---|---|---|---|---|---|---|
| | | | | | | Longest duration (month) | Period of longest drought | Drought severity | Average intensity | Drought category |
| SDI-1 | 396 | 51 | 40 | 12.88 | 1.28 | 6 | Jun to Nov, 1992 | -8.41 | -1.40 | 1 |
| SDI-3 | 394 | 61 | 46 | 15.48 | 1.33 | 8 | Dec, 2003 to Jul, 2004 | -11.15 | -1.39 | 1 |
| SDI-6 | 391 | 67 | 56 | 17.14 | 1.20 | 14 | Dec, 2001 to Jan, 2003 | -17.78 | -1.27 | 1 |
| SDI-9 | 388 | 66 | 60 | 17.01 | 1.10 | 23 | Aug, 1991 to Jun, 1993 | -44.88 | -1.95 | 2 |
| SDI-12 | 385 | 58 | 56 | 15.06 | 1.04 | 35 | Dec, 2001 to Oct, 2004 | -49.93 | -1.43 | 1 |
| SDI-24 | 373 | 55 | 53 | 14.75 | 1.04 | 36 | Jul, 2002 to Jun, 2005 | -67.32 | -1.87 | 2 |

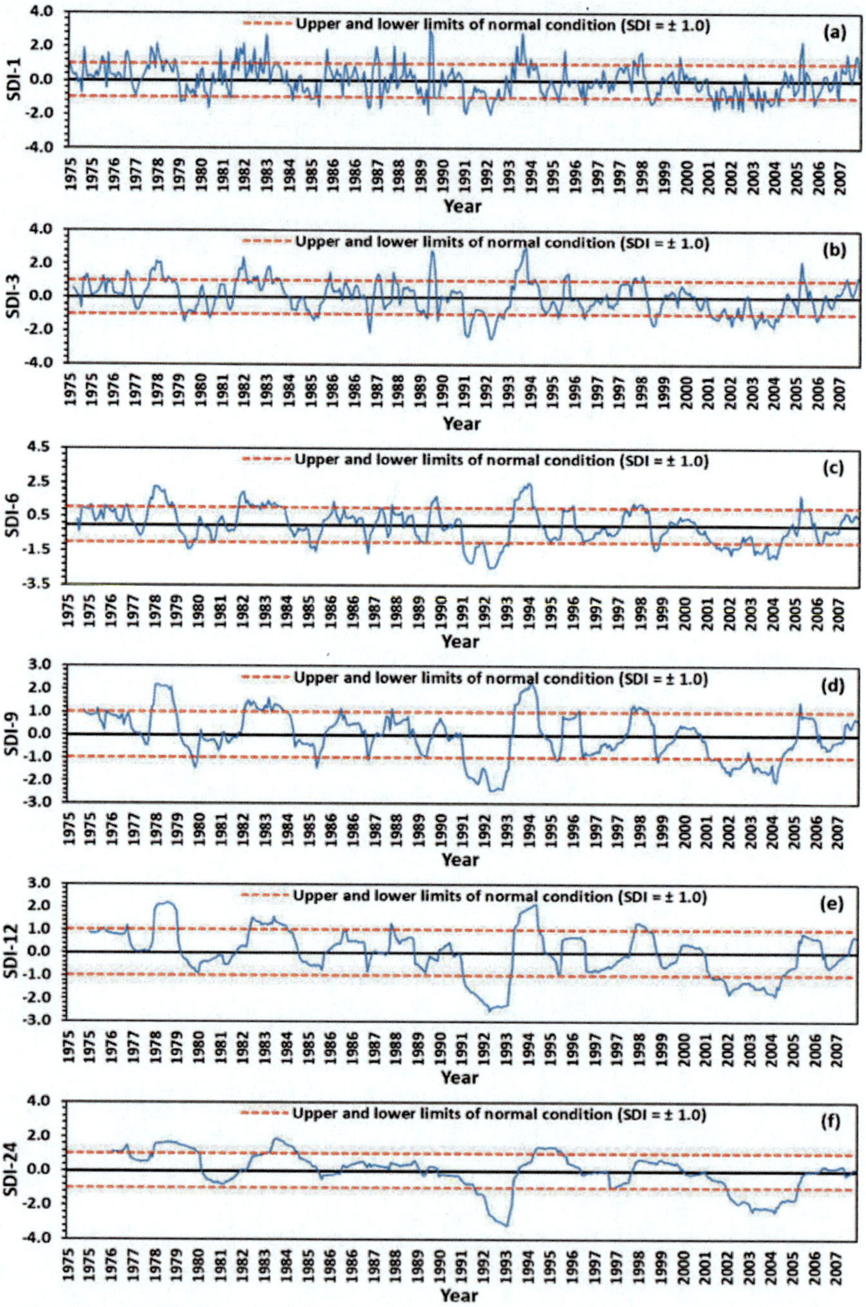


**Fig. 5** **(a–f)** Drought and wet patterns for SDI-1, SDI-3, SDI-6, SDI-9, SDI-12 and SDI-24 at Kedar station

Table 4.7 (b) SDI-based wet pattern of streamflow series for Kedar station

| SDI Category | Total months | No. of wet months | No. of wet Incidents | Wet percentage (%) | Average duration (month) | Longest wet pattern | | | | |
|---|---|---|---|---|---|---|---|---|---|---|
| | | | | | | Longest duration (month) | Period of longest wet | Wet severity | Average intensity | Wet category |
| SDI-1 | 396 | 54 | 40 | 13.64 | 1.35 | 3 | May to Jul, 1977<br>Sep to Nov, 1978<br>Mar to May, 1982<br>Oct to Dec, 1987<br>Nov, 1993 to Jan, 1994 | 4.33<br>4.72<br>5.60<br>4.27<br>5.51 | 1.44<br>1.57<br>1.87<br>1.40<br>1.84 | A<br>B<br>B<br>A<br>B |
| SDI-3 | 394 | 60 | 46 | 15.23 | 1.30 | 7 | Jun to Dec, 1978 | 12.32 | 1.76 | B |
| SDI-6 | 391 | 65 | 55 | 16.62 | 1.18 | 11 | Mar, 1983 to Jan, 1984<br>Sep, 1993 to Jul, 1994 | 13.02<br>20.42 | 1.18<br>1.86 | A<br>B |
| SDI-9 | 388 | 63 | 59 | 16.24 | 1.07 | 22 | Jun, 1982 to Mar, 1984 | 27.06 | 1.23 | A |
| SDI-12 | 385 | 53 | 49 | 13.77 | 1.08 | 19 | Sep, 1982 to Mar, 1984 | 24.74 | 1.30 | A |
| SDI-24 | 373 | 65 | 61 | 17.43 | 1.07 | 24 | Sep, 1978 to Aug, 1980 | 35.39 | 1.47 | A |

The general pattern of wet events for Kedar station is given in Table 7(b) which indicates that out of total available months, the number of wet months was 54, 60, 65, 63, 53, and 65 for 1-, 3-, 6-, 9-, 12-, and 24-SDI scales, respectively. The percentage of wet occurrence was found to be 13.64, 15.23, 16.62, 16.24, 13.77, and 17.43% for 1-, 3-, 6-, 9-, 12-, and 24-SDI scales, respectively. The number of wet incidents (e.g., single or consecutive) was identified as 40, 46, 55, 59, 49, and 61 for 1-, 3-, 6-, 9-, 12-, and 24-SDI scales, respectively. The average duration of wet events was found to be 1.35, 1.30, 1.18, 1.07, 1.08, and 1.07 months for 1-, 3-, 6-, 9-, 12-, and 24-SDI scales, respectively. The period of the longest duration of wet events was also identified with its duration, period, severity, average intensity, and corresponding drought category. The longest wet duration was found to be of 3, 7, 11, 22, 19, and 24 months for 1-, 3-, 6-, 9-, 12-, and 24-SDI scales, respectively. Wet severity and average intensity were found to be 4.33, 4.72, 5.60, 4.27, 5.51 and 1.44, 1.57, 1.87, 1.40, 1.84, 12.32 and 1.76, 13.02, 20.42 and 1.18, 1.86, 27.06 and 1.23, 24.74 and 1.30, and 35.39 and 1.47 for 1-, 3-, 6-, 9-, 12-, and 24-SDI scales, respectively. Accordingly, the wet category varied from moderate wet (1.23–1.47) for 1-, 6-, 9-, 12-, and 24-SDI scales to severe wet (1.57–1.87) for 1-, 3-, and 6-SDI scales. Figure 5(a to f) shows the distribution of wet events for 1-, 3-, 6-, 9-, 12-, and 24-SDI scales at Kedar station for the study period (1975–2007).

### 4.3.4 Probability of Occurrence of Drought and Wet Events at Kedar Station

The probability of occurrence of drought and wet (i.e., moderate, severe, and extreme) events at Kedar station is summarized in Table 8 for SDI-1 to SDI-24. Table 8 reveals that out of total drought months, the probability of occurrence of moderate, severe, and extreme drought events was calculated as 0.71, 0.29, and 0.00 for SDI-1; 0.75, 0.13, and 0.11 for SDI-3; 0.60, 0.27, and 0.13 for SDI-6; 0.53, 0.29, and 0.18 for SDI-9; 0.36, 0.40, and 0.24 for SDI-12; and 0.13, 0.36, and 0.51 for SDI-24, respectively. Similarly, out of total wet months, the probability of occurrence of moderate, severe and extreme wet events was produced as 0.46, 0.35, and 0.19 for SDI-1; 0.62, 0.22, and 0.17 for SDI-3; 0.68, 0.15, and 0.17 for SDI-6; 0.68, 0.11, and 0.21 for SDI-9; 0.57, 0.21, and 0.23 for SDI-12; and 0.66, 0.34, and 0.00 for SDI-24, respectively.

Figure 6(a and b) demonstrates the probability of occurrence of drought and wet events for 1.47 of 1-, 3-, 6-, 9-, 12-, and 24-SDI scales at Kedar station. As observed from Fig. 6(a), the probability of occurrence of moderate drought was higher than severe and extreme droughts for SDI-1, SDI-3, SDI-6, and SDI-9, but the probability of occurrence of severe and extreme droughts was 0.40 higher than SDI-12, and 0.36 and 0.51 for SDI-24 with severe and extreme, respectively. Similarly, Fig. 6(b) depicts that the probability of occurrences of moderate wet was higher than severe and extreme wet conditions for 1.47 of 1-, 3-, 6-, 9-, 12-, and 24-SDI scales. It was

**Table 4.8** Probability of occurrence of drought and wet events at different SDI time scales for Kedar station

| SDI time scale | Drought events | | | Wet events | | |
|---|---|---|---|---|---|---|
| | Moderate | Severe | Extreme | Moderate | Severe | Extreme |
| | 1 | 2 | 3 | A | B | C |
| **SDI-1** | | | | | | |
| Event months | 36 | 15 | 0 | 25 | 19 | 10 |
| Total months | 51 | 51 | 51 | 54 | 54 | 54 |
| Probability | 0.71 | 0.29 | 0.00 | 0.46 | 0.35 | 0.19 |
| **SDI-3** | | | | | | |
| Event months | 46 | 8 | 7 | 37 | 13 | 10 |
| Total months | 61 | 61 | 61 | 60 | 60 | 60 |
| Probability | 0.75 | 0.13 | 0.11 | 0.62 | 0.22 | 0.17 |
| **SDI-6** | | | | | | |
| Event months | 40 | 18 | 9 | 44 | 10 | 11 |
| Total months | 67 | 67 | 67 | 65 | 65 | 65 |
| Probability | 0.60 | 0.27 | 0.13 | 0.68 | 0.15 | 0.17 |
| **SDI-9** | | | | | | |
| Event months | 35 | 19 | 12 | 43 | 7 | 13 |
| Total months | 66 | 66 | 66 | 63 | 63 | 63 |
| Probability | 0.53 | 0.29 | 0.18 | 0.68 | 0.11 | 0.21 |
| **SDI-12** | | | | | | |
| Event months | 21 | 23 | 14 | 30 | 11 | 12 |
| Total months | 58 | 58 | 58 | 53 | 53 | 53 |
| Probability | 0.36 | 0.40 | 0.24 | 0.57 | 0.21 | 0.23 |
| **SDI-24** | | | | | | |
| Event months | 7 | 20 | 28 | 43 | 22 | 0 |
| Total months | 55 | 55 | 55 | 65 | 65 | 65 |
| Probability | 0.13 | 0.36 | 0.51 | 0.66 | 0.34 | 0.00 |

also evident from Fig. 6(b) that 0.35 and 0.34 for probability of occurrence of wet conditions (e.g., inundation/flood) for SDI-1 and SDI-24 were calculated.

## 4.4 Conclusions and Remarks

This chapter has analyzed hydrological drought and wet conditions using multi-scalar streamflow drought index (SDI). The study area was Naula and Kedar stations, located in the upper Ramganga River catchment in Uttarakhand, India. The following specific conclusions are drawn from this study:

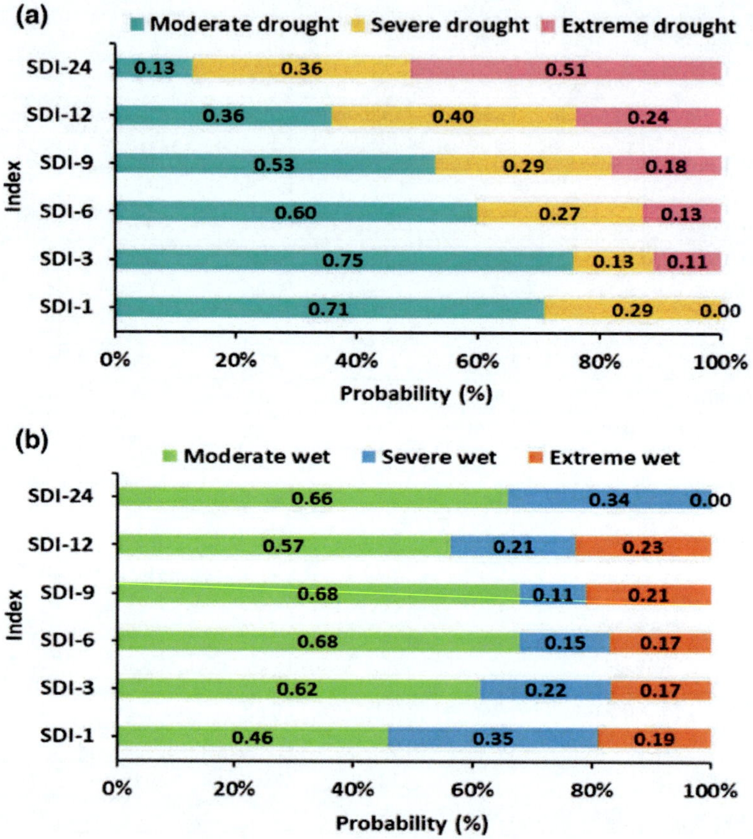


**Fig. 6** (**a & b**) Comparison of probability of occurrence of drought and wet categories demarcated using SDI-1, SDI-3, SDI-6, SDI-9, SDI-12 and SDI-24 for Kedar station

- The K-S test shows all the distributions fit well to streamflow data series for 1-, 3-, 6-, 9-, 12-, and 24-SDI scales for all the months at 1% and 5% significance levels. However, gamma distribution is used for analysis of hydrological drought and wet conditions for Naula and Kedar stations.
- At Naula station, on an average, there is 0.72 probability of having normal condition. Under the influence of hydrological drought, probability of occurrence of moderate drought condition is 0.72, followed by 0.27 for severe drought conditions with merely 0.01 for extreme condition at 1-, 3-, 6-, 9-, 12-, and 24-SDI scales.
- At Naula station, on an average, there is 0.58 probability of having moderate wet condition, followed by 0.25 for severe wet conditions with only 0.17 for extreme condition at 1-, 3-, 6-, 9-, 12-, and 24-SDI scales.
- At Kedar station, on an average, there is 0.69 probability of having normal condition. Under the influence of hydrological drought, probability of occurrence of

moderate drought condition is 0.51, followed by 0.29 for severe conditions with 0.20 for extreme condition at 1-, 3-, 6-, 9-, 12-, and 24-SDI scales.

- At Kedar station, on an average, there is 0.61 probability of having moderate wet condition, followed by 0.23 for severe wet conditions with only 0.16 for extreme condition at 1-, 3-, 6-, 9-, 12-, and 24-SDI scales.
- At Kedar station, probability of occurrence of extreme drought condition is 0.51 for SDI at 24-month timescale.
- Occurrence of moderate hydrological drought and wet conditions can be taken care of by adopting appropriate water/moisture conservation measures such as in situ rainwater harvesting, mulching, terracing, contour bunding for enhancing agricultural productivity in the study region.

**Conflicts of Interest** The authors declare no conflict of interest.

## References

Angelidis P, Maris F, Kotsovinos N, Hrissanthou V (2012) Computation of drought index SPI with alternative distribution functions. Water Resour Manage. https://doi.org/10.1007/s11269-012-0026-0

Beyaztas U, Yaseen ZM (2019) Drought interval simulation using functional data analysis. J Hydrol 124141

Bhattacharjya RK (2004) Optimal design of unit hydrographs using probability distribution and genetic algorithms. Sadhana Acad Proc Eng Sci. https://doi.org/10.1007/BF02703257

Borji M, Malekian A, Salajegheh A, Ghadimi M (2016) Multi-time-scale analysis of hydrological drought forecasting using support vector regression (SVR) and artificial neural networks (ANN). Arab J Geosci. https://doi.org/10.1007/s12517-016-2750-x

Choi W, Byun HR, Cassardo C, Choi J (2018) Meteorological and streamflow droughts: characteristics, trends, and propagation in the Milwaukee river basin. Prof Geographer. https://doi.org/10.1080/00330124.2018.1432368

David V, Davidová T (2017) Relating hydrological and meteorological drought indices in order to identify causes of low flows in the catchment of blanice river. Environ Processes. https://doi.org/10.1007/s40710-017-0223-1

Dobrovolski SG (2015) World droughts and their time evolution: agricultural, meteorological, and hydrological aspects. Water Resour. https://doi.org/10.1134/s0097807815020049

Dracup JA, Lee KS, Paulson EG (1980) On the statistical characteristics of drought events. Water Resour Res. https://doi.org/10.1029/WR016i002p00289

Gumus V, Algin HM (2017) Meteorological and hydrological drought analysis of the Seyhan—Ceyhan River Basins, Turkey. Meteorological Applications. https://doi.org/10.1002/met.1605

Hassani H, Silva E, Kolmogorov-Smirnov A (2015) Based test for comparing the predictive accuracy of two sets of forecasts. Econometrics. https://doi.org/10.3390/econometrics3030590

Hayes MJ, Wilhelmi OV, Knutson CL (2004) Reducing drought risk: bridging theory and practice. Nat Hazards Rev. https://doi.org/10.1061/(ASCE)1527-6988(2004)5:2(106)

Lindgren B (1968) The statistical theory, 2nd edn. The Macmillan Company, London, p 521

Lloyd-Hughes B, Saunders MA (2002) A drought climatology for Europe. Int J Climatol 22:1571–1592. https://doi.org/10.1002/joc.846

Mandal KG, Padhi J, Kumar A, et al (2015) Analyses of rainfall using probability distribution and Markov chain models for crop planning in Daspalla region in Odisha, India. Theoretical and Applied Climatology. https://doi.org/10.1007/s00704-014-1259-z

Mckee TB, Doesken NJ, Kleist J (1993) The relationship of drought frequency and duration to time scales. In: AMS 8th conference on applied climatology 179–184. doi: citeulike-article-id:10490403

Mishra AK, Singh VP (2009) Analysis of drought severity-area-frequency curves using a general circulation model and scenario uncertainty. J Geophys Res Atmos. https://doi.org/10.1029/2008JD010986

Mishra AK, Singh VP (2010) A review of drought concepts. J Hydrol 391:202–216

Moivre A de (1738) The doctrine of chances: or, a method of calculating the probabilities of events in play

Myronidis D, Ioannou K, Fotakis D, Dörflinger G (2018) Streamflow and hydrological drought trend analysis and forecasting in cyprus. Water Resour Manage. https://doi.org/10.1007/s11269-018-1902-z

Nabaei S, Sharafati A, Yaseen ZM, Shahid S (2019) Copula based assessment of meteorological drought characteristics: regional investigation of Iran. Agric For Meteorol 276:107611

Nalbantis I (2008) Evaluation of a hydrological drought index. Euro Water 23:67–77

Nalbantis I, Tsakiris G (2009) Assessment of hydrological drought revisited. Water Resour Manage. https://doi.org/10.1007/s11269-008-9305-1

Nikbakht J, Tabari H, Talaee PH (2013) Streamflow drought severity analysis by percent of normal index (PNI) in northwest Iran. Theoret Appl Climatol. https://doi.org/10.1007/s00704-012-0750-7

Olea RA, Pawlowsky-Glahn V (2009) Kolmogorov-Smirnov test for spatially correlated data. Stoch Env Res Risk Assess. https://doi.org/10.1007/s00477-008-0255-1

Pathak AA, Channaveerappa DBM (2016) Comparison of two hydrological drought indices. Perspect Sci. https://doi.org/10.1016/j.pisc.2016.06.039

Qutbudin I, Shiru MS, Sharafati A et al (2019) Seasonal drought pattern changes due to climate variability: case study in Afghanistan. Water 11:1096. https://doi.org/10.3390/w11051096

Razmkhah H (2017) Comparing threshold level methods in development of stream flow drought severity-duration-frequency curves. Water Resour Manage. https://doi.org/10.1007/s11269-017-1587-8

Riebsame W, Changnon S, Karl T (1991) Drought and natural resource management in the United States: impacts and implications of the 1987–89 drought. Westview Press, Boulder, CO, p 174

Sardou F, Bahremand A (2014) Hydrological drought analysis using SDI index in Halilrud basin of Iran. Environ Resour Res 2:48–56

Sayl KN, Muhammad NS, Yaseen ZM, El-shafie A (2016) Estimation the physical variables of rainwater harvesting system using integrated GIS-based remote sensing approach. Water Resour Manage 30:3299–3313. https://doi.org/10.1007/s11269-016-1350-6

Stephens MA (1974) EDF statistics for goodness of fit and some comparisons. J Am Statistical Association. https://doi.org/10.1080/01621459.1974.10480196

Subramanya K (2005) Engineering hydrology. Tata McGraw Hill, New Delhi, p 392

Tabari H, Nikbakht J, Hosseinzadeh Talaee P (2013) Hydrological drought assessment in Northwestern Iran based on Streamflow Drought Index (SDI). Water Resour Manage. https://doi.org/10.1007/s11269-012-0173-3

Wilhite DA (2000) Drought as a natural hazard. In: Drought: A Global Assessment

Wilhite DA, Glantz MH (1985) Understanding: the drought phenomenon: the role of definitions

Yaseen ZM, Sulaiman SO, Deo RC, Chau K-W (2018) An enhanced extreme learning machine model for river flow forecasting: state-of-the-art, practical applications in water resource engineering area and future research direction. J Hydrol 569:387–408. https://doi.org/10.1016/j.jhydrol.2018.11.069

Yevjevich V (1967) An objective approach to definitions and investigations of continental hydrologic droughts. Hydrol Papers, Colorado State University, Fort Collins, Colorado, USA, p 23

# Chapter 5
# Intelligent Data Analytics Approaches for Predicting Dissolved Oxygen Concentration in River: Extremely Randomized Tree Versus Random Forest, MLPNN and MLR

**Salim Heddam**

## 5.1 Introduction

Control of water quality by monitoring water variables is still of major importance for the protection of human life (El Najjar et al. 2019). Rivers and streams are the major's component of the freshwater ecosystems and constitute the source of life for both humans and animals and they become the most "*endangered ecosystems*" in the world (Kumar and Jayakumar 2020; Kebede et al. 2020; Emenike et al. 2020). From year to year, a considerable work was carried out to maintain a good water quality status (Jerves-Cobo et al. 2020). Dissolved oxygen concentration (DO) in rivers, lakes and streams freshwater is a key variable in the assessment and control of the aquatic life and freshwater health (Banerjee et al. 2019). It also plays tremendous role in the control of water quality, and has many important effects on aquatic hydrochemistry and aquatic toxicology, and used as an "*early warning*" and classifying the eutrophication status of lakes (Hoang et al. 2019).

Dissolved oxygen should be sufficiently and accurately estimated and monitored, and the variability of DO in freshwaters ecosystems can stem from various reasons and the amount of variability is particularly great when considering the interaction between several other water quality variables (Hutchins and Hitt 2019; Yaseen et al. 2018a). Over the years, it was demonstrated that there is substantial evidence that DO concentration is highly related to water quality variables (e.g. water temperature, water pH, specific conductance), which can be linked together. For example, it was demonstrated that decrease of DO concentration was mainly related to the increase

S. Heddam (✉)
University 20 Août 1955 Skikda, Faculty of Science, Agronomy Department, Hydraulics Division, Laboratory of Research in Biodiversity Interaction Ecosystem and Biotechnology, Route El Hadaik, BP 26, Skikda, Algeria
e-mail: heddamsalim@yahoo.fr

© Springer Nature Singapore Pte Ltd. 2021

R. C. Deo et al. (eds.), *Intelligent Data Analytics for Decision-Support Systems in Hazard Mitigation*, Springer Transactions in Civil and Environmental Engineering,
https://doi.org/10.1007/978-981-15-5772-9_5

of algal growth, especially excessive concentration of nitrogen and phosphorus (Crossman et al. 2019). In this regard, a combination of several in situ measurements of historical data was largely used to build robust models for predicting and forecasting DO concentration (Heddam 2017; Suarez et al. 2019). Past and present works conducted worldwide proved that models based on intelligent data analytic techniques have accurately estimated DO, and over the years alternative models have been developed. For example, see the studies of (Banerjee et al. 2019; Elkiran et al. 2019; Mitrović et al. 2019; Cao et al. 2019; Shi et al. 2019; Yang et al. 2019; Liu et al. 2019; Ross and Stock 2019; Csábrági et al. 2019; Suarez et al. 2019; Rahman et al. 2019; Yahya et al. 2019; Tao et al. 2019; Zounemat-Kermani et al. 2019; Antanasijević et al. 2019).

Banerjee et al. (2019) employed deep artificial neural network (DNN) and standard multiple linear regression (MLR) for predicting DO using sixteen water quality variables as predictors using data from three stations in the Bakreswar Reservoir, India. Elkiran et al. (2019) compared three intelligent data analytic techniques, namely multilayer perceptron neural networks (MLPNN), adaptive neuro-fuzzy inference system (ANFIS) and support vector machine (SVM) for forecasting DO in the Yamuna River, India, according to three scenarios: (i) simple average ensemble, (ii) weighted average ensemble (WAE) and (iii) neural network ensemble. Mitrović et al. (2019) proposed the Monte Carlo simulations (MCS) approach for optimizing an artificial neural network model, which applied for the spatial interpolation of DO concentration and other water quality parameters at the Danube River, Serbia. Cao et al. (2019) employed an improved version of the extreme learning machines (ELM) model called regularized extreme learning machine (RELM) for predicting DO in an aquaculture pond ecosystem. The RELM was combined with the ensemble empirical mode decomposition (EEMD) algorithm and the most significant input variables (e.g. water temperature and pH) were determined using the grey relational degree method. Shi et al. (2019) proposed an original ELM model called Softplus extreme learning machine method (CSELM) for modelling DO time series, using data measure at 10 min interval of time. Compared to the standard ELM, the authors demonstrated that CSELM was more accurate with high accuracy and less running time. In another's studies, the recurrent neural networks (RNN) regression model was applied for predicting DO concentration in recirculating aquaculture system (Liu et al. 2019; Yang et al. 2019).

Keshtegar et al. (2019) compared the polynomial chaos expansions (PCE), the MLPNN and the MLPNN optimized particle swarm optimization, for predicting daily DO concentration. Ross and Stock (2019) applied a model tree (MTree) for predicting minimum and mean DO concentration using water temperature and salinity as predictors, measured at one minute interval. Csábrági et al. (2019) compared the MLR, the radial basis function neural network (RBFNN) and the general regression neural network (GRNN). The models were developed using the runoff, water temperature, electric conductivity and water pH, collected at 13 stations in the River Tisza, Hungary. Rahman et al. (2019) investigated the capabilities and the robustness of three data-driven models, namely the MLR, the MLPNN and the SVM, applied for predicting DO concentration in two aquaculture prawn ponds in Queensland,

Australia. Zhu and Heddam (2019) employed the ELM model with different activation functions for predicting DO in urban rivers at the Three Gorges Reservoir, China. Yahya et al. (2019) employed the SVM with fold cross-validation for predicting DO using several water quality variables, namely chemical oxygen demand (COD), ammonia nitrogen (AN), pH, suspended solids (SS) and biochemical oxygen demand (BOD). Tao et al. (2019) compared the hybrid response surface method (HRSM) and the SVM models for predicting monthly DO concentration in the Euphrates River, Iraq. Antanasijević et al. (2019) introduced an innovative approach by combining the location similarity index (LSI) with the ward neural networks (WNNs) for predicting the DO at multiple sites. Zounemat-Kermani et al. (2019) applied the cascade correlation neural network (CCNN) with discrete wavelet transform (DWT) and variational mode decomposition (VMD), for predicting DO concentration in the St. Johns River, Florida, USA.

The main goal of this chapter is to present a new intelligent data analytic model for predicting DO concentrations using well-known water quality variables as predictors. To achieve our goals, we applied and compared two new data-driven models, namely the extremely randomized tree (ERT) and the random forest (RF).

## 5.2 Materials and Method

### *5.2.1 Study Area Description*

High quality in situ measurements of water variables are essential for developing robust models. In the present study, the dissolved oxygen concentration (DO) and the four water quality variables were obtained from two monitoring stations, where the data is provided at 15 min interval of time: (i) the Charles River Buoy (CR-Buoy) (https://www.epa.gov/charlesriver/live-water-quality-data-lower-charles-river) and (ii) the Mystic River Buoy (MR-Buoy) (https://www.epa.gov/mysticriver/live-water-quality-data-mystic-river). At the two stations, a self-contained solar powered buoy takes measurements for water temperature (TE), water pH, specific conductance (SC), phycocyanin pigment concentration (PC) and DO concentration. The sensors on the buoy are located 1 m below the water's surface. Consequently, TE, pH, SC and PC were selected as input variables for predicting DO. The data for the year 2019 has been used in the present study. The total length of the data set was: 12,640 patterns for the CR-Buoy station and 10,303 patterns for the MR-Buoy station. For the two stations, we used 70% of the data set for training and the remaining 30% for validation. All variables were normalized using the Z-score method:

$$Z_n = \frac{x_n - x_m}{\sigma_x} \tag{5.1}$$

where: $Z_n$ is the normalized value of the observation $n$; $x_n$ is the measured value of the observation $n$; $x_m$ and $\sigma_x$ are the mean value and standard deviation of the variable $x$. The correlation plot between DO concentration and the four water quality variables using scatterplot matrix is shown in Fig. 5.1. Based on the correlation matrix, we evaluated several combinations of the inputs variables and in total nine input combination were compared (Table 5.1).

**Performance Assessment of the Models**

Following several literature researches, the current study was evaluated using several statistical metrics, including mean absolute error, root mean square error, Nash–Sutcliffe index and determination coefficient (Yaseen et al. 2018b; Kisi and Yaseen 2019; Khosravi et al. 2018):

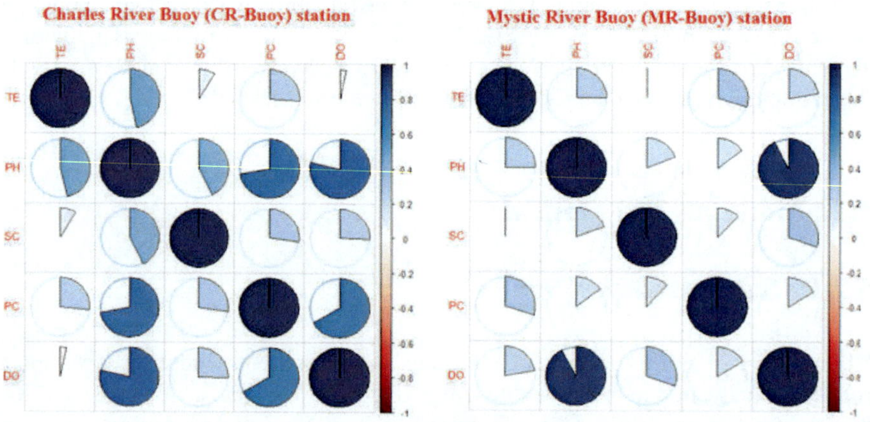


**Fig. 5.1** Correlation plot using scatterplot matrix for visual representation of the relations between dissolved oxygen and water quality variables for the two stations

**Table 5.1** Input combinations of different models

| ERT | RF | MLPNN | MLR | Input combination | Input |
|---|---|---|---|---|---|
| ERT1 | RF1 | MLPNN1 | MLR1 | TE, pH, SC, PC | 1234 |
| ERT2 | RF2 | MLPNN2 | MLR2 | TE, SC, PC | 134 |
| ERT3 | RF3 | MLPNN3 | MLR3 | pH, SC, PC | 234 |
| ERT4 | RF4 | MLPNN4 | MLR4 | TE, pH, PC | 124 |
| ERT5 | RF5 | MLPNN5 | MLR5 | TE, pH, SC | 123 |
| ERT6 | RF6 | MLPNN6 | MLR6 | pH, PC | 24 |
| ERT7 | RF7 | MLPNN7 | MLR7 | pH, SC | 23 |
| ERT8 | RF8 | MLPNN8 | MLR8 | TE, pH | 12 |
| ERT9 | RF9 | MLPNN9 | MLR9 | SC, PC | 34 |

$$\mathrm{MAE} = \frac{1}{N}\sum_{i=1}^{N} |(\mathrm{DO}_0)_i - (\mathrm{DO}_p)_i| \tag{5.2}$$

$$\mathrm{RMSE} = \sqrt{\frac{1}{N}\sum_{i=1}^{N} [(\mathrm{DO}_0)_i - (\mathrm{DO}_p)_i]^2} \tag{5.3}$$

$$\mathrm{NES} = 1 - \frac{\sum_{i=1}^{N} [(\mathrm{DO}_0)_i - (\mathrm{DO}_p)_i]^2}{\sum_{i=1}^{N} [(\mathrm{DO}_0)_i - \overline{\mathrm{DO}_0}]^2}, \quad -\infty < \mathrm{NES} \leq 1 \tag{5.4}$$

$$R = \left[ \frac{\frac{1}{\mathrm{N}}\sum_{i=1}^{N} \left((\mathrm{DO}_0)_i - \overline{\mathrm{DO}_0}\right)\left((\mathrm{DO}_p)_i - \overline{\mathrm{DO}_P}\right)}{\sqrt{\frac{1}{\mathrm{N}}\sum_{i=1}^{n} \left((\mathrm{DO}_0)_i - \overline{\mathrm{DO}_0}\right)^2}\sqrt{\frac{1}{\mathrm{N}}\sum_{i=1}^{n} \left((\mathrm{DO}_p)_i - \overline{\mathrm{DO}_p}\right)^2}} \right] \tag{5.5}$$

In which, $N$ is the number of data, $\mathrm{DO}_0$, $\mathrm{DO}_p$, $\overline{\mathrm{DO}_0}$ $\overline{\mathrm{DO}_p}$ are the measured, calculated, mean measured and mean calculated dissolved oxygen concentration, respectively.

### *5.2.2 Modelling Approaches*

**Random Forest for Regression**

Random forest (RF) is one of the ensemble machines learning algorithms introduced by Breiman (2001), and as with all other ensemble learning, e.g. bagging (Breiman 1996), boosting (Breiman et al. 1984) and classification and regression Trees (CART); the RF creates an ensemble of randomized CART using the impurity Gini index method as a measure of the best split selection (Breiman 2001), where each tree contributes with a single vote for the assignment of the most frequent class to the input data (Deng et al. 2020). RF possesses the capability of random feature selection at each node and no pruning or stopping rule during the training process (Tan et al. 2020). Each CART is trained on a bootstrapped sample of the original training data by selecting many bootstrap observations from the original data (Tao et al. 2020). The RF uses a random subset of predictive variables in the division of every node, which reduces the generalization error (Chen et al. 2020), and after a large number of regression trees have been generated, they are used to predict the class of new data, the best split at each node of the tree is searched only amongst a randomly selected subset of predictors, using the so-called out-of-bag (OOB) data (Hanna et al. 2020). Building a RF model needs three parameters: the number of trees in the forest, the minimum number of data points in each terminal node and the number of features tried at each node (Breiman 2001).

**Extremely Randomized Tree**

Extremely randomized tree also called extra trees were introduced by Geurts et al. (2006), and it belongs to the decision tree-based models. Since it was proposed, the

ERT has largely been considered as an improved version of the RF model. However, several authors have highlighted two majors difference between the two (Manavalan et al. 2019): (i) during the construction of the tree, the ERT uses all the training patterns with varying parameters, while the RF adopts a bagging procedure and (ii) the best split was used by the RF model, while the ERT randomly chooses the node split upon the construction of each tree. Regarding the major improvement achieved using the ERT model, it was demonstrated that the ERT substantially reduces the variance and slightly increases the bias of the prediction model with low computational cost (Basith et al. 2018). From a mathematical point of view, the ERT is composed of a set of decision trees ($T$) and each one ($t \in \{1 \ldots T\}$) uses all the training patterns separately during the training process (Pinto et al. 2018) for the construction of either decision or regression trees (Nattee et al. 2017). In the present study, the ERT models were fitted using the MATLAB code available at https://github.com/rtaormina/MATLAB_ExtraTrees.

**Multilayer Perceptron Neural Networks**

Artificial neural networks (ANN) are an information processing paradigm composed of highly interconnected simple elements called neurons, and structured into a sequence of layer operating in parallel (Orimoloye et al. 2020). Multilayer perceptron neural network (MLPNN) is the well-know ANN model reported in the literature, and usually consists of an input layer that comprises all the inputs variables (TE, pH, SC and PC), one hidden layer and an output layer (Moustris et al. 2020). Each neuron in the hidden layer calculates a weighted sum from each input neuron and the product is then passed by means of an activation function, generally the sigmoid (Jang and Xing 2020; Ozonoh et al. 2020). The output layer contains only one neuron that corresponded to the DO concentration, and a linear transfer function was adopted between the hidden and the output layers. The ANN like all the other intelligent data analytic models needs a data set of available input and output patterns for training the model, involving an adjustment of the connections weights and biases, from the input to the hidden layer and from the hidden to the output layer (Jang and Xing 2020). The process of adapting the models parameters is called the "*learning by example*", generally achieved using the supervised backpropagation training algorithm, in order to progressively reduce the mean square error performance function (MSE) given by Eq. (5.6):

$$\mathrm{MSE} = \frac{1}{N}\sum_{i=1}^{N}(e_i)^2 = \frac{1}{N}\sum_{i=1}^{N}(\mathrm{DO}_P - \mathrm{DO}_\mathrm{O})^2 \tag{5.6}$$

where MSE is the mean square error; $N$ is the number of patterns forming the training subset; $i$ is the index for pattern; $e_i$ is the error of the $i$th element; $DO_\mathrm{O}$ is the measured value of dissolved oxygen for $i$th element; $DO_P$ is the calculated value for $i$th element (Dickel et al. 2020).

**Multiple Linear Regression**

The multiple linear regression analysis between DO concentration and the four water quality variables, namely water TE, pH, SC and the PC was done with using the general expression as follow:

$$DO = \lambda_0 + \lambda_1\,TE + \lambda_2\,pH + \lambda_3\,SC + \lambda_4\,PC \quad (5.7)$$

The parameters $\lambda_i$ are the model coefficients determined using the least square method (LSM).

## 5.3 Results and Discussion

Dissolved oxygen (DO) was predicted using three intelligent data analytic models, namely extremely randomized tree (ERT), random forest (RF) and MLPNN, and the obtained results were compared to those obtained using the MLR model. The models were developed for assessing DO by using four water quality variables (e.g. TE, SC, pH and PC). To further explore the sensitivity of the proposed models caused by inclusion or exclusion of different variables, we examined several input combinations (Table 5.1). According to Table 5.1, the models were developed whether using all the input variables, using only three input variables or using only two input variables, and the models demonstrated ability to provide high correlation between measured and calculated DO even with fewer input variables or all input variables (Tables 5.2 and 5.3). However, in both CR-Buoy and MR-Buoy stations, the relationship of measured DO versus predicted DO was extremely close with coefficient of correlation above 0.98 by including all the input variables. In order to demonstrate the best possible input combination, we performed two calculations and present here two sets of results, for each station separately. This section provides a detailed analysis of the models accuracy, their sensitivity and responses to different water quality variables.

**Predicting DO at the Charles River Buoy (CR-Buoy) Station**

The results obtained from the proposed models at CR-Buoy station are presented in Table 5.2, and the estimated DO concentration was compared with the DO concentration measured in situ. Firstly, using all the four water quality variables as input (combination 1, Table 5.2), the all three intelligent data analytic models (ERT1, RF1 and MLPNN1) yielded remarkably accurate estimation of DO concentration (Figs. 5.2 and 5.3), with very low magnitudes of the RMSE and MAE, and very high $R$ and NSE values. RF1 model had the lowest MAE of 0.086 mg/L and RMSE of 0.123 mg/L. The ERT1 ranked second and had a slightly higher MAE (0.107 mg/L) and RMSE (0.147 mg/L). The MLPNN1 had slightly higher RMSE and MAE values compared to the ERT1 and RF1 but were still significantly lower than the MLR1(RMSE = 0.594 mg/L, MAE = 0.448 mg/L). According to Table 5.2, the ERT1 and RF1 have comparable results in that the obtained $R$ and NSE between measured and calculated DO were above 0.99 and 0.98, respectively, slightly more

**Table 5.2** Intelligent data analytic model performances for dissolved oxygen prediction at CR-Buoy station

| Models | Training | | | | Validation | | | |
|---|---|---|---|---|---|---|---|---|
| | *R* | NSE | RMSE | MAE | *R* | NSE | RMSE | MAE |
| MLPNN1 | 0.990 | 0.981 | 0.179 | 0.136 | 0.989 | 0.978 | 0.190 | 0.142 |
| MLPNN2 | 0.906 | 0.821 | 0.543 | 0.406 | 0.910 | 0.828 | 0.534 | 0.395 |
| MLPNN3 | 0.978 | 0.957 | 0.266 | 0.197 | 0.978 | 0.956 | 0.270 | 0.199 |
| MLPNN4 | 0.983 | 0.966 | 0.237 | 0.176 | 0.984 | 0.967 | 0.233 | 0.176 |
| MLPNN5 | 0.982 | 0.965 | 0.241 | 0.179 | 0.983 | 0.965 | 0.240 | 0.179 |
| MLPNN6 | 0.938 | 0.881 | 0.444 | 0.322 | 0.938 | 0.879 | 0.447 | 0.320 |
| MLPNN7 | 0.954 | 0.911 | 0.383 | 0.282 | 0.955 | 0.913 | 0.380 | 0.281 |
| MLPNN8 | 0.972 | 0.945 | 0.301 | 0.235 | 0.973 | 0.947 | 0.296 | 0.232 |
| MLPNN9 | 0.865 | 0.749 | 0.643 | 0.488 | 0.874 | 0.763 | 0.626 | 0.468 |
| MLR1 | 0.887 | 0.786 | 0.594 | 0.450 | 0.887 | 0.787 | 0.594 | 0.448 |
| MLR2 | 0.680 | 0.462 | 0.942 | 0.763 | 0.700 | 0.490 | 0.919 | 0.741 |
| MLR3 | 0.801 | 0.642 | 0.768 | 0.588 | 0.807 | 0.651 | 0.760 | 0.585 |
| MLR4 | 0.875 | 0.766 | 0.621 | 0.477 | 0.878 | 0.771 | 0.616 | 0.470 |
| MLR5 | 0.882 | 0.778 | 0.605 | 0.461 | 0.882 | 0.778 | 0.606 | 0.461 |
| MLR6 | 0.797 | 0.635 | 0.776 | 0.585 | 0.805 | 0.647 | 0.764 | 0.575 |
| MLR7 | 0.790 | 0.623 | 0.788 | 0.590 | 0.795 | 0.631 | 0.781 | 0.588 |
| MLR8 | 0.870 | 0.757 | 0.633 | 0.484 | 0.872 | 0.760 | 0.630 | 0.478 |
| MLR9 | 0.664 | 0.441 | 0.960 | 0.790 | 0.684 | 0.467 | 0.939 | 0.770 |
| ERT1 | 0.995 | 0.990 | 0.131 | 0.095 | 0.994 | 0.987 | 0.147 | 0.107 |
| ERT2 | 0.962 | 0.923 | 0.355 | 0.250 | 0.955 | 0.910 | 0.386 | 0.272 |
| ERT3 | 0.988 | 0.975 | 0.202 | 0.144 | 0.986 | 0.971 | 0.219 | 0.155 |
| ERT4 | 0.989 | 0.977 | 0.196 | 0.143 | 0.987 | 0.975 | 0.204 | 0.149 |
| ERT5 | 0.991 | 0.981 | 0.177 | 0.128 | 0.989 | 0.978 | 0.192 | 0.140 |
| ERT6 | 0.951 | 0.903 | 0.399 | 0.282 | 0.946 | 0.894 | 0.419 | 0.293 |
| ERT7 | 0.968 | 0.938 | 0.320 | 0.227 | 0.963 | 0.927 | 0.347 | 0.248 |
| ERT8 | 0.982 | 0.964 | 0.243 | 0.183 | 0.981 | 0.961 | 0.253 | 0.192 |
| ERT9 | 0.904 | 0.816 | 0.550 | 0.405 | 0.896 | 0.802 | 0.573 | 0.420 |
| RF1 | 0.998 | 0.997 | 0.073 | 0.050 | 0.996 | 0.991 | 0.123 | 0.086 |
| RF2 | 0.979 | 0.955 | 0.272 | 0.182 | 0.961 | 0.920 | 0.363 | 0.251 |
| RF3 | 0.993 | 0.986 | 0.154 | 0.107 | 0.987 | 0.974 | 0.207 | 0.143 |
| RF4 | 0.994 | 0.988 | 0.140 | 0.098 | 0.989 | 0.978 | 0.192 | 0.136 |
| RF5 | 0.996 | 0.991 | 0.120 | 0.081 | 0.991 | 0.981 | 0.177 | 0.124 |

(continued)

**Table 5.2** (continued)

| Models | Training | | | | Validation | | | |
|---|---|---|---|---|---|---|---|---|
| | *R* | NSE | RMSE | MAE | *R* | NSE | RMSE | MAE |
| RF6 | 0.960 | 0.921 | 0.360 | 0.247 | 0.945 | 0.893 | 0.420 | 0.289 |
| RF7 | 0.975 | 0.950 | 0.287 | 0.194 | 0.964 | 0.929 | 0.343 | 0.236 |
| RF8 | 0.990 | 0.979 | 0.184 | 0.132 | 0.981 | 0.961 | 0.253 | 0.180 |
| RF9 | 0.924 | 0.852 | 0.494 | 0.357 | 0.894 | 0.799 | 0.576 | 0.415 |

accurate than the MLPNN1 model ($R = 0.989$, NSE = 0.978). The results for MLR1 model were comparatively poor, with $R = 0.887$ and NSE = 0.787, and the obtained RMSE and MAE were the largest for MLR1. The large RMSE and MAE values of MLR1 indicate that it achieves low accuracy than the other three intelligent data analytic models. Secondly, removing the pH from the input variables (combination 2), the models performances were significantly decreased, and the models without pH were substantially less accurate than the models with pH (combination 1).

Thirdly, for the remaining models having only three input variables (combination 3, 4 and 5), it is clear from Table 5.2 that comparing MLPNN3, MLPNN4 and MLPNN5 models with each other reveals only small and negligible differences in corresponding statistical indices. A slight difference becomes evident between MLPNN4 and MLPNN5, on one hand, and MLPNN3, on the other hand, and the bets accuracy was achieved using the MLPNN4 marginally less than the MLPNN1 having all the water quality variables as input. The MLPNN1 decrease the RMSE and the MAE of the MLPNN4 by 18.45% and 19.32%, respectively, while, when the two models (MLPNN1, MLPNN4) were compared according to the *R* and NSE, the difference between the two was negligible.

From Table 5.2, it can be seen that the values of *R* and NSE of the ERT3, ERT4 and ERT5 are greater than those obtained using the MLPNN models: the *R* and NSE have been improved and the RMSE and MAE have been reduced. Moreover, the ERT1 decrease the RMSE and the MAE of the ERT5 by 23.43% and 23.57%, respectively. The RF5 model had the lower errors indices in predicting DO (RMSE = 0.177, MAE = 0.124) compared to the other models with three input variables: (i) 7.81 and 11.42% less than the values achieved using the ERT5 and (ii) 24.03 and 29.54% less than the values achieved using the MLPNN4.

This study found the RF intelligent data analytic models to be more robust in predicting DO compared to the ERT and MLPNN, whether using four input or three input variables, and the MLR models were significantly less accurate than the intelligent data analytic models with high errors indices (RMSE and MAE) and lower goodness of fit indices (*R* and NSE). The results from the models having only two input variables are also listed in Table 5.2.

RF achieved the best results with *R* and NSE in the range of 0.894–0.981, and 0.799–0.961, respectively, equally with the ERT models, which were followed by the MLPNN models with R and NSE in the range of 0.874–0.973, and 0.763–0.947, respectively. The MLR produced the worst results with R and NSE varying from

**Table 5.3** Intelligent data analytic models performance for dissolved oxygen prediction at MR-Buoy station

| Models | Training | | | | Validation | | | |
|---|---|---|---|---|---|---|---|---|
| | *R* | NSE | RMSE | MAE | *R* | NSE | RMSE | MAE |
| MLPNN1 | 0.972 | 0.945 | 0.541 | 0.414 | 0.973 | 0.948 | 0.547 | 0.420 |
| MLPNN2 | 0.741 | 0.549 | 1.548 | 1.175 | 0.740 | 0.547 | 1.610 | 1.224 |
| MLPNN3 | 0.970 | 0.942 | 0.556 | 0.424 | 0.971 | 0.942 | 0.575 | 0.441 |
| MLPNN4 | 0.965 | 0.931 | 0.604 | 0.464 | 0.967 | 0.935 | 0.610 | 0.470 |
| MLPNN5 | 0.968 | 0.937 | 0.576 | 0.434 | 0.969 | 0.939 | 0.588 | 0.444 |
| MLPNN6 | 0.956 | 0.915 | 0.673 | 0.532 | 0.958 | 0.917 | 0.688 | 0.546 |
| MLPNN7 | 0.958 | 0.918 | 0.658 | 0.505 | 0.959 | 0.919 | 0.679 | 0.530 |
| MLPNN8 | 0.956 | 0.914 | 0.674 | 0.532 | 0.958 | 0.919 | 0.682 | 0.537 |
| MLPNN9 | 0.502 | 0.252 | 1.993 | 1.580 | 0.510 | 0.260 | 2.056 | 1.646 |
| MLR1 | 0.930 | 0.866 | 0.845 | 0.684 | 0.933 | 0.869 | 0.864 | 0.700 |
| MLR2 | 0.381 | 0.145 | 2.131 | 1.758 | 0.378 | 0.142 | 2.214 | 1.837 |
| MLR3 | 0.930 | 0.866 | 0.845 | 0.684 | 0.933 | 0.869 | 0.864 | 0.700 |
| MLR4 | 0.921 | 0.849 | 0.897 | 0.740 | 0.926 | 0.857 | 0.905 | 0.751 |
| MLR5 | 0.930 | 0.865 | 0.845 | 0.683 | 0.933 | 0.869 | 0.865 | 0.700 |
| MLR6 | 0.921 | 0.848 | 0.898 | 0.740 | 0.926 | 0.856 | 0.907 | 0.751 |
| MLR7 | 0.930 | 0.865 | 0.845 | 0.683 | 0.933 | 0.869 | 0.865 | 0.700 |
| MLR8 | 0.921 | 0.848 | 0.899 | 0.738 | 0.926 | 0.856 | 0.908 | 0.750 |
| MLR9 | 0.325 | 0.105 | 2.180 | 1.799 | 0.335 | 0.112 | 2.253 | 1.876 |
| ERT1 | 0.986 | 0.972 | 0.387 | 0.285 | 0.985 | 0.969 | 0.419 | 0.311 |
| ERT2 | 0.897 | 0.786 | 1.067 | 0.778 | 0.878 | 0.757 | 1.179 | 0.871 |
| ERT3 | 0.980 | 0.961 | 0.458 | 0.341 | 0.979 | 0.957 | 0.495 | 0.375 |
| ERT4 | 0.977 | 0.954 | 0.496 | 0.382 | 0.976 | 0.953 | 0.519 | 0.403 |
| ERT5 | 0.982 | 0.963 | 0.444 | 0.326 | 0.980 | 0.959 | 0.483 | 0.355 |
| ERT6 | 0.964 | 0.928 | 0.617 | 0.487 | 0.962 | 0.925 | 0.654 | 0.526 |
| ERT7 | 0.970 | 0.941 | 0.558 | 0.413 | 0.967 | 0.935 | 0.610 | 0.459 |
| ERT8 | 0.966 | 0.933 | 0.598 | 0.465 | 0.965 | 0.931 | 0.630 | 0.490 |
| ERT9 | 0.715 | 0.501 | 1.628 | 1.251 | 0.673 | 0.449 | 1.775 | 1.392 |
| RF1 | 0.997 | 0.993 | 0.192 | 0.126 | 0.992 | 0.984 | 0.303 | 0.205 |
| RF2 | 0.946 | 0.880 | 0.800 | 0.543 | 0.888 | 0.781 | 1.119 | 0.773 |
| RF3 | 0.991 | 0.981 | 0.318 | 0.224 | 0.984 | 0.967 | 0.435 | 0.317 |
| RF4 | 0.989 | 0.978 | 0.343 | 0.250 | 0.982 | 0.964 | 0.454 | 0.334 |
| RF5 | 0.994 | 0.986 | 0.269 | 0.184 | 0.986 | 0.972 | 0.402 | 0.277 |

(continued)

**Table 5.3** (continued)

| Models | Training | | | | Validation | | | |
|---|---|---|---|---|---|---|---|---|
| | *R* | NSE | RMSE | MAE | *R* | NSE | RMSE | MAE |
| RF6 | 0.975 | 0.950 | 0.513 | 0.398 | 0.963 | 0.928 | 0.641 | 0.502 |
| RF7 | 0.981 | 0.963 | 0.444 | 0.312 | 0.968 | 0.938 | 0.597 | 0.434 |
| RF8 | 0.980 | 0.961 | 0.455 | 0.337 | 0.969 | 0.938 | 0.594 | 0.443 |
| RF9 | 0.791 | 0.613 | 1.433 | 1.061 | 0.658 | 0.433 | 1.801 | 1.381 |

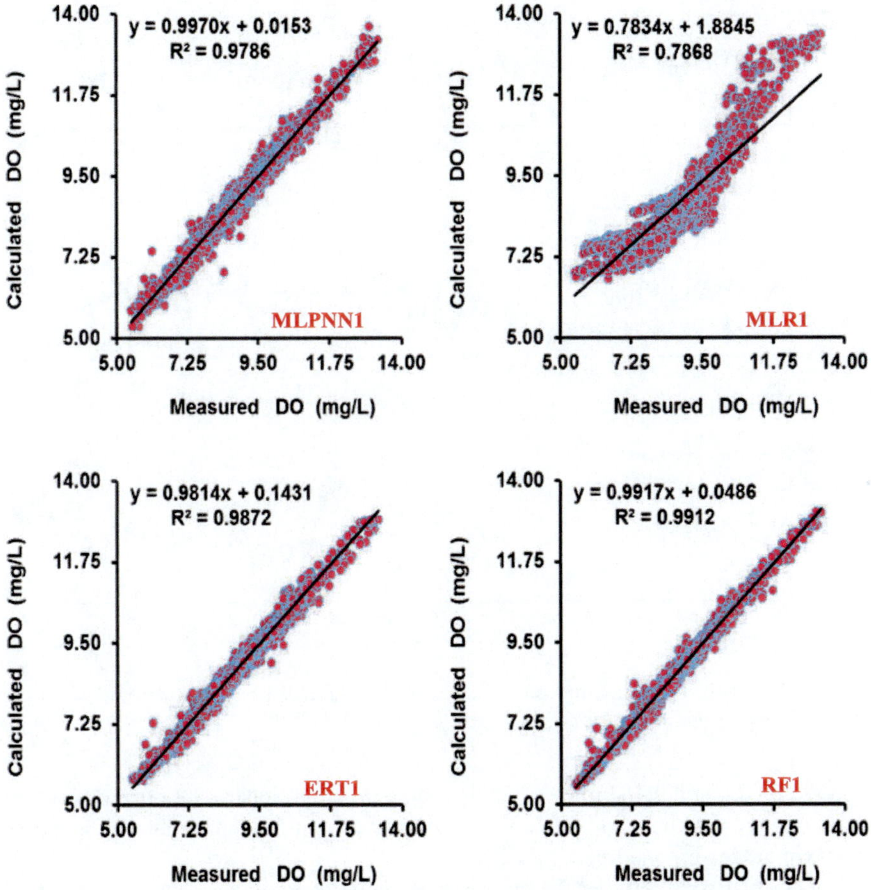


**Fig. 5.2** Scatterplots of measured against calculated dissolved oxygen concentration at the CR-BUPY station

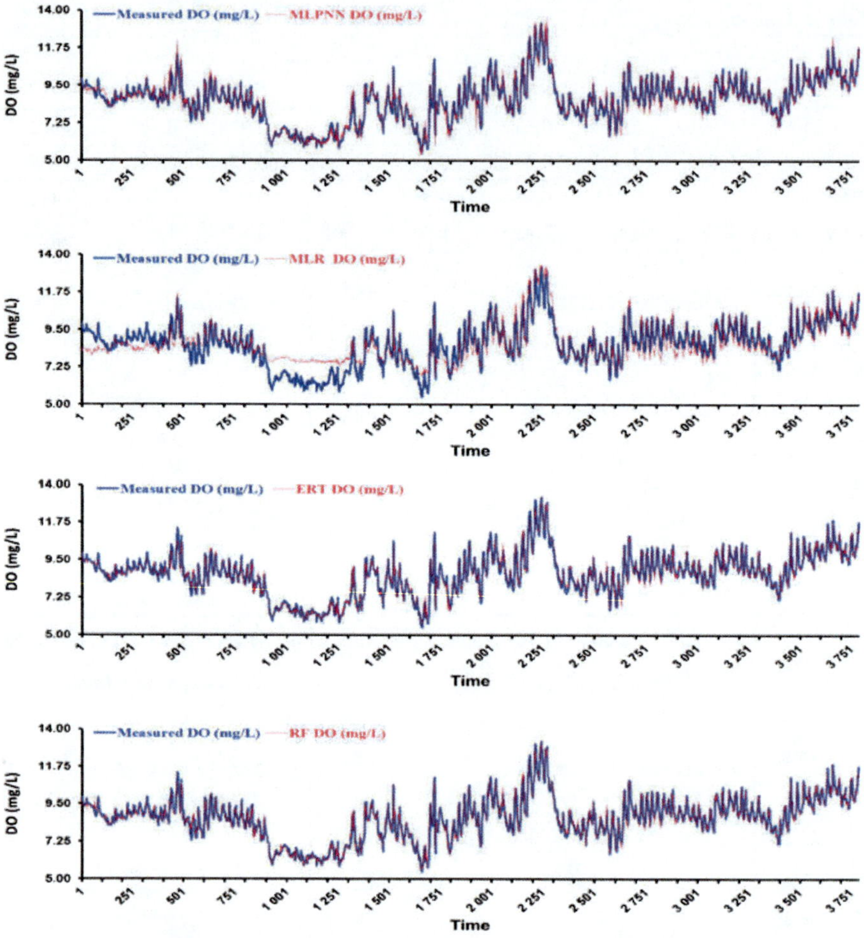


**Fig. 5.3** Comparison between measured and calculated dissolved oxygen concentration at the CR-BUPY station

0.684 to 0.872, and 0.467 to 0.760 indicating a high nonlinear relationship between water quality variables and DO concentration may not be appropriate to be easily approximated using the linear models.

For comparisons between the models with only two input variables (combination 6–9), the results showed that the models having TE and pH as input variables were better than the other models by jointly observing *R*, NSE, RMSE and MAE in the use of the RF8, ERT8, MLPNN8 and MLR8.

**Predicting DO at the Mystic River Buoy (MR-Buoy) Station**

For the MR-Buoy station, Table 5.3 reports the training and validation results showing the variation of the four statistical indices in relation to the input combination.

According to Table 5.3, during the validation phase, when all water quality variables were included in the models input (combination 1), the numerical models performances (R, NSE, RMSE and MAE) show excellent efficiency on the estimation of DO concentration for the three intelligent data analytic models, with high R and NSE, reaching their maximum (i.e. $R = 0.992$ and NSE = 0.982) using the RF1 model, followed by the ERT1 (i.e. $R = 0.985$ and NSE = 0.969), the MLPNN1 in the third place (i.e. $R = 0.973$ and NSE = 0.948).

The good agreement between modelled and measured DO concentration using RF1 at MR-Buoy station is also shown in Fig. 5.4 (scatterplot), and further supported by a RMSE value of about 0.303 (mg/L) and a MAE equal to 0.205 (mg/L) (see Table 5.3). The performances indices (i.e. $R = 0.933$, NSE = 0.869, RMSE = 0.864 mg/L and MAE = 0.700 mg/L) of the MLR1 model are significantly lower and therefore not comparable with those obtained using the RF1, thus highlighting the robustness of the RF1 model predictions.

Concerning configuration with only three input variables, consistently with the previous analysis, the results in terms of model efficiency highlighting the superiority of the RF5 model having the TE, pH, as SC as input variables compared to the all other models. In particular, the $R$ and NSE values are maximal (i.e. $R = 0.986$, NSE = 0.972), and the RMSE and MAE are the minimal (i.e. RMSE = 0.402, MAE = 0.277).

Referring to the ERT models, the best performances (i.e. $R = 0.980$, NSE = 0.959) were achieved using the ERT5 model, slightly lower than the RF5, for which the RF5 decrease the RMSE and MAE of the ERT5 by 16.77% and 21.98%, respectively. However, the lowest RMSE and MAE values were achieved using the MLPNN3 with values equal to 0.575 mg/L and 0.441 mg/L, respectively.

In addition to the above results, it is clear from Table 5.3 that the MLR3 and MLR5 provided the same accuracy as achieved by the MLR1, which leads to conclude that using the multiple linear regression models, three input variables are sufficient for providing the best accuracy.

Further analysis of the results reported in Table 5.3, by showing the comparison between the models accuracy using only three input variables, it is clear that removing the pH from the input variables (combination 2: TE, SC, PC), the models accuracy were dramatically decreased which leads to the lowest accuracy. In this case, removing the pH form the input variables of the MLPNN1, the RMSE and MAE were increased by 66.02%, 65.68%, while the R and NSE values were decreased by 23.3% and 40.1%, respectively (MLPNN1 vs. MLPNN2). Similarly, the RMSE and MAE of the MLR1 were increased by 60.97%, 61.89%, while the R and NSE values were decreased by 55.5% and 72.7%, respectively (MLR1 vs. MLR2). In addition, the RMSE and MAE of the ERT1 were increased by 64.46%, 64.29%, while the R and NSE values were decreased by 10.7% and 21.2%, respectively (ERT1 vs. ERT2). Finally, if the pH variable was removed from the input variables of the RF1 model, the validation RMSE and MAE decreased significantly (i.e. 72.92 and 73.48%), becoming the largest percentage decrease compared to the other models, highlighting how RF model proves to be more accurate and suitable for capturing all available information in the input variables. According to the results obtained for

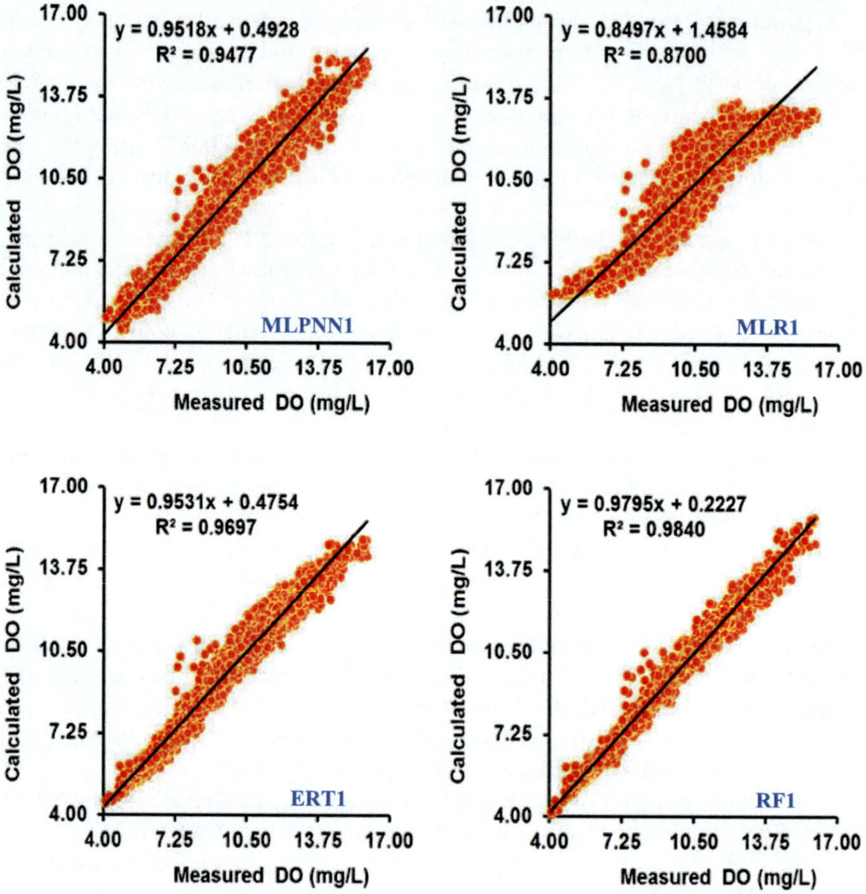


**Fig. 5.4** Scatterplots of measured against calculated DO concentration at the MR-BUPY station

combination 6–9 (Table 5.3) using only two input variables, all the performances have worsened. In particular, the MLPNN6, MLPNN7 and MLPNN8 provided relatively the same accuracy slightly lower than the ERT and RF models. The R and NSE values were still satisfactory and the best accuracy was obtained using the RF8 having the TE and pH as input variables, with R and NSE values of 0.969 and 0.938, respectively.

Figures 5.2 and 5.4 display the scatter plots of the modelled versus calculated values of DO concentration using the four applied models for the two stations. A significant correlation was found between modelled and in situ values for all the three intelligent data analytic models analysed above (ERT1, RF1 and MLPNN1), with a high degree of scatter in the data is apparent for all the models, except for the MLR. The RF1 model, not surprisingly, gave the closest results compared to measured values; it showed a high coefficient of determination R-squared

($R^2 = 0.99$). Figures 5.3 and 5.5 show modelled DO concentration using all proposed models plotted against corresponding in situ measurement for the two stations. It is remarkable that the performance of the MLR model was significantly lower than the machine learning models and the MLPNN1 was slightly lower than the RF1 and ERT1, however, RF1, ERT1 and MLPNN1 models perform well and the RF1 was the unique model that possesses the capacity for correctly fitting the maximum DO values. The boxplot of the DO data during the validation (Fig. 5.6) shows that RF1 provides the best estimation of DO slightly higher than the ERT1 and MLPNN1, and the MLR1 was largely less accurate than the intelligent data analytic models. Finally, the models were further compared according to their performances, using

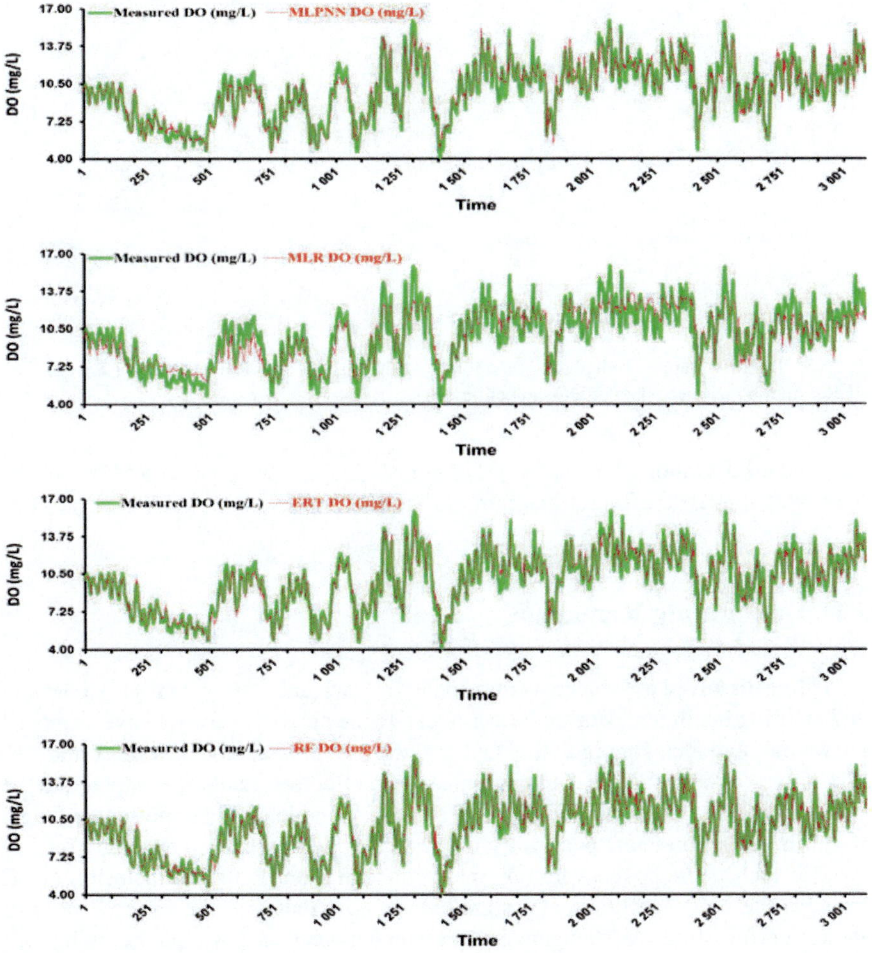


**Fig. 5.5** Comparison between measured and calculated DO concentration at the MR-BUPY station

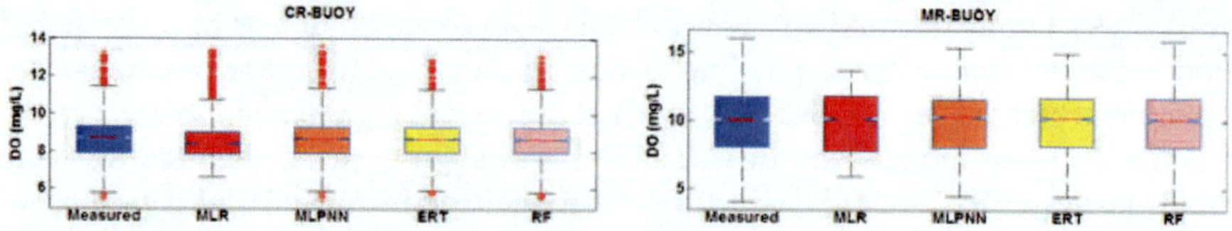


**Fig. 5.6** Boxplot with whiskers from minimum to maximum data for all models. The box stretches from the 25th percentile to the 75th percentile

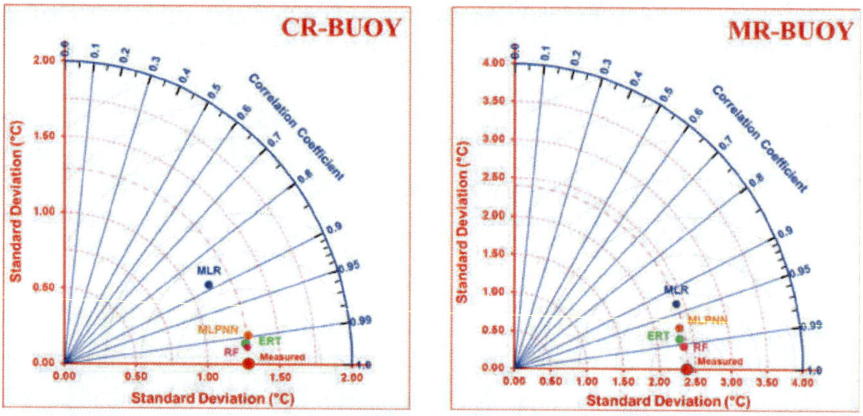


**Fig. 5.7** Taylor's diagrams displaying a statistical comparison between MLPNN, ERT, RF and MLR with the measured data of DO concentration

the standard deviation, the coefficient of correlation and the centred RMSE; all the three plotted using the Taylor diagram in Fig. 5.7.

## 5.4 Concluding Remarks

Modelling dissolved oxygen concentration in freshwater ecosystems was extensively studied using intelligent data analytic models. In the present study, we have proposed a new kind of models for predicting DO, and the proposed approaches (ERT and RF) have proved to be workable, and guarantee more effective accuracy compared to the well-known neural network and multiple regression models. One important outcome of our investigation is the possibility of predicting DO based on new water quality variable such as the phycocyanin pigment concentration (PC). The inclusion of PC provides the better accuracy and significantly contributes to the improvement of models performances. Although a comparison between various input combinations was done, our results suggest that the ERT and the RF models provide more accuracy results than the MLPNN models, even with fewer input variables. The ERT and

RF models tend to capture the high nonlinearities between DO and water quality variables more efficiently that the MLPNN and, therefore they are particularly suited for predicting DO allowing high precision and strong correlation between measured and calculated data. Furthermore, further analysis using other water quality variables could indeed help to drawing more and in-depth conclusions.

## References

Antanasijević D, Pocajt V, Perić-Grujić A, Ristić M (2019) Multilevel split of high-dimensional water quality data using artificial neural networks for the prediction of dissolved oxygen in the Danube River. Neural Comput Appl 1–10. https://doi.org/10.1007/s00521-019-04079-y

Breiman L (2001) Random forests. Mach Learn 45(1):5–32. https://doi.org/10.1023/A:1010933404324

Breiman L (1996) Bagging predictors. Mach Learn 24(2):123–140. https://doi.org/10.1007/BF00058655

Breiman L, Friedman J, Stone CJ, Olshen RA (1984) Classification and regression trees, 1st edn. Chapman and Hall/CRC, Belmont, CA

Basith S, Manavalan B, Shin TH, Lee G (2018) IGHBP: Computational identification of growth hormone binding proteins from sequences using extremely randomized tree. Comput Struct Biotechnol J 16:412–420. https://doi.org/10.1016/j.csbj.2018.10.007

Banerjee A, Chakrabarty M, Rakshit N, Bhowmick AR, Ray S (2019) Environmental factors as indicators of dissolved oxygen concentration and zooplankton abundance: deep learning versus traditional regression approach. Ecol Ind 100:99–117. https://doi.org/10.1016/j.ecolind.2018.09.051

Crossman J, Futter MN, Elliott JA, Whitehead PG, Jin L, Dillon PJ (2019) Optimizing land management strategies for maximum improvements in lake dissolved oxygen concentrations. Sci Total Environ 652:382–397. https://doi.org/10.1016/j.scitotenv.2018.10.160

Cao W, Huan J, Liu C, Qin Y, Wu F (2019) A combined model of dissolved oxygen prediction in the pond based on multiple-factor analysis and multi-scale feature extraction. Aquacult Eng 84:50–59. https://doi.org/10.1016/j.aquaeng.2018.12.003

Chen L, Su W, Feng Y, Wu M, She J, Hirota K (2020) Two-layer fuzzy multiple random forest for speech emotion recognition in human-robot interaction. Inf Sci 509:150–163. https://doi.org/10.1016/j.ins.2019.09.005

Csábrági A, Molnár S, Tanos P, Kovács J, Molnár M, Szabó I, Hatvani IG (2019) Estimation of dissolved oxygen in riverine ecosystems: comparison of differently optimized neural networks. Ecol Eng 138:298–309. https://doi.org/10.1016/j.ecoleng.2019.07.023

Deng X, Liu Z, Zhan Y, Ni K, Zhang Y, Ma W, Shao S, Lv X, Yuan Y, Rogers KM (2020) Predictive geographical authentication of green tea with protected designation of origin using a random forest model. Food Control 107:106807. https://doi.org/10.1016/j.foodcont.2019.106807

Dickel D, Francis DK, Barrett CD (2020) Neural network aided development of a semi-empirical interatomic potential for titanium. Comput Mater Sci 171:109157. https://doi.org/10.1016/j.commatsci.2019.109157

Elkiran G, Nourani V, Abba SI (2019) Multi-step ahead modelling of river water quality parameters using ensemble artificial intelligence-based approach. J Hydrol 577:123962. https://doi.org/10.1016/j.jhydrol.2019.123962

Emenike PC, Neris JB, Tenebe IT, Nnaji CC, Jarvis P (2020) Estimation of some trace metal pollutants in River Atuwara southwestern Nigeria and spatio-temporal human health risks assessment. Chemosphere 239:124770. https://doi.org/10.1016/j.chemosphere.2019.124770

El Najjar P, Kassouf A, Probst A, Probst JL, Ouaini N, Daou C, El Azzi D (2019) High-frequency monitoring of surface water quality at the outlet of the Ibrahim River (Lebanon): a multivariate assessment. Ecol Ind 104:13–23. https://doi.org/10.1016/j.ecolind.2019.04.061

Geurts P, Ernst D, Wehenkel L (2006) Extremely randomized trees. Mach Learn 63(1):3–42. https://doi.org/10.1007/s10994-006-6226-1

Hoang THT, Nguyen VD, Van AD, Nguyen HT (2019) Decision tree techniques to assess the role of daily DO variation in classifying shallow eutrophicated lakes in Hanoi, Vietnam. Water Qual Res J. https://doi.org/10.2166/wqrj.2019.105

Heddam S (2017) Fuzzy neural network (EFuNN) for modelling dissolved oxygen concentration (DO). In: Kahraman C, Sari IU (eds) Intelligence systems in environmental management: theory and applications, intelligent systems reference library 113. https://doi.org/10.1007/978-3-319-42993-9_11

Hanna BN, Dinh NT, Youngblood RW, Bolotnov IA (2020) Machine-learning based error prediction approach for coarse-grid computational fluid dynamics (CG-CFD). Prog Nucl Energy 118:103140. https://doi.org/10.1016/j.pnucene.2019.103140

Hutchins MG, Hitt OE (2019) Sensitivity of river eutrophication to multiple stressors illustrated using graphical summaries of physics-based river water quality model simulations. J Hydrol 577:123917. https://doi.org/10.1016/j.jhydrol.2019.123917

Jerves-Cobo R, Forio MAE, Lock K, Van Butsel J, Pauta G, Cisneros F, Nopens I, Goethals PL (2020) Biological water quality in tropical rivers during dry and rainy seasons: a model-based analysis. Ecol Ind 108:105769. https://doi.org/10.1016/j.ecolind.2019.105769

Jang HS, Xing S (2020) A model to predict ammonia emission using a modified genetic artificial neural network: analyzing cement mixed with fly ash from a coal-fired power plant. Constr Build Mater 230:117025. https://doi.org/10.1016/j.conbuildmat.2019.117025

Khosravi K, Mao L, Kisi O, Yaseen ZM, Shahid S (2018) Quantifying hourly suspended sediment load using data mining models: case study of a glacierized andean catchment in Chile. J Hydrol

Keshtegar B, Heddam S, Hosseinabadi H (2019) The employment of polynomial chaos expansions approach for modeling dissolved oxygen concentration in River. Environ Earth Sci 78:34. https://doi.org/10.1007/s12665-018-8028-8

Kisi O, Yaseen ZM (2019) The potential of hybrid evolutionary fuzzy intelligence model for suspended sediment concentration prediction. CATENA 174:11–23

Kumar AU, Jayakumar KV (2020) Hydrological alterations due to anthropogenic activities in Krishna River Basin, India. Ecol Indicators 108:105663. https://doi.org/10.1016/j.ecolind.2019.105663

Kebede G, Mushi D, Linke RB, Dereje O, Lakew A, Hayes DS, Farnleitner AH, Graf W (2020) Macro invertebrate indices versus microbial fecal pollution characteristics for water quality monitoring reveals contrasting results for an Ethiopian river. Ecol Ind 108:105733. https://doi.org/10.1016/j.ecolind.2019.105733

Liu Y, Zhang Q, Song L, Chen Y (2019) Attention-based recurrent neural networks for accurate short-term and long-term dissolved oxygen prediction. Comput Electron Agric 165:104964. https://doi.org/10.1016/j.compag.2019.104964

Manavalan B, Basith S, Shin TH, Wei L, Lee G (2019) AtbPpred: a robust sequence-based prediction of anti-tubercular peptides using extremely randomized trees. Comput Struct Biotechnol J 17:972–981. https://doi.org/10.1016/j.csbj.2019.06.024

Moustris K, Kavadias KA, Zafirakis D, Kaldellis JK (2020) Medium, short and very short-term prognosis of load demand for the Greek Island of Tilos using artificial neural networks and human thermal comfort-discomfort biometeorological data. Renew Energy 147:100–109. https://doi.org/10.1016/j.renene.2019.08.126

Mitrović T, Antanasijević D, Lazović S, Perić-Grujić A, Ristić M (2019) Virtual water quality monitoring at inactive monitoring sites using Monte Carlo optimized artificial neural networks: a case study of Danube River (Serbia). Sci Total Environ 654:1000–1009. https://doi.org/10.1016/j.scitotenv.2018.11.189

Nattee C, Khamsemanan N, Lawtrakul L, Toochinda P, Hannongbua S (2017) A novel prediction approach for antimalarial activities of trimethoprim, pyrimethamine, and cycloguanil analogues using extremely randomized trees. J Mol Graph Model 71:13–27. https://doi.org/10.1016/j.jmgm.2016.09.010

Ozonoh M, Oboirien BO, Higginson A, Daramola MO (2020) Performance evaluation of gasification system efficiency using artificial neural network. Renew Energy 145:2253–2270. https://doi.org/10.1016/j.renene.2019.07.136

Orimoloye LO, Sung MC, Ma T, Johnson JE (2020) Comparing the effectiveness of deep feedforward neural networks and shallow architectures for predicting stock price indices. Expert Syst Appl 139:112828. https://doi.org/10.1016/j.eswa.2019.112828

Ross AC, Stock CA (2019) An assessment of the predictability of column minimum dissolved oxygen concentrations in Chesapeake Bay using a machine learning model. Estuarine Coast Shelf Sci 221:53–65. https://doi.org/10.1016/j.ecss.2019.03.007

Rahman A, Dabrowski J, McCulloch J (2019) Dissolved oxygen prediction in prawn ponds from a group of one step predictors. Inf Process Agric. https://doi.org/10.1016/j.inpa.2019.08.002

Suarez VVC, Brederveld RJ, Fennema M, Moreno-Rodenas A, Langeveld J, Korving H, Schellart NA, Shucksmith J (2019) Evaluation of a coupled hydrodynamic-closed ecological cycle approach for modelling dissolved oxygen in surface waters. Environ Model Softw 119:242–257. https://doi.org/10.1016/j.envsoft.2019.06.003

Shi P, Li G, Yuan Y, Huang G, Kuang L (2019) Prediction of dissolved oxygen content in aquaculture using Clustering-based Softplus Extreme Learning Machine. Comput Electron Agric 157:329–338. https://doi.org/10.1016/j.compag.2019.01.004

Tao H, Bobaker AM, Ramal MM, Yaseen ZM, Hossain MS, Shahid S (2019) Determination of biochemical oxygen demand and dissolved oxygen for semi-arid river environment: application of soft computing models. Environ Sci Pollut Res 26(1):923–937. https://doi.org/10.1007/s11356-018-3663-x

Tan K, Wang H, Chen L, Du Q, Du P, Pan C (2020) Estimation of the spatial distribution of heavy metal in agricultural soils using airborne hyperspectral imaging and random forest. J Hazard Mater 382:120987. https://doi.org/10.1016/j.jhazmat.2019.120987

Tao H, Chen R, Xuan J, Xia Q, Yang Z, Zhang X, He S, Shi T (2020) Prioritization analysis and compensation of geometric errors for ultra-precision lathe based on the random forest methodology. Precision Eng 61:23–40. https://doi.org/10.1016/j.precisioneng.2019.09.012

Yang H, Csukás B, Varga M, Kucska B, Szabó T, Li D (2019) A quick condition adaptive soft sensor model with dual scale structure for dissolved oxygen simulation of recirculation aquaculture system. Comput Electron Agric 162:807–824. https://doi.org/10.1016/j.compag.2019.05.025

Yahya A, Saeed A, Ahmed AN, Binti Othman F, Ibrahim RK, Afan HA, El-Shafie A, Fai CM, Hossain MS, Ehteram M, Elshafie A (2019) Water quality prediction model based support vector machine model for Ungauged River Catchment under dual scenarios. Water 11(6):1231. https://doi.org/10.3390/w11061231

Yaseen ZM, Ramal MM, Diop L, Jaafar O, Demir V, Kisi O (2018a) Hybrid adaptive neuro-fuzzy models for water quality index estimation. Water Resour Manage 32:2227–2245. https://doi.org/10.1007/s11269-018-1915-7

Yaseen Z, Ehteram M, Sharafati A, Shahid S, Al-Ansari N (2018b) The integration of nature-inspired algorithms with least square support vector regression models: application to modeling river dissolved oxygen concentration. Water 10:1–21

Zhu S, Heddam S (2019) New formulation for predicting dissolved oxygen in urban rivers at the Three Gorges Reservoir, China: extreme learning machines (ELM) versus artificial neural network (ANN). Water Qual Res J Can. https://doi.org/10.2166/wqrj.2019.053

Zounemat-Kermani M, Seo Y, Kim S, Ghorbani MA, Samadianfard S, Naghshara S, Kim NW, Singh VP (2019) Can decomposition approaches always enhance soft computing models? Predicting the dissolved oxygen concentration in the St. Johns River, Florida. Appl Sci 9(12): 2534. https://doi.org/10.3390/app9122534

# Chapter 6
# Evolving Connectionist Systems Versus Neuro-Fuzzy System for Estimating Total Dissolved Gas at Forebay and Tailwater of Dams Reservoirs

**Salim Heddam and Ozgur Kisi**

## 6.1 Introduction

The importance of regularly monitoring water quality in fresh water ecosystem is well highlighted and broadly discussed in literature. Dam's reservoirs are one of the most important hydraulic structures that have received great importance, especially in regard to the control of the stored water quality. At several high dams reservoir, especially in USA and China, water is released through the spillways of the dams causing the concentration of the total dissolved gas (*TDG*) to be very high (Feng et al. 2014a, b). The concentration of *TDG* in water is a key indicator of the biophysical quality of the water, and when compared to the standard level of concentration (110%), most of the reach of the river water downstream of a high-dam spillway is affected by the supersaturation of the total dissolved gas (*TDG*) levels, which complicates the aquatic life of the fish. Excess *TDG* concentrations cause "*gas bubble disease GBD*" (Weitkamp and Katz 1980) and now considered as the most important and serious problem associated by the elevation of *TDG*. For example, (Feng et al. 2018) argued that during the summer of 2014, TDG supersaturations have caused more 100,000 kg of dead fish in China.

Assessments for the formation of *TDG* supersaturation and its relation with the other factor governing its formation remain critically important and can help to a reliable and accurate understanding of *TDG* process. Numerical models are one of the important tools used by researchers to improve the process understanding in several

S. Heddam (✉)
Agronomy Department, Hydraulics Division, Laboratory of Research in Biodiversity Interaction Ecosystem and Biotechnology, University 20 Août 1955 Skikda, Route El Hadaik, BP 26, Skikda, Algeria
e-mail: heddamsalim@yahoo.fr

O. Kisi
School of Technology, Ilia State University, 0162 Tbilisi, Georgia

© Springer Nature Singapore Pte Ltd. 2021

R. C. Deo et al. (eds.), *Intelligent Data Analytics for Decision-Support Systems in Hazard Mitigation*, Springer Transactions in Civil and Environmental Engineering,
https://doi.org/10.1007/978-981-15-5772-9_6

areas of environmental and ecological sciences. Consequently, several algorithms that use several numerical, fluid mechanic, and hydrodynamic equations have been developed and shown to yield accurate simulation of the *TDG* process (Feng et al. 2014a, b; Ou et al. 2016; Ma et al. 2016, 2018; Weber et al. 2004; Politano et al. 2007, 2009, 2012, 2017; Wilhelms and Schneider 2006; Witt et al. 2017; Deng et al. 2017; Shen et al. 2019; Picket et al. 2004; Tawfik and Diez 2014; Wang et al. 2018; Yuan et al. 2018; Heddam 2017). Looking to the importance assigned to *TDG* supersaturation, an important part of *TDG* formation and variation was explained by the increase of spill from dam (*SFD*) through the spillway until the tailrace; and this observation is supported by numerous studies (Politano et al. 2009, 2012). Historically, previous investigations focused on the application of data-driven approaches have been tested for solving a variety of environmental problem. However, to date, less importance has been attributed to modeling *TDG* using data-driven models. To the best of the author's knowledge, only the generalized regression neural network (*GRNN*) model proposed by (Heddam 2017) and applied for modelling *TDG* concentration using several variables, namely water temperature, barometric pressure, spill from dam, sensor depth, and total flow, no other studies are available in the literature.

The focus of the present chapter is on a wide range of aspects related to *TDG* modelling. Key points addressed in this investigation are (i) a new intelligent data analytic approach for *TDG* prediction from easily measured data is proposed based on data-driven models. The proposed models belong to the categories of neuro-fuzzy approaches and in total, five models were proposed; (ii) comparing the accuracy of the proposed models based on several input combination and an attempt was made for predicting *TDG* using only fewer variables, especially, the accuracy of the models using only *SFD* variable was highlighted; (iii) the results obtained using the data-driven models were compared to those obtained using the standard multiple linear regression (*MLR*); (iv) the best models obtained were tested by interchanging the training and validation data sets by setting the training data set (30%) and validation data set (70%); and (v) the best models were applied and compared for predicting *TDG* using the component of the Gregorian calendar as input variables.

## 6.2 Materials and Method

### *6.2.1 Study Area Description*

The *TDG* data and four input variables collected during two years (2015 and 2016) were used for developing the *TDG* models.

For the two-year period, we used the data of the spilling season only, i.e., from April to September (six months). For each month, data was available at hourly time steps. The stations chosen are operated by the US Army Corps of Engineers (*USACE*). The two stations are located at the Lower Granite Dam at the Snake River: the Lower

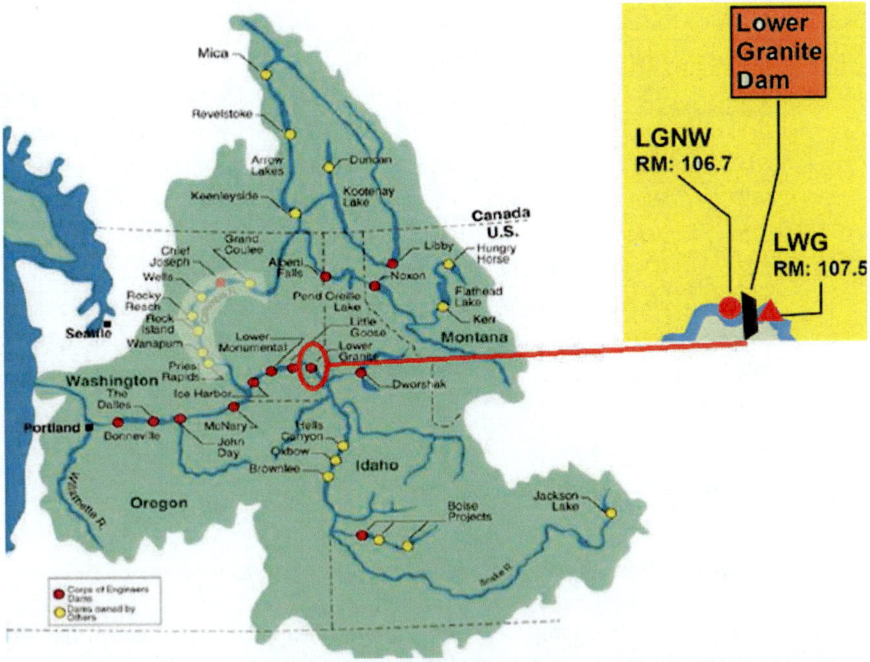


**Fig. 6.1** Location of the two stations at the Lower Granite Dam located at the Snake River, USA: Lower Granite Tailwater (*LGNW*) and the Lower Granite Forebay (*LWG*) (Stewart et al. 2015)

Granite Tailwater (*LGNW*) (Latitude 46° 39′58.1″; Longitude 117° 26′19.3″) and the Lower Granite Forebay (*LWG*) (Latitude 46° 39′34.9″; Longitude 117° 26′34.9″).

Figure 6.1 shows the location of the two stations. Data for the two stations can be found at the website: www.nwd-wc.usace.army.mil/ftppub/water_quality/tdg/. For developing the models, we used four input variables, namely water temperature (*TE* °C), barometric pressure (BP mm Hg), spill from dam (*SFD* kcfs), and the total flow (*DIS* kcfs).Total dissolved gas measured as the percent of saturation (*TDG %*) is the predicted variable. The total length of the data set was 8770 patterns for the *LGNW* station and 7190 patterns for the *LWG* station. In the present study, we split the data into 70% for training and 30% for validation.

The *TDG* and the four variables were normalized using the *Z*-score method:

$$Z_n = \frac{x_n - x_m}{\sigma_x} \tag{6.1}$$

where $Z_n$ is the normalized value of the observation *n*; $x_n$ is the measured value of the observation *n;* $x_m$ and $\sigma_x$ are the mean value and standard deviation of the variable *x*. In Table 6.1, we report the mean, maximum, minimum, standard deviation, and coefficient of variation values, and the coefficient of correlation with *TDG*, i.e., $X_{mean}$, $X_{max}$, $X_{min}$, $S_x$, $C_v$, and *R*, respectively.

**Table 6.1** Statistical parameters of the used data sets for the two stations

| Station | Data set | Unit | $X_{mean}$ | $X_{max}$ | $X_{min}$ | $S_x$ | $C_v$ | $R$ |
|---|---|---|---|---|---|---|---|---|
| *LGNW* | *TE* | °C | 15.995 | 21.400 | 7.600 | 3.653 | 0.228 | −0.227 |
| | *BP* | mm Hg | 743.098 | 757.300 | 732.100 | 3.680 | 0.005 | −0.124 |
| | *DIS* | kcfs | 44.752 | 140.500 | 12.300 | 26.386 | 0.590 | 0.482 |
| | *SFD* | feet | 14.343 | 68.300 | 0.000 | 8.304 | 0.579 | 0.907 |
| | *TDG* | % sat. | 107.966 | 122.800 | 95.500 | 4.896 | 0.045 | 1.000 |
| *LWG* | *TE* | °C | 16.514 | 21.700 | 7.800 | 3.505 | 0.212 | −0.500 |
| | *BP* | mm Hg | 740.238 | 754.600 | 729.700 | 3.458 | 0.005 | -0.285 |
| | *DIS* | kcfs | 47.108 | 140.500 | 12.500 | 27.529 | 0.584 | 0.584 |
| | *SFD* | feet | 15.463 | 68.300 | 0.000 | 7.472 | 0.483 | 0.629 |
| | *TDG* | % sat. | 101.992 | 106.900 | 96.400 | 1.885 | 0.018 | 1.000 |

## 6.2.2 *Modeling Approaches*

### 6.2.2.1 Adaptive Neuro-Fuzzy Inference System (*ANFIS*)

*ANFIS*, first suggested by (Jang 1993), is a universal approximator and it has the ability in approximating any continuous function on a compact set. *ANFIS*' structure comprises a number of nodes connected to each other with directional links similar to neural networks (Jang et al. 1997).

Let assume a fuzzy system involving three inputs $x$, $y$, and $z$ and one output $f$. In this case, the rule base has 2 fuzzy if–then rules

$$\text{Rule 1: if } x \text{ is } \mathrm{A}_1,\ y \text{ is } \mathrm{B}_1 \text{ and } z \text{ is } \mathrm{C}_1 \text{ then } \quad f_1 = p_1 x + q_1 y + r_1 z + s_1 \tag{6.2}$$

$$\text{Rule 2: if } x \text{ is } \mathrm{A}_2,\ y \text{ is } \mathrm{B}_2 \text{ and } z \text{ is } \mathrm{C}_2 \text{ then } \quad f_2 = p_2 x + q_2 y + r_2 z + s_2 \tag{6.3}$$

Here, $f_1$ and $f_2$ indicate the output function corresponding to the rule 1 and rule 2, respectively. Figure 6.2 illustrates the structure of *ANFIS*. The description of the node functions is as follow for each layer.

First layer: This layer includes adaptive nodes ($i$) as

$$O_{l,i} = \phi A_i(x), \quad \text{for } i = 1, 2 \tag{6.4}$$

where $x$ = input for the node $i$ and $A_i$ indicates the label (e.g., "small" or "big"). $O_{l,i}$ shows the is the membership function of a fuzzy set $A$ (=$A_1$, $A_2$, $B_1$, $B_2$, $C_1$, or $C_2$). Gaussian function is generally preferred ($\phi A_i(x)$)

$$\phi A_i(x) = \exp\left(-\left(\frac{x - a_i}{b_i}\right)^2\right) \tag{6.5}$$

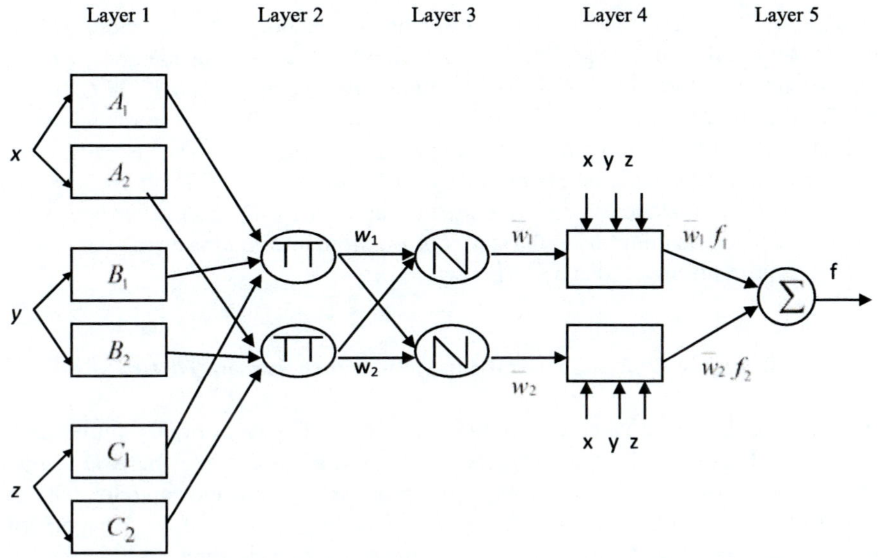


**Fig. 6.2** Structure of *ANFIS*

where $a_i$, $b_i$ are the parameters of membership function which are known as premise parameters (Jang et al. 1997) (Jang 1993).

Second layer: The nodes in this layer multiply the incoming data and give the product out.

$$w_i = \phi A_i(x)\phi B_i(y)\phi C_i(z), \quad i = 1, 2 \tag{6.6}$$

Third layer: The nodes in this layer calculates the ratio of the $i$th rule's firing strength to the sum of all rules' firing strengths as

$$\overline{w}_i = \frac{w_i}{w_1 + w_2}, \quad i = 1, 2 \tag{6.7}$$

Fourth layer: The node function in this layer is

$$O_{4,i} = \overline{w}_i f_i = \overline{w}_i(p_i x + q_i y + r_i z + s_i) \tag{6.8}$$

where $\overline{w}_i$ = the output of the third layer and the parameter set is $\{p_i, q_i, r_i, s_i\}$ and they are named as consequent parameter.

Fifth layer: The node in this layer computes the final output as

$$O_{5,i} = \sum_{i=1} \overline{w}_i f = \frac{\sum_i w_i f_i}{\sum_i w_i} \tag{6.9}$$

The *ANFIS* output involves constant or linear functions. Details about *ANFIS* can be acquired from (Jang 1996). In this chapter, three *ANFIS* methods, *ANFIS* with subtractive clustering (*ANFIS-SC*), *ANFIS* with grid partition (*ANFIS-GP*), and *ANFIS* with fuzzy c-means (*ANFIS-FCM*) were employed. The *GP* commonly used in related literature is an input partitioning method (Jang 1996). Data is partitioned into clusters and a smaller number of fuzzy rules are gotten in SC method compared to *GP*. In *FCM* method, an iterative method used in *FCM* and cluster centers is calculated based on minimizing the square error. More information on these methods can be acquired from (Kisi et al. 2017).

#### 6.2.2.2 Dynamic Evolving Neural-Fuzzy Inference System (*DENFIS*)

In this section, the architecture of the proposed *DENFIS* is briefly introduced, with multiple input and single output, consisting with a fuzzy rules base. Dynamic evolving neural-fuzzy inference system was proposed by (Kasabov and Song 2002) based on the original neuro-fuzzy system (*NF*) and belongs to the category of evolving connectionist systems. Contrary to the standard well-known *ANFIS* model, for which several clustering methods can be used, i.e., fuzzy c-mean clustering (*FC*) algorithm and subtractive clustering (*SC*) method; *DENFIS* is based on the so-called evolving clustering method (*ECM*) (Kasabov 2007). The fuzzy rules base of the *DENFIS* based on the *ECM* can be presented as follow (Kasabov 2007):

$$\begin{cases} \text{if } x_1 \text{ is } R_{11} \text{ and } x_2 \text{ is } R_{12} \text{ and} \ldots \text{and } x_q \text{ is } R_{1q}, \text{ then } y \text{ is } f_1(x_1, x_2, \ldots, x_q) \\ \text{if } x_1 \text{ is } R_{21} \text{ and } x_2 \text{ is } R_{22} \text{ and} \ldots \text{and } x_q \text{ is } R_{2q}, \text{ then } y \text{ is } f_2(x_1, x_2, \ldots, x_q) \\ \text{if } x_1 \text{ is } R_{m1} \text{ and } x_2 \text{ is } R_{m2} \text{ and} \ldots \text{and } x_q \text{ is } R_{mq}, \text{ then } y \text{ is } f_m(x_1, x_2, \ldots, x_q) \end{cases} \tag{6.10}$$

where "$x_j$ *is* $R_{ij}$", $i = 1, 2 \ldots m$; $j = 1, 2, \ldots q$, are $m \times q$ fuzzy propositions that form $m$ antecedents for $m$ fuzzy rules, respectively; $x_j, j = 1, 2, \ldots, q$, are antecedent variables defined over universes of discourse $X_j, j = 1, 2, \ldots, q$, and $R_{ij}$, $i = 1, 2, \ldots m$; $j = 1, 2, \ldots, q$, are fuzzy sets defined by their fuzzy membership functions $\mu_{Rij}$: $X_j \rightarrow [0, 1]$, $i = 1, 2, \ldots, m$; $j = 1, 2, \ldots, q$ (Kasabov 2007). Generally, there are two methods for learning the *DENFIS* model, i.e., online learning called *DENFIS_ON* and offline learning called *DENFIS_OF*, and the two models were based on the Takagi–Sugeno neuro-fuzzy (*TS*) approach (Kasabov and Song 2002; Kasabov 2007). During the last few years, *DENFIS* has been applied successfully for solving various problems. Modelling hourly dissolved oxygen (*DO*) in river (Heddam 2014); modelling coagulant dosage in water treatment plant (Heddam and Dechemi 2015); modelling daily reference evapotranspiration ($ET_0$) (Heddam et al. 2018); prediction of the solar radiation (Kisi et al. 2019). Controlling the broadening of the signal from mode division multiplexer (Noori et al. 2019); and prediction of the Hydrodynamics of river-channel confluence (Kisi et al. 2019b). In the present investigation, *DENFIS* was developed using the NeuCom toolbox.

### 6.2.3 Performance Assessment

To evaluate the accuracy of the developed models, this chapter uses five performance indices. These indices are: the coefficient of correlation ($R$), the Nash–Sutcliffe efficiency (NSE), the root mean squared error (RMSE), and the mean absolute error (MAE).

$$R = \left[ \frac{\frac{1}{\mathrm{N}} \sum (\mathrm{O}_i - \mathrm{O}_m)(\mathrm{P}_i - \mathrm{P}_m)}{\sqrt{\frac{1}{\mathrm{N}} \sum_{i=1}^{n} (\mathrm{O}_i - \mathrm{O}_m)^2} \sqrt{\frac{1}{\mathrm{N}} \sum_{i=1}^{n} (\mathrm{P}_i - \mathrm{P}_m)^2}} \right] \tag{6.11}$$

$$\mathrm{NSE} = 1 - \frac{\sum_{i=1}^{N} [O_i - P_i]^2}{\sum_{i=1}^{N} [O_i - O_m]^2} \tag{6.12}$$

$$\mathrm{RMSE} = \sqrt{\frac{1}{N} \sum_{i=1}^{N} (O_i - P_i)^2} \tag{6.13}$$

$$\mathrm{MAE} = \frac{1}{N} \sum_{i=1}^{N} |O_i - P_i| \tag{6.14}$$

where $N$ is the data number, $O_i$ is the measured *TDG* value, and $P_i$ is the predicted *TDG*. $\mathrm{O_m}$ and $P_m$ indicate the average of $O_i$ and $P_i$.

## 6.3 Results and Discussion

In this study, *TDG* measured at the Forebay (*LWG*) and Tailwater (*LGNW*) stations located at the Lower Granite Dam at the Snake River, USA, was predicted using five data-driven models, in addition to the standard multiple linear regression (*MLR*). The proposed models described in the previous sections were: (i) the *DENFIS_OF*; (ii) the *DENFIS_ON*; (iii) the *ANFIS_FC*; (iv) the *ANFIS_GP* and (v) the *ANFIS_SC*. The correspondence between the measured and model estimates of *TDG* are reported in Tables 6.3, 6.4, 6.5 and 6.6, in terms of *RMSE, MAE, R,* and *NSE*, during the training and validations phases.

We evaluated several combinations of the inputs variables and in total, six input combination were compared (Table 6.2). The present investigation is conducted according to the following four different scenarios: (i) predicting *TDG* concentration at the Lower Granite Tailwater (*LGNW*) station according to the six input combinations (scenario 1, Table 6.3); (ii) predicting *TDG* concentration at the Lower Granite Forebay (*LWG*) station according to the six input combinations (scenario 2, Table 6.4); (iii) the capabilities and usefulness of the best obtained models having the four variables as input at the *LWG* and the *LGNW* stations were applied and compared

**Table 6.2** Input combinations of different models

| Models | | | | | | Input combinations |
|---|---|---|---|---|---|---|
| *DENFIS_ON* | *DENFIS_OF* | *ANFIS_GP* | *ANFIS_SC* | *ANFIS_FC* | *MLR* | |
| *DENFIS_ON1* | *DENFIS_OF1* | *ANFIS_GP1* | *ANFIS_SC1* | *ANFIS_FC1* | *MLR1* | *TE, BP, SFD, DIS* |
| *DENFIS_ON2* | *DENFIS_OF2* | *ANFIS_GP2* | *ANFIS_SC2* | *ANFIS_FC2* | *MLR2* | *TE, BP, DIS* |
| *DENFIS_ON3* | *DENFIS_OF3* | *ANFIS_GP3* | *ANFIS_SC3* | *ANFIS_FC3* | *MLR3* | *TE, BP* |
| *DENFIS_ON4* | *DENFIS_OF4* | *ANFIS_GP4* | *ANFIS_SC4* | *ANFIS_FC4* | *MLR4* | *TE, DIS* |
| *DENFIS_ON5* | *DENFIS_OF5* | *ANFIS_GP5* | *ANFIS_SC5* | *ANFIS_FC5* | *MLR5* | *BP, DIS* |
| *DENFIS_ON6* | *DENFIS_OF6* | *ANFIS_GP6* | *ANFIS_SC6* | *ANFIS_FC6* | *MLR6* | *SFD* |

by interchanging the training and validation data sets by setting the training data set (30%) and validation data set (70%) (scenario 3, Table 6.5); and (iv) prediction of *TDG* without the well-known input variables, rather the models were developed using the component of the Gregorian calendar that are (i) year number, (ii) month number from 1 to 12, (iii) day number of the month from 1 to 31, and (iv) hour number from 0:00 to 24:00 (scenario 4, Table 6.6). Hereafter, we focused our discussion about the obtained results during the validation phase.

Results obtained using the proposed models at the Lower Granite Tailwater (*LGNW*) station (scenario 1) were reported in Table 6.3. When only spill from dam (*SFD*) was used as predictor, the best accuracy was obtained using the *ANFIS_FC6* with larger *R* and *NSE* values, and smaller *RMSE* and *MAE* values, while the lowest accuracy was obtained using the MLR6. Among the five data-driven models, *DENFIS_ON6* performed worst with low *R* and *NSE* values, and high *RMSE* and *MAE* values. *ANFIS_SC6* and *ANFIS_GP6* possessed relatively the same accuracy with marginal difference and the *ANFIS_GP6* showed lower *RMSE* and *MAE* values.

In addition, the *DENFIS_OF6* performed significantly better than the *DENFIS_ON6*. For numerical comparison, results from Table 6.3 clearly indicated that, the *MAE* and *RMSE* of the *ANFIS_FC6* was decreased by (93.73% and 94.40%), respectively, compared to the *MLR6*, and by (97.082% and 95.89%) compared to the *DENFIS_ON6*. The *ANFIS_FC6* improved the predictive accuracy of the *DENFIS_OF6* by decreasing the *RMSE* and *MAE* values by 29.07% and 36.136%, respectively. Finally, *ANFIS_FC6* exhibited an increase in *R* and *NSE* values by 1.4% and 3% compared to the *ANFIS_GP6*, and by 1.9% and 3.7% compared the *ANFIS_SC6*, respectively. According to Table 6.3, using the four input variables leads to models with the best accuracy among the six input combinations reported in Table 6.2. Using all the four input variables (*TE, BP, DIS* and *SFD*), the *ANFIS_FC1* model provided better prediction results ($R = 0.977$ and NSE = 0.954) and the lowest errors indexes (RMSE = 1.084 and MAE = 0.773), slightly better than the *ANFIS_SC1* ($R = 0.972$ and NSE = 0.945).

Overall accuracy results were high for all the proposed models and the *MLR1* produced the lowest overall accuracy ($R = 0.944$ and NSE = 0.891) on an equal accuracy with the *DENFIS_ON1*. Finally, the *ANFIS_GP1* and the *DENFIS_OF1*

**Table 6.3** Performances of different models in modelling *TDG* at *LGNW* station

| Models | Training | | | | Validation | | | |
|---|---|---|---|---|---|---|---|---|
| | RMSE | MAE | *R* | NSE | RMSE | MAE | *R* | NSE |
| *DENFIS_ON1* | 0.823 | 0.455 | 0.986 | 0.973 | 1.672 | 0.988 | 0.944 | 0.890 |
| *DENFIS_ON2* | 1.288 | 0.719 | 0.966 | 0.933 | 3.296 | 2.001 | 0.795 | 0.572 |
| *DENFIS_ON3* | 1.929 | 1.149 | 0.923 | 0.851 | 5.993 | 4.109 | 0.413 | 0.114 |
| *DENFIS_ON4* | 0.859 | 0.503 | 0.985 | 0.970 | 4.570 | 2.983 | 0.674 | 0.178 |
| *DENFIS_ON5* | 1.537 | 0.887 | 0.952 | 0.905 | 4.635 | 2.987 | 0.661 | 0.154 |
| *DENFIS_ON6* | 30.739 | 10.318 | 0.770 | 0.590 | 45.060 | 22.376 | 0.627 | 0.310 |
| *DENFIS_OF1* | 1.363 | 0.935 | 0.962 | 0.925 | 1.271 | 0.868 | 0.968 | 0.936 |
| *DENFIS_OF2* | 2.731 | 1.941 | 0.837 | 0.701 | 2.720 | 1.941 | 0.842 | 0.709 |
| *DENFIS_OF3* | 4.382 | 3.443 | 0.485 | 0.229 | 4.433 | 3.502 | 0.484 | 0.226 |
| *DENFIS_OF4* | 2.747 | 1.908 | 0.835 | 0.697 | 2.731 | 1.893 | 0.842 | 0.706 |
| *DENFIS_OF5* | 3.037 | 2.098 | 0.794 | 0.630 | 3.023 | 2.080 | 0.800 | 0.640 |
| *DENFIS_OF6* | 1.861 | 1.456 | 0.936 | 0.861 | 1.854 | 1.439 | 0.941 | 0.865 |
| *ANFIS_GP1* | 1.381 | 0.954 | 0.961 | 0.923 | 1.278 | 0.881 | 0.968 | 0.936 |
| *ANFIS_GP2* | 2.536 | 1.844 | 0.861 | 0.742 | 2.518 | 1.817 | 0.866 | 0.750 |
| *ANFIS_GP3* | 3.941 | 2.953 | 0.613 | 0.376 | 4.009 | 3.002 | 0.606 | 0.367 |
| *ANFIS_GP4* | 2.475 | 1.703 | 0.868 | 0.754 | 2.439 | 1.664 | 0.875 | 0.766 |
| *ANFIS_GP5* | 2.991 | 2.011 | 0.800 | 0.641 | 2.976 | 1.982 | 0.807 | 0.651 |
| *ANFIS_GP6* | 1.646 | 1.235 | 0.944 | 0.891 | 1.581 | 1.173 | 0.950 | 0.902 |
| *ANFIS_SC1* | 1.258 | 0.900 | 0.968 | 0.936 | 1.185 | 0.850 | 0.972 | 0.945 |
| *ANFIS_SC2* | 2.489 | 1.786 | 0.867 | 0.751 | 2.499 | 1.789 | 0.868 | 0.754 |
| *ANFIS_SC3* | 4.148 | 3.216 | 0.556 | 0.309 | 4.191 | 3.253 | 0.555 | 0.308 |
| *ANFIS_SC4* | 2.429 | 1.708 | 0.874 | 0.763 | 2.419 | 1.692 | 0.877 | 0.770 |
| *ANFIS_SC5* | 2.950 | 1.975 | 0.806 | 0.650 | 2.948 | 1.962 | 0.811 | 0.658 |
| *ANFIS_SC6* | 1.695 | 1.298 | 0.941 | 0.885 | 1.632 | 1.243 | 0.946 | 0.895 |
| *ANFIS_FC1* | 1.112 | 0.776 | 0.975 | 0.950 | 1.084 | 0.773 | 0.977 | 0.954 |
| *ANFIS_FC2* | 2.249 | 1.549 | 0.893 | 0.797 | 2.283 | 1.572 | 0.892 | 0.795 |
| *ANFIS_FC3* | 3.758 | 2.783 | 0.658 | 0.433 | 3.928 | 2.921 | 0.627 | 0.392 |
| *ANFIS_FC4* | 2.181 | 1.398 | 0.899 | 0.809 | 2.185 | 1.404 | 0.901 | 0.812 |
| *ANFIS_FC5* | 2.850 | 1.904 | 0.821 | 0.674 | 2.902 | 1.919 | 0.818 | 0.669 |
| *ANFIS_FC6* | 1.360 | 0.948 | 0.962 | 0.926 | 1.315 | 0.919 | 0.965 | 0.932 |
| *MLR1* | 1.706 | 1.219 | 0.940 | 0.883 | 1.667 | 1.164 | 0.944 | 0.891 |
| *MLR2* | 3.888 | 3.160 | 0.627 | 0.393 | 3.891 | 3.170 | 0.636 | 0.404 |
| *MLR3* | 4.804 | 3.763 | 0.270 | 0.073 | 4.878 | 3.826 | 0.252 | 0.063 |
| *MLR4* | 3.906 | 3.182 | 0.622 | 0.387 | 3.904 | 3.185 | 0.633 | 0.400 |
| *MLR5* | 4.257 | 3.423 | 0.522 | 0.272 | 4.295 | 3.472 | 0.523 | 0.274 |
| *MLR6* | 2.153 | 1.660 | 0.902 | 0.814 | 20.988 | 16.412 | 0.501 | 0.253 |

**Table 6.4** Performances of different models in modelling *TDG* at *LWG* station

| Models | Training | | | | Validation | | | |
|---|---|---|---|---|---|---|---|---|
| | RMSE | MAE | *R* | NSE | RMSE | MAE | *R* | NSE |
| *DENFIS_ON1* | 0.535 | 0.342 | 0.959 | 0.919 | 1.229 | 0.856 | 0.786 | 0.574 |
| *DENFIS_ON2* | 0.608 | 0.393 | 0.947 | 0.896 | 1.236 | 0.844 | 0.768 | 0.569 |
| *DENFIS_ON3* | 0.777 | 0.516 | 0.911 | 0.830 | 1.809 | 1.358 | 0.554 | 0.077 |
| *DENFIS_ON4* | 0.546 | 0.378 | 0.958 | 0.916 | 1.687 | 1.238 | 0.600 | 0.198 |
| *DENFIS_ON5* | 0.709 | 0.494 | 0.927 | 0.858 | 1.367 | 1.030 | 0.707 | 0.473 |
| *DENFIS_ON6* | 1.562 | 0.874 | 0.696 | 0.312 | 2.779 | 2.221 | 0.433 | 0.162 |
| *DENFIS_OF1* | 1.153 | 0.854 | 0.795 | 0.625 | 1.188 | 0.898 | 0.782 | 0.602 |
| *DENFIS_OF2* | 1.177 | 0.877 | 0.783 | 0.610 | 1.215 | 0.920 | 0.766 | 0.583 |
| *DENFIS_OF3* | 1.423 | 1.072 | 0.655 | 0.429 | 1.474 | 1.121 | 0.624 | 0.388 |
| *DENFIS_OF4* | 1.366 | 1.052 | 0.691 | 0.474 | 1.365 | 1.068 | 0.693 | 0.475 |
| *DENFIS_OF5* | 1.199 | 0.907 | 0.775 | 0.595 | 1.229 | 0.943 | 0.761 | 0.574 |
| *DENFIS_OF6* | 1.397 | 1.110 | 0.671 | 0.450 | 1.416 | 1.123 | 0.661 | 0.435 |
| *ANFIS_GP1* | 1.089 | 0.838 | 0.816 | 0.666 | 1.148 | 0.893 | 0.793 | 0.628 |
| *ANFIS_GP2* | 1.160 | 0.887 | 0.788 | 0.621 | 1.201 | 0.930 | 0.771 | 0.593 |
| *ANFIS_GP3* | 1.404 | 1.055 | 0.666 | 0.444 | 1.458 | 1.117 | 0.634 | 0.401 |
| *ANFIS_GP4* | 1.309 | 0.996 | 0.719 | 0.517 | 1.311 | 1.021 | 0.718 | 0.515 |
| *ANFIS_GP5* | 1.149 | 0.874 | 0.792 | 0.628 | 1.194 | 0.921 | 0.774 | 0.598 |
| *ANFIS_GP6* | 1.369 | 1.070 | 0.687 | 0.472 | 1.394 | 1.091 | 0.676 | 0.452 |
| *ANFIS_SC1* | 0.812 | 0.603 | 0.902 | 0.814 | 0.995 | 0.749 | 0.851 | 0.721 |
| *ANFIS_SC2* | 1.017 | 0.763 | 0.842 | 0.709 | 1.111 | 0.844 | 0.808 | 0.652 |
| *ANFIS_SC3* | 1.309 | 0.993 | 0.719 | 0.517 | 1.367 | 1.055 | 0.688 | 0.473 |
| *ANFIS_SC4* | 1.241 | 0.965 | 0.752 | 0.566 | 1.279 | 1.004 | 0.734 | 0.539 |
| *ANFIS_SC5* | 1.102 | 0.835 | 0.811 | 0.658 | 1.183 | 0.908 | 0.778 | 0.605 |
| *ANFIS_SC6* | 1.368 | 1.069 | 0.687 | 0.473 | 1.390 | 1.089 | 0.677 | 0.455 |
| *ANFIS_FC1* | 0.864 | 0.652 | 0.889 | 0.789 | 0.979 | 0.737 | 0.855 | 0.730 |
| *ANFIS_FC2* | 1.010 | 0.752 | 0.844 | 0.712 | 1.091 | 0.827 | 0.816 | 0.664 |
| *ANFIS_FC3* | 1.386 | 1.054 | 0.677 | 0.459 | 1.440 | 1.107 | 0.645 | 0.415 |
| *ANFIS_FC4* | 1.226 | 0.925 | 0.759 | 0.576 | 1.253 | 0.963 | 0.747 | 0.558 |
| *ANFIS_FC5* | 1.100 | 0.830 | 0.812 | 0.659 | 1.171 | 0.891 | 0.784 | 0.613 |
| *ANFIS_FC6* | 1.368 | 1.069 | 0.687 | 0.472 | 1.391 | 1.090 | 0.677 | 0.455 |
| *MLR1* | 1.234 | 0.949 | 0.755 | 0.571 | 1.249 | 0.982 | 0.749 | 0.560 |
| *MLR2* | 1.369 | 1.058 | 0.687 | 0.472 | 1.392 | 1.089 | 0.674 | 0.454 |
| *MLR3* | 1.450 | 1.105 | 0.638 | 0.407 | 1.485 | 1.145 | 0.615 | 0.378 |
| *MLR4* | 1.532 | 1.197 | 0.582 | 0.339 | 1.521 | 1.202 | 0.590 | 0.348 |
| *MLR5* | 1.377 | 1.077 | 0.682 | 0.466 | 1.395 | 1.104 | 0.672 | 0.451 |
| *MLR6* | 1.470 | 1.156 | 0.625 | 0.391 | 5.912 | 4.871 | 0.591 | 0.345 |

**Table 6.5** Evaluation of the optimal models in modelling *TDG* trained with validation data set (30%) and tested with training data set (70%)

| Models | Training (30%) | | | | Validation (70%) | | | |
|---|---|---|---|---|---|---|---|---|
| | RMSE | MAE | *R* | NSE | RMSE | MAE | *R* | NSE |
| *Lower granite tailwater (LGNW) station* | | | | | | | | |
| *DENFIS_ON1* | 0.976 | 0.553 | 0.981 | 0.963 | 1.329 | 0.872 | 0.964 | 0.929 |
| *DENFIS_OF1* | 1.274 | 0.876 | 0.968 | 0.936 | 1.38 | 0.954 | 0.961 | 0.924 |
| *ANFIS_GP1* | 1.218 | 0.85 | 0.97 | 0.942 | 1.337 | 0.933 | 0.964 | 0.928 |
| *ANFIS_SC1* | 1.168 | 0.828 | 0.973 | 0.946 | 1.282 | 0.912 | 0.967 | 0.934 |
| *ANFIS_FC1* | 0.92 | 0.642 | 0.983 | 0.967 | 1.126 | 0.759 | 0.974 | 0.949 |
| *MLR1* | 1.666 | 1.159 | 0.944 | 0.891 | 1.707 | 1.214 | 0.94 | 0.883 |
| *Lower granite forebay (LWG)* | | | | | | | | |
| *DENFIS_ON1* | 0.785 | 0.518 | 0.912 | 0.826 | 1.191 | 0.884 | 0.785 | 0.6 |
| *DENFIS_OF1* | 1.133 | 0.858 | 0.803 | 0.638 | 1.111 | 0.845 | 0.811 | 0.652 |
| *ANFIS_GP1* | 1.128 | 0.869 | 0.801 | 0.641 | 1.097 | 0.841 | 0.813 | 0.661 |
| *ANFIS_SC1* | 1.062 | 0.811 | 0.826 | 0.682 | 1.032 | 0.783 | 0.837 | 0.7 |
| *ANFIS_FC1* | 0.876 | 0.664 | 0.885 | 0.784 | 1.036 | 0.768 | 0.838 | 0.697 |
| *MLR1* | 1.245 | 0.975 | 0.75 | 0.563 | 1.239 | 0.945 | 0.754 | 0.567 |

**Table 6.6** Performances of the proposed models in modeling *TDG* at the two stations using the component of the Gregorian calendar

| Models | Training | | | | Validation | | | |
|---|---|---|---|---|---|---|---|---|
| | RMSE | MAE | *R* | NSE | RMSE | MAE | *R* | NSE |
| *Lower granite tailwater (LGNW) station* | | | | | | | | |
| *DENFIS_ON1* | 51.371 | 34.599 | 0.13 | 0.11 | 59.014 | 41.498 | 0.12 | 0.1 |
| *DENFIS_OF1* | 3.849 | 2.194 | 0.737 | 0.405 | 5.08 | 3.123 | 0.566 | 0.31 |
| *ANFIS_GP1* | 2.254 | 1.67 | 0.892 | 0.796 | 2.287 | 1.691 | 0.891 | 0.794 |
| *ANFIS_SC1* | 2.218 | 1.702 | 0.896 | 0.802 | 10.323 | 10.063 | 0.429 | 0.184 |
| *ANFIS_FC1* | 1.594 | 1.134 | 0.948 | 0.898 | 18.726 | 12.731 | 0.569 | 0.32 |
| *MLR1* | 3.589 | 2.991 | 0.695 | 0.483 | 3.598 | 3.021 | 0.7 | 0.49 |
| *Lower granite forebay (LWG)* | | | | | | | | |
| *DENFIS_ON1* | 61.548 | 44.499 | 0.12 | 0.101 | 69.13 | 51.398 | 0.11 | 0.1 |
| *DENFIS_OF1* | 1.397 | 1.11 | 0.671 | 0.45 | 1.416 | 1.123 | 0.661 | 0.435 |
| *ANFIS_GP1* | 1.282 | 1.032 | 0.733 | 0.537 | 1.289 | 1.04 | 0.729 | 0.531 |
| *ANFIS_SC1* | 1.275 | 1.028 | 0.736 | 0.542 | 1.447 | 1.15 | 0.68 | 0.41 |
| *ANFIS_FC1* | 0.87 | 0.675 | 0.887 | 0.787 | 1.631 | 1.231 | 0.688 | 0.249 |
| *MLR1* | 1.386 | 1.111 | 0.677 | 0.458 | 1.367 | 1.093 | 0.688 | 0.473 |

possess the same accuracy with equal *R* and NSE values. It is clear from the obtained results that the proposed models were indeed more sensitive to the exclusion of the SFD from the input variables (combination 2, Table 6.2). First, the *ANFIS_FC2*, *ANFIS_SC2,* and the *ANFIS_GP2* models provided relatively the same accuracy, with slight superiority of the *ANFIS_FC2* ($R = 0.892$ and $\text{NSE} = 0.798$), while the *MLR2* possess the lowest accuracy ($R = 0.636$ and $\text{NSE} = 0.404$).

Exclusion of the *SFD* from the model inputs decreased the performances of the models; however, the *MLR* model is more sensitive to the exclusion of *SFD* compared with the data-driven models. The *RMSE* and *MAE* of the *DENFIS_ON1* and *DENFIS_OF1* have increased by (49.27%, 50.62%) and (53.27% and 55.28%), respectively. In addition, the *RMSE* and *MAE* of the *ANFIS_GP1*, *ANFIS_SC1* and *ANFIS_FC1* have increased by (53.27%, 55.28%), (49.24%, 51.51%) and (52.58% and 52.48%), respectively.

Finally, using only two input variables, the best accuracy was obtained using the models having the *TE* and *DIS* as input variables, and the *ANFIS_FC4* was more accurate compared to all other models. *MLR4* produces relatively large *RMSE* and *MAE* errors significantly higher than the values calculated using the five data-driven models. Nonetheless, the use of the data-driven models with only fewer input variables (*TE* and *DIS*) resulted in substantial differences with respect to the statistical indexes, and only the data-driven models was characterized by the strongest accuracy. The comparison between measured and calculated values of the *TDG* at the Lower Granite Tailwater (*LGNW*) station is plotted in Fig. 6.3.

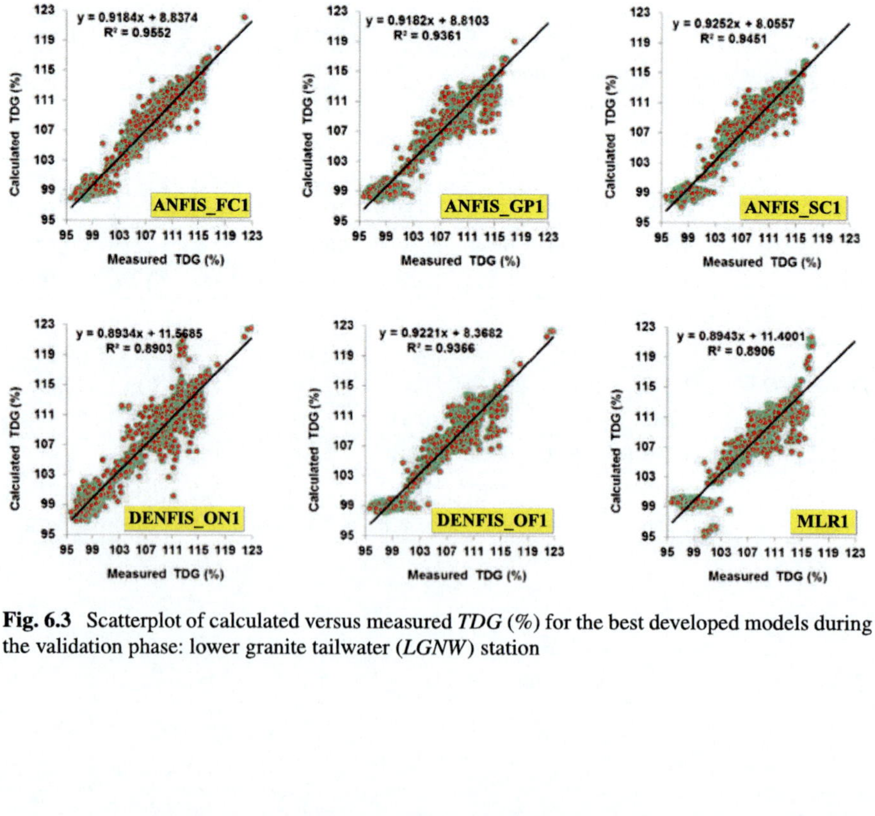


**Fig. 6.3** Scatterplot of calculated versus measured *TDG* (%) for the best developed models during the validation phase: lower granite tailwater (*LGNW*) station

At the Lower Granite Forebay (*LWG*) station (Table 6.4), the estimation of *TDG* was less accurate than the estimation at the *LGNW* station and all the proposed models were less accurate, having a *RMSE* and *MAE* values higher than the values obtained at the *LGNW* station. Estimates based on *NSE* and *R* indexes were almost less accurate, showing a significant decrease in the *NSE* and *R* values. Table 6.4 shows that high variability of models performances has been observed between the six input combinations and the best accuracy was obtained using the models with the first input combination having the four variables as input (*TE, BP, DIS* and *SFD*). In addition, it is clear from the results reported in Table 6.4; the models were indeed less sensitive to the exclusion of the *SFD* from the input variables. Among all the proposed *MLPNN* models (Table 6.4), the *ANFIS_FC1* has the highest accuracy with lower *RMSE* and *MAE*, and high NSE and d values. The *ANFIS_FC1* had an *R* value of 0.855 and *NSE* value of 0.730, when evaluated using the validation data. Predictor variables selected by this model included the four original variables (*TE, BP, DIS,* and *SFD*).

The *R* and *NSE* values were slightly decreased when the *SFD* was removed from the input: the *R* and *NSE* values of the *DENFIS_ON1*, *DENFIS_OF1* and *ANFIS _GP1* were decreased by (1.8%, 0.5%), (1.6%, 1.9%) and (2.2%, 3.5%), respectively. In addition, the *R* and *NSE* values of the *ANFIS _SC1*, *ANFIS _FC1* and *MLR1* were decreased by (4.3%, 6.9%), (3.9%, 6.6%) and (7.5%, 10.6%), respectively. It is clear from the results reported in Table 6.4 that the *MLR1* model is more sensitive to the exclusion of the *SFD* variable compared to the five data-driven models. In addition, it is clear from Table 6.4 that, using fewer input variables, the *ANFIS_FC5* model having the *BP* and *DIS* as input variables was able to provide high accuracy compared to the other models. Results clearly indicated that, the *RMSE* and *MAE* of the *ANFIS_FC5* were decreased by 1.014% and 1.782%, 1.926% and 3.257%, 4.719.6% and 5.514%, compared to the *ANFIS_SC5*, *ANFIS_GP5*, and *DENFIS_OF5*, respectively.

Overall, when comparing the *ANFIS_FC5* with the other models, it is clear that there was a strong relationship between the number of input variables and the accuracy of the models. The relationship varied, however, which means that *ANFIS_FC5* was more suitable for building models using only fewer inputs. The *ANFIS_FC5* with only *BP* and *DIS* as input variables, decreasing the values of the *RMSE* and *MAE* by 14.33% and 13.49%, 16.057% and 19.293% compared to the *DENFIS_ON5* and *MLR5* models, respectively. Finally, using only the *SFD* as input variable, it is clear from the results reported in Table 6.4 that the *DENFIS_ON6* was the less accurate model, significantly less than the *MLR6* model, and the *ANFIS_SC6* had the high accuracy equally with the *ANFIS_FC6*, *ANFIS_GP6,* and *DENFIS_OF6* models. The comparison between measured and calculated values of the *TDG* at the Lower Granite Tailwater (*LGNW*) station is plotted in Fig. 6.4.

According to Table 6.5, for the third scenario, when the models were trained using the validation data set (30%) and tested using the training data set, there is a slightly difference between the five data-driven models and the *MLR* model. At *LGNW* station, the best accuracy was obtained using the *ANFIS_FC1* model ($R$ = 0.974, NSE = 0.949) slightly higher than the other data-driven models, and

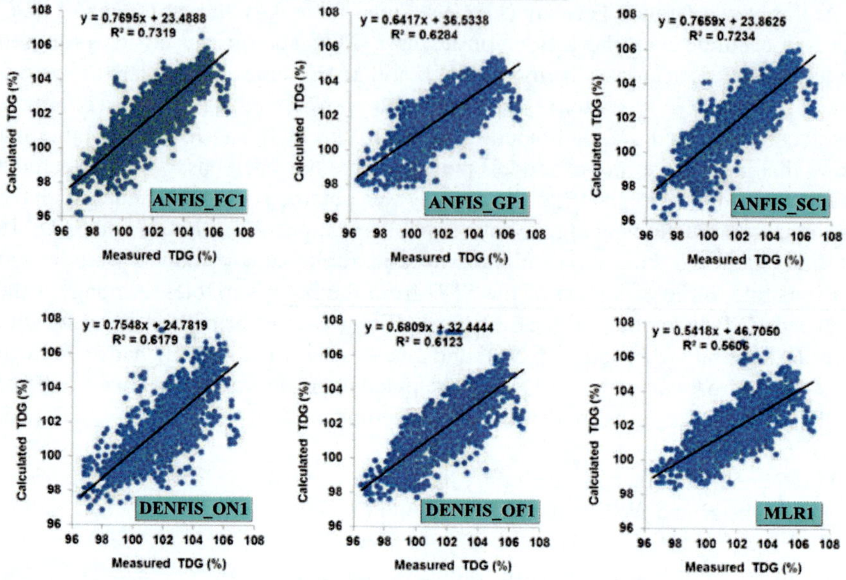


**Fig. 6.4** Scatterplot of calculated versus measured *TDG* (%) for the best developed models during the validation phase: lower granite Forebay (*LWG*)

significantly higher than the *MLR1* model ($R = 0.940$, NSE $= 0.883$). At *LWG* station, the best accuracy was obtained simultaneously using the *ANFIS_FC1* and *ANFIS_SC1* models, while the *MLR1* model possesses the lowest accuracy. The comparison between measured and calculated values of the *TDG* at the *LGNW* and *LWG* stations is plotted in Figs. 6.5 and 6.6. Finally, for the fourth scenario, an attempt is made to predict *TDG* using the component of the Gregorian calendar, and the results are reported in Table 6.6. It is clear that the models had low-to-moderate accuracy at the two stations. At *LGNW* station, the *ANFIS_GP1* produces high nonlinear mapping of *TDG* which are substantially higher than all the three other models. This can be nearly always advantageous, especially in the situation where the input variables were missing. *ANFIS_GP1* exhibited a decrease in *RMSE* and *MAE* values compared to the *DENFIS_OF1*, *ANFIS_SC1*, and *MLR1*, respectively. The good accuracy of the *ANFIS_GP1* model is obvious, especially when comparing its accuracy to those of *DENFIS_ON1* and *ANFIS_SC1*. At *LWG* station, *ANFIS_SC1*, *ANFIS_FC1*, and *MLR* yielded similar accuracy in terms of all the statistical indexes, and slightly less than the *ANFIS_GP1*. Performances characteristics also differ between *DENFIS_OF1* and *DENFIS_ON1*, with a very low accuracy obtained using the *DENFIS_ON1*.

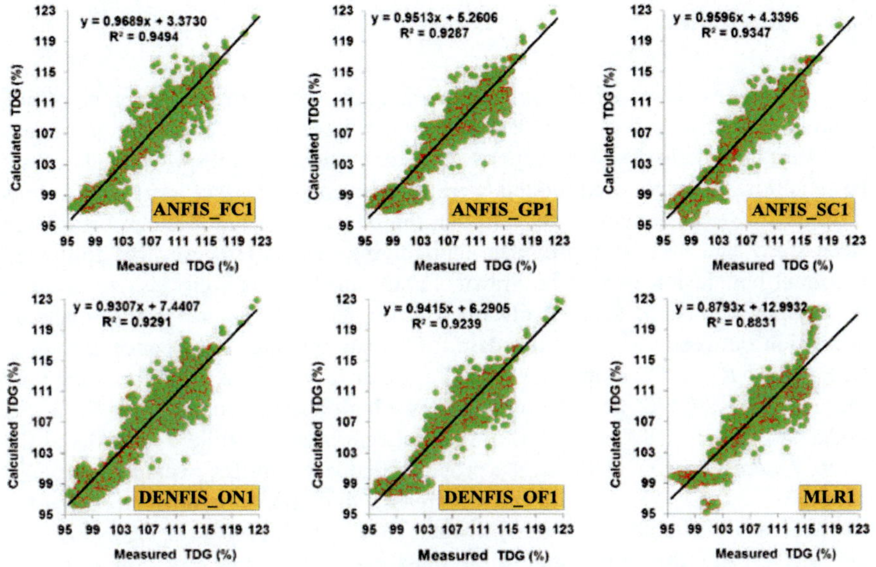


**Fig. 6.5** Scatterplot of calculated versus measured *TDG* (%) for the best models during the validation phase at the lower granite tailwater (*LGNW*) station: 30–70%

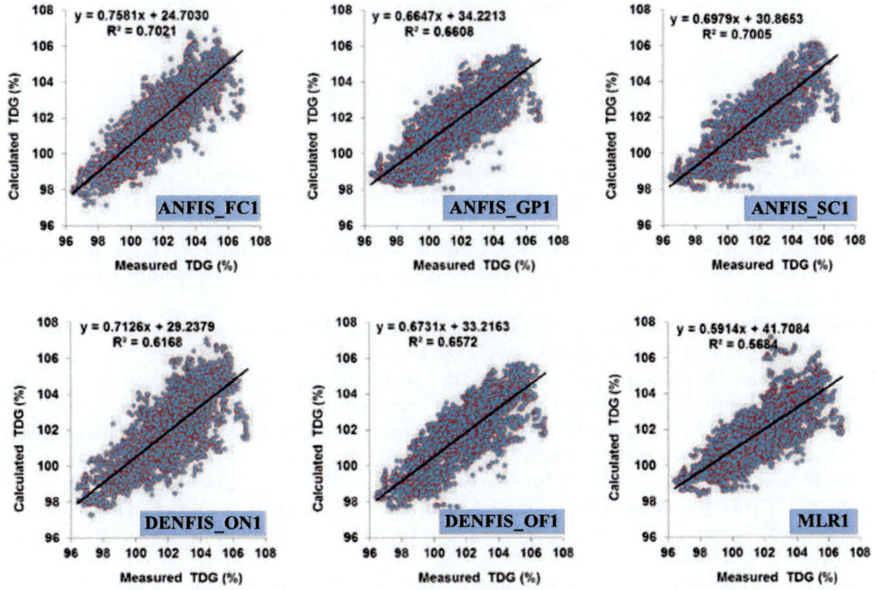


**Fig. 6.6** Scatterplot of calculated versus measured *TDG* (%) for the best models during the validation phase at the Lower Granite Forebay (*LWG*) station: 30–70%

## 6.4 Conclusion

This chapter has presented new intelligent data analytic approaches to predict *TDG* at the Forebay and Tailwater of dams reservoirs based on the combination of four input variables, namely water temperature, barometric pressure, spill from dam, and the total flow. The proposed models were tested using *TDG* data from two stations in the *USA*. Results obtained showed that the proposed techniques can be applied successfully and they have provided encouraging results. However, the following important conclusion should be drawn. Firstly, the models were less accurate at the Forebay compared to the Tailwater and this is certainly due to the moderate correlation between the spill from dam (*SFD*) variable and *TDG* concentration at the Forebay ($R = 0.62$) compared to the Tailwater ($R = 0.91$). Secondly, using only *SFD* as input variable, we have obtained very high accuracy using the *ANFIS_FC* model ($R = 0.965$, NSE $= 0.932$) at the Tailwater station, while the results at the Forebay were very low using all the proposed models. Thirdly, excluding the *SFD* from the input variables significantly decreased the models performances and the best accuracy was obtained at the Tailwater ($R = 0.892$, NSE $= 0.795$). Fourthly, and finally, using the component of the Gregorian calendar as input variables, the best accuracy was obtained at the Tailwater ($R = 0.891$, NSE $= 0.794$) using the *ANFIS_GP*, while at the Forebay station, all the proposed models have provided very low accuracy and the *NSE* value does not excessed 0.53.

## References

Deng ZD, Duncan JP, Arnold JL, Fu T, Martinez J, Lu J, Titzler PS, Zhou D, Mueller RP (2017) Evaluation of boundary dam spillway using an autonomous sensor fish device. J Hydro-Environ Res 14:85–92

Feng JJ, Li R, Liang RF, Shen X (2014a) Eco-environmentally friendly operational regulation: an effective strategy to diminish the TDG supersaturation of reservoirs. Hydrol Earth Syst Sci Discuss 18:1213–1223

Feng JJ, Li R, Ma Q, Wang LL (2014b) Experimental and field study on dissipation coefficient of supersaturated total dissolved gas. J Cent South Univ 21(5):1995–2003

Feng JJ, Wang L, Li R, Li K, Pu X, Li Y (2018) Operational regulation of a hydropower cascade based on the mitigation of the total dissolved gas supersaturation. Ecol Ind 92:124–132

Heddam S (2014) Modelling hourly dissolved oxygen concentration (DO) using dynamic evolving neural-fuzzy inference system (DENFIS) based approach: case study of Klamath River at Miller Island Boat Ramp, Oregon, USA. Environ Sci Pollut Res 21:9212–9227

Heddam S (2017) Generalized regression neural network based approach as a new tool for predicting total dissolved gas (TDG) downstream of spillways of dams: a case study of Columbia River Basin Dams, USA. Environ Process 4:235–253

Heddam S, Dechemi N (2015) A new approach based on the dynamic evolving neural-fuzzy inference system (DENFIS) for modelling coagulant dosage: case study of water treatment plant of Algeria Country. Desalination Water Treat Taylor Francis 53–4:1045–1053

Heddam S, Watts MJ, Houichi L, Djemili L, Sebbar A (2018) Evolving connectionist systems (ECoSs): a new approach for modeling daily reference evapotranspiration ($ET_0$). Environ Monit Assess 190(9):516

Jang J-SR (1993) ANFIS: adaptive-network-based fuzzy inference system. IEEE Trans Sys Manage Cybern 23(3):665–685
Jang J-S (1996) Input selection for ANFIS learning. In: Proceedings of the fifth IEEE international conference on fuzzy systems, pp 1493–1499
Jang J-SR, Sun C-T, Mizutani E (1997) Neuro-fuzzy and soft computing: a computational approach to learning and machine intelligence. Prentice Hall, Upper Saddle River, New Jersey, USA
Kasabov N (2007) Evolving connectionist systems: the knowledge engineering approach. Springer, New York, p 465. ISBN 978-1-84628-345-1
Kasabov N, Song Q (2002) DENFIS: dynamic, evolving neural-fuzzy inference systems and its application for time-series prediction. IEEE Trans Fuzzy Syst 10:144–154
Kisi O, Demir V, Kim S (2017) Estimation of long-term monthly temperature by three different adaptive neuro-fuzzy approaches using geographical inputs. J Irrig Drainage Eng 143:1–18
Kisi O, Heddam S, Yaseen ZM (2019a) The implementation of univariable scheme-based air temperature for solar radiation prediction: New development of dynamic evolving neural-fuzzy inference system model. Appl Energy 241:184–195
Kisi O, Khosravinia P, Nikpour MR, Sanikhani H (2019b) Hydrodynamics of river-channel confluence: toward modeling separation zone using GEP, MARS, M5 Tree and DENFIS techniques. Stochast Environ Res Risk Assess: 1–19
Ma Q, Li R, Zhang Q, Hodges BR, Feng JJ, Yang H (2016). Two-phase flow simulation of supersaturated total dissolved gas in the plunge pool of a high dam. Environ Prog Sustain Energy
Ma Q, Li R, Feng J, Lu J, Zhou Q (2018) Cumulative effects of cascade hydropower stations on total dissolved gas supersaturation. Environ Sci Pollut Res 25(14):13536–13547. https://doi.org/10.1007/s11356-018-1496-2
Noori A, Amphawan A, Ghazi A, Ghazi SA (2019) Dynamic evolving neural fuzzy inference system equalization scheme in mode division multiplexer for optical fiber transmission. Bull Electr Eng Inform 8(1):127–135
Ou Y, Li R, Tuo Y, Niu J, Feng JJ, Pu X (2016) The promotion effect of aeration on the dissipation of supersaturated total dissolved gas. Ecol Eng 95:245–251
Picket J, Rueda H, Herold M (2004) Total maximum daily load for total dissolved gas in the Mid-Columbia River and Lake Roosevelt. Submittal Report. No. 04-03-002, Washington State Department of Ecology, Olympia, WA
Politano M, Carrica PM, Turan C, Weber L (2007) A multidimensional two phase flow model for the total dissolved gas downstream of spillways. J Hydraul Res 45(2):165–177
Politano M, Carrica P, Weber L (2009) A multiphase model for the hydrodynamics and total dissolved gas in tailraces. Int J Multiphase Flow 35:1036–1050
Politano M, Arenas Amado A, Bickford S, Murauskas J, Hay D (2012) Evaluation of operational strategies to minimize gas supersaturation downstream of a dam. Comput Fluids 68:168–185
Politano M, Castro A, Hadjerioua B (2017) Modeling total dissolved gas for optimal operation of multireservoir systems. J Hydraul Eng 143(6):04017007
Shen X, Li R, Hodges BR, Feng J, Cai H, Ma X (2019) Experiment and simulation of supersaturated total dissolved gas dissipation: Focus on the effect of confluence types. Water Res 155:320–332
Stewart K, Witt A, Hadjerioua B, Politano M, Magee T, DeNeale S, Bender M, Maloof A (2015) Total dissolved gas prediction and optimization in riverware. Oak Ridge National Laboratory ORNL/TM-2015/551. Environ Sci Div. info.ornl.gov/sites/publications/files/Pub59285.pdf
Tawfik ME, Diez FJ (2014) On the relation between onset of bubble nucleation and gas supersaturation concentration. Electrochim Acta 146:792–797
Wang Y, Politano M, Weber L (2018) Spillway jet regime and total dissolved gas prediction with a multiphase flow model. J Hydraul Res:1–13
Weber L, Huang H, Lai Y, McCoy A (2004) Modeling total dissolved gas production and transport downstream of spillways-three-dimensional development and applications. Int J River Basin Manage 2(3):1–11
Weitkamp DE, Katz M (1980) A review of dissolved gas supersaturation literature. Trans Am Fish Soc 109(6):659–702

Wilhelms S, Schneider M (2006) TDG at lower monumental dam for alternative spill operations. In: ASCE proceedings: operating reservoirs in changing conditions, pp 391–399
Witt A, Stewart K, Hadjerioua B (2017) Predicting total dissolved gas travel time in hydropower reservoirs. J Environ Eng 143(12):06017011
Yuan Y, Feng J, Li R, Huang Y, Huang J, Wang Z (2018) Modelling the promotion effect of vegetation on the dissipation of supersaturated total dissolved gas. Ecol Model 386:89–97

# Chapter 7
# Modulation of Tropical Cyclone Genesis by Madden–Julian Oscillation in the Southern Hemisphere

**Kavina S. Dayal, Bin Wang, and Ravinesh C. Deo**

## Abbreviation

| | |
|---|---|
| ECMWF | European Centre for Medium-Range Weather Forecasts |
| ENSO | El-Niño Southern Oscillation |
| ITCZ | Intertropical Convergence Zone |
| JTWC | Joint Typhoon Warning Center |
| HYCOM | Hybrid Coordinate Ocean Model |
| MJO | Madden–Julian Oscillation |
| NCEP | National Center for Environmental Prediction |
| NH | Northern Hemisphere |
| OLR | Outgoing Longwave Radiation |
| RH | Relative Humidity |
| SH | Southern Hemisphere |
| SIO | South Indian Ocean |
| SPCZ | South Pacific Convergence Zone |
| SPO | South Pacific Ocean |
| SST | Sea Surface Temperature |
| TC | Tropical Cyclone |
| TRMM | Tropical Rainfall Measuring Mission |

K. S. Dayal (✉)
Commonwealth Scientific and Industrial Research Organisation (CSIRO), Sandy Bay 7005, Hobart, TAS, Australia
e-mail: kavinadayal@gmail.com

B. Wang
Department of Atmospheric Science, School of Ocean and Earth Science and Technology, University of Hawaii at Manoa, Honolulu, HI, USA

R. C. Deo
School of Sciences, University of Southern Queensland, Springfield Central, QLD 4300, Australia

© Springer Nature Singapore Pte Ltd. 2021

R. C. Deo et al. (eds.), *Intelligent Data Analytics for Decision-Support Systems in Hazard Mitigation*, Springer Transactions in Civil and Environmental Engineering,
https://doi.org/10.1007/978-981-15-5772-9_7

| | |
|---|---|
| U | Zonal wind component |
| V | Meridional wind component |
| WH04 | Wheeler and Hendon 2004 |

## 7.1 Introduction

Tropical cyclones are naturally driven calamities or more technically, an extreme meteorological event of paramount interest to people of tropical Pacific especially that they have tremendous impacts on the economy of island nations. Genesis of TCs has been noted over relatively warmer waters with sea surface temperatures (SST) greater than 26 °C, low magnitudes of vertical wind shear and large magnitude of low-level cyclonic potential vorticity and middle tropospheric relative humidity which provides dynamical explanations for TC motions (Yun et al. 2012). Climatologically favorable conditions are related to enhancement or reduction in large-scale, low amplitude atmospheric circulation anomalies that occur on intraseasonal, inter-seasonal, and inter-decadal timescales (Liebmann et al. 1994). One particular atmospheric anomaly known to influence heavy precipitation events (e.g., 90th and 95th percentiles) including favorable conditions for TC genesis is the Madden–Julian Oscillation (MJO) phenomenon (Jones and Carvalho 2014). Consequently, the MJO, which has with a period of 40–50 days, is recognized as a leading mode of climate variability and climate extremes in the tropics (Madden and Julian 1971, 1972).

The role of MJO in modulating meteorological events has been studied extensively for over four decades. In such studies, it has mainly been categorized as an eastward propagating disturbance where a wave or disturbance event leads to the development of the next one (Knutson and Weickmann 1987; Rui and Wang 1990; Hendon and Salby 1994; Kiladis and Weickmann 1992). However, in terms of the impacts of the MJO on cyclones, Gray (1979) was among the earliest to demonstrate that TC formation clustered into a 2–3 week of active period followed by a similar period of no convective activity. After Gray, a number of studies investigated on how the MJO acted to modulate the TC activity in global regions, including the Western North Pacific (von Storch and Xu 1990, Nakazawa 1988), the eastern North Pacific and the Gulf of Mexico (Maloney and Hartmann 2000a, b). Importantly, these studies demonstrated a pivotal role of the MJO phenomenon in modulating the TC genesis in geographically diverse regions.

In comparison to TC genesis in abovementioned regions, focus in the South Indian Ocean (SIO) and

South Pacific Ocean (SPO) has mainly been on climatological behaviors of TCs (Sinclair 2002; Terry and Gienko 2010), frequency or intensity change as a result of climatic change factors (IPCC 2007; Kuleshov et al. 2010; Nott 2011; Peduzzi et al. 2012; Webster et al. 2005) and possible relationships with El-Niño Southern Oscillation (Basher and Zheng 1995; Callaghan and Power 2011; Kuleshov et al. 2008). Accordingly, little attention has been given on how TC genesis and its development or progression may vary depending on the phase of MJO in the SIO and SPO regions. This underexplored geographic region of TC genesis and its modulation is the subject of our chapter.

A search of literature reveals a number of studies that examined MJO-TC relationships in the Southern Hemisphere (SH) [e.g., (Bessafi and Wheeler 2006; Chand and Walsh 2010; Hall et al. 2001)]. In particular, the work of Hall et al. (2001) checked the impacts of MJO on TC genesis in Australia (80° E–170° E) using best tracks of TCs, central pressure using Dvorak technique, satellite-derived OLR and reanalysis wind fields for the period of November 1974–April 1998. Using the MJO basic state defined by leading Empirical Orthogonal Functions (EOFs) of the filtered OLR over 20–200 days, they spotted the first EOF to exhibit an MJO convective dipole pattern with negative OLR anomaly over the Indian Ocean with corresponding positive value of OLR anomaly over the western Pacific while the second EOF had an enhanced convection activity over Indonesia region along the South Pacific Convergence Zone (SPCZ) and reduced activity over Africa, Central Pacific, and South America. Interestingly, their results exemplified an enhanced TC activity in the phase B (enhanced convection) and reduced activity in phase D (suppressed convection) over (Hall et al. 2001) North-west Australia region where TC genesis was seen to be outnumbered by 4:1 ratio between the most active to inactive MJO phase. Likewise, the genesis of TCs in Northeastern Australia yielded a ratio of 3:1 between phase C (enhanced convection) and phase A (reduced convection). This modulation of TC activity was attributable to the low-level relative vorticity and vertical wind shear, associated with equatorial wave response to the MJO convective anomalies.

Bessafi and Wheeler (2006) studied the impacts of large-scale atmospheric waves including the MJO modulation on TCs in the SIO using TC tracks from the La Réunion Regional Specialized Meteorological Center (RSMC), OLR from the National Oceanic and Atmospheric Administration (NOAA) and 850- and 200-hPa wind fields from the European Center for Medium-Range Weather Forecasts (ECMWF) over 1979–2004. OLR filtering was conducted for eastward propagating planetary wavenumbers 1 through 5 and periods of 30–96 days. To define the eastward propagating MJO, they performed EOF analysis on the covariance matrix of the data in the domain 20° S–20° N, 0°–357° E. For MJO to represent its state over a domain at any particular time, they defined six equal-sized (number of days) phases and binned TC genesis in each respective phase, a method similar to the one described earlier in Hall et al. (2001). A modulation of 2.6–1 between phase 2 with enhanced TC genesis and phase 5 with fewer TC geneses between 30° S–0, and 30° E–100° E was found, where the daily genesis rate (DGR, TC genesis number divided by total number of days) for phase 2 as 7.1% and for phase 5 as 2.7%. Chand and Walsh (2010) examined TC genesis modulation by MJO on the local domain of Fiji-Samoa-Tonga (25° S–5° S and 170° E–170° W) between November 1970 and April 2006. The procedure of identifying different phases of MJO was similar to Wheeler and Hendon (2004), hereafter referenced as WH04) and Hall et al. (2001) except that the eight MJO phases were considered with TC data into each phase and time of its genesis. It was noteworthy that a statistically significant, strong MJO-TC relationship was evident with TCs forming ~5 times more frequently during the active phases of MJO (phases 7 and 8) than during the inactive period (phasees 2 and 3).

In relation to ENSO effects on MJO-TC relationship, the modulation was enhanced during El-Niño periods with dynamic conditions (low-level relative vorticity and vertical shear of horizontal winds and upper-level divergence associated with TC genesis) exhibiting large variation between the active and inactive MJO phases.

Based on the literature review, we aver that intraseasonal clustering of TCs on timescales of 1–2 weeks, the active MJO phase is followed by a period of 2–3 weeks of inactive MJO phase (Gray 1979) may be related to the actual phase of the MJO. As also shown in earlier studies, e.g., (Hall et al. 2001; Maloney and Hartmann 2000a, b), the low-level vorticity and vertical wind shear is a crucial determinant of subseasonal TC variability. Earlier studies based on shorter records compared to ours, however, neglected the role of thermodynamic conditions in relation to TC genesis, although such conditions are crucially important for cyclogenesis (Frank 1987; McBride and Zehr 1981). Also importantly, the low-level relative vorticity and vertical shear of the horizontal wind plays a dominant role in TC formation to the west of 100° E in the SIO (Bessafi and Wheeler 2006). The modulation of TC genesis by MJO in the SIO appears to be of a lesser magnitude than the observed value in the Australian region (Hall et al. 2001). Klotzbach (2014) showed that above (below) average TC frequency for convectively enhanced (suppressed) phases of MJO in all ocean basins as were affected by the vertical wind shear, mid-level moisture, vertical motion, and sea-level pressure. Likewise, the frequency of rapidly intensifying TCs also increased when MJO was in its enhanced convective phase, clearly showing the crucial role of MJO in the modulation of TCs.

Although investigations in basins across the globe have advanced our knowledge of MJO-TC relationships, they only focused on their specific domains of interest, hence lacked to relate the features and impact of MJO between SIO and SPO—regions where MJO occurrence is prominent. Moreover, the role of low-level background mean flow, which plays an important role in the MJO events, has not been addressed. Accordingly, it is of interest that an analysis of MJO modulation of TC genesis for active and inactive phases and an emphasis on convections associated with varying OLR anomalies assumed to be effective in determining the circulation anomalies in SIO and SPO, be performed.

This chapter adopts a modified MJO index with an emphasis on convective anomalies associated with two phases of MJO. That is, in this chapter, the OLR has been chosen over the conventional use of zonal wind fields to examine MJO activity in the two study regions vulnerable to cyclone activities, namely the South Indian Ocean (SIO: 0–30° S, 30° E–130° E) and the South Pacific Ocean (SPO: 0–30° S, 130° E–130° W) without an assumption that the location of variability centers remain on the equator (e.g., Wheeler and Hendon 2004). It focuses on the variation of TC genesis due to the large-scale dynamic and thermodynamic conditions including the requirement of sufficient ocean thermal energy (SST > 26 °C up to 60 m depth), enhanced mid-tropospheric relative humidity (RH), conditional instability, enhanced low-level relative vorticity, weak vertical shear of the horizontal winds at the genesis site, and displacement at least by 5° latitude away from the equator. Out of the six environmental conditions conducive to cyclogenesis, this chapter addresses the condition related to 850-hPa relative vorticity, vertical shear of the horizontal wind between 850- and 200-hPa, SST, and the 700-hPa RH.

In Data and Methods section, we describe the data processing techniques, preliminary assessment of the distribution of TCs in the study basins, and the methodology developed for computing the modified MJO index. In Results and Discussion section, we detail the findings of this work particularly the modulation of MJO on TC genesis in the SOI and SPO regions. In Conclusion section, the findings are presented and suggestions for potential topics for future work are made. The conclusion is that low-level relative vorticity is a primary modulator of TC genesis in both the SIO and SPO regions, but the MJO has little effect in the western SIO through the cyclonic season where the signal is very weak.

## 7.2 Materials and Method

### *7.2.1 Data Pre-processing*

For the two study basins (SIO and SPO) used in the present chapter, the daily interpolated OLR values from the National Oceanic and Atmospheric Administration (NOAA) and reanalysis data of pressure levels corresponding to zonal (U) and meridional (V) wind components at 850-hPa and 200-hPa from the National Centre for Environment Prediction (NCEP) with grid resolution of 2.5° × 2.5° were acquired (Kalnay et al. 1996; Liebmann 1996). In addition, the relative humidity (RH) based on ERA-Interim data produced by the European Centre for Medium-Range Weather Forecasts (ECMWF) with global reanalysis of climate observations at a grid resolution of 1.5° × 1.5° at 700-hPa were obtained. In order to examine atmospheric circulation effects of MJO on TC genesis, the Southern Hemispheric (SH) best-track TC data from Joint Typhoon Warning Center (JTWC) recorded every 6-hours were also obtained. The TC data comprised of the time, geographic locations (latitude and longitude), and maximum sustained surface wind speeds of tropical cyclones in the region of study [JTWC, available online at http://www.usno.navy.mil/JTWC/].

In the present chapter, only the austral summer TC season which spans from November to April within the SIO (0–30° S, 30° E–130° E) and SPO (0–30° S, 130° E–130° W) basins are considered. November 1980 to April 1981 period is referenced as the 1981 TC season, resulting in 32 TC seasons altogether. We identified the TC genesis by considering an extreme weather event with wind speeds approximately 35 knots or higher, and the hurricane genesis when the wind speeds were equal to or greater than 64 knots. Based on the thresholds, a total of 453 (for SIO) and 293 (for SPO) TCs over the 32-year period are considered in the present research.

In order to deduce the effects of atmospheric circulation (and hence, the two phases of the MJO) on cyclogenesis in the study basin, we pre-filtered our original data in order to achieve a smooth climatic signal, primarily by averaging the daily anomalies of OLR, 850 hPa winds, SST and 700-hPa RH for a five-day period (hereafter, called the "pentads"). This chapter also utilized the ERA-Interim reanalysis RH dataset that had less quality and reliance issues (Camargo et al. 2009; Wang and Yang 2008). For filtering purposes, we have applied a ninth order ("Butterworth") filter from MATLAB in order to isolate the MJO signal for consideration. Note that this bandpass provided passage of frequencies equivalent to a period of 30–80 days (Madden and Julian 1971), deemed sufficient for capturing the typical periodicity or the natural variation in the MJO phases (Zhang 2005).

### 7.2.2 *Assessment of TC Distribution*

Before performing an investigation on the MJO modulation of TC genesis in the study basin it is important that we first assess the seasonal and the geographical distribution of TC genesis. As shown in Fig. 7.1 we constructed a seasonal map (November to April) average climatology of the OLR, 850-hPa winds, and TC genesis locations

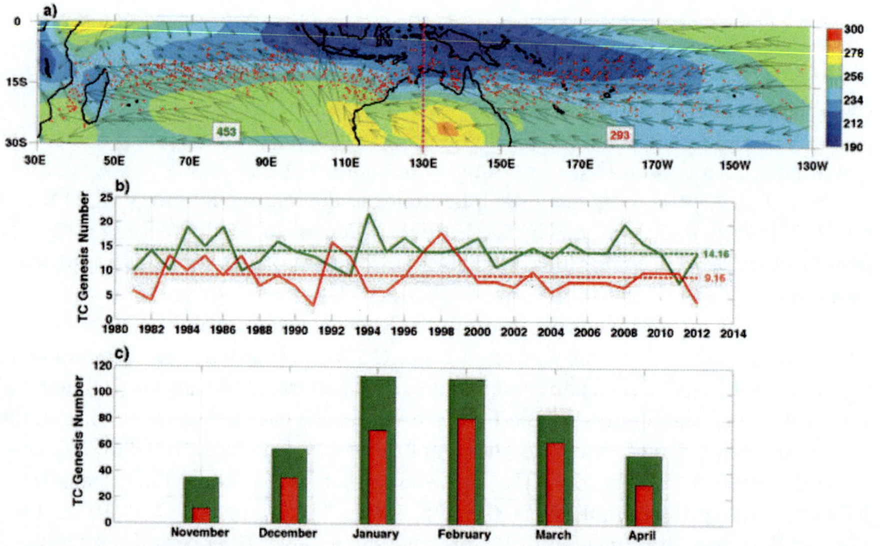


**Fig. 7.1** **a** The 1981–2012 November to April climatologically averaged TC genesis (red dots), outgoing longwave radiation (OLR, $Wm^{-2}$) (contours), and circulation patterns at 850-hPa (vectors). Pink dashed line at 130° E separates the two ocean basins considered in this chapter, namely the South Indian Ocean (SIO: 30° S–0°, 30° E–130° E) and South Pacific Ocean (SPO: 30° S–0° S, 130° E–130° W). The number 453 and 293 are actual TC genesis numbers that were analyzed. **b** The shows interannual variation of TC genesis numbers for the SIO (green) and the SPO (red) regions, respectively, where the dashed lines show the average TC numbers over the study period. On average, ~14 and ~9 TCs form every TC season in SIO and SPO, respectively. **c** The number of TC genesis recorded from 1981 to 2012 within the SIO (green bars) and SPO (red bars) regions, respectively, stratified by the corresponding months of genesis

for the period 1980–2012. Here, the pink dashed line at 130° E is used to show the separation of the two ocean basins considered in the chapter (SIO and SPO) with TC genesis locations in red. Note that this location represents the onset origin of an extreme weather system where the wind speed exceeded the 35-knot threshold for more than 24 h. Based on Fig. 7.1, it was obvious that during the cyclone season from November to April summed for the entire course of 32 seasons, there were 453 and 293 TC formations in SIO and SPO basins, respectively. Interestingly, the distribution of TC genesis appeared to be quite even in the SIO region, while in the SPO, most TCs formed to the west of the International Dateline. Very few TCs occurred in the eastern SPO, perhaps due to strong subsidence caused by the Walker Circulation (Walker and Bliss 1930, 1937) that creates a stable environment that inhibited TC formation.

Figure 7.1b shows the interannual variation of TC genesis numbers in the ocean basins for November to April TC season in the SH. Notably; the SIO had more TCs produced with an average of ~14 while the SPO has ~9. In general, the pattern of interannual variation between the two basins was out-of-phase, with a small downward correlation of −0.22. For instance, the years when the TCs are high in numbers in the SIO, the SPO region appeared to experience the counter effect. This out-of-phase pattern was partly due to the El-Niño Southern Oscillation (ENSO) dominant in the Pacific Ocean. For example, in the years 1992 and 1998, the SPO region appeared to exhibit an enhanced TC genesis, whereas the SIO region had below to near its average value in the years 1992 and 1998, respectively. However, the years with the lowest TC genesis in the SPO region (e.g., 1982, 1991, 1994, and 2004) were ENSO neutral years. This illustrated that the TC genesis was enhanced in the SPO region during an El-Niño period while the number of events stayed quite close to average value during the La Niña years and mostly below average during the ENSO neutral phase.

To further the understanding of the climatology of TC genesis it is imperative to demonstrate how TC numbers fluctuated in the respective seasons in the SH. Figure 7.1c shows the monthly variation of TC genesis numbers. A large number of TCs occurred in the months of January, February, and March. That is, the TC number was directly proportional to the progression of the summer season, which picked up in magnitude by January, reached its peak in February, and weakened as the month of March progressed. By comparison, over the January–March period, the SIO and SPO regions experienced ~68% and ~74% of all TC occurrences, respectively. Clearly, the SPO appeared to encompass more favorable convective activities that were conducive to TC formation in the region.

### *7.2.3 Definition of MJO Indices*

There are several ways of defining the MJO index [e.g., (Bessafi and Wheeler 2006; Chand and Walsh 2010; Hall et al. 2001; Maloney and Hartmann 2000a, b; Wheeler and Hendon 2004)]. In this chapter, a new MJO index has been introduced where the emphasis was placed on convective anomalies associated with the MJO phases. Although the Real-time Multivariate (RMM) MJO index used in earlier studies is the acceptable form for defining the MJO (Wheeler and Hendon 2004), we have adopted the alternate approach in order to capture the true amplitudes of the "MJO like" convective signals, as stipulated recently by Kiladis et al. (2014). Consequently, in our work, the outgoing longwave radiation (OLR) was chosen over the traditional use of zonal wind fields. In contrast, earlier studies considered multivariate Empirical Orthogonal Function (EOF) to decompose the MJO signal on time-series and spatial maps using OLR and the 850- and 200-hPa zonal wind fields. Our modified MJO index was different in three ways. First, we defined MJO based on an index from the OLR data alone. Second, it was the index extracted from regions with the largest OLR variance (standard deviation) averaged over the November to April TC season for 32 year period. Finally, we defined two sets of MJO indices based on regional differences; one for the SIO and the other for the SPO basin. Consequently, the alternative MJO indices have been accustomed to yield a greater understanding of seasonal variation of MJO in the SH, particularly for the geographically distinct basins.

To compose the MJO-based indices, a map of the standard deviations of the OLR was created for the SIO and SPO regions to determine the region with the largest variance in OLR (Fig. 7.2a). In this map, the regions are located over 5° N–10° S, 80° E–95° E and 5° S–15° S, 175° E–175° W. The spatially averaged OLR were

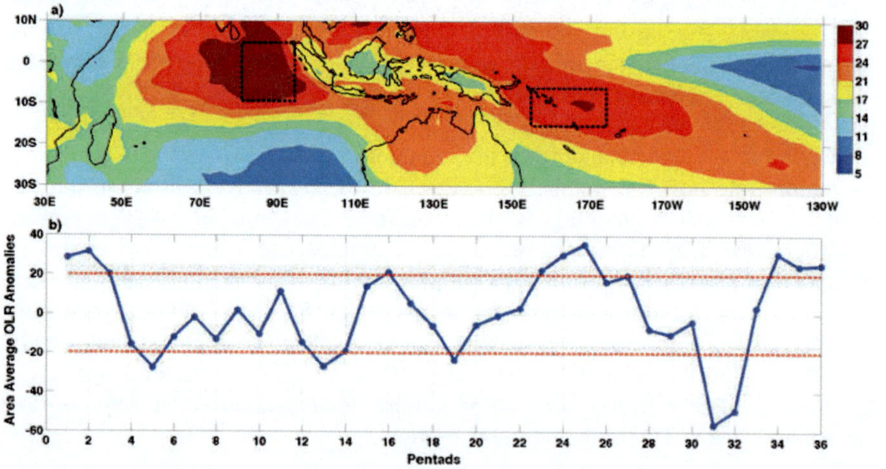


**Fig. 7.2** **a** The map of standard deviation of 5-day mean filtered outgoing longwave radiation (OLR; $Wm^{-2}$). The two rectangles mark the regions with largest OLR variability and domain from which the Madden–Julian Oscillation (MJO) indices are extracted. SIO MJO index is chosen over 10° S–5° N, 80° E–95° E, and SPO MJO index is chosen over 15° S–5° S, 155° E–175° W. **b** The area-averaged OLR anomalies versus the pentad values within the SIO MJO index region for the TC season 1981. There are 36 pentads in one TC season. The dashed red lines are ±1 standard deviation, where pentad values >1 and <−1 standard deviations are used for making composites

taken to be the MJO index for the SIO and SPO, respectively. Only the pentads (five-day non-overlapping average) OLR standard deviations greater than 1 and less than −1 in the spatially averaged locations were used to construct the composite maps. An example is shown in Fig. 7.2b for the 1981 TC season with the values above and below the ± dashed lines are used for making composite maps. The MJO indices were categorized into active and inactive phases in which the former phase corresponded to standard deviations less than −1 and the latter phase corresponded to standard deviations greater than 1.

Based on the preliminary analysis, we saw that for the SIO region, the active phase of the MJO consists of approximately 203 pentads and inactive phase with 210 pentads out of a total of 1152 pentads over the 32 seasons. Likewise, the SPO region had approximately 202 pentads for the active phase and 193 pentads for the inactive phase. It is imperative to mention that the MJO indices in this chapter are similar to those of WH04 in a way that this chapter focuses on the convective activity in the Indian Ocean and western Pacific as illustrated in WH04. In other words, our MJO index was simply a combination of the WH04's phases 2 and 3 that define the SIO MJO index, and phases 6 and 7 that define the SPO MJO index.

As the SIO and SPO-based MJO indices have been extracted from a spatial box as indicated earlier, we checked the evolution and propagation of convective and non-convective centers based on the MJO indices within ±2-pentad lead-lag range (Fig. 7.3). It was unambiguous that an eastward propagating disturbance existed

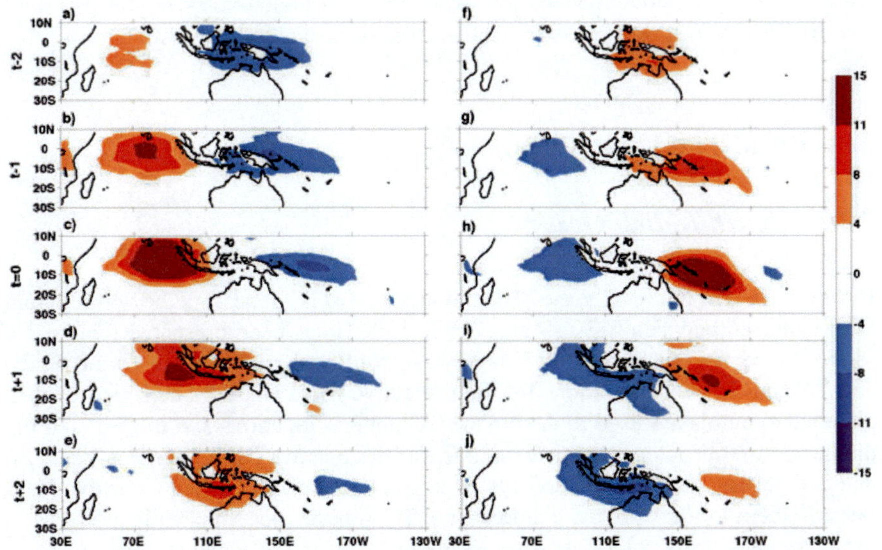


**Fig. 7.3** A lead-lag regression of filtered OLR ($Wm^{-2}$) as a contour from ±2 pentads using the SIO region MJO index (left: **a–e**) and using SPO MJO index (right: **f–j**). It shows the evolution and propagation of MJO using the two MJO indices. One active (convective phase) MJO leads the other active phase by 25-days

where the active and the inactive MJO periods were spaced out by approximately 25 days (or 5 pentads). When the correlation coefficient of the SIO and SPO MJO indices was computed, largest value of −0.26 occurred at lag 0. This indicated that the SPO region simultaneously experienced the inactive phase when the MJO in SIO region was in its active phase.

As a qualitative measure of the atmospheric circulation anomaly, we used the 850-hPa wind fields to compute the low-level relative vorticity in order to deduce the nature of convective activities within the upper atmosphere. The wind fields at 850-hPa were then subtracted from winds at 200-hPa to deduce the vertical shear, which is also believed to impact TC genesis and its characteristics [e.g., (Zhang and Tao 2013)]. TC and hurricane genesis were binned into their active and inactive phases whereby each genesis that occurred during the same time as the inactive phase was placed into the inactive bin and vice versa. For consistency, the hurricane genesis, which upgraded from the same TC location during its respective MJO phase were considered. For example, if a TC formed during an inactive MJO phase upgraded into a hurricane during the inactive MJO phase, then that hurricane genesis was considered for analysis. On the contrary, when a TC formed during the inactive phase but became hurricane during the active phase of MJO, then that hurricane was disregarded. The pentad dates that categorized the inactive and active MJO phases were used to construct composite maps of OLR, low-level winds, low-level relative vorticity, vertical wind shear, RH, SST, TC, and hurricane genesis locations. The statistical significance of results was assessed by a parametric approach (*t*-test) and crosschecked with an equivalent non-parametric test (Wilcoxon's test). It is found that the Wilcoxon signed-rank test yielded similar results to the *t*-test.

## 7.3 Results and Discussion

### *7.3.1 South Indian Ocean*

In the first part of our results, the MJO modulation on TC genesis in the South Indian Ocean located between 0–30° S and 30° E–130° E has been presented. Figure 7.4 shows the composite map of the 5-day averaged filtered OLR, 850-hPa wind fields, and TC and hurricane locations for the respective MJO phases. The superimposed diamond symbols are used to denote the TC genesis locations and in Fig. 7.4b the diamonds are hurricane genesis locations, for the inactive MJO phase. In Fig. 7.4c and d, the OLR shading, and 850-hPa wind vectors are composited for active MJO phase, except in Fig. 7.4c the diamonds are TC genesis locations while in Fig. 7.4d the diamonds are hurricane genesis locations. During the inactive phase of MJO over the SIO region, the positive OLR anomalies were concentrated in the region 0–15° S and 60° E–120° E (a, b). This indicated a suppressed convective activity. An anticyclone gyre appeared within this region while anomalous easterly winds developed to the west of this region. A large cyclonic feature was centered on the

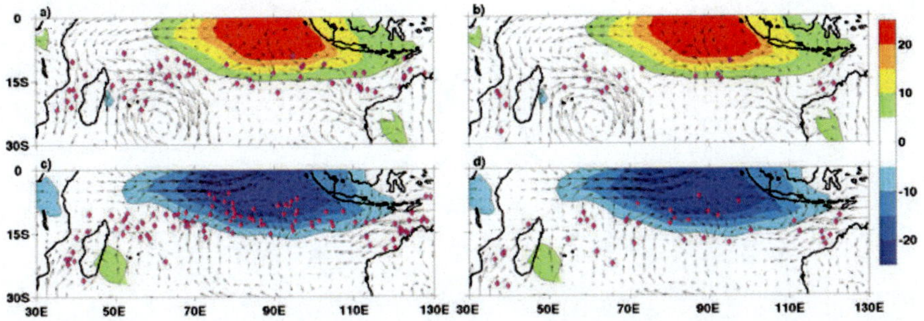


**Fig. 7.4** The 850-hPa wind vectors for: **a**, **b** the inactive MJO phase, **c**, **d** the active MJO phase. Note that the contours (shading) in all four panels show the composite of 5-day mean filtered OLR ($Wm^{-2}$) and the magenta diamond symbols are the TC genesis locations in (**a**) and hurricane genesis locations in (**b**) for inactive MJO phase, and the TC genesis locations in (**c**) and hurricane genesis locations in (**d**) for active MJO phases

Southeast of Madagascar and another off of the Southwest coast of Australia. The TC genesis inhibiting anomalous circulations appeared to suppress the TC formation to 60 TCs (Fig. 7.4a) of which only 24 have intensified into the hurricane stage (Fig. 7.4b) in the entire basin. More TCs and hurricanes originated approximately north of the large cyclonic circulation off of Madagascar and also in the Mozambique Channel where northerly winds are responsible for the entrainment of warm air and moisture from the equatorial region to create more favorable conditions for TC genesis. In contrast, the TC and hurricane genesis numbers in the entire SIO (Fig. 7.4c and d) during the active phase were approximately 105 and 38, respectively. Interestingly, the enhanced TC genesis during the active phase was associated with westerly wind anomalies toward the negative OLR anomaly center.

Interestingly, we observed a significant impact of the MJO phases in the region east of 70° E, where the TC modulation was relatively strong. That is, there was an increase in TC and hurricane genesis by ~2 folds for the active phase compared to the inactive phase. Furthermore, there was also an apparent shift in the TC genesis locations between the two MJO phases. For instance, on average the TC formation occurred along 15° S latitude during the inactive phase while during the active MJO phase, TCs were clustered equator-ward of 15° S. This latitudinal shift was not statistically significant at 95% level of confidence as the average latitudinal difference was less than 5°. In general, the MJO modulation of TC genesis was found to be stronger where the MJO signal was also reasonably strong.

As our results have demonstrated a strong modulation of TC genesis to the right of 70° E longitude, it was important to verify the relevance of these results. For this purpose, the region between 5° S–15° S and 70° E–110° E where enhanced TC or hurricane genesis was observed, has been selected and the area-averaged OLR anomalies were computed to determine the number of TCs that were formed during the negative ($<-1$ standard deviations) anomalies and positive anomalies ($>+1$ standard deviation) for up to $\pm 2$ pentads. Accordingly, a time-series of the

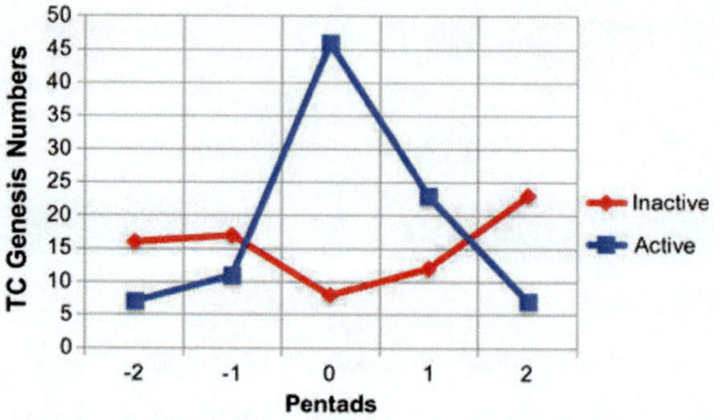


**Fig. 7.5** The TC genesis numbers for active (blue) and inactive (red) phases of the MJO with respect to area-averaged OLR in the region: 5° S–15° S and 70° E–95° E, where the MJO modulation is strongest. Pentad 0 is when the MJO is active with maximum or minimum standard deviation in the region. The TC genesis numbers are taken for up to ±2 pentads

TC genesis versus the OLR with the OLR represented in form of time lags was constructed. Figure 7.5 plots the number of TC genesis that occurred during the active MJO (pentad 0) and up to ±2 pentads phases. The pentad 0 consisted of all area-averaged OLR values >+1 and <−1 standard deviation occurring at the time of the MJO event. Interestingly, an approximate Gaussian distribution was obtained for the active phase of MJO with maximum (active MJO phase: blue line) and minimum (inactive MJO phase: red line) number of TCs at pentad 0. For clarity, pentad 0 is when the MJO was active, i.e., the convection was either enhanced or suppressed. Here, the modulation at pentad 0 for active/inactive MJO phase is ~6:1, indicating TC modulation by MJO events is very significant in this particular region.

It is imperative to realize that consistent with previous works [e.g., (Hall et al. 2001)], the primary cause for observed TC modulation is probably linked to changes in large-scale atmospheric fields with both dynamic and thermodynamic origin. In this chapter, the atmospheric dynamic parameters (850-hPa relative vorticity and vertical shear of the horizontal wind fields) and thermodynamic parameters (700-hPa RH and SST) were examined. It is noteworthy that these four parameters were also used in the original work of Gray (1968, 1979, 1998). Previous studies on MJO-TC relationships, however, have neglected thermodynamic parameters as these conditions are believed to be more appropriate for investigating longer periods, say more than a year [e.g., (McBride and Zehr 1981)] whereas MJO occurs on an intraseasonal timescale—shorter time period. Regardless, in our chapter, the 700-hPa RH and SST fields were used to determine whether the MJO-induced perturbations based on MJO indices having an influence on the number of TC genesis.

Next, we checked the role of vorticity, which also describes the state of the MJO on TC genesis (Fig. 7.6). Note that the negative relative vorticity is actually cyclonic in the SH region. In agreement with previous work (e.g., Kuleshov et al. 2009), a relatively large correspondence between negative relative vorticity and TC genesis locations in the central and southwest Indian Ocean was obtained. However, the

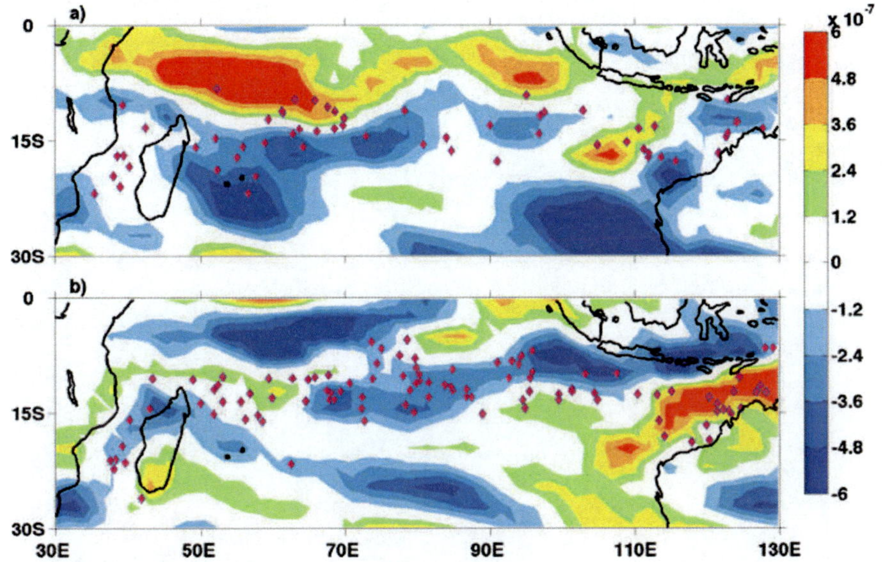


**Fig. 7.6** Composite of 850-hPa relative vorticity ($s^{-1}$) in shading, superimposed with the TC genesis (shown in magenta diamonds) for the inactive (top), and the active (bottom) MJO phases. The negative vorticity is cyclonic in the Southern Hemisphere

MJO-induced perturbations of low-level relative vorticity were enhanced to the right of 70° E during the active phase. Interestingly, we also observe that to the right 70° E, TCs are still seen to originate albeit with some degree of positive relative vorticity. During the inactive MJO phase, there were also large values of negative vorticity that seemed to favor TC genesis to the east of Madagascar. This was consistent with strong cyclonic flows; convective OLR anomalies and enhanced TC genesis (see Fig. 7.4).

In terms of the relationship between TC genesis and vertical wind shear, we found that the changes with respect to horizontal wind field between 850- and 200-hPa in the inactive and active phase of the MJO was relatively small (Fig. 7.7). For instance, TCs were formed in wind shear conditions where the wind speed was less than 5 $ms^{-1}$ (shown in black contour line) during both MJO phases. Also, the 5 $ms^{-1}$ contour line was located along ~20° S in both phases. While the change in wind shear was very small between the two phases in the entire SIO, there was no apparent influence of the MJO-induced perturbation on TC genesis during the active phase. Our finding was consistent with the deduction of Hall et al. (2001) for MJO modulation of TC genesis in the Australian region. That is, the climatological mean wind shear appeared to dominate the MJO-induced anomalies, so in the present chapter, we conclude that there is an apparent lack of the influence of MJO-induced vertical wind shear on TC genesis in the SIO region.

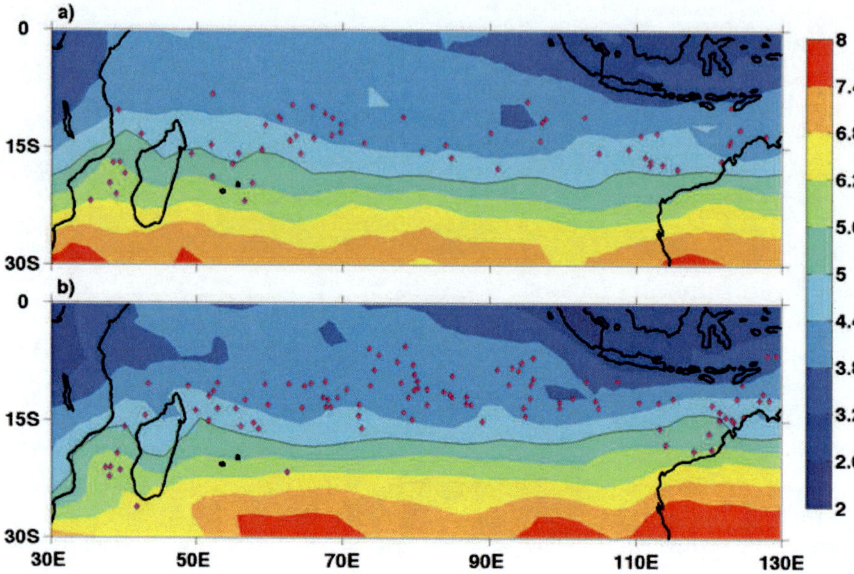


**Fig. 7.7** Composite of vertical shear of horizontal wind fields between 850- and 200-hPa levels ($ms^{-1}$) in shading, superimposed with the TC genesis shown in magenta-colored diamond shapes for the inactive (top) and the active (bottom) MJO phases. Note that the low vertical shear favors cyclogenesis

Next, we assessed the impacts of relative humidity (RH) as a thermodynamic factor, particularly for examining the MJO-induced perturbations based on mid-tropospheric moisture contents. However, fluctuations in moisture content between the inactive and active phases appeared to be relatively small (figure not shown). Also importantly, there was no apparent MJO-induced perturbation based on RH in the active phase that favored TC genesis. However, during the inactive phase, the anomalous value of RH fluctuations near Madagascar appeared to induce a larger proportion of atmospheric moisture into existing moist conditions, evidenced more clearly in monthly averaged plots. This condition appeared to generate a more favorable environment for TC genesis. While the actual threshold of RH required for TC genesis is unknown, an important point to note is that the fluctuations in RH were based on the order of the instrument error, which is presumably smaller than the error of analysis.

As the SST is a primary heat source for the upper atmosphere and may impact the conditions for TC genesis, we checked its correspondence with wind fields and solar radiation data in the study basin. Evidently, our results showed that to the west of the convectively active region, anomalously strong surface winds and enhanced evaporation, together with reduced solar insolation was seen to produce a decrease in SST, while to the east, weaker surface winds and enhanced solar insolation appeared to increase the SST (not shown here). In agreement with Woolnough et al. (2000), the negative SST anomaly during the active MJO phase was perhaps caused by the passage of convective clouds that reduce solar radiation reaching the surface and the strong winds associated with MJO near the surface lowered the SST.

In previous works performed using OLR and gridded reanalysis data, the intraseasonal SST variation appeared to be driven by anomalous latent heat flux and surface insolation. The physical process that accounts for intraseasonal SST variability has been examined extensively. For example, Shinoda and Hendon (1998) argued that the solar insolation was more important than latent heat flux in producing an intraseasonal variability in the SST along the 5° S and 75° E and 175° E regions, where the amplitude of MJO-induced SST variance exhibited a peak value. It is noteworthy that the mean SST generally exceeds 29 °C threshold in the equatorial warm pool and the Inter-tropical Convergence Zone (ITCZ) where strong intraseasonal variations in convective activities occur (Han et al. 2007). Using a bandpass filter of 30–90 days, the Tropical Rainfall Measuring Mission (TRMM) Hybrid Coordinate Ocean Model (HYCOM) SST appeared to span across the 0.23–0.34 °C (0.27–0.42 °C) range in the equatorial Indian Ocean. Although the mean SST exceeded the minimum threshold for TC formation [e.g., (Gray 1979)], the MJO-induced perturbation was often very small. This perturbation generally produces a cooling effect on SST during the active MJO phase and is primarily linked to the strong winds acting on the thin surface mixed layer (Duvel et al. 2004; Saji et al. 2006). In general, although cooling of SST is expected from the passage of convective clouds and MJO-induced wind fields, our analysis indicates that the MJO-induced SST perturbations do not produce a significant influence on TC genesis.

### 7.3.2 South Pacific Ocean

In this section, we considered the MJO-driven modulation of TC genesis in the South Pacific Ocean region located between 0–30°S and 130° E–130° W. Figure 7.8 plots the composite map of the OLR, 850-hPa wind fields, and TC/hurricane genesis. During the inactive MJO phase, the positive OLR anomaly region was centered between the location 0–20° S and 145° E–170° W and extended southeastward. Recall that this is an area of suppressed convection and was accompanied by anomalous easterly winds from the southeast (Fig. 7.8a and b). During the active MJO phase (Fig. 7.8c and d) the negative OLR anomalies that represented convective disturbances were accompanied by anomalous westerly winds. In the entire SPO basin, the TC (hurricane) genesis was approximately 3.6 (4.3) times greater during the active phase. The largest modulation of the MJO was observed to the left of 170° W, where TC (hurricane) genesis number was approximately 6.6 (6.3) times more during the active phase than during the inactive phase. To the east of 170° W, the inactive phase had more TC and hurricane genesis. A direct physical interpretation of this is that the MJO modulation of TC genesis was probably much stronger to the west of 170° W while being almost non-existent to the east of this location. A shift in the location of TC genesis was also apparent between the two MJO phases. For instance, during the inactive phase, the TCs and hurricanes were produced at a geographic distance further away from

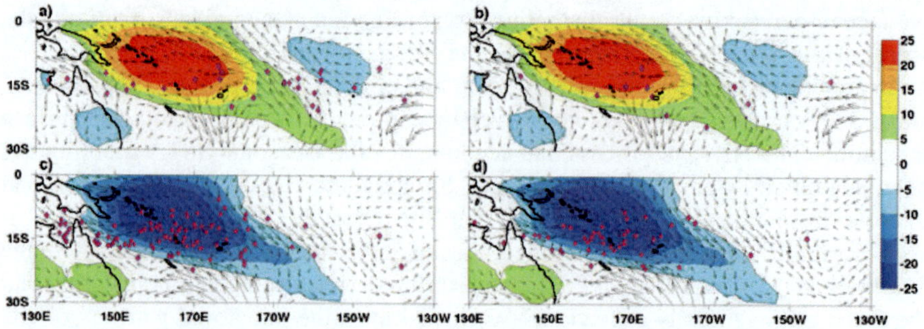


**Fig. 7.8** The shading in all four panels (**a–d**) is the composite of 5-day mean filtered OLR measured in $Wm^{-2}$ and vectors are composite of 850-hPa winds, for inactive MJO phase (**a, b**) and active MJO phase (**c, d**). In **a** the magenta-colored diamond shapes are the TC genesis locations and in **b** hurricane genesis locations for the inactive MJO phase. In **c** magenta-colored diamond shapes are TC genesis locations and in **d** hurricane genesis locations for active MJO phase

the suppressed convection area; while during the active phase, nearly all geneses occurred closer to the equator and within the enhanced convective region represented by negative OLR anomalies.

Since our results demonstrated a strong modulation on TC genesis to the west of 170° W, it was important to verify the significance of our results. For this purpose, the geographic region (22° S–7° S and 145° E–180° E) where enhanced TC or hurricane genesis occurs was examined closely. The area-averaged OLR anomalies were used to examine how the number of TCs formed during negative anomalies $<-1$ standard deviation and positive anomalies of OLR $>+1$ standard deviation at time zero and up to $\pm 2$ pentads. Pentad 0 is taken only for the extreme minimum and maximum standard deviation values. It was notable from Fig. 7.9 that the TC genesis number was enhanced (denoted by blue line) at pentad 0 when the MJO was in its convective

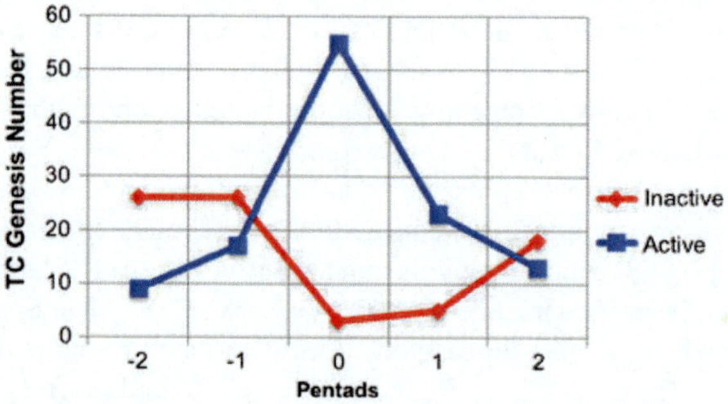


**Fig. 7.9** The TC genesis numbers for the active (blue) and inactive (red) phases of the MJO with respect to the area-averaged OLR in the region: 7° S–22° S and 145° E–180° E, where the MJO modulation is strongest. Pentad 0 is when the MJO is active with maximum or minimum standard deviation in the region. The TC genesis numbers are taken for up to $\pm 2$ pentads

phase. A decrease in TC genesis numbers was observed during ±5 and ±10 days, which was consistent with the 25-day lead-lag of the active MJO as evidenced earlier. By contrast, during the inactive phase, the number of TC genesis was the lowest at pentad 0 when the MJO was in the suppressed convective phase, although it increased for ±5 and ±10 day periods. This comparison yielded a modulation ratio of ~18:1 at pentad 0. Importantly, the enhanced (reduced) TC genesis during the active (inactive) phase confirmed that the MJO modulation of TC genesis was significant in this particular region. During the active MJO phase, very few TCs or hurricanes were produced to the east of 170° W longitude. This was probably due to the existence of a stable environment caused by strong subsidence to the east of 170° W longitude that weakened the MJO-induced perturbation, hence the conditions were unfavorable for cyclogenesis (Walker and Bliss 1930, 1937).

When compared with the SIO region, the observed MJO modulation of TC genesis in the SPO appeared to be much larger especially at the east of 70° E. A plausible explanation for this is the comparatively stronger background mean flow ($\approx$8 $ms^{-1}$) in the SIO where MJO-induced perturbations of the wind field were small (less than 2 $ms^{-1}$). This indicated that the strong background mean state had greater control on TC formations in this basin. In the SPO, the MJO-induced perturbations in the wind fields were also up to 2 $ms^{-1}$ but these perturbations were relatively large compared to the climatological mean flow, which was up to ~5 $ms^{-1}$. Quite clearly, this showed that MJO-induced circulation anomalies in the SPO were much stronger; hence the modulation of TC genesis was also comparably large. In addition, during the active MJO phase, nearly all TCs form in the SPCZ region, which was enhanced with convective activity. Furthermore, all TCs were produced within a region where the MJO basic state was also convectively enhanced during the active phase, as indicated by the large negative OLR anomalies.

In Fig. 7.10, we plot a composite map of the relative vorticity computed at 850-hPa during the inactive (Fig. 7.10a) and active (Fig. 7.10b) MJO phases. Strikingly, there was a very strong correspondence between cyclonic vorticity and TC genesis locations during the active MJO phase. That is, all TCs were produced in the region where the relative vorticity was highly negative (as well as cyclonic). By contrast, during the inactive phase, the TC genesis does not generally occur in the cyclonic vorticity region. In agreement with previous findings [e.g., (Chand and Walsh 2010; Hall et al. 2001)], relative vorticity was highly correspondent to MJO-induced perturbations in the SPO, as was the case for the SIO region.

Figure 7.11 plots the composite maps of vertical shear of the horizontal wind fields between the location of 850- and 200-hPa levels during the inactive (Fig. 7.11a) and active (Fig. 7.11b) phases of the MJO. There was little evidence of any MJO-induced perturbations during the active phase of MJO. This apparent lack of influence of vertical shear on TC formations has been indicated by the 5 $ms^{-1}$, contour lines that seemed to shift equator-ward during the active MJO phase. Also, all geneses occurred when the vertical wind shear was less than 6.8 $ms^{-1}$ regardless of the phase of the MJO. Clearly, this illustrates that the climatological mean value of the vertical wind shear during the TC season dominated the shear anomalies induced by the MJO in the SPO region, which was also similar to SIO region. Although our finding is

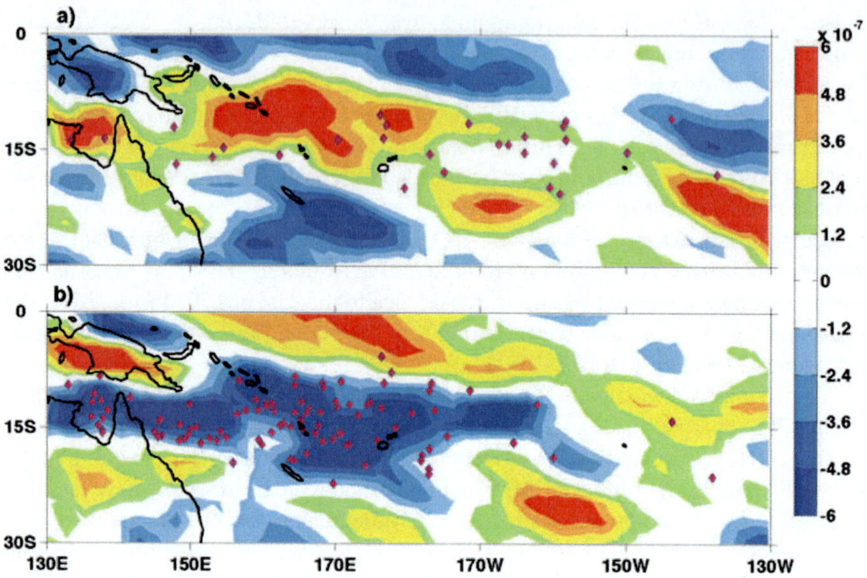


**Fig. 7.10** The composite of 850-hPa relative vorticity ($s^{-1}$) in shading, superimposed with the TC genesis shown in magenta-colored diamond shapes for the inactive (top) and the active (bottom) MJO phases

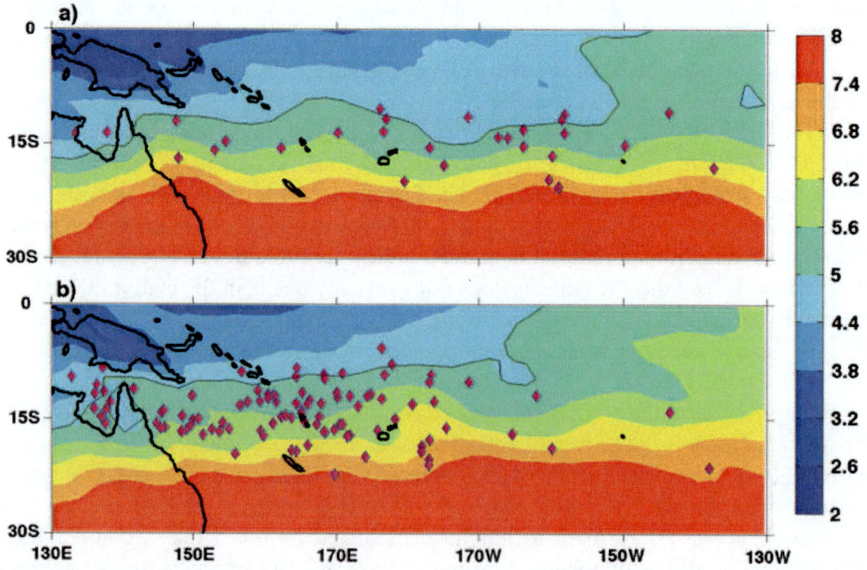


**Fig. 7.11** The composite of the vertical shear of the horizontal wind fields between 850- and 200-hPa (in $ms^{-1}$) in shading, superimposed with the TC genesis in magenta-colored diamond shapes for the inactive (top) and the active (bottom) MJO phases. Note that the low vertical shear favors cyclogenesis

consistent with Hall et al (2001), it does contradict with Chand and Walsh (2010) where a net southward shift of the zero contour line during the enhanced MJO phase was found for the Fiji-Samoa-Tonga region.

In terms of TC genesis modulation based on relative humidity, the climatological mean RH showed more than 50% moisture content in a region where a large number of TCs were formed. Although mid-tropospheric RH is an important thermodynamic condition, the actual threshold value of this parameter for the formation of cyclones is unknown. When compared by the SST, the MJO-induced perturbation appeared to exhibit a cooling effect for the active phase in the SPO (figure not shown). This was the result of the MJO related strong winds near the surface driving down the SST and the blockage of solar radiation by clouds over the convective anomaly region. Although the mean SST in the tropical Pacific exceeded the minimum threshold for TC formation (Gray 1968), the MJO-induced perturbation was very small.

Although the OLR variability was small, modulation of TC genesis in the SIO was much stronger than the SPO region. A primary reason for the observed weak modulation in the SIO region is due to the strong background mean flow in this basin. Hence the mean background flow was calculated using the wind convergence at 850-hPa (Fig. 7.12). Interestingly, the climatological November to April mean wind convergence was prominent along the SPCZ and the South Indian convergence zone. Furthermore, the magnitude of MJO-induced perturbations of convergence

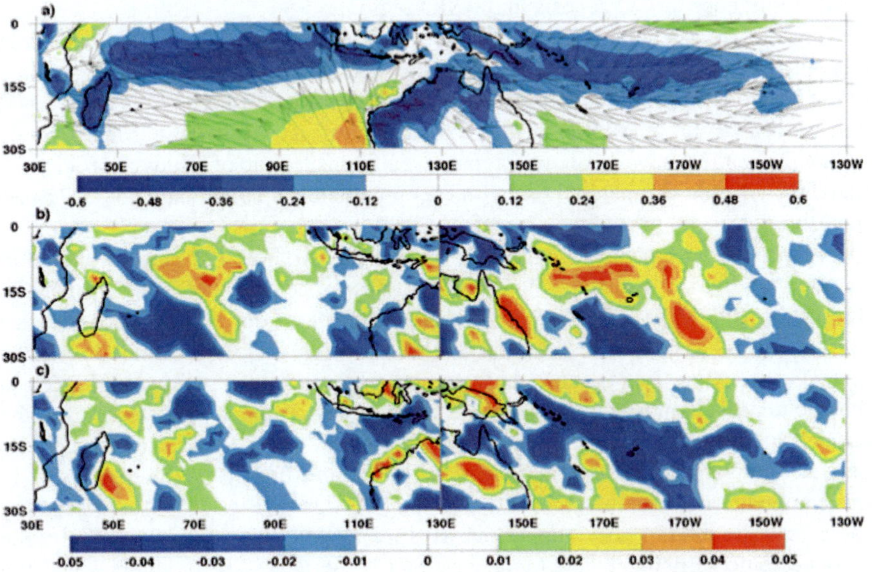


**Fig. 7.12** **a** November to April climatology of 850-hPa winds (vectors) and the corresponding divergence (positive) and convergence (negative) of the wind fields. The MJO-induced anomalous divergence and convergence for the inactive MJO phase is given in (**b**) and the active MJO phase given in (**c**). The unit for convergence/divergence is $\times 10^{-5}\ s^{-1}$

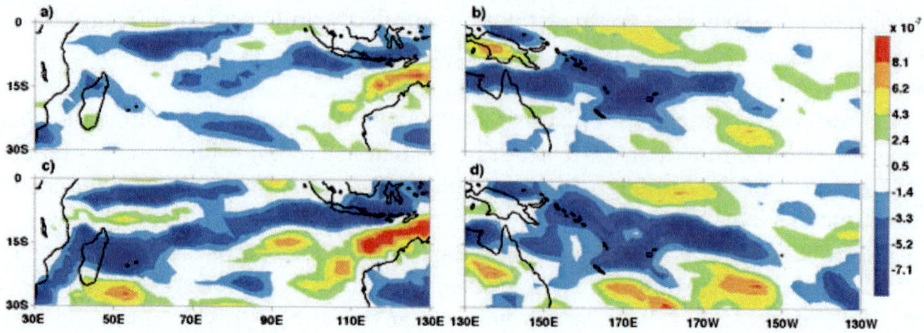


**Fig. 7.13** The composite of low-level relative vorticity for all TC and non-TC cases in **a** South Indian Ocean (SIO), **b** South Pacific Ocean (SPO) during the active phase of the MJO. In **c** and **d** the composites are shown for only non-TC cases in the SIO and SIO regions, respectively. The number of pentads used for constructing these composites in (**a**) are 203, 202 in (**b**), 121 in (**c**), and 117 in (**d**). The negative vorticity (blue) is cyclonic in the Southern Hemisphere with units as $s^{-1}$

was relatively small in the SIO region during the active MJO phase (Fig. 7.12c) compared to the climatological mean value (Fig. 7.12a). By contrast, a relatively strong wind convergence anomaly was evident in the SPO during the active MJO phase. This contrasted in wind fields as depicted by the convergence and divergence, and thus explained why the MJO-induced wind perturbations had a weak, albeit notable influence on TC genesis in the SIO region.

There is an acceptable scientific consensus that TCs are responsible for some form of climate variability in the regions where they are manifested. Following some of the previous studies [e.g., (Bessafi and Wheeler 2006; Hendon and Salby 1994)] that advocated the possibility of the contribution of TCs to intraseasonal climate variability in the study region, we prepared a composite map of relative vorticity for non-TC cases and compared those with the composite map of TC and non-TC cases inclusive.

Figure 7.13 presents the composite maps for the two ocean basins considered in this chapter. Evidently, just over half of the number of pentads had no TC formations during the active MJO phase. In the SIO region, it appeared that TCs have no influence or contribution to the relative vorticity. In other words, a closer examination of the non-TC cases (Fig. 7.13c) showed higher cyclonic vorticity compared to Fig. 7.13a that includes a composite map of relative vorticity for TC and non-TC cases. For the case of the SPO region, the modulation of 7:1 for TC and 6:1 for hurricane genesis between active and inactive phase appears to exhibit higher cyclonic vorticity in the composite map inclusive of TC and non-TC cases (Fig. 7.13b) compared to the composite map of only non-TC cases (Fig. 7.13d). Apparently, there are three peak cyclonic vorticity centers to the west of 170° W (Fig. 7.13b), which is also the region where strong modulation of TC genesis by the MJO is observed. Although the difference in vorticity (Fig. 7.13b and d) in this study region is relatively small, it does indicate that the TCs have some contribution from the deep convections near the center of cyclonic vorticity (Arnault and Roux 2011) during the active MJO phase.

In contrast, the work of Chand and Walsh (2010) found no significant differences in the magnitude of TC contribution to the enhanced MJO phases between the TC and the non-TC cases, as was the case with earlier investigations [e.g., (Camargo et al. 2009)].

## 7.4 Concluding Remarks

The goal of this chapter was to investigate the intraseasonal variability of TC genesis in austral summer for the SIO (0–30° S, 30° E–130° E) and the SPO (0–30° S, 130° E–130° W) regions. The chapter used outgoing longwave radiation (OLR), reanalysis wind fields, and relative humidity (RH) data in addition to satellite-derived sea surface temperature (SST) from November 1981 to April 2012. This analysis also examined large-scale MJO modulations on TC genesis for the SH TC season as well as the dynamic and thermodynamic parameters that were related to TC genesis. In this chapter, a relatively new MJO index based on the standard deviation of area-averaged OLR where OLR variance was the maximum for the two ocean basins was formulated. The modified MJO index was different in three ways. (1) By defining the MJO based on an index from the OLR data alone, (2) the index extracted from regions with largest OLR variance averaged over the November to April TC season for 32 year period, and (3) two sets of MJO indices defined based on regional differences; one for the SIO and the other for the SPO basin. The main reason for adopting the alternate approach was to capture the true amplitudes of the "MJO like" convective signals. Consequently, the outgoing longwave radiation (OLR) was chosen over the traditional use of zonal wind fields. Hence, the MJO indices yielded a better understanding of the seasonal variation of MJO signal in the SH. Using the SIO and SPO MJO indices, the MJO indices that described the inactive phase (suppressed convection) and the active phase (enhanced convection) were considered.

The results showed that TC genesis was enhanced by ~2 folds to the east of 70° E in SIO and ~7 folds to west of 170° W in the SPO region in the convectively enhanced phase of the MJO. The latter was much stronger than previously observed. The OLR and wind fields during different phases of the MJO reveal the dynamics responsible for modulation of TC genesis. TC genesis location tends to shift during different phases of MJO, i.e., they were strongly enhanced toward the negative OLR anomaly region where anomalous winds were strong westerlies, while reduced from the positive OLR anomaly region where anomalous winds were strong easterlies. The large longitudinal shift between inactive and active phases of MJO further confirmed a shift in TC genesis locations.

MJO-induced modulation of TC genesis was weaker in SIO compared to the SPO region despite larger amplitude of OLR variance in the former region. The discernment in the magnitude of MJO-induced modulation between the two basins is probably due to the strong background mean flow in the SIO where the MJO-induced perturbations were relatively small. In addition, the strong modulation in the SPO region also makes TC genesis in this region potentially more predictable than the SIO region. Such modulation occurs due to changes in large-scale flows. The active phase of the MJO coincided very well with the most favorable combination of dynamic and thermodynamic conditions. In the SIO and SPO regions, only low-level cyclonic relative vorticity strongly contributed to the convective phase of the MJO. Our analysis found that a large number of TCs formed in the southwest Indian Ocean (near Madagascar). Since the occurrence of MJO was mainly confined to the tropics (15° N–15° S) (Madden and Julian 1971; Zhang 2005), the TC formations in this region occurred away from the equator, and thus were unaffected by the MJO.

Based on this chapter's results, we can conclude that the observed MJO modulation of TC genesis is relatively significant especially in the SPO region. Given the notable magnitude of the modulation ratio of TC genesis between two MJO phases, it is possible to utilize the OLR and other atmospheric data for the development of systems, software, or models for forecasting TC activities using MJO indices used in our chapter. The prediction skill of TC genesis can be improved if the specific region with the strongest modulations is considered instead of the entire basin with a wide range of forecasting techniques (Roy and Kovordányi 2012). Our chapter, however, was limited in that it only analyzed the large-scale atmospheric conditions that although oceanic contributions to TC modulation could also be considered as a useful follow-up study in the South Pacific and South Indian Oceans.

## References

Arnault J, Roux F (2011) Characteristics of African easterly waves associated with tropical cyclogenesis in the Cape Verde Islands region in July–August–September of 2004–2008. Atmos Res 100(1):61–82

Basher R, Zheng X (1995) Tropical cyclones in the southwest Pacific: spatial patterns and relationships to southern oscillation and sea surface temperature. J Clim 8(5):1249–1260

Bessafi M, Wheeler MC (2006) Modulation of South Indian Ocean tropical cyclones by the Madden-Julian Oscillation and convectively coupled equatorial waves. Mon Weather Rev 134(2):638–656

Callaghan J, Power SB (2011) Variability and decline in the number of severe tropical cyclones making land-fall over eastern Australia since the late nineteenth century. Clim Dyn 37(3–4):647–662

Camargo SJ, Wheeler MC, Sobel AH (2009) Diagnosis of the MJO modulation of tropical cyclogenesis using an empirical index. J Atmos Sci 66(10):3061–3074

Chand SS, Walsh KJ (2010) The influence of the Madden-Julian oscillation on tropical cyclone activity in the Fiji region. J Clim 23(4):868–886

Duvel JP, Roca R, Vialard J (2004) Ocean mixed layer temperature variations induced by intraseasonal convective perturbations over the Indian Ocean. J Atmos Sci 61(9):1004–1023

Frank WM (1987) Tropical cyclone formation. A global view of tropical cyclones, pp 53–90

Gray WM (1968) Global view of the origin of tropical disturbances and storms. Mon Weather Rev 96(10):669–700

Gray WM (1979) Hurricanes: their formation, structure and likely role in the tropical circulation. Meteorol Over Tropical Oceans 77:155–218

Gray WM (1998) The formation of tropical cyclones. Meteorol Atmos Phys 67(1–4):37–69

Hall JD, Matthews AJ, Karoly DJ (2001) The modulation of tropical cyclone activity in the Australian region by the Madden-Julian Oscillation. Mon Weather Rev 129(12):2970–2982

Han W, Yuan D, Liu WT, Halkides D (2007). Intraseasonal variability of Indian Ocean sea surface temperature during boreal winter: Madden–Julian Oscillation versus submonthly forcing and processes. J Geophys Res Oceans 112(C4)
Hendon HH, Salby ML (1994) The life cycle of the Madden-Julian oscillation. J Atmos Sci 51(15):2225–2237
IPCC (2007) Summary for Policymakers. In: Climate change 2007: the physical science basis. contribution of working group i to the fourth assessment report of the intergovernmental panel on climate change. Cambridge University Press, United Kingdom and New York, NY, USA
Jones C, Carvalho LM (2014) Sensitivity to Madden–Julian Oscillation variations on heavy precipitation over the contiguous United States. Atmos Res 147:10–26
Kalnay E et al (1996) The NCEP/NCAR 40-year reanalysis project. Bull Am Meteor Soc 77(3):437–471
Kiladis GN et al (2014) A comparison of OLR and circulation-based indices for tracking the MJO. Mon Weather Rev 142(5):1697–1715
Kiladis GN, Weickmann KM (1992) Circulation anomalies associated with tropical convection during northern winter. Mon Weather Rev 120(9):1900–1923
Klotzbach PJ (2014) The Madden–Julian oscillation's impacts on worldwide tropical cyclone activity. J Clim 27(6):2317–2330
Knutson TR, Weickmann KM (1987) 30–60 day atmospheric oscillations: Composite life cycles of convection and circulation anomalies. Mon Weather Rev 115(7):1407–1436
Kuleshov Y et al (2010) Trends in tropical cyclones in the South Indian Ocean and the South Pacific Ocean. J Geophys Res Atmos 115(D1)
Kuleshov Y, Qi L, Fawcett R, Jones D (2008) On tropical cyclone activity in the Southern Hemisphere: trends and the ENSO connection. Geophys Res Lett 35(14)
Kuleshov Y, Ming FC, Qi L, Choaibou I, Hoareau C, Roux F (2009) Tropical cyclone genesis in the southern hemisphere and its relationship with ENSO. Ann Geophys 27:2423–2538
Liebmann B (1996) Description of a complete (interpolated) outgoing longwave radiation dataset. Bull Amer Meteor Soc 77:1275–1277
Liebmann B, Hendon HH, Glick JD (1994) The relationship between tropical cyclones of the western Pacific and Indian Oceans and the Madden-Julian oscillation. J Meteorol Soc Jpn 72(3):401–412
Madden RA, Julian PR (1971) Detection of a 40-50 day oscillation in the zonal wind in the tropical Pacific. J Atmos Sci 28(5):702–708
Madden RA, Julian PR (1972) Description of global-scale circulation cells in the tropics with a 40–50 Day Period. J Atmos Sci 29:1109–1123. https://doi.org/10.1175/1520-0469(1972)029<1109:DOGSCC>2.0.CO;2
Maloney ED, Hartmann DL (2000a) Modulation of eastern North Pacific hurricanes by the Madden-Julian oscillation. J Clim 13(9):1451–1460
Maloney ED, Hartmann DL (2000b) Modulation of hurricane activity in the Gulf of Mexico by the Madden-Julian oscillation. Science 287(5460):2002–2004
McBride JL, Zehr R (1981) Observational analysis of tropical cyclone formation. Part II: Comparison of non-developing versus developing systems. J Atmos Sci 38(6): 1132–1151
Nakazawa T (1988) Tropical super clusters within intraseasonal variations over the western Pacific. J Meteor Soc Japan 66:823–839
Nott J (2011) Tropical cyclones, global climate change and the role of Quaternary studies. J Quat Sci 26(5):468–473
Peduzzi P et al (2012) Global trends in tropical cyclone risk. Nat Clim Change 2(4):289–294
Rui H, Wang B (1990) Development characteristics and dynamic structure of tropical intraseasonal convection anomalies. J Atmos Sci 47(3):357–379
Roy C, Kovordányi R (2012) Tropical cyclone track forecasting techniques-a review. Atmos Res 104:40–69
Saji N, Xie SP, Tam CY (2006) Satellite observations of intense intraseasonal cooling events in the tropical south Indian Ocean. Geophys Res Lett 33(14)

Shinoda T, Hendon HH (1998) Mixed layer modeling of intraseasonal variability in the tropical western Pacific and Indian Oceans. J Clim 11(10):2668–2685

Sinclair MR (2002) Extratropical transition of southwest Pacific tropical cyclones. Part I: climatology and mean structure changes. Monthly Weather Rev 130(3): 590–609

Terry JP, Gienko G (2010). Climatological aspects of South Pacific tropical cyclones, based on analysis of the RSMC-Nadi (Fiji) regional archive

von Storch H, Xu J (1990) Principal Oscillation Pattern analysis of the tropical 30–60 day oscillation. Part I: definition of an index and its prediction. Clim Dyn 4:179–190

Walker GT, Bliss EW (1930) World weather IV. Memoirs Roy Meteorol Soc 3(24):81–95

Walker GT, Bliss EW (1937) World weather VI. Memoirs Roy Meteorol Soc 4(39):119–139

Wang B, Yang H (2008) Hydrological issues in lateral boundary conditions for regional climate modeling: Simulation of East Asian summer monsoon in 1998. Clim Dyn 31(4):477–490

Webster P, Holland G, Curry J, Chang H (2005) Changes in tropical cyclone number, duration, and intensity in a warming environment. Science 309(5742):1844–1846

Wheeler MC, Hendon HH (2004) An all-season real-time multivariate MJO index: development of an index for monitoring and prediction. Mon Weather Rev 132(8):1917–1932

Woolnough SJ, Slingo JM, Hoskins BJ (2000) The relationship between convection and sea surface temperature on intraseasonal timescales. J Clim 13(12):2086–2104

Yun KS, Chan JC, Ha KJ (2012) Effects of SST magnitude and gradient on typhoon tracks around East Asia: Acase study for Typhoon Maemi (2003). Atmos Res 109:36–51

Zhang C (2005) Madden–Julian oscillation. Rev Geophys 43(2)

Zhang F, Tao D (2013) Effects of vertical wind shear on the predictability of tropical cyclones. J Atmos Sci 70:975–983. https://doi.org/10.1175/JAS-D-12-0133.1

# Chapter 8
# Intelligent Data Analytics for Time Series, Trend Analysis and Drought Indices Comparison

**Kavina S. Dayal, Ravinesh C. Deo, and Armando A. Apan**

## 8.1 Introduction

This chapter employs Standardised Precipitation-Evapotranspiration Index (SPEI), formulated as an improved version of WMO-approved SPI. Unlike in the case of the Standardised Precipitation Index (SPI), the SPEI has an ability to encapsulate the contributory influence of temperatures on the demand for water, and therefore, it appears to be more suitable for the monitoring of hydrological and agricultural impacts. Also, unlike the case of the Palmer Drought Severity Index (PDSI), the SPEI is able to operate on multiple timescales (1–48 months), acting as an essential tool for assessment of the hydrologic cycles and for accounting for different category of drought (meteorological, hydrological and agricultural). The SPEI can replicate the sensitivity embedded in the PDSI for monitoring of hydrological status in terms of the estimated evaporation and transpiration driven by warm temperatures, while assessing the multi-temporal nature of drought afforded by SPI.

An idealistic characteristic of the SPEI is its ability to capture the evaporative demand of the hydrosphere (i.e., via reference evapotranspiration; *ETo*) and the indicative aberrations in overall water resource conditions. SPEI holds the advantages of less data requirement, flexibility, and simple computation. These accord to the viewpoint of Keyantash and Dracup (2002) that a drought metric must be simple,

K. S. Dayal (✉)
Commonwealth Scientific and Industrial Research Organisation (CSIRO), Sandy Bay 7005, Hobart, TAS, Australia
e-mail: kavinadayal@gmail.com

R. C. Deo
School of Sciences, University of Southern Queensland, Springfield Central, QLD 4300, Australia

A. A. Apan
School of Civil Engineering, University of Southern Queensland, Toowoomba, QLD 4350, Australia

© Springer Nature Singapore Pte Ltd. 2021 
R. C. Deo et al. (eds.), *Intelligent Data Analytics for Decision-Support Systems in Hazard Mitigation*, Springer Transactions in Civil and Environmental Engineering,
https://doi.org/10.1007/978-981-15-5772-9_8

clear, comprehensible and statistically robust, and also be independent of the climatic characteristics (i.e., standardised) to be comparable in the wider temporal and spatial domains across geographically diverse regions.

Despite its infancy in hydrologic research community, many case studies performed outside of Australia have applied SPEI for drought assessment and demonstrated its strong statistical correlation with hydro-meteorological variables that affect drought impacts in such diverse climatic regions. For example, SPEI has been used for drought variability studies (Das et al. 2016; Li et al. 2012; Paulo et al. 2012; Potop 2011), hydrological impact assessments, agricultural drought studies, impact of drought on ecological systems (Barbeta et al. 2013; Cavin et al. 2013; Martin-Benito et al. 2013; Toromani et al. 2011; Vicente-Serrano et al. 2013) as well as for the monitoring of drought events (Fuchs et al. 2012). To contribute to the growth of knowledge in the subject area, this study has performed several analyses using SPEI as the primary drought-monitoring index.

The main objectives of this chapter are to (1) compare SPEI with precipitation-based SPI, RDDI and RAI drought indices using statistical metrics and wavelet analysis and (2) assess any change in the trend of the SPEI time series over the period 1915–2016 using the change-point analysis.

## 8.2 Materials and Method

### *8.2.1 Hydrological Data and Study Area*

Four case study locations in southeast Queensland (SEQ): R1, R2, R3 and R4 with distinct climatological characteristics are selected. Figure 8.1 plots their location. Table 8.1 lists their geographic coordinates with average annual precipitation for base period (1971–2000). The monthly precipitation data (*P*; in millimetres), maximum temperature ($T_{\max}$; in degree Celsius) and FAO-56 Penman–Monteith-based Reference Evapotranspiration (*ETo*; in millimetres) for any grid point, interpolated from gridded datasets on 5 km × 5 km resolution, are acquired from SILO database for the period 1915–2016. The upper-layer soil moisture (*WRel1*; fractional [0 1]) is acquired from AWAP.

The sensitivity of potential evapotranspiration in any calculations involving the SPEI requires caution since previous studies have demonstrated that the potential evapotranspiration should not be forced by temperature data alone [e.g., (Hobbins et al. 2012; Roderick et al. 2007; Sheffield et al. 2012)]. Therefore, this study utilises the FAO-56 Penman–Monteith-based *ETo* data to compute the SPEI. Note that instead of using the generic term 'potential evapotranspiration', hereafter this study refers to 'reference evapotranspiration' specifically (denoted as *ETo*). The *ETo* data available in the SILO database has been estimated via the FAO-56 Penman–Monteith formula (Allen et al. 1998). The *ETo* values correspond to the short crop cases whereby a hypothetical reference with the assumed grass top height of 0.12 m, a fixed surface

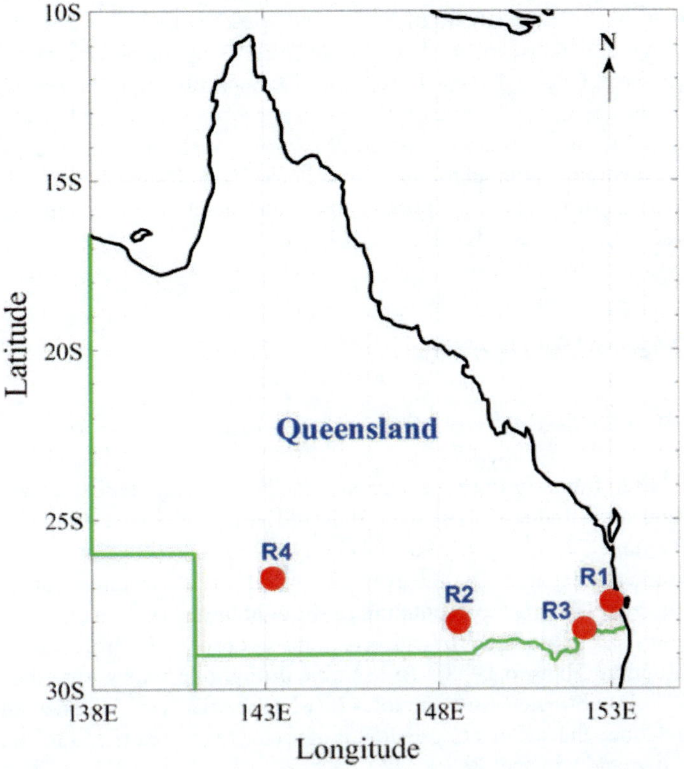


**Fig. 8.1** Map of point-based study locations: R1—Subtropical, R2—Grassland, R3—Temperate, and R4—Desert

**Table 8.1** Study locations and their descriptive statistics

| Location label | Climatic regions | Geographical location | Elevation above sea-level (*m*) | Annual mean precipitation (*P*; mm) |
|---|---|---|---|---|
| R1 | Subtropical | 153.05° E, 27.45° S | 61 | 1154.61 |
| R2 | Grassland | 148.60° E, 28.05° S | 250 | 551.93 |
| R3 | Temperate | 152.25° E, 28.25° S | 561 | 739.70 |
| R4 | Desert | 143.25° E, 26.75° S | 211 | 307.82 |

resistance of 70 $sm^{-1}$, an albedo of 0.23, wind speed value at 2 m height, radiation derived from cloud oktas, and the hours of sunshine using the procedure documented in Zajaczkowski et al. (2013) has been used. The monthly upper-layer soil moisture (*WRel1*; fractional [0, 1]) and end of the month aggregated soil moisture (*WRel1End*; fractional [0, 1]) are obtained from the AWAP historical runs constructed from the *WaterDyn* hydrological model for the period 1915–2016 (Raupach et al. 2009, 2012). The AWAP and SILO data are all obtained for the matching geographical locations for this case study (i.e., R1, R2, R3 and R4).

## 8.2.2 *Theoretical Overview*

### 8.2.2.1 Standardised Precipitation-Evapotranspiration Index

The point-based monthly time series, comprised of the Standardised Precipitation-Evapotranspiration Index (SPEI) over 1915–2016 period based on the cumulative effects of *P* and *ETo* are generated where *ETo* is used to depict the evaporative demand of the atmosphere, i.e., the evapotranspiration that would occur if sufficient water was available. This representation of drought aimed to incorporate the role of *ETo* that could act to moderate or exacerbate the underlying hydrological cycles in a drought situation (Hanson 1988). To examine drought periods within the historical data, the *ETo* is subtracted from the total *P* (where $\text{SDB}_i = P_i - ETo_i$ and $i =$ the month) to deduce the surplus or the deficit of water resources (i.e., the computation of supply–demand balance; *SDB*).

As precipitation data generally exhibits seasonality within the case study regions and the distribution of these data are especially pronounced in different regimes, it is necessary to transform the *SDB* time series via an equal probability framework to a normal distribution with a mean of zero ($\mu = 0$) and standard deviation of one ($\alpha = 1$). This allows the water deficits and surpluses to be comparable in space and time, and the SPEI to be standardised so that it is free from seasonality and the data distribution effects when assessing the different drought event. To achieve this, the *SDB* time series are fitted to the three-parameter log-logistic, Gamma and Pearson III distributions based on the goodness-of-fit tests, i.e., Kolmogorov–Smirnov (K–S) statistic. With the null hypothesis that the *SDB* time series follows a specified distribution at significance level ($\alpha = 0.05$), the best (smallest) K-S statistic is obtained for log-logistic distribution, concurring with Vicente-Serrano et al. (2010). Thus, the transformation of the *SDB* time series employed a probability density function of a three-parameter ($\alpha$, $\beta$ and $\gamma$) log-logistic distribution, $F(x)$ according to Vicente-Serrano et al. (2010):

$$F(x) = \left[1 + \left(\frac{\beta}{x - \gamma}\right)^{\alpha}\right]^{-1} \tag{8.1}$$

The SPEI is then computed as (Vicente-Serrano et al. 2010):

$$\text{SPEI} = W - \frac{C_o + C_1 W + C_2 W^2}{1 + d_1 W + d_2 W^2 + d_3 W^3} \tag{8.2}$$

In Eq. (8.2), the term $W = \sqrt{-2ln(P)}$ for $P \leq 0.5$, and $P$ is the exceedance probability of a determined *SDB* value, $P = 1 - F(x)$ while $C_0 = 2.515517$, $C_1 = 0.802853$, $C_2 = 0.010328$, $d_1 = 1.432788$, $d_2 = 0.189269$ and $d_3 = 0.001308$ are the empirical constants. The SPEI values corresponding to deficits or surpluses of water resources at six timescales ($T = 1, 3, 6, 9, 12$ and 24 months) are computed; where for instance, the $\text{SPEI}_9$ is constructed by a sum of *SDB* values from eight months before to the current month. In all SPEI calculations, the base period is set to be 30 years (i.e., 1971–2000), which is a common practice for drought studies in Australia (Dayal et al. 2018; Deo and Şahin 2015; Deo et al. 2009). While the computational aspect of deriving SPEI is generally simple, the data requirement however has to be of at least 30 years for better distributional fit and true representation of the regional droughts.

#### 8.2.2.2 Rainfall Decile-Based Drought Index

The rainfall decile drought index (RDDI) is a measure of rainfall deficiency (Gibbs and Maher 1967). It is calculated from monthly rainfall values. While maintaining the traditional ten classes (deciles) and instead of sorting rainfall data into slices of 10%, this study has sorted rainfall data into slices of 5%, (20 quantiles) to obtain a better accuracy in detecting the drought signatures. First, the rainfall values of each Julian calendar month, for the base period, are sorted in ascending order and ranked from lowest to highest to construct a cumulative frequency distribution. The distribution is then split into 20 quantiles (slices of 5%). Using the monthly rainfall amount for each quantile obtained for the base period, the monthly rainfall amounts for the study period are then assigned the corresponding quantile. The first quantile that has the lowest rainfall values indicates driest months in the series while the 20th quantile indicates the wettest months. The other quantiles show the range from driest to wettest months. The ten classes (deciles) are assigned to each rainfall value as integers from 0.5 to 10 at intervals of 0.5.

#### 8.2.2.3 Standardised Precipitation Index

Developed by McKee et al. (1993), the Standardised Precipitation Index (SPI) primarily defines and monitors the drought events. To compute SPI for the desired station, a frequency distribution is constructed for a long-term precipitation record. It is then fitted to a theoretical probability distribution, i.e., Gamma distribution, and then transformed into a normal distribution so that the mean SPI is zero with a unit variance. The transformed distribution helped determine the extent of rainfall deficit, thus facilitated comparison and monitoring of spatial drought conditions at various temporal scales. The following equations are useful for calculating SPI. A Gamma distribution defined by its probability density function is given by:

$$g(x) = \frac{1}{\beta^{\alpha} \Gamma(a)} x^{\alpha-1} e^{-\frac{x}{\beta}}, \quad \text{for } x > 0 \tag{8.3}$$

The $\alpha$ (shape) and $\beta$ (scale) parameters are calculated as:

$$\alpha = \frac{1}{4A}\left(1 + \sqrt{\frac{4A}{3}}\right) \text{ and } \beta = \frac{\bar{x}}{\alpha} \tag{8.4}$$

where $\bar{x}$ represent the rainfall average over the base period and $n$ is the number of observations, i.e., 1224 months, for each study location, and;

$$A = \ln(\bar{x}) - \frac{\sum \ln(x)}{n} \tag{8.5}$$

The above parameters are then used to derive the cumulative probability distribution, given as:

$$G(x) = \int_0^x g(x)\mathrm{d}x = \frac{1}{\beta^{\alpha} \Gamma(\alpha)} \int_0^x x^{\alpha-1} e^{-\frac{x}{\beta}} \mathrm{d}x \tag{8.6}$$

Since the Gamma distribution is undefined for zero rainfall amount, the cumulative probability, $H(x)$, of zero and nonzero rainfall is calculated as:

$$H(x) = q + (1 - q)G(x) \tag{8.7}$$

where $q$ is the probability of zero rainfall. For instance, if there are $m$ number of months with zero rainfall, then $q$ is estimated as $m/n$ and the cumulative probability is transformed into a standardised normal distribution that gives mean SPI and variance to be 0 and 1, respectively.

#### 8.2.2.4 Rainfall Anomaly Index

The monthly Rainfall Anomaly Index (RAI) is computed according to Van Rooy (1965):

$$\text{RAI} = x_i - \bar{x} \tag{8.8}$$

where $x_i$ is the monthly precipitation and $\bar{x}$ is the mean precipitation for the base period, 1971–2000.

#### 8.2.2.5 Wavelet Analysis

A wavelet is a function localised in both frequency ($\Delta\omega$ or bandwidth) and time ($\Delta t$) with zero mean (Torrence and Compo 1998). The Morlet wavelet is selected in this study because it comprises both real and imaginary parts in the function that enables it to investigate a signal's coherence and phase angle, after Chang et al. (2017) and Grinsted et al. (2004). It is expressed as:

$$\psi_0(\eta) = \pi^{\frac{-1}{4}} e^{i\omega_0\eta} e^{\frac{-1}{2}\eta^2} \tag{8.9}$$

where $\omega_0$ and $\eta$ are dimensionless frequency and time, respectively. For feature extraction purposes, Morlet wavelet with $\omega_0 = 6$ is a recommended choice since it provides a good balance between time and frequency localisation (Grinsted et al. 2004).

#### 8.2.2.6 The Continuous Wavelet Transform (CWT)

The CWT has been used to apply a bandpass filter to the time series. Unlike Fourier transformation, the CWT has the ability to construct time–frequency localisation of a signal. By varying its scale ($s$), the wavelet is stretched in time so that, $\eta = s \cdot t$ and subsequently normalising it to have a unit energy. The CWT of a time series ($x_n, n = 1, \ldots, N$) with a uniform time steps $\delta t$ is defined as the convolution of $x_n$ with scaled and normalised wavelet (Grinsted et al. 2004) as:

$$W_n^X(s) = \sqrt{\frac{\delta t}{s}} \sum_{n'=1}^{N} x_{n'} \psi_0\left[(n' - n)\frac{\delta t}{s}\right] \tag{8.10}$$

#### 8.2.2.7 Cross-Wavelet Transform (XWT)

While CWT divides a continuous-time function into wavelets, the similarity between two series in the same period is generally hard to identify. To overcome this, the XWT has been used. The XWT of two time series $x_n$ and $y_n$ is defined as $W^{XY} = W^X W^{Y*}$ (Grinsted et al. 2004) where * refers to complex conjugation and the absolute value of $W^{XY}$ is the cross-wavelet power. Detailed theoretical distribution of the XWT is given in Torrence and Compo (1998).

The CWT and XWT are powerful methods for testing proposed linkages between the two data time series.

#### 8.2.2.8 Change-Point Analysis (CPA)

A CPA is a powerful tool that determines whether a change has taken place in a series. There are numerous ways to perform a CPA on a times series. This study has used the approach implemented in Taylor (2000). It is capable of detecting subtle changes and better characterises the detected changes by providing confidence levels and confidence intervals. The confidence levels indicate the likelihood that a change has occurred while the confidence interval indicates when the change has occurred.

The change-point analysis iteratively uses a combination of cumulative sum charts (CUSUM) and bootstrapping to detect the changes. In this study, the CUSUM charts are constructed by calculating and plotting a cumulative sum based on the data, viz. Taylor (2000). Suppose $X_1, X_2, \ldots, X_N$ are $N$ number of data points. The cumulative sums (CUSUM) $S_0, S_1, \ldots, S_N$ are then calculated in three steps:

1. Calculate the average of the data series: $\overline{X} = \frac{X_1+X_2+\cdots+X_N}{N}$;
2. Set the initial condition, $S_0 = 0$;
3. Compute $S$ recursively, $S_i = S_{i-1} + \left(X_i - \overline{X}\right), i = 1, 2, \ldots, N$

The interpretation of a CUSUM chart requires close attention to the pattern of $S_i$. For the case of SPEI time series, consider a period of time where SPEI is above the average value. Most values adding up to the cumulative sum can produce positive values so the trend line is expected to rise steadily. In this case, a segment with increasing $S_i$ indicates a period where SPEI values are above average. Likewise, a segment with a decreasing $S_i$ indicates a period where values are below average. A sudden change in the direction of the time series, detected by the change in the sign of the gradient of $S_i$ at a stationary point, $x$, is likely to indicate a sudden, abrupt shift in the average value of the time series. However, $S_i$ following a relatively straight path indicate a period where the average does not change. Based on this, we can detect any abrupt changes over the given study period.

To be certain that a change has occurred, a confidence interval can be used for the obvious change by performing a bootstrap analysis. For bootstrap analysis, an estimator of the magnitude of change is required, and in this case, it is the difference between the maximum and minimum cumulative sums, $S_i$, i.e., $S_{\text{diff}} = S_{i\max} - S_{i\min}$. A single bootstrap analysis is then performed in four steps:

1. Generate a bootstrap sample of $N$ values, denoted as $X_1^0, X_2^0, \ldots, X_N^0$, by randomly reordering the original $N$ values in the series. This is called sampling without replacement;
2. Calculate CUSUM based on the bootstrap sample, denoted as $S_1^0, S_2^0, \ldots, S_N^0$;
3. Calculate $S_{\min}^0$, $S_{\max}^0$ and $S_{\text{diff}}^0$;
4. Determine whether the bootstrap $S_{\text{diff}}^0$ is less than the original $S_{\text{diff}}$.

Bootstrapped samples represent a random reordering of the data that 'mimic' the behaviour of the CUSUM if no change has occurred. With a large number of bootstrap samples, we can estimate by how much $S_{\text{diff}}$ would vary if no change has taken place, *i.e.*, the confidence interval. To determine the confidence level, let $M$ be the number of bootstrap samples performed and $Y$ be the number of bootstraps for which $S^0_{\text{diff}} < S_{\text{diff}}$, then:

$$\text{Confidence Level} = \frac{Y}{M} \tag{8.11}$$

The corresponding exceedance probability of bootstrapped samples $p$-value is computed viz:

$$p\text{-value} = 1 - \text{Confidence Level} \tag{8.12}$$

The $p$-value is then compared against $\alpha = 0.001$, 0.05 and 0.1.

The other estimators of when the change occurred are the mean square error (MSE). If the point $m$ estimated the last point before the change occurred, then the MSE($m$) is defined as (Taylor 2000):

$$\text{MSE}(m) = \sum_{i=1}^{m} (X_i - \overline{X}_1)^2 + \sum_{i=,m+1}^{N} (X_i - \overline{X}_2)^2 \tag{8.13}$$

where

$$\overline{X}_1 = \frac{\sum_{i=1}^{m} X_i}{m} \text{ and } \overline{X}_2 = \frac{\sum_{i=m+1}^{N} X_i}{N - m} \tag{8.14}$$

## 8.3 Results and Discussion

### *8.3.1 Comparison Between Drought Indices*

The comparison between SPEI and only precipitation-based DIs, i.e., SPI, RDDI and RAI, is carried out.

Figures 8.2 and 8.3 show the area plot of monthly SPEI and SPI, respectively. The negative range of values refers to the water deficit periods. Since SPEI and SPI are the standardised indices with mean equal to zero and standard deviation equal to 1, the two are comparable. It is apparent that the SPEI extremes are more compared to SPI extremes, especially distinguishable for location R4 for the Millennium Drought period (1996–2010). Such difference could be explained by the location of R4, which is in an arid to the semi-arid region where water deficit through evapotranspiration is consequential. This difference can also be assessed numerically viz. Pearson correlation values listed in Table 8.2.

The correlation of SPEI with SPI, RDDI and RAI is slightly less in R4, as compared to the other study regions. Additionally, while SPEI has a high correlation

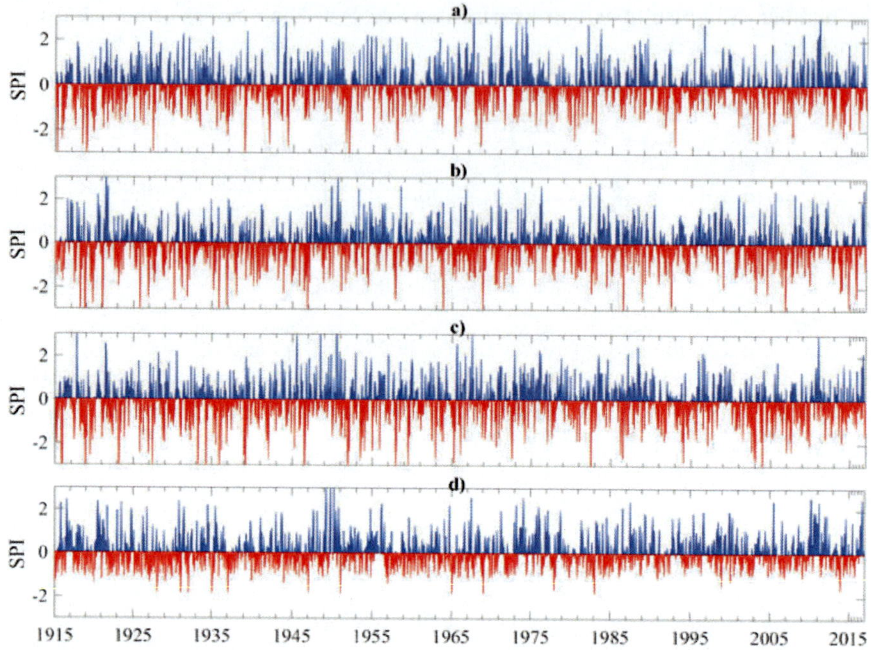


**Fig. 8.2** Area plot of the monthly SPEI from 1915 to 2016 for **a** R1, **b** R2, **c** R3 and **d** R4

with SPI for their similar computational procedure, the former also has a reasonably high correlation with RDDI that is predominantly used by BoM in Australia. This preliminary comparison suggests that SPEI is a high-calibre drought index for characterising drought events.

A comparative study using wavelet analysis on DIs is the first study of its kind for assessment of droughts in Australia. Figure 8.4 shows the wavelet power spectrum for *WRel1*, SPEI, SPI, RDDI and RAI time series for the location R1. The bold contour lines indicate 95% confidence level. The period is measured in months. Clearly, there are common features in the wavelet power of all time series between 128 and 256 months band for the Millennium Drought during mid-1990s to 2010. However, the significance level varies where RAI appears to have the smallest region during this drought period. While there are various smaller bands with significant wavelet spectrum for all time series, the SPEI, however, clearly captures the significant power spectra for the World War II drought between ~32 and 80 months band. Nonetheless, the CWT reveals the common feature between the DIs; however, it is hard to tell if there is merely any confidence. As such, the XWT is used to identify the common features.

Figure 8.5 shows the XWT computed between *WRel1* and DIs. Note that *WRel1* is used for comparison as it is an important indicator of the agricultural drought and is expected to have a high correlation with DIs. It is clearly evident that the DIs have a significant correlation with *WRel1* for all major droughts occurred at

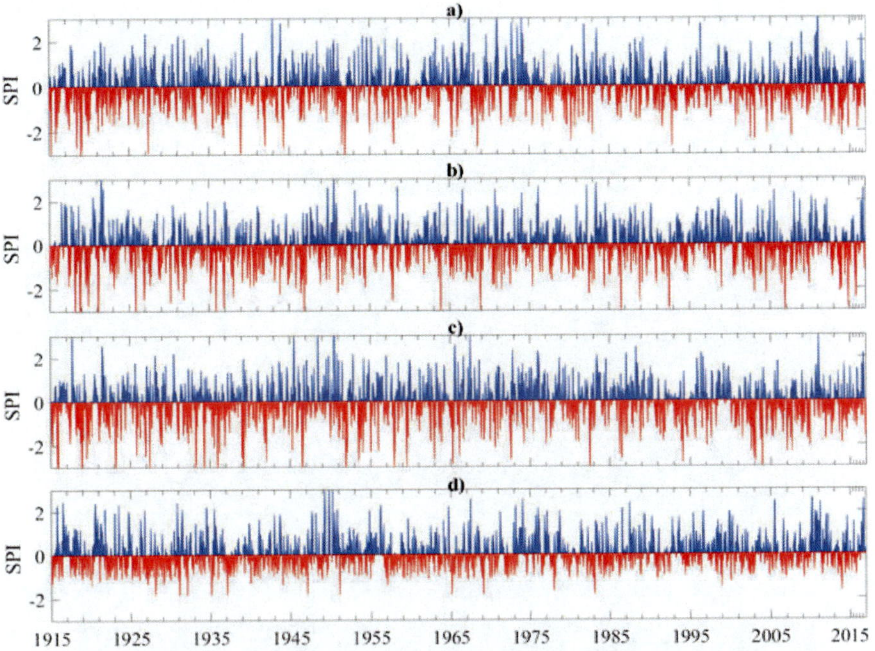


**Fig. 8.3** Area plot of the monthly SPI from 1915 to 2016 for **a** R1, **b** R2, **c** R3 and **d** R4

**Table 8.2** Pearson correlation between drought indices

| Drought indices | Pearson correlation | | | |
|---|---|---|---|---|
| | R1 | R2 | R3 | R4 |
| SPEI versus SPI | 0.9604 | 0.9148 | 0.9367 | 0.8950 |
| SPEI versus RDDI | 0.9569 | 0.9375 | 0.9492 | 0.8778 |
| SPEI versus RAI | 0.8290 | 0.8450 | 0.9109 | 0.7608 |
| SPI versus RDDI | 0.9402 | 0.9400 | 0.9285 | 0.9402 |
| SPI versus RAI | 0.8561 | 0.8366 | 0.8914 | 0.8238 |
| RDDI versus RAI | 0.8104 | 0.8488 | 0.8941 | 0.7323 |

the location R1. However, the significant correlation of *WRel1* with SPEI is more conspicuous for the World War II drought in the band range ~32–80 from the period 1930–1960, compared to that with SPI, RDDI and RAI. The arrows pointing to the east implies both time series are in phase. From the significant power spectrum and phase relationship, particularly for the major drought events, we can speculate that there is a strong link between soil moisture content and DIs, where SPEI stands out to be more prominent. The correlation assessment of SPEI with other DIs has demonstrated that SPEI is in fact very effective at identifying the historical drought events and thus can be advocated for monitoring of drought events in real time.

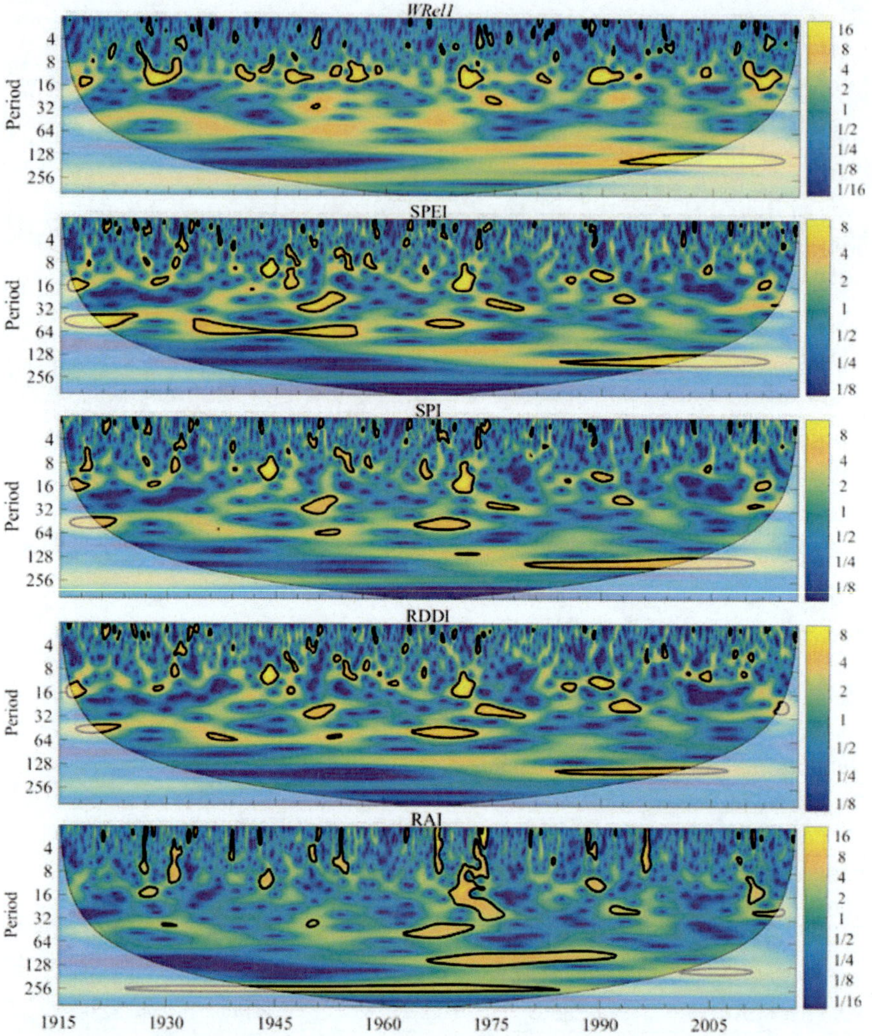


**Fig. 8.4** Continuous wavelet transform (CWT) power spectrum for the time series of upper-layer soil moisture (*WRel1*), SPEI, SPI, RDDI and RAI drought indices

## 8.3.2 Trend Changes in SPEI

In studying drought climatology, it is important to investigate whether there has been any change in the trend of the SPEI, which reveals the periods of water deficiency in its standardised form. A change-point analysis has been carried out on the monthly SPEI time series for the 1915–2016 periods for each study location. Tables 8.3, 8.4 and 8.5 show the results for locations R2, R3 and R4. No significant change has been

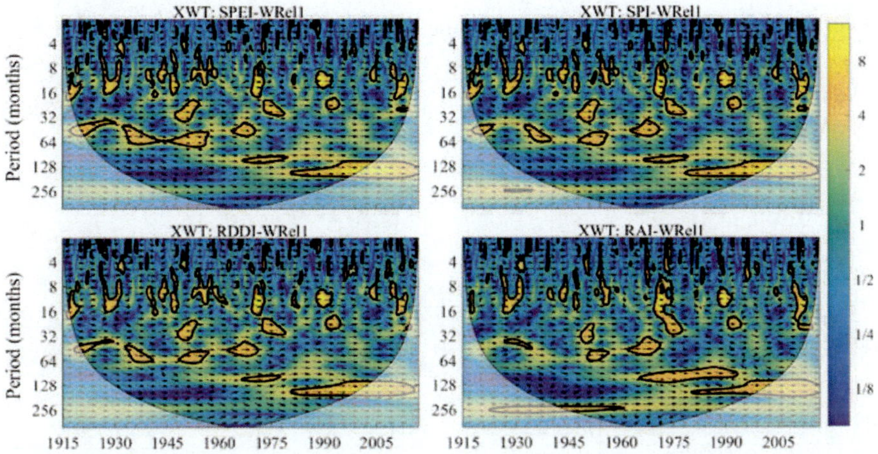


**Fig. 8.5** Cross-wavelet spectrum (XWT) between the upper-layer soil moisture (*WRel1*) and drought indices

**Table 8.3** Table of significant changes in the time series of SPEI for location R2

| Row | Confidence interval | Conf. Level (%) | From | To | Level |
|---|---|---|---|---|---|
| 16 | (11, 185) | 98 | −1.2308 | −0.27877 | 7 |
| 387 | (294, 440) | 100 | −0.27877 | 0.19399 | 1 |
| 476 | (391, 494) | 94 | 0.19399 | 0.66059 | 5 |
| 506 | (494, 553) | 98 | 0.66059 | −0.019886 | 6 |
| 1041 | (958, 1072) | 95 | −0.019886 | −0.5149 | 11 |
| 1116 | (1084, 1129) | 93 | −0.5149 | 0.24262 | 8 |
| 1136 | (1129, 1137) | 98 | 0.24262 | −1.4479 | 6 |
| 1141 | (1141, 1143) | 93 | −1.4479 | 0.62607 | 4 |
| 1172 | (1166, 1187) | 100 | 0.62607 | 0.39774 | 3 |

Confidence level for candidate changes = 50%, confidence level for inclusion in table = 90%, confidence interval = 95%, bootstraps = 1000, without replacement, MSE estimates

**Table 8.4** Table of significant changes in the time series of the SPEI for location R3

| Row | Confidence interval | Conf. Level (%) | From | To | Level |
|---|---|---|---|---|---|
| 916 | (829, 917) | 96 | −0.023271 | −1.2465 | 2 |
| 925 | (923, 956) | 98 | −1.2465 | −0.46026 | 3 |
| 970 | (940, 1001) | 98 | −0.46026 | 0.033142 | 2 |
| 1003 | (983, 1010) | 96 | 0.033142 | 0.75432 | 4 |
| 1023 | (1021, 1032) | 100 | 0.75432 | −0.48461 | 3 |
| 1141 | (1082, 1209) | 96 | −0.48461 | 0.059474 | 5 |

Confidence level for candidate changes = 50%, confidence level for inclusion in table = 90%, confidence interval = 95%, bootstraps = 1000, without replacement, MSE estimates

observed in the location R1. For the discussion purpose, only those change points are considered that has a confidence level of 100%. At the location R2, of the nine changes, there are two periods when a change has occurred with 100% confidence. Those are 387th and 1172nd month in the time series, corresponding to March 1947 and August 2012. At 95% confidence interval, the first change with 100% level of confidence occurred between June 1939 and August 1951, while the second change occurred between February 2012 and November 2013. The fact that the confidence interval for the first change is wider indicates that the time for this change cannot be as accurately pinpointed compared to the second change. In the table, we can also see the average SPEI values prior to and after the change has occurred, i.e., prior to the March 1947 change the average SPEI value was −0.27877 while after the change the SPEI was 0.19399. Similarly, the average SPEI values for the August 2012 change are 0.62607 and −0.39774. The table also gives a level associated with each change that gives an indication of the importance of the change. For instance, the level 1 change denotes first change detected while level 2 changes are detected on a second pass through the data, and so on.

The question as to whether all nine changes are significant or not is cross-checked using 1000 bootstraps without replacement and MSE. To graphically illustrate the results, Figure 8.6 shows the CUSUM chart with significant changes shown in the background for the location R2. The number of significant changes is indicated by the number of times the background colour has changed. It appears that the significant changes in the SPEI have occurred nine times at location R2, concurring with Table 8.3.

Similarly, the change with 100% level of confidence occurred at the 1023rd month (i.e., March 2000) as given in Table 8.4 for the location R3. There are six periods in total where significant change has occurred. The graphical representation of the change in terms of CUSUM is shown in Fig. 8.7. As opposed to the locations R1,

**Table 8.5** Table of significant changes in the time series of the SPEI for location R4

| Row | Confidence interval | Conf. Level (%) | From | To | Level |
|---|---|---|---|---|---|
| 19 | (15, 20) | 100 | −0.63934 | 1.2351 | 9 |
| 28 | (26, 40) | 93 | 1.2351 | 0.24884 | 10 |
| 47 | (37, 52) | 98 | 0.24884 | −0.58503 | 8 |
| 67 | (65, 68) | 100 | −0.58503 | 1.1087 | 11 |
| 84 | (82, 90) | 100 | 1.1087 | −0.23955 | 6 |
| 164 | (85, 172) | 94 | −0.23955 | −0.74341 | 14 |
| 182 | (178, 200) | 90 | −0.74341 | 0.17805 | 7 |
| 263 | (210, 322) | 98 | 0.17805 | 0.22256 | 10 |
| 409 | (396, 421) | 100 | 0.22256 | 0.79671 | 9 |
| 434 | (415, 443) | 96 | 0.79671 | −0.017313 | 10 |
| 479 | (465, 487) | 99 | −0.017313 | 0.77868 | 11 |
| 503 | (499, 514) | 98 | 0.77868 | −0.28625 | 8 |
| 699 | (682, 713) | 100 | −0.28625 | 0.5016 | 9 |
| 772 | (762, 774) | 100 | 0.5016 | −0.6594 | 12 |
| 794 | (787, 805) | 94 | −0.6594 | 0.058864 | 15 |
| 809 | (799, 811) | 94 | 0.058864 | −0.96257 | 14 |
| 820 | (818, 867) | 91 | −0.96257 | −0.067982 | 13 |
| 893 | (865, 893) | 91 | −0.067982 | 1.5758 | 15 |
| 896 | (896, 896) | 96 | 1.5758 | −0.024475 | 19 |
| 904 | (903, 904) | 99 | −0.024475 | 1.1383 | 18 |
| 909 | (909, 912) | 95 | 1.1383 | −0.63758 | 15 |

(continued)

**Table 8.5** (continued)

| Row | Confidence interval | Conf. Level (%) | From | To | Level |
|---|---|---|---|---|---|
| 934 | (925, 962) | 96 | −0.63758 | 0.14244 | 18 |
| 1035 | (1008, 1058) | 100 | 0.14244 | −0.4749 | 4 |
| 1142 | (1139, 1144) | 100 | −0.4749 | 1.4303 | 5 |
| 1157 | (1155, 1162) | 99 | 1.4303 | −0.37256 | 7 |
| 1219 | (1205, 1223) | 100 | −0.37256 | 1.0161 | 2 |

Confidence level for candidate changes = 50%, confidence level for inclusion in table = 90%, confidence interval = 95%, bootstraps = 1000, without replacement, MSE estimates

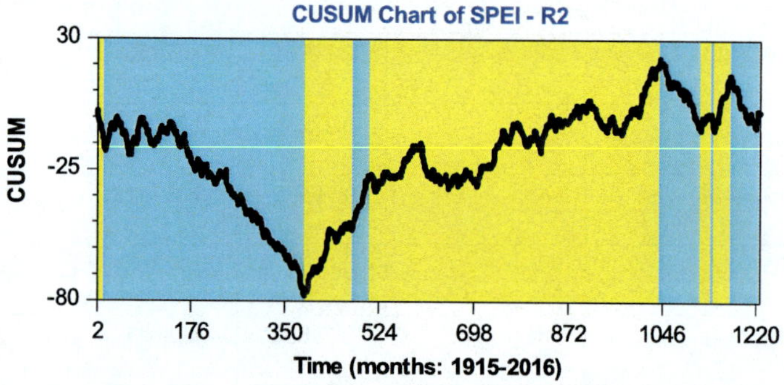


**Fig. 8.6** CUSUM chart of the SPEI data for location R2 with significant changes shown in the background

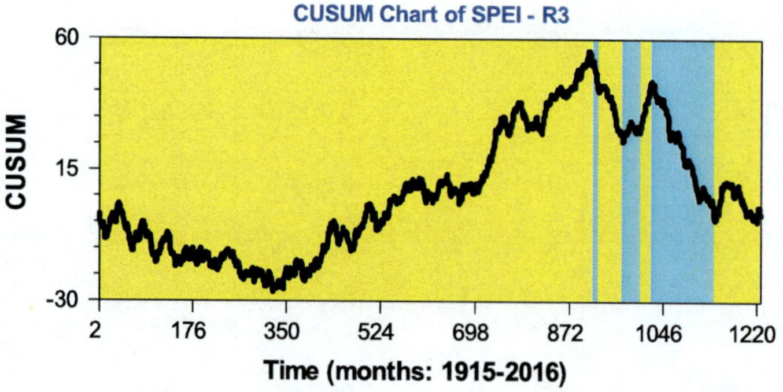


**Fig. 8.7** CUSUM chart of the SPEI data for location R3 with significant changes shown in the background

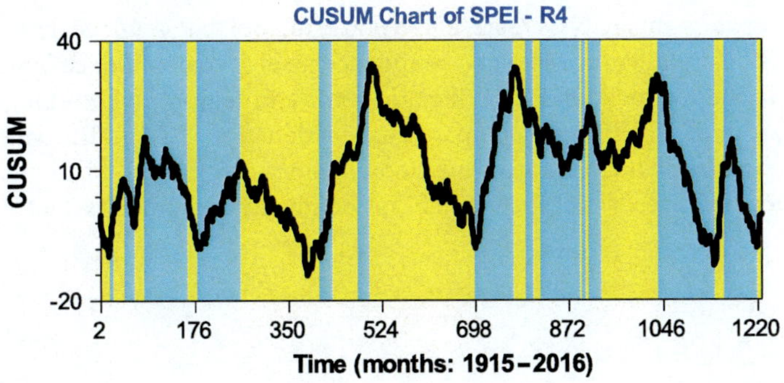


**Fig. 8.8** CUSUM chart of SPEI data for location R4 with significant changes shown in the background

R2 and R3, the location R4 has twenty-six periods where a significant change in the SPEI trend has occurred, see Table 8.5 and corresponding CUSUM in Fig. 8.8. Of the twenty-six, nine changes have occurred with 100% confidence. These points are (numerically in month/year format) 07/1916, 07/1920, 12/1921, 01/1949, 03/1973, 04/1979, 03/2001, 02/2010 and 07/2016. Most of these changes occurred when the phase of ENSO changed as well. The significant difference at R4, compared to the locations R1, R2 and R3, could be due to its location in the arid/semi-arid region where meteorological parameters are highly variable.

## 8.4 Conclusion

The main conclusions, promoting the use of intelligent data analytic methods, drawn from this chapter are as follows.

1. The SPEI is able to identify extreme droughts better than the SPI;
2. The SPEI highly correlates with precipitation-based DIs, especially with SPI and RDDI but can additionally provide complementary information about hydrological effects of drought;
3. Illustrated by the wavelet analysis, the SPEI concurs with all major drought events to a greater extent, significant at 95% confidence interval, compared to SPI, RDDI and RAI;
4. The DI cross-wavelet analysis with the upper-layer soil moisture (i.e., *WRel1*) indicated SPEI to be in phase more significantly at 95% confidence interval compared to the SPI, RDDI and RAI.

5. The change-point analysis represents a powerful tool that is able to detect changes in the SPEI trend with associated confidence levels and confidence intervals. The study found the location R4 (in the arid/semi-arid region) to have undergone 26 changes in the SPEI trend compared to locations R1, R2 and R3 with 0, 9 and 6, respectively. The location of the study matter where inland from the coastline experiences more variability in the environmental parameters that define the SPEI.

## References


Allen RG, Pereira LS, Raes D, Smith M (1998) Crop evapotranspiration-Guidelines for computing crop water requirements-FAO Irrigation and drainage paper 56. FAO Rome 300(9):D05109

Barbeta A, Ogaya R, Peñuelas J (2013) Dampening effects of long-term experimental drought on growth and mortality rates of a Holm oak forest. Glob Change Biol 19(10):3133–3144

Cavin L, Mountford EP, Peterken GF, Jump AS (2013) Extreme drought alters competitive dominance within and between tree species in a mixed forest stand. Funct Ecol 27(6):1424–1435

Chang T-P, Liu F-J, Ko H-H, Huang M-C (2017) Oscillation characteristic study of wind speed, global solar radiation and air temperature using wavelet analysis. Appl Energy 190:650–657

Das PK, Dutta D, Sharma J, Dadhwal V (2016) Trends and behaviour of meteorological drought (1901–2008) over Indian region using standardized precipitation–evapotranspiration index. Int J Climatol 36(2):909–916

Dayal K, Deo R, Apan A (2018) Investigating drought duration-severity-intensity characteristics using the Standardised Precipitation-Evapotranspiration Index: case studies in drought-prone southeast Queensland. J Hydrol Eng 23(1)

Deo RC, Syktus J, McAlpine C, Lawrence P, McGowan H, Phinn SR (2009) Impact of historical land cover change on daily indices of climate extremes including droughts in eastern Australia. Geophys Res Lett 36(8)

Deo RC, Şahin M (2015) Application of the extreme learning machine algorithm for the prediction of monthly Effective Drought Index in eastern Australia. Atmos Res 153:512–525

Fuchs B, Svoboda M, Nothwehr J, Poulsen C, Sorensen W, Guttman N (2012) A new national drought risk Atlas for the US from the National Drought Mitigation Center

Gibbs WJ, Maher JV (1967) Rainfall deciles as drought indicators. Bureau Meteorol 48

Hanson RL (1988) Evapotranspiration and droughts. In: Paulson RW, Chase EB, Roberts RS, Moody DW (eds) Compilers, national water summary, pp 99–104

Hobbins M, Wood A, Streubel D, Werner K (2012) What drives the variability of evaporative demand across the conterminous United States? J Hydrometeorol 13(4):1195–1214

Grinsted A, Moore JC, Jevrejeva S (2004) Application of the cross wavelet transform and wavelet coherence to geophysical time series. Nonlinear Process Geophys 11(5/6):561–566

Li W, Hou M, Chen H, Chen X (2012) Study on drought trend in south China based on standardized precipitation evapotranspiration index. J Nat Disasters 21:84–90

Keyantash J, Dracup JA (2002) The quantification of drought: an evaluation of drought indices. Bull Am Meteor Soc 83(8):1167–1180

Martin-Benito D, Beeckman H, Canellas I (2013) Influence of drought on tree rings and tracheid features of Pinus nigra and Pinus sylvestris in a mesic Mediterranean forest. Eur J Forest Res 132(1):33–45

McKee TB, Doesken NJ, Kleist J (1993) The relationship of drought frequency and duration to time scales. In: Proceedings of the 8th conference on applied climatology, american meteorological society Boston, MA, pp 179–183

Paulo A, Rosa R, Pereira L (2012) Climate trends and behaviour of drought indices based on precipitation and evapotranspiration in Portugal. Nat Hazards Earth Syst Sci 12:1481–1491
Potop V (2011) Evolution of drought severity and its impact on corn in the Republic of Moldova. Theoret Appl Climatol 105(3–4):469–483
Raupach M, Briggs P, Haverd V, King E, Paget M, Trudinger C (2009) Australian water availability project (AWAP): CSIRO marine and atmospheric research component: final report for phase 3. In: Centre for Australian weather and climate research (bureau of meteorology and CSIRO). Melbourne, Australia
Raupach M, Briggs P, Haverd V, King E, Paget M, Trudinger C (2012) Australian Water Availability Project. CSIRO Marine and Atmospheric Research. Canberra, Australia
Roderick ML, Rotstayn LD, FarquharGD, Hobbins MT (2007) On the attribution of changing pan evaporation. Geophys Res Lett 34(17)
Sheffield J, Wood EF, Roderick ML (2012) Little change in global drought over the past 60 years. Nature 491(7424):435–438
TaylorWA (2000) Change-point analysis: a powerful new tool for detecting changes
Torrence C, Compo GP (1998) A practical guide to wavelet analysis. Bull Am Meteor Soc 79(1):61–78
Toromani E, Sanxhaku M, Pasho E (2011) Growth responses to climate and drought in silver fir (Abies alba) along an altitudinal gradient in southern Kosovo. Can J For Res 41(9):1795–1807
Van Rooy M (1965) A rainfall anomaly index independent of time and space. Notos 14:43–48
Vicente-Serrano SM, Beguería S, López-Moreno JI (2010) A multiscalar drought index sensitive to global warming: the standardized precipitation evapotranspiration index. J Climatol 23(7):1696–1718
Vicente-Serrano SM, Gouveia C, Camarero JJ, Beguería S, Trigo R, López-Moreno JI, Azorín-Molina C, Pasho E, Lorenzo-Lacruz J, Revuelto J (2013) Response of vegetation to drought time-scales across global land biomes. Proc Natl Acad Sci 110(1):52–57
Zajaczkowski J, Wong K, Carter J (2013) Improved historical solar radiation gridded data for Australia. Environ Model Softw 49:64–77

# Chapter 9
# Conjunction Model Design for Intermittent Streamflow Forecasts: Extreme Learning Machine with Discrete Wavelet Transform

**Ozgur Kisi, Meysam Alizamir, and Jalal Shiri**

## 9.1 Introductory Note

Correct estimation and prediction of streamflow are essentially crucial in hydrology and water resource management due to its direct effect on a dam's performance, groundwater level fluctuations, scoring and sedimentation of rivers, water catchment management, etc. Precipitation, evapotranspiration, groundwater table level and soil moisture content might be considered as the most influential parameters of streamflow magnitude (Kisi 2008). Given that the deterministic models need lots of parameters to predict streamflow values, employing auto-regressive techniques that use previously recorded streamflow values for prediction issues might be considered as suitable alternatives for deterministic models. Smith et al. (1998) adopted a discrete wavelet transform to quantify the streamflow variations. Coulibaly and Burn (2004) applied wavelet transform to analyze the variability of stream lows in annual time scales. Zhou et al. (2008) suggested a wavelet-based model of river flow prediction in monthly time window. Shiri and Kisi (2010) introduced a wavelet-neuro-fuzzy approach for forecasting short-term and long-term streamflow values. Shiri et al. (2012) compared different heuristic data-driven approaches for predicting daily streamflows. Pandhiani and Shabri (2013) applied wavelet-support vector machines and wavelet-regression models to predict monthly streamflow values. Karimi et al. (2016) established a

O. Kisi
Department of Civil Engineering, Ilia State University, 0162 Tbilisi, Georgia
e-mail: ozgur.kisi@iliauni.edu.ge

M. Alizamir (✉)
Department of Civil Engineering, Hamedan Branch, Islamic Azad University, Hamedan, Iran
e-mail: meysamalizamir@gmail.com

J. Shiri
Water Engineering Department, Faculty of Agriculture, University of Tabriz, Tabriz, Iran
e-mail: j_shiri2005@yahoo.com

© Springer Nature Singapore Pte Ltd. 2021

R. C. Deo et al. (eds.), *Intelligent Data Analytics for Decision-Support Systems in Hazard Mitigation*, Springer Transactions in Civil and Environmental Engineering,
https://doi.org/10.1007/978-981-15-5772-9_9

wavelet-gene expression programming model for predicting streamflows. Karimi et al. (2018) applied genetic programming and support vector machine techniques with different data management scenarios for predicting streamflows in successive hydrometric stations. The present chapter aims at predicting intermittent streamflow values employing extreme learning machine (ELM) algorithm coupled with discrete wavelet transform (DWT) as a data pre-processing technique.

## 9.2 Materials and Method

### 9.2.1 *Case Study Region*

The study uses daily intermittent streamflow time series data from Uzunkopru (105) and Babaeski (101) stations situated on Ergene River and Seytan Streams, respectively (Kisi 2009). The stations can be observed from Fig. 9.1.

The drainage areas of the stations are 10,195 $km^2$ for Uzunkopru and 478 $km^2$ for Babaeski. The measured data cover 15 years (5475 values) long with the period of 1980–1994 for Uzunkopru and 35 years (12,775 values) long with the period of 1957–1992 for Babaeski. For both stations, 80% of the entire data was used in the

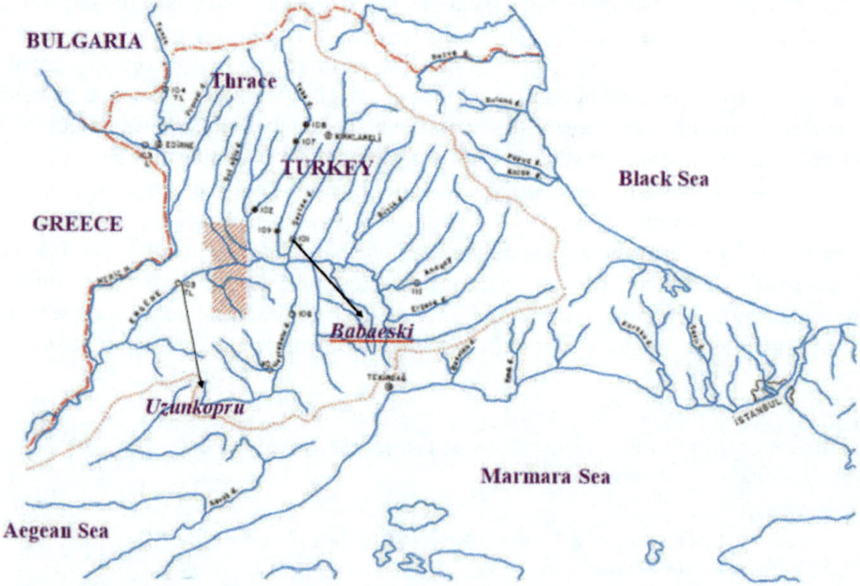


**Fig. 9.1** Uzunkopru (105) and Babaeski (101) stations on Ergene River and Seytan Streams, respectively (Kisi 2009)

training stage of the employed methods and remaining 20% was used in testing stage. The brief statistical properties of the used data can be reached from Kisi (2009).

### 9.2.2 Extreme Learning Machine

ELM has been introduced by Huang et al. (2004) for the single layer feed-forward neural networks (SLFN) (2006). ELM selects the input weights of SLFN randomly, but identifies the output weights analytically. This algorithm does not include too much human intervention since it determines all the network parameters analytically, and can be established much faster than the traditional algorithms (Alizamir et al. 2017; Kisi and Alizamir 2018).

### 9.2.3 Principles of ELM

**Theorem 1** (Liang et al. 2006) *Let an SLFN with L additive hidden nodes and an activation function g(x) that is boundlessly differentiable at any interim of R. For a capricious L distinct input vectors* $\{x_i | x_i \in R^n, i = 1, \ldots, L\}$ *and* $\{(a_i, b_i)\}_{i=1}^{L}$ *randomly produced by any continuous probability distribution, respectively, the hidden layer output matrix is invertible with probability one, the hidden layer output matrix H of the SLFN is invertible and* $\|H\beta - T\| = 0$.

**Theorem 2** (Liang et al. 2006) *Given any small positive value of* $\varepsilon > 0$ *and activation function of* $g(x)$: $R \to R$ *that is boundlessly differentiable at any interim, there exists* $L \leq N$ *so that for N capricious distinct input vectors* $\{x_i | x_i \in R^n, i = 1, \ldots, L\}$ *for any* $\{(a_i, b_i)\}_{i=1}^{L}$ *randomly produced based upon any continuous probability distribution* $\|H_{N \times L} \beta_{L \times m} - T_{N \times m}\| < \varepsilon$ *with probability one.*

### 9.2.4 Discrete Wavelet Transform (DWT)

Wavelet function $\psi(t)$ that is called as the mother wavelet could be presented as $\int_{-\infty}^{+\infty} \psi(t)dt = 0$. $\psi_{a,b}(t)$ might be determined by compressing and expanding $\psi(t)$:

$$\psi_{a,b}(t) = |a|^{-1/2} \psi\left(\frac{t-b}{a}\right) \quad b \in R, a \in R, a \neq 0 \tag{9.1}$$

where $\psi_{a,b}(t)$ is called as the successive wavelet, $a$ and $b$ show the scale and time factors, respectively; and $R$ stands for the real numbers domain.

If $\psi_{a,b}(t)$ satisfies Eq. (9.1), for the time series $f(t) \in L^2(R)$, successive wavelet transform of $f(t)$ is given as:

$$W_\psi f(a,b) = |a|^{-1/2} \int_R f(t)\, \bar{\psi}\left(\frac{t-b}{a}\right) dt \quad (9.2)$$

Here, $\bar{\psi}(t)$ presents the complex conjugate functions of $\psi(t)$. From Eq. (9.2), it is clear seen that the wavelet transform is the decomposition of $f(t)$ under different resolution scales.

The successive wavelet is frequently discrete in real applications. Let us assume $a = a_0^j$, $b = kb_0a_0^j$, $a_0 > 1$, $b_0 \in$ R, $k$, $j$ are integer numbers. Discrete wavelet transform (DWT) of $f(t)$ reads:

$$W_\psi f(j,k) = a_0^{-j/2} \int_R f(t)\, \overline{\psi}\left(a_0^{-j} t - kb_0\right) dt \quad (9.3)$$

The most common (and simplest) selection for the $a_0$ and $b_0$ parameters would be 2 and 1 time steps, respectively (Mallat 1989), so Eq. (9.3) is re-written as:

$$W_\psi f(j,k) = 2^{-j/2} \int_R f(t)\, \overline{\psi}\left(2^{-j} t - k\right) dt \quad (9.4)$$

The characteristics of the original time series in frequency ($a$ or $j$) and time domain ($b$ or $k$) at the same time are presented by $W_\psi f(a,b)$ or $W_\psi f(j,k)$.

For a discrete time series $f(t)$, where occurs at different time t (i.e., here integer time steps are used), the DWT would be considered as

$$W_\psi f(j,k) = 2^{-j/2} \sum_{t=0}^{N-1} f(t)\, \overline{\psi}\left(2^{-j} t - k\right) \quad (9.5)$$

where $W_\psi f(j,k)$ denotes the wavelet coefficient for the discrete wavelet of scale $a = 2^j$, $b = 2^j k$.

DWT operates two sets of functions, e.g., high-pass and low-pass filters. The original time series are passed through these filters and separated at different levels. The time series is decomposed into one comprising its trend (the approximation) and one comprising the high frequencies and the fast events (the detail) (Shiri and Kişi 2010; Mallat 1989). In this research, the detail coefficients and approximation (A) sub-time series were computed through Eq. (9.5).

## 9.3 Results and Discussion

Training and test results of single ANN, ELM and wavelet conjunction models, WANN and WELM, are provided in forecasting monthly intermittent streamflow of Babaeski station in Table 9.1.

**Table 9.1** Training and test results of single and wavelet conjunction models in forecasting monthly intermittent streamflow of Babaeski station

| Model | Methods | Training | | | | Testing | | | |
|---|---|---|---|---|---|---|---|---|---|
| | | RMSE | NS | $r$ | $R^2$ | RMSE | NS | $r$ | $R^2$ |
| $Q_{t-1}$ | WELM-haar | 3.7684 | 0.5761 | 0.76 | 0.5781 | 2.1198 | 0.5115 | 0.724 | 0.5245 |
| | **WELM-db3** | **2.7917** | **0.7673** | **0.876** | **0.7687** | **1.7051** | **0.684** | **0.827** | **0.6842** |
| | ELM | 3.897 | 0.5466 | 0.74 | 0.548 | 2.1569 | 0.4943 | 0.715 | 0.5118 |
| | WANN-haar | 3.8155 | 0.5654 | 0.752 | 0.5661 | 2.1636 | 0.4911 | 0.706 | 0.4995 |
| | WANN-db3 | 3.2253 | 0.6895 | 0.832 | 0.6925 | 1.8138 | 0.6424 | 0.801 | 0.6425 |
| | ANN | 3.8031 | 0.5682 | 0.754 | 0.5694 | 2.2026 | 0.4727 | 0.696 | 0.4855 |
| $Q_{t-1}$, $Q_{t-2}$ | WELM-haar | 2.7717 | 0.7706 | 0.878 | 0.7717 | 1.6946 | 0.6878 | 0.832 | 0.6937 |
| | **WELM-db3** | **2.359** | **0.8339** | **0.913** | **0.8343** | **1.3199** | **0.8106** | **0.9** | **0.8117** |
| | ELM | 3.48 | 0.6384 | 0.799 | 0.6385 | 1.8866 | 0.613 | 0.797 | 0.6366 |
| | WANN-haar | 3.0572 | 0.721 | 0.85 | 0.7232 | 1.7486 | 0.6676 | 0.819 | 0.671 |
| | WANN-db3 | 2.7406 | 0.7758 | 0.886 | 0.786 | 1.4724 | 0.7643 | 0.894 | 0.7994 |
| | ANN | 3.5264 | 0.6287 | 0.793 | 0.6293 | 1.8882 | 0.6123 | 0.794 | 0.6315 |
| $Q_{t-1}$, $Q_{t-2}$, $Q_{t-3}$ | WELM-haar | 1.9561 | 0.8858 | 0.941 | 0.8864 | 1.1502 | 0.8562 | 0.926 | 0.8593 |
| | **WELM-db3** | **1.4605** | **0.9363** | **0.968** | **0.9377** | **0.7908** | **0.932** | **0.966** | **0.9336** |
| | ELM | 3.4639 | 0.6417 | 0.801 | 0.6417 | 1.8829 | 0.6145 | 0.794 | 0.6319 |
| | WANN-haar | 2.5034 | 0.8129 | 0.903 | 0.8166 | 1.334 | 0.8066 | 0.903 | 0.8157 |
| | WANN-db3 | 1.9465 | 0.8869 | 0.942 | 0.8874 | 0.8512 | 0.9212 | 0.959 | 0.9215 |
| | ANN | 3.5094 | 0.6322 | 0.795 | 0.6329 | 1.9184 | 0.5998 | 0.784 | 0.6149 |
| $Q_{t-1}$, $Q_{t-2}$, $Q_{t-3}$, $Q_{t-4}$ | WELM-haar | 1.0291 | 0.9684 | 0.984 | 0.9694 | 0.5456 | 0.9676 | 0.984 | 0.9693 |
| | **WELM-db3** | **1.1971** | **0.9572** | **0.979** | **0.9592** | **0.4425** | **0.9787** | **0.991** | **0.9834** |
| | ELM | 3.4152 | 0.6517 | 0.807 | 0.6518 | 1.826 | 0.6374 | 0.806 | 0.6507 |
| | WANN-haar | 1.4536 | 0.9369 | 0.98 | 0.9606 | 1.0073 | 0.8897 | 0.978 | 0.9565 |
| | WANN-db3 | 1.2306 | 0.9548 | 0.977 | 0.9563 | 0.4712 | 0.9759 | 0.988 | 0.978 |
| | ANN | 3.5193 | 0.6302 | 0.793 | 0.6304 | 1.9251 | 0.5971 | 0.787 | 0.62 |
| $Q_{t-1}$, $Q_{t-2}$, $Q_{t-3}$, $Q_{t-4}$, $Q_{t-5}$ | WELM-haar | 0.9918 | 0.9706 | 0.985 | 0.972 | 0.4366 | 0.9793 | 0.989 | 0.9794 |
| | **WELM-db3** | **0.5897** | **0.9896** | **0.994** | **0.9898** | **0.355** | **0.9863** | **0.993** | **0.9863** |
| | ELM | 2.4797 | 0.8164 | 0.907 | 0.8228 | 0.9065 | 0.9107 | 0.954 | 0.912 |
| | WANN-haar | 1.8922 | 0.8931 | 0.946 | 0.895 | 0.8802 | 0.9158 | 0.963 | 0.9278 |
| | WANN-db3 | 0.7751 | 0.9821 | 0.992 | 0.9841 | 0.375 | 0.9847 | 0.992 | 0.9855 |
| | ANN | 2.6595 | 0.7889 | 0.893 | 0.7977 | 1.1289 | 0.8615 | 0.939 | 0.8824 |

It is clear from the table that two different wavelet functions, db3 and haar, were utilized for the WANN and WELM models. Five input combinations from $Q_{t-1}$, to $Q_{t-1}$, $Q_{t-2}$, $Q_{t-3}$, $Q_{t-4}$, $Q_{t-5}$ were tried. Table clearly shows that the wavelet-based models outperform the single ones for all input cases. For example, the RMSE accuracy of the WELM-db3 (WANN-db3) ranges from 1.7051 $m^3$/s (1.8138 $m^3$/s) to 0.355 $m^3$/s (0.375 $m^3$/s) while the ELM and ANN have the ranges of 2.1569 $m^3$/s – 0.9065 $m^3$/s and 2.2026 – 1.1289 $m^3$/s, respectively.

Increasing input lags positively affects wavelet conjunction models; decrease in RMSE of WELM-db3 from 1.7051 to 0.355 $m^3$/s (input $Q_{t-1}$, $Q_{t-2}$, $Q_{t-3}$, $Q_{t-4}$, $Q_{t-5}$). ELM and WELM models perform superior to the ANN and WANN models, respectively. For example, for the best input case (the last input combination with 5 discharge lags), the ELM has lower RMSE (0.9065 $m^3$/s) than the ANN model. The improvements of the single ELM (ANN) models obtained applying wavelet are by 61% (69%) for the last input combination.

Table 9.2 sums up the forecasting results of single and wavelet models for intermittent streamflow of Uzunkopru station.

In this station, considerable improvement is observed by employing wavelet decomposition method. The RMSEs of the WELM-db3 (WANN-db3) have the range 10.1167 – 5.6462 $m^3$/s (11.3567 – 6.5836 $m^3$/s) while the ELM (ANN) vary from 12.9152 $m^3$/s (21.4387 $m^3$/s) to 9.0572 $m^3$/s (13.7085 $m^3$/s) corresponding to inputs from $Q_{t-1}$, to $Q_{t-1}$, $Q_{t-2}$, $Q_{t-3}$, $Q_{t-4}$, $Q_{t-5}$, respectively.

By increasing input lags, considerable improvements are observed for the wavelet conjunction models; an increase in the accuracy of the WELM-db3 with respect to RMSE is by 44% (RMSE decreased from 10.1167 $m^3$/s to 5.6462 $m^3$/s). ELM model performs superior to the ANN model in all input combinations. The relative RMSE differences between ELM and ANN are 40, 31, 32, 33 and 34% for the time lags from 1 to 5 in the test stage, respectively. Similarly, the corresponding differences between the WELM-db3 and WANN-db3, respectively, are 6, 10, 7, 6 and 5%. There is a high difference between these two methods compared to previous station (Babaeski), especially for the single models.

The main reason of this might be the difference behavior and distribution of the discharge data of two stations. The Uzunkopru station is located on the main river Ergene, and the Babaeski is on a tributary upstream of the Ergene. The WELM-db3 (WANN-db3) model improved the test accuracy of the ELM (ANN) with respect to RMSE by 61% (67%) for the last input combination.

Training and test predictions of the applied models are compared in Fig. 9.2 for the Babaeski station.

It is clear from the scatterplots, wavelet conjunction models (WELM and WANN) have less scattered discharges forecasts compared to single models (ELM and ANN). The superior accuracy of the ELM (WELM) models over ANN (WANN) models is clearly observed from the graphs.

Figure 9.3 illustrates the observed and predicted streamflow by the WELM, WANN, ELM and ANN models for the Uzunkopru station. Here, also the similar

**Table 9.2** Training and test results of the single and wavelet conjunction models in forecasting monthly intermittent streamflow of Uzunkopru station

| Model | Methods | Training | | | | Testing | | | |
|---|---|---|---|---|---|---|---|---|---|
| | | RMSE | NS | $r$ | $R^2$ | RMSE | NS | $r$ | $R^2$ |
| $Q_{t-1}$ | WELM-haar | 39.4792 | 0.6686 | 0.828 | 0.6856 | 11.3739 | 0.2472 | 0.784 | 0.6155 |
| | **WELM-db3** | **44.8303** | **0.573** | **0.764** | **0.5844** | **10.1167** | **0.4047** | **0.706** | **0.4998** |
| | ELM | 46.1001 | 0.5481 | 0.766 | 0.5871 | 12.9152 | 0.0294 | 0.733 | 0.5376 |
| | WANN-haar | 46.7843 | 0.5349 | 0.736 | 0.543 | 18.8674 | −1.0707 | 0.724 | 0.5255 |
| | WANN-db3 | 47.2234 | 0.5261 | 0.732 | 0.5368 | 11.3567 | 0.2498 | 0.639 | 0.4086 |
| | ANN | 68.5808 | 0.1182 | 0.363 | 0.1323 | 21.4387 | −1.6314 | 0.247 | 0.0615 |
| $Q_{t-1}, Q_{t-2}$ | WELM-haar | 41.6395 | 0.6316 | 0.811 | 0.6584 | 11.2641 | 0.262 | 0.829 | 0.6873 |
| | **WELM-db3** | **36.51** | **0.7168** | **0.876** | **0.7685** | **9.6163** | **0.4621** | **0.834** | **0.6967** |
| | ELM | 45.3321 | 0.5631 | 0.778 | 0.6068 | 12.501 | 0.0906 | 0.785 | 0.6168 |
| | WANN-haar | 41.2276 | 0.6388 | 0.8141 | 0.6628 | 17.8266 | −0.8485 | 0.64 | 0.4104 |
| | WANN-db3 | 38.0351 | 0.6926 | 0.859 | 0.7386 | 12.2877 | 0.1217 | 0.785 | 0.6173 |
| | ANN | 66.1208 | 0.0705 | 0.361 | 0.1307 | 18.209 | −0.9294 | 0.509 | 0.2591 |
| $Q_{t-1}, Q_{t-2}, Q_{t-3}$ | WELM-haar | 36.1391 | 0.7225 | 0.876 | 0.7678 | 10.8911 | 0.31 | 0.832 | 0.6923 |
| | **WELM-db3** | **31.735** | **0.786** | **0.906** | **0.8224** | **7.327** | **0.6877** | **0.873** | **0.7637** |
| | ELM | 35.5059 | 0.732 | 0.86 | 0.7412 | 11.0745 | 0.2863 | 0.829 | 0.688 |
| | WANN-haar | 40.1672 | 0.6572 | 0.838 | 0.7026 | 16.3089 | −0.5472 | 0.743 | 0.5524 |
| | WANN-db3 | 37.1945 | 0.706 | 0.86 | 0.74 | 13.5236 | −0.0638 | 0.8 | 0.6406 |
| | ANN | 52.1842 | 0.421 | 0.656 | 0.4305 | 16.3645 | −0.5583 | 0.741 | 0.55 |
| $Q_{t-1}, Q_{t-2}, Q_{t-3}, Q_{t-4}$ | WELM-haar | 32.2541 | 0.7789 | 0.906 | 0.8212 | 9.2912 | 0.4979 | 0.843 | 0.7118 |
| | **WELM-db3** | **29.3982** | **0.8162** | **0.903** | **0.8162** | **5.9478** | **0.7941** | **0.912** | **0.833** |

(continued)

**Table 9.2** (continued)

| Model | Methods | Training | | | | Testing | | | |
|---|---|---|---|---|---|---|---|---|---|
| | | RMSE | NS | $r$ | $R^2$ | RMSE | NS | $r$ | $R^2$ |
| | ELM | 38.2753 | 0.6885 | 0.839 | 0.7055 | 10.0042 | 0.4176 | 0.816 | 0.6668 |
| | WANN-haar | 42.7786 | 0.6109 | 0.809 | 0.6556 | 11.4664 | 0.2349 | 0.778 | 0.6067 |
| | WANN-db3 | 33.9841 | 0.7546 | 0.891 | 0.794 | 6.9081 | 0.7224 | 0.909 | 0.8266 |
| | ANN | 39.7959 | 0.6633 | 0.829 | 0.6875 | 14.9361 | −0.2981 | 0.762 | 0.5813 |
| $Q_{t-1}, Q_{t-2}, Q_{t-3}, Q_{t-4}, Q_{t-5}$ | WELM-haar | 35.4938 | 0.7324 | 0.883 | 0.7802 | 8.0199 | 0.6258 | 0.871 | 0.7588 |
| | **WELM-db3** | **28.5088** | **0.8272** | **0.909** | **0.8272** | **5.6462** | **0.8145** | **0.916** | **0.8405** |
| | ELM | 39.5185 | 0.6682 | 0.84 | 0.7072 | 9.0572 | 0.5228 | 0.862 | 0.7437 |
| | WANN-haar | 46.0746 | 0.549 | 0.766 | 0.5875 | 11.0524 | 0.2894 | 0.845 | 0.7149 |
| | WANN-db3 | 36.5058 | 0.7167 | 0.862 | 0.7438 | 6.5836 | 0.7478 | 0.883 | 0.7811 |
| | ANN | 36.6745 | 0.7143 | 0.874 | 0.7648 | 13.7085 | −0.0932 | 0.768 | 0.5901 |

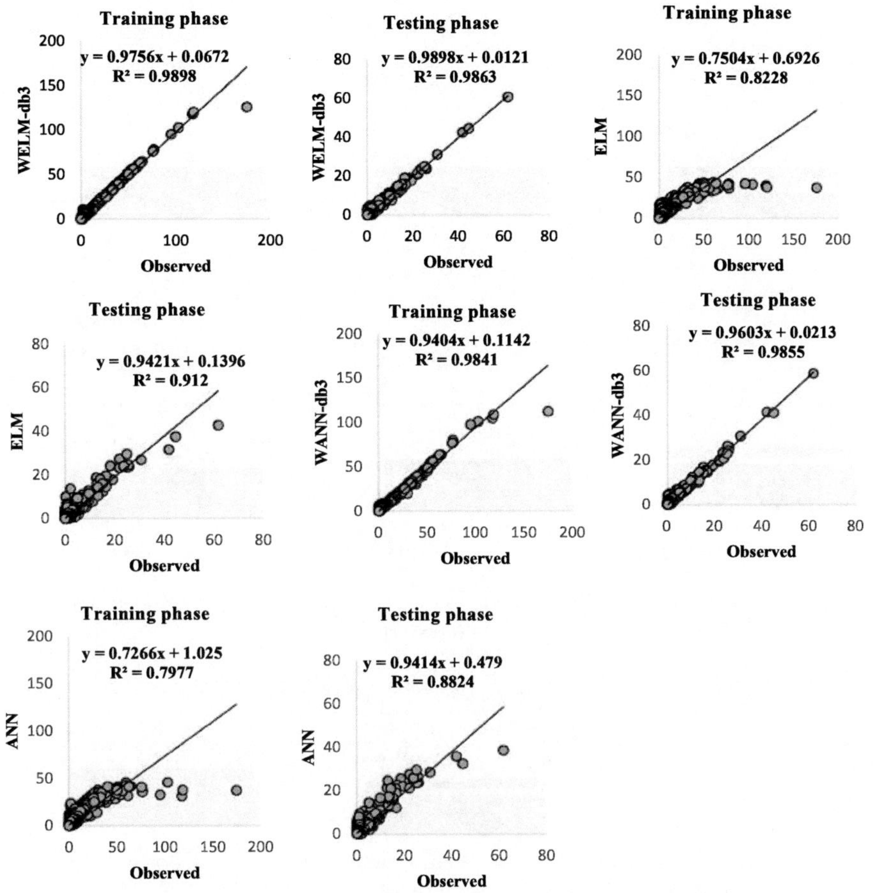


**Fig. 9.2** Training and test predictions of the applied models for the Babaeski station

results with the previous station are seen from the scatterplots. In both stations, the slope and bias coefficients of the fit line equations are closer to the 1 and 0 with higher correlation for the wavelet-based models compared to single models.

## 9.4 Concluding Remarks

The ability of a new intelligent data analytic method: ELM-DWT was investigated in this chapter to forecast intermittent streamflow. Daily time series data from two stations, Uzunkopru and Babaeski, Turkey were used as a case study. The results of ELM-DWT were compared with ANN-DWT, single ELM and ANN methods.

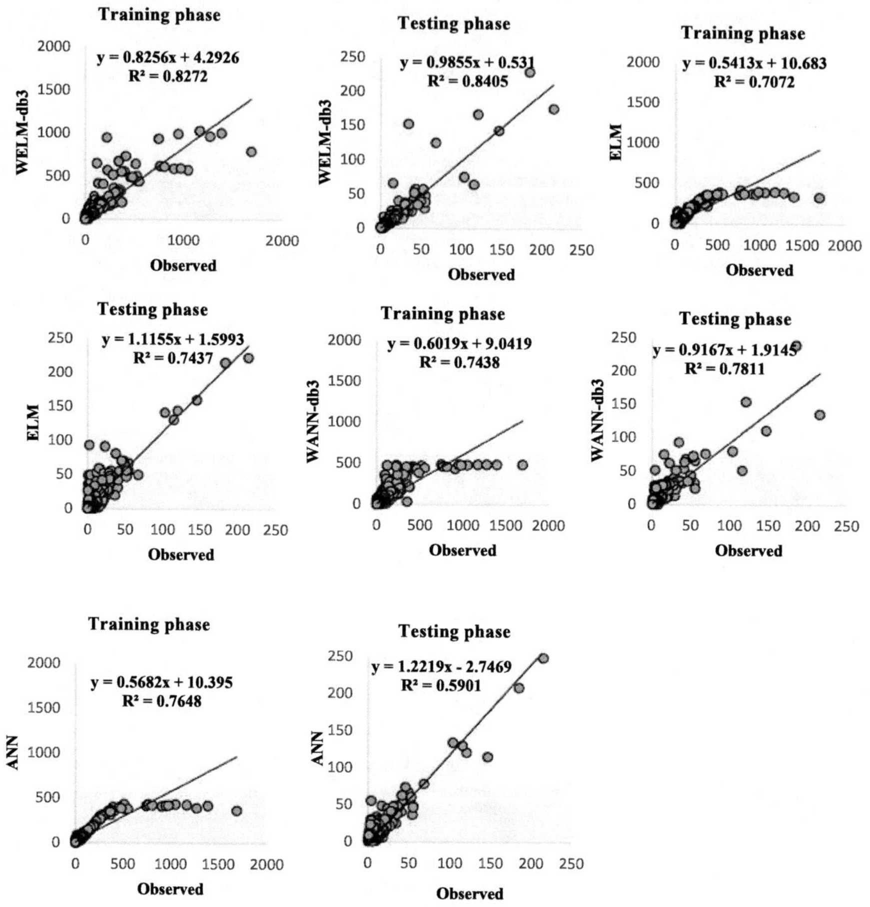


**Fig. 9.3** Training and test predictions of the applied models for the Uzunkopru station

Three commonly used statistics, RMSE, NSE and R, were employed for assessment of the applied methods. The ELM-DWT model with five previous streamflow values showed the best performance compared to other corresponding models. It was observed that DWT considerably increased ability of single models in forecasting intermittent streamflow of both stations; a decrease in RMSE for the ELM and ANN models were 38 (61)% and 59 (69)% using the ELM-DWT for the Uzunkopru (Babaeski) stations, respectively. The methodology proposed can be adopted as a useful data intelligent method for stream prediction, to aid in mitigation of water-related disasters and other forms of natural hazards (e.g., drought events).

## References

Alizamir M, Kisi O, Zounemat-Kermani M (2017) Modelling long-term groundwater fluctuations by extreme learning machine using hydro-climatic data. Hydrol Sci J 63:63–67

Coulibaly P, Burn HD (2004) Wavelet analysis of variability in annual Canadian streamflows. Water Resour Res 40:W03105

Huang GB, Zhu QY, Siew CK (2004) Extreme learning machine: a new learning scheme of feedforward neural networks. Int Joint Conf Neural Netw 2:985–990

Huang GB, Zhu QY, Siew CK (2006) Extreme learning machine: theory and applications. Neurocomputing 70:489–501

Karimi S, Shiri J, Kisi O, Shiri AA (2016) Short-term and long-term streamflow prediction by using 'wavelet–gene expression' programming approach. ISH J Hydraul Eng 22(2):148–162

Karimi S, Shiri J, Kisi O, Xu T (2018) Forecasting daily streamflow values: assessing heuristic models. Hydrol Res 49(3):658–669

Kisi O (2008) River flow forecasting and estimation using different artificial neural network techniques. Hydrol Res 39(1):27–40

Kisi O (2009) Neural networks and wavelet conjunction model for intermittent streamflow forecasting. J Hydrol Eng 14(8):773–782

Kisi O, Alizamir M (2018) Modelling reference evapotranspiration using a new wavelet conjunction heuristic method: wavelet extreme learning machine versus wavelet neural networks. Agric Meteorol 263:41–48

Liang NY, Huang GB, Rong HJ, Saratchandran P, Sundararajan N (2006) A fast and accurate on-line sequential learning algorithm for feedforward networks. IEEE Trans Neural Netw 17:1411–1423

Mallat SG (1989) A theory for multi resolution signal decomposition: the wavelet representation. IEEE Trans Pattern Anal Mach Intell 11(7):674–693

Pandhiani SM, Shabri AB (2013) Time series forecasting suing wavelet-least squares support vector machines and wavelet regression models for monthly stream flow data. Open J Statist 3(3):183–194

Shiri J, Kişi Ö (2010) Short term and long term streamflow forecasting using a wavelet and neuro-fuzzy conjunction model. J. Hydrology 394:486–493

Shiri J, Kişi O, Makarynskyy O, Abbas-Ali Shiri AA, Nikoofar B (2012) Forecasting daily stream flows using artificial intelligence approaches. ISH J Hydraul Eng 18(3):204–214

Smith LC, Turcotte DL, Isacks B (1998) Stream flow characterization and feature detection using a discrete wavelet transform. Hydrol Process 12:233–249

Zhou HC, Peng Y, Liang G-H (2008) The research of monthly discharge predictor-corrector model based on wavelet decomposition. Water Resour Manag 22(2):217–227

# Chapter 10
# Systematic Integration of Artificial Intelligence Toward Evaluating Response of Materials and Structures in Extreme Conditions

**M. Z. Naser**

## 10.1 Introduction

The last few years have witnessed a wave of natural disasters and manmade events of devastating magnitude, i.e., wildfires, tsunamis and terrorist attacks (NYDOT 2008).

This sudden rise in frequency and intensity of natural occurrences, combined with escalation of manmade threats, is posing substantial threats to civil infrastructure (i.e., high-rise buildings, bridges). Since much of our infrastructure were built during the post-World War II era, these infrastructures are expected to experience increasing demands arising from growing population and to undergo a variety of hazards including climate change-triggered environmental conditions and/or occasional occurrences (i.e., earthquakes) (Kodur and Naser 2013).

Out of all infrastructure population, only 10–15% is thought of to be vital to the continual functionality of the society. Combining the aforementioned factors paints, a concerning scenario in which the vulnerability of aging infrastructure continues to hinder our progress and may in fact be detrimental to the safety and prosperity of our societies (Lounis and McAllister 2016). It is then necessary to minimize (or mitigate) destruction or incapacitation of critical infrastructure, whether during or in the aftermath of a disaster (Hudson et al. 2012). This has been duly noted in past and recent works (Landrigan et al. 2004).

The American Society of Civil Engineers (ASCE) also shares a similar view to that discussed above. This international organization continues to document and rank the *health of our infrastructure* in a report card that is shared publicly every four years (ASCE 2017). In the latest report card, published in 2017, the ASCE noted that

M. Z. Naser (✉)
Glenn Department of Civil Engineering, Clemson University, Clemson, SC 29634, USA
e-mail: mznaser@clemson.edu; m@mznaser.com
URL: http://www.mznaser.com

© Springer Nature Singapore Pte Ltd. 2021

R. C. Deo et al. (eds.), *Intelligent Data Analytics for Decision-Support Systems in Hazard Mitigation*, Springer Transactions in Civil and Environmental Engineering,
https://doi.org/10.1007/978-981-15-5772-9_10

the overall score of infrastructure in the US is "*D+*",[1] with some specific types of infrastructure scoring a much lower grade (i.e., *D*: aviation, dams and roads; *D-*: transit). This card also shows few surprising statistics. For a start, about 56,007 and 83,556 out of the 614,387 bridges that were in service are categorized as *structurally deficient*[2] or *functionally obsolete*,[3] respectively. Secondly, 17% of dams are considered as *high hazard potential* that require $45 billion to be repaired. Thirdly, more than 1,029,980 km of high-voltage transmission lines in 48 states bypassed their expected life span and are operating at full capacity (which led to an annual average of 3571 power outages). The reader is encouraged to review other key statistics which are avoided herein for brevity but can be found elsewhere (ASCE 2017).

The main outcome of ASCE's report card stresses the fact that our infrastructure is aging and at maximum if not full capacity. As such, in the event where an infrastructure (say a bridge) was exposed to an extreme event, then this bridge must undergo a proper inspection to identify possible degree of damage and needed repairs (if any). This inspection process would require the bridge to be shut down to ensure safety of commuters and on-site engineers as inspection process often takes few hours to complete. In the case a bridge does not undergo any damage, then the bridge is reopened for traffic (DesRoches 2006). On the other hand, an extended shut down of a bridge triggers a *domino effect* that starts by detouring ongoing traffic to nearby routes. This disrupts supply chain operations and impose significant traffic delays. As a result of this closure, the flow of newly detoured traffic adversely affects commuters' convenience and business operations (Kiremidjian et al. 2007; Kleindorfer and Saad 2009).

While the example of the bridge discussed above shows the expected degree of disruption on a limited scale (i.e. nearby routes, portion of a city), this damage is expected to be significantly scaled up in the case of a major critical infrastructure (i.e., power plant) undergoes partial/complete collapse due to tsunami, meltdown or sabotage. A disaster of this magnitude could potentially affect a much larger population and for a prolonged period of time (Ladkin 2012; Steinhauser et al. 2014). Knowing that it is virtually impractical nor economical to prevent all disasters or to design a fully disaster-resistant infrastructure has led researcher to seek other solutions as means to minimize disaster-related damages. One such solution that has been noted in the published literature over the past few years calls for adopting structural resilience (Bruneau and Reinhorn 2006; Kim et al. 2015).

Holling (2003) introduced resilience in early 1970s as a multidimensional concept to describe resistance to changes in natural and ecological systems. Since then, this concept has been adopted by a number of fields including structural and construction engineering. From this chapter's point of view, the resilience of an infrastructure that is subjected to an extreme loading event requires such infrastructure to not

[1] A score of "*D*" implies a poor state.

[2] Structurally deficient describe bridges with restricted service, or bridges that require rehabilitation (Vdot).

[3] Functionally obsolete describe bridges built to standards that are not used today (Vdot).

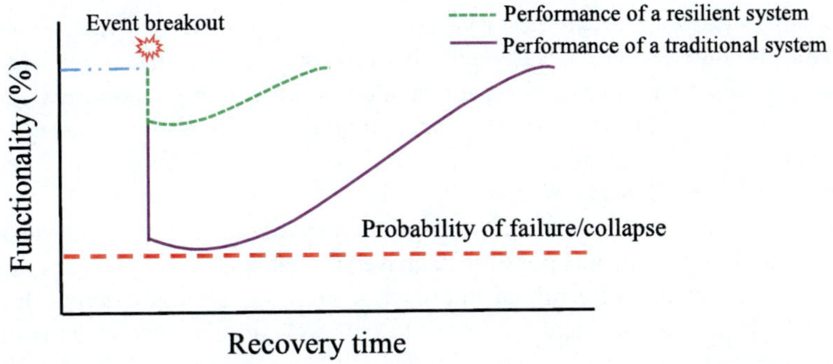


**Fig. 10.1** Comparison between a resilient and a traditional infrastructure in the aftermath of an extreme event

only successfully withstand such an event, but to also minimize expected damages, consequences and time to achieve full functionality and recovery.

Figure 10.1 illustrates the loss of functionality in the aftermath of a particular extreme event between two identical infrastructures wherein one is resilient while the other is not.

Evidently, the resilient infrastructure has an improved structural response with a much shorter down time as compared to the conventional infrastructure. This improvement in structural response is quantified by observing how a resilient structure undergoes less degree of damage which could be translated as shorter time for repairs, etc. It is worth noting that a key component of resilience that is often neglected is that resilient structures allow occupants/commuters to safely evacuate and provide first responders with sufficient time to overcome the adversity of an extreme event (Naser and Kodur 2018; Naser 2019a).

Structural resilience can be attained by implementing few design solutions such as adopting appropriate detailing and protection measures to ensure ductile response, enhance structural safety and resistance to damaging effects. However, one should remember that adopting such solutions may meet resilience requirements for low-to-medium-sized events, and still these solutions may not be sufficient to satisfy performance requirements under extreme loading events or under simultaneous loading events (e.g., aircraft impact followed by fire) as observed during the collapse of World Trade Center twin towers (Eagar and Musso 2001). On a similar note, the above solutions may not be practical to install in some infrastructure (i.e., bridges and tunnels) due to their unique characteristics arising from complex structural systems, demanding service requirement, etc. (O'Rourke 2007).

When resilience principles may not be fully adopted into infrastructure, then identifying such structures becomes of importance (1) as to be structurally strengthened before the start of the disaster season (i.e., hurricanes) or even closed for public (in some scenarios) and (2) can be of importance to authorities while planning for emergency response and evacuation. As one can imagine, identifying vulnerable structures is expected to be a hectic and tedious task. This is given the fact that the expected damage arising from an extreme event is governed by a variety of factors some of which may include: structural characteristics, type of construction materials, incident intensity, etc. Even in the case where information on all of these factors are collected, few questions arise: (1) *What is the proper methodology to analyze these factors?* *and* (2) *how such an analysis be scaled for infrastructure of different characteristics and/or for other extreme conditions?*

While finding answers to the aforementioned questions is not an easy task, this chapter hypothesizes that it is possible to arrive at reliable answers to these questions by adopting principles of artificial intelligence (AI). Given that machine learning (ML) techniques are specifically tailored to comprehend complex systems of high dimensionality, these tools have been successfully used to unlock hidden mechanisms in various applications such as materials science (Abdalla and Hawileh 2011; Ding et al. 2019; Naser 2019b, c), decision making (Lu 1991; Chen and Tan 1994), investigative engineering (Kushida et al. 1997; Naser 2019d) and planning (Seitllari 2014; Jahangiri and Rakha 2015). In fact, a number of prominent works have lobbied toward integrating AI into modern structural engineering (Mohan 1990; Levitt et al. 2008). This is the main drive behind this chapter.

In order to display the meritorious use of AI in the field of engineering for extreme conditions, this chapter shares the development of an AI-driven framework aimed to accurately trace the response of materials and structures at elemental and systemal levels. This chapter starts with a brief review on notable structural failures that occurred around the world in the last two decades to demonstrate the magnitude and impact of structural failures due to extreme loading conditions with emphasis on large-scale infrastructure.

To provide a concise presentation, this chapter will primarily cover response and behavior of construction materials and structural elements when subjected to extreme temperatures as in the case of building or bridge fires. This will provide the reader with a comprehensive understanding on material and membral behavior under simultaneous/concurrent thermal and mechanical loading. Then, this chapter will showcase the development of an AI-based framework and decision-making tool that can aid engineers and authorities in predicting magnitude of damage in bridge infrastructure once subjected to a variety of extreme loadings such as fire, wind and flooding.

## 10.2 Structural Failure Due to Extreme Loading Events

This section reviews principles of structural engineering and highlights noteworthy structural failures. These incidents cover the collapse of World Trade Center twin towers in the USA, Cathédrale Notre-Dame de Paris in France and Viadotto Polcevera bridge in Italy.

### 10.2.1 Why Structures Fail?

The structural response of a given structure or a load bearing member (say a beam) is a reflection of its inherent features and how these features interact with respect to changes arising due to internal or external actions. *From a pure structural engineering point of view, a structure fails once the applied loading (or effects of loading),* $P_{app}$, *exceeds the structure's capacity,* $R_{cap}$, *to resist such loading/effects.* These features are briefly covered herein and a more in-depth discussion can be found elsewhere (Levy and Salvadori 2002; Bisby 2016):

*Material features*: This covers the type of the construction material the beam is made of and from load bearing point of view, include concrete, steel, wood, along with their derivatives. These features are highly influenced by the type of loading the material is subjected to (i.e., for the same specimen size, concrete cracks/fails under low levels of tensile forces, but this specimen would require much higher load levels to crush in compression. On the other hand, steel behave comparatively well under compression and tensile forces types of loading, etc.).

*Geometric features*: This group covers aspects such as beam's cross section, slenderness, length/span, aspect ratio, etc. Naturally, a thicker or larger beam would have higher/better structural response than a slender beam (given that all other features are the same). The geometric features are highly dependent upon the selected construction material. For example, steel beams are often made of W-shaped sections, while concrete and wood beams are made of rectangular shapes.

*Restraints features*: Beams with full restraints often achieve improved performance due to their ability to redistribute some of the applied loadings to regions of hogging/negative moments, as opposed to beams with simply supported restraints, etc. Overall, the higher the restraint conditions, the better the structural response (from bending and axial capacity points of view).

*Loading features*: Loadings are often forces or actions (i.e., heat) that need to be carried out (resisted) by the structural member in order to transfer such forces from one point to another in a *structural system.*[4] Loading features differ in type (e.g., compressive, shear), intensity (heavy, light, sudden, repetitive, compound) and duration (short, long, seasonal). For example of a beam, if this beam is to be located in a classroom, then this beam is expected to be moderately stressed during class hours as opposed to during nighttime/weekends, etc.

*Other features*: These may include environmental conditions, aging, deterioration-related issues with regard to construction materials (i.e., cracking in concrete, corrosion in steel, rotting in wood), settlement due to sudden/heavy loading, etc. Features that may belong to this group also include errors in design and fabrication, poor inspection and upkeep, etc.

The above factors are in continuous state of seeking balance, and this balance gets violated once any of the above features overpower the rest. For instance, if a beam made of concrete is heated, temperature across the cross section of this beam starts

[4]A structural system is a collection of structural members, i.e., the joining of a beam and two columns yields a frame.

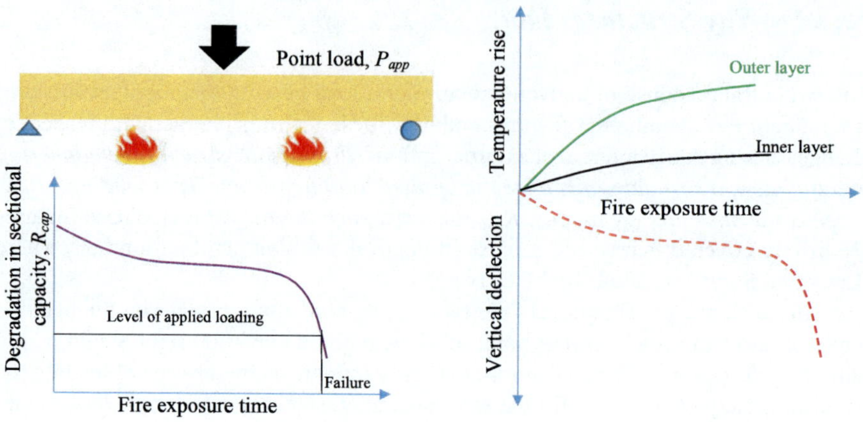


**Fig. 10.2** Typical response of a RC beam under fire conditions

to slowly rise due to the low thermal conductivity and high specific heat of concrete. Thus, in the initial stage of heating (i.e., fire), a thermal gradient is expected to develop as a result of hotter temperatures at the outer layers of concrete as opposed to those in the inner layers of concrete (see Fig. 10.2). This rise in temperature adversely affects concrete material[5] properties causing degradation (i.e., loss in strength and modulus). Due to this degradation, the moment capacity of this beams starts to weaken (reduces with fire exposure time, see Fig. 10.2), and this is physically reflected by a downward deflection. With extended duration of exposure to fire, the beam continues to deflect as a result of degrading properties/capacity and eventually fails once the level of moment capacity falls below that of existing applied loading.

While this example covers the basics of how a typical reinforced concrete (RC) beam behaves under fire, the reader should note that other effects such as fire-induced creep (accelerated deformation) and fire-induced spalling may also occur. These effects are governed by the composition of concrete mix, use of admixture, peak temperature, etc., and a more detailed discussion on these effects can be found elsewhere (Wang et al. 2012; Kodur 2014; Buchanan and Abu 2016). Similar principles to those highlighted herein for the case of fire can also be applied to understand the behavior of similar structural members under other loading events (i.e., impact, flood). The following presents three real case studies on actual structures that failed/collapsed due to some of the aforementioned events.

[5]Due to the smaller size of rebars, as compared to the beam's cross section, temperature in rebars is assumed to be similar to that in surrounding concrete (Hawileh et al. 2011).

### *10.2.2 Case Studies of Falling Structure*

***World Trade Center***

The twin towers of the World Trade Center (WTC) were square in plan with an edge size of 63 m and stood at 110 storeys high.

The north tower (often referred to as WTC 1) and south tower (WTC 2) had a full height of 417 m and 415 m, respectively. These towers were built in 1973 and were designed with *tube-in-tube* structural system (see Fig. 10.3). Each face of these towers had 59 perimeter composite steel columns and a flooring system comprising of 100-mm lightweight concrete slabs. These slabs were attached to steel decks carried by a network of 18-m-long steel trusses. In general, both towers were designed to have high levels of structural redundancy as to achieve improved performance and resilience.

An airplane hits both of these towers as part of the terrorist attacks that occurred on September 11, 2001.

The impact of the plane on each tower has fractured the bulk of the columns across the impacted face and started a fire. As a result of these simultaneous/combined loading events, the towers collapsed. The collapse of the towers was examined though a joint investigation between the American Society of Civil Engineers (ASCE), the Federal Emergency Management Agency (FEMA) and the National Institute of Standards and Technology (NIST).

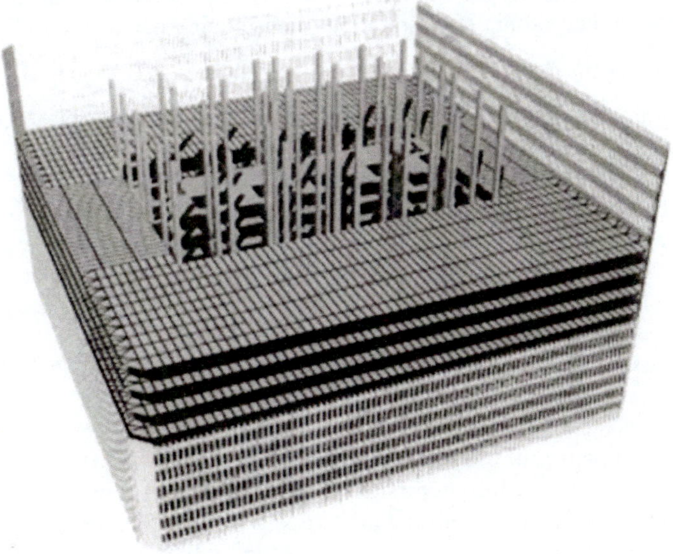

**Fig. 10.3** Layout of floor plan in WTC (Eagar and Musso 2001)

The outcome of this investigation noted the high resilience of the tube-in-tube structural system which was able to withstand the impact of the plane despite the large number of fractured columns (i.e., the impact of the planes did not cause an immediate collapse of the towers). This investigation identified how the burning of gasoline fuel that leaked from the airplane upon impact has started multiple fires across the towers' floors. The failure/non-operationality of active fire protection systems (i.e., sprinklers) and passive protection systems (i.e., debonding of insulation) have led to spread of these fires, which induced additional thermal damage to the already-weakened and overstressed steel trusses and columns, eventually leading to collapse. It is worth noting that the collapse of WTC towers claimed the lives of more than 2190 people and caused at least $10 billion in property loss.

***Cathédrale Notre-Dame de Paris***

Notre-Dame de Paris is a medieval Catholic cathedral in Paris that was built during 1163–1345.

This cathedral has survived for more than 670 years, including World War I and II. Unfortunately, a major fire broke out on April 15, 2019, under the roof of the Cathedral (see Fig. 10.4). This fire caused the roof to collapse along with causing severe damage to the interior walls and vaulted ceiling arising from high temperatures exceeding 800 °C (Décugis et al. 2019).

Over 400 firefighters were dispatched to fight the fire. Due to the massive size of this structure, an what appears to be poor planning, the fire could not be controlled immediately and lasted for 15 h. In the aftermath of this event, a major donation campaign started and managed to collect over €1 billion to restore the cathedral. It is worth noting that a complete restoration could require +20 years.

***Viadotto Polcevera Bridge***

The Viadotto Polcevera bridge is a cable-stayed bridge that connects A10 motorway in Genoa, Italy.

The construction of this bridge was completed in 1967. This bridge had a total span of 1182 m. The bridge had three 90-m-high-reinforced concrete pylons. The bridge carried four lanes of traffic. On August 14, 2018, a 210-m segment collapsed due to the combined effects of severe storm and high levels of corrosion in the steel cables (see Fig. 10.5). About 30 vehicles were on top of the segment at the time of the collapse, and this has led to the death of 43 people.

The discussion on the above notable structural failures highlights few observations.

For a start, these incidents show the devastating effects of extreme conditions upon society (in terms of human, historical and economic losses) and environment (in terms of generating debris, toxic smokes, etc.). Secondly, despite development of advanced construction materials and modern knowledge of structural behavior and promoting novel codal provisions, we do not seem to fully grasp an understanding on how structural systems respond to extreme loading conditions. Thirdly, one could argue that since most infrastructures are not frequently exposed to extreme loadings, and the fact that there is in fact a limited number of structural failures that (1)

Fig. 10.4 Burning of cathedral ceiling. Courtesy of LeLaisserPasserA38 (2019)

Fig. 10.5 Collapsed span of the Viadotto Polcevera bridge. Courtesy of Salvatore Fabbrizio (2019)

occur spread around the world, and (2) lack proper documentations, then applying traditional methods to analyze such failures may not be sufficient to unlock hidden mechanisms or relations between various influencing factors. An attractive solution to the above observations is to leverage modern technologies such as AI in order to explore key aspects of structural behavior under extreme loading conditions.

## 10.3 Development of an Intelligent Data Analytic (AI) Framework

The first serious discussion of AI as a mean to tackle complex problems was noted at a conference in Dartmouth College in the mid-1950s (Salehi and Burgueño 2018).

In this mention, techniques were proposed to develop frameworks with the ability to mimic human cognition in order to realize implicit relations between influencing factors to discover quantifiable conclusions to a particular phenomenon through analyzing datasets. A review of open literature identifies some of the primary AI techniques as neural networks (NN), genetic programming (GP), etc. (Goldberg and Holland 1988; Ferreira 2001).

From this chapter's point of view, AI is considered as the realm of modern technologies that can be leveraged in favor of structural and construction engineering. For those who are enthusiast about this particular field of engineering, they know how structural and construction engineering requires a thorough knowledge of analysis/design principles, materials science, fluid dynamics, etc. The design, detailing, construction, maintenance and upkeep for infrastructure are governed by a number of variables spanning geometric features, material properties, load conditions, etc. The actual representation of these variables is often not clear, but rather estimated through a series of procedures and provisions.

In some instances, where an engineer is required to design for an extreme loading condition (i.e., flooding), this engineer is often faced with immature/complex codal provisions to follow, further hindering his/her ability of achieving a proper design (Kodur et al. 2012; Watson and Adams 2012; Michael Grayson et al. 2013; Khorasani et al. 2016; Naser 2018). In all reality, some infrastructures (such as bridges) are not even designed to withstand some events (i.e., fire) as design codes do not specifically require such structures to be designed for such events. The above discussion demonstrates how this community is tackling phenomena in which traditional methodologies struggle to resolve. An attempt to integrate AI into this field could be promising.

AI, and unlike traditional analysis approaches, does not rely on pure mathematical models or well-established sets of criteria to start an analysis. But rather, AI attempts to simulate the cognition process of the brain in order to analyze a phenomenon by employing heuristic and evolutionary search methods and algorithms to carry out adaptive learning and systematic analysis of factors governing a particular phenomenon. The accuracy of an AI model largely depends on how such a model was developed and also upon the amount of training and fine-tuning it underwent. Such accuracy is often measured in terms of fitness/error metrics.

This chapter highlights the use of three AI techniques that belong to the supervised learning family, namely artificial neural networks (ANNs), genetic algorithms (GAs) and gene expression programming (GEP). These techniques are applied to carryout regression and classification operations and are further examined herein.

### 10.3.1 Artificial Neural Networks (ANNs)

An ANN consists of visible and hidden layers that incorporate processing units or neurons (see Fig. 10.6). Neurons are organized to form a network for processing—of a similar topology to the brain. These neurons maintain a constant state of interaction and link other functioning neurons and layers. The neurons in each hidden layer process input data points and relay the information to the following layer through specific connections/weightages. Thus, by using neurons in the hidden layer, the network can learn and recognize the relevant data patterns and approximate complex nonlinear mapping (transformation) between the inputs and output(s). These hidden layers connect to the final layer to displays the outcome of the ANN.

Once input parameters are added into the first layer, these parameters flow toward the hidden and output layers passing through neurons. This process of forward flowing of data is known as the *feed-forward network*. This process continues for a predefined number of iterations and/or as long as a prespecified metric fitness or error tolerance is achieved between experimental and ANN-predicted output. Oftentimes, this training process is performed using *back-propagation algorithm* which aims at minimizing the error between the input and output layers (Kisi and Çobaner 2009). One of the most commonly used optimization method is *Levenberg–Marquardt* which evaluates the error in terms of mean squared error (MSE). In this method, if $z$ is the experimental dataset, then MSE can be calculated using Eq. 10.1.

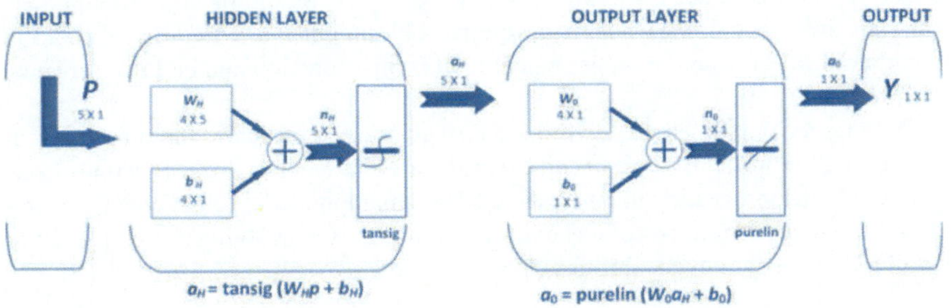


**Fig. 10.6** Typical ANN structure ($W_H$ and $W_O$ represent the interconnection weights for hidden layer and output layer, respectively. Likewise, $b_H$ and $b_O$ are the biases for hidden layer and output layers, respectively)

$$\text{MSE} = \frac{1}{z}\sum_{i=0}^{z}(e_i)^2 = \frac{1}{z}\sum_{i=0}^{z}(m_i - p_i)^2 \tag{10.1}$$

where $z$ = the total number of datasets, $e_i$ = the error for each input set, $m_i$ = the measured output and $p_i$ = the estimated output.

The best fitting model can also be statistically evaluated in terms of MSE as well as coefficient of determination ($R^2$) or mean absolute relative error (MARE) (see equations below).

$$R^2 = \frac{\Sigma(m - p)^2}{\Sigma\left(p - p_{\text{avg}}\right)^2}. \tag{10.2}$$

$$\text{MARE} = \frac{1}{z}\sum_{i=1}^{z}\left|\frac{m - p}{m}\right| \times 100 \tag{10.3}$$

where $z$ = the total number of datasets, $e_i$ = the error for each input set, $m_i$ = the measured output, $p_i$ = the estimated output and $p_{\text{avg}}$ = the average estimate output.

### *10.3.2 Genetic Algorithm (GA)*

GA is an evolutionary search technique that was introduced by Koza (1992). This technique belongs to the supervised learning family and attempts to solve the problem on hand by following Darwinian natural selection process. In GA, a random population of candidate solutions, often referred to as "expression trees," is created to house possible solutions. These solutions are structured in strings comprising of functions and terminals (i.e., mathematical symbols). For example, a function (*F*) may contain trivial operations (addition "+", multiplication "×" etc.), power functions (logarithm "log", exponential "exp") or logic functions ("AND", "OR", "NOR", "NAND", etc.), to name a few. Conversely, a terminal (*T*) may comprise of arguments and/or numerical constants/variables. Hence, a developed GA model has a tree-like formation (configuration) in which branches can extend from a function and end in a terminal as shown in Fig. 10.7.

Once a set of candidate expressions is arrived, the GA evaluates the fitness (accuracy) of such expression (i.e., a fit expression is one that accurately traces actual observations). The fittest models are then selected and manipulated through genetic operations, i.e., reproduction, cross-over and mutation to ensure evolving elite expressions with highest prediction capabilities (Koza 1992).

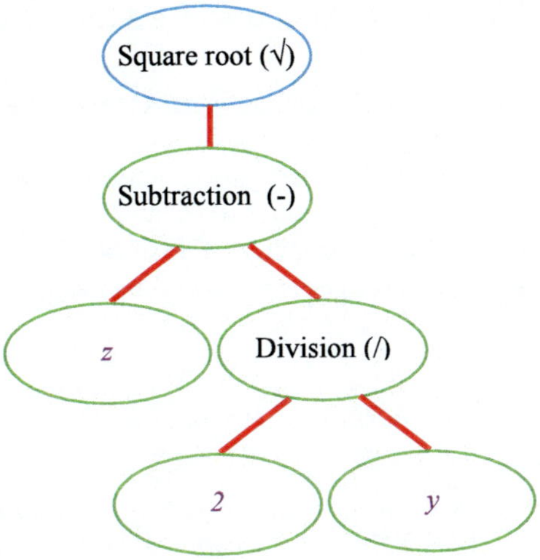


**Fig. 10.7** Typical tree representation for expression: $\sqrt{z - \frac{2}{y}}$ in GA

### *10.3.3 Gene Expression Programming (GEP)*

GA is an umbrella that encompasses regression techniques (which include genetic programming (GP) and a newly developed subset referred to as (GEP)). GEP, introduced by Ferreira (2001), is a modern supervised learning processes that also mimics the natural selection process (i.e., Darwinian evolution) to express hidden relations. GEP is particularly beneficial in situations where statistical methods turn complicated or require high computational capacity. Other scenarios may also include those where the exact physics of a problem is not well established in detail (Alavi et al. 2010; Mousavi et al. 2012; Jafari and Mahini 2017). A main advantage of GEP is its capability to produce computer codes to express predictive expressions without relying on past formulas or relationships.

The key variance between GA and GEP lies in their depiction of the final relation between the selected inputs and outputs. While GA creates a binary string that lists actions and values (i.e., equation) that represent the solution, on the other hand, GEP develops computer codes often in a tree structure of actions and values that are expressed in a functional programming language (e.g., C++, Fortran). In other words, GA searches a data space, while GEP searches a program space. Once developed, these programs can be run in respective software to solve a phenomenon (i.e., for a given problem, GEP can develop a macro that can be run using Fortran or C++ software, while GA can derive a formula that can be substituted into by hand calculation or a spreadsheet, i.e., Excel). Readers are encouraged to review the following references to arrive at a comprehensive understanding of GA and GEP (Goldberg and Holland 1988; Koza 1992; Ferreira 2001).

## 10.4 Collection and Development of Databases

The successful application of AI techniques requires compiling comprehensive and proper databases.

A literature survey was carried out to identify well-documented studies and reports that cover three dimensions: temperature-dependent material properties, thermal and structural behavior of load bearing members and prominent cases of failed infrastructure. These three will be further detailed in the following subsections.

### *10.4.1 Material Properties*

As discussed earlier, *the response of structures under a given extreme event is governed by how the constituent materials of such structure respond to this specific event*. A particular case for the event of fire is selected herein as it provides a unique opportunity to understand material behavior under thermal and mechanical loading. Under such effects, thermal and mechanical properties oscillate as temperature rises in response to the series of physio-chemical changes that occur at various temperatures. Since concrete is widely used in constructions around the world, this material is selected for analysis herein (Naik 2008).

To start with the thermal properties. These properties govern temperature rise and propagation within a concrete component/structural element and consist of density ($\rho$), thermal conductivity ($k$) and specific heat ($C_\rho$). The behavior of these properties is governed by specifics of mixed composition and temperature rise (duration, etc.) (Neville 2012). The density, mass of a unit volume, is mostly governed by the amount of water present in concrete mix and density of coarse aggregates. The density of normal weight concrete ranges between 2100 and 2300 kg/m$^3$ and slightly reduces beyond 100 °C as a result of water loss to evaporation. The thermal conductivity ($k$) reflects the rate at which concrete transmits heat and is comparatively low (1.4 – 3.6 W/m.K) at ambient conditions. Since this property is sensitive to the amount of moisture and crystallinity of aggregates (Naser 2019b), the thermal conductivity decreases with rise in temperature rise due loss of moisture and increased porosity (Kodur 2014). It should be noted that siliceous concretes have higher conductivity than carbonate concrete (see Fig. 10.8a).

The specific heat, $C_\rho$, describes the extent of energy required to raise a unit mass of concrete a unit temperature. The specific heat can vary between 840 and 1800 J/kg K at ambient temperature as a result to variation in production process, mix proportions, etc. (Kodur 2014). $C_\rho$ is highly sensitive to physio-chemical changes that occur at temperatures of 100 °C, 400–500 °C and 600–700 °C as these reflect an increase in thermal energy demand toward loss of free water, breakdown of $Ca(OH)_2$ into ($CaO$) and ($H_2O$), and decomposition of carbonate aggregates beyond 600 °C, respectively (see Fig. 10.9a, b).

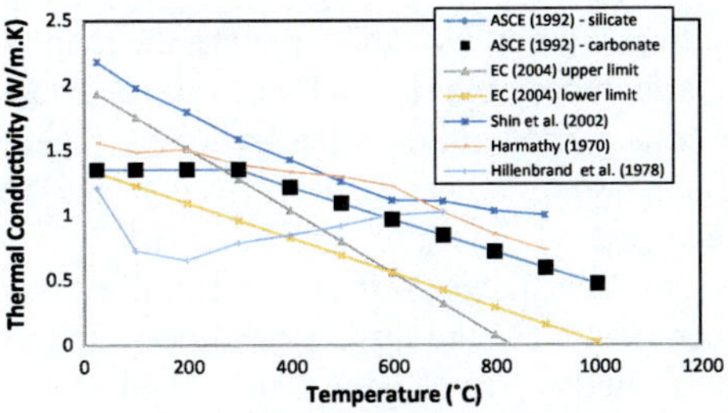


(a) Normal strength concrete

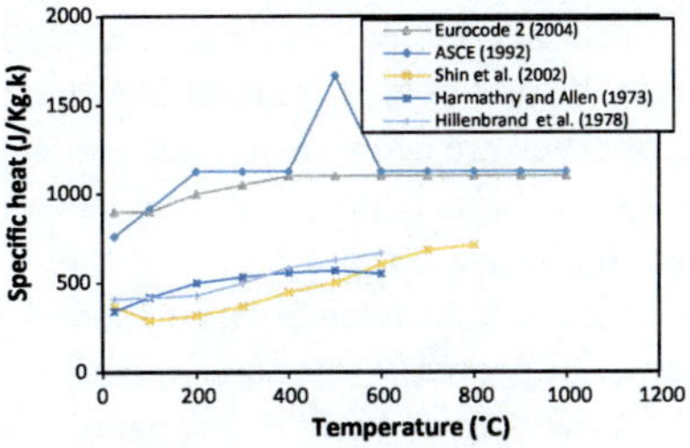


(b) Silicate-based concrete

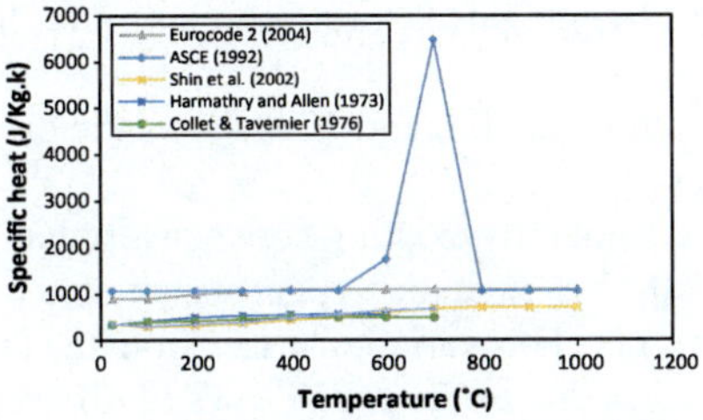


(c) Carbonate-based concrete

**Fig. 10.8** Variation of thermal properties of normal weight concrete with temperature rise

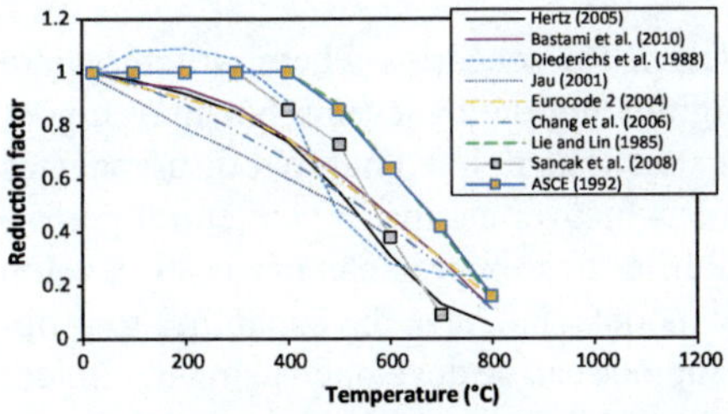


(a) Normal strength concrete – siliceous

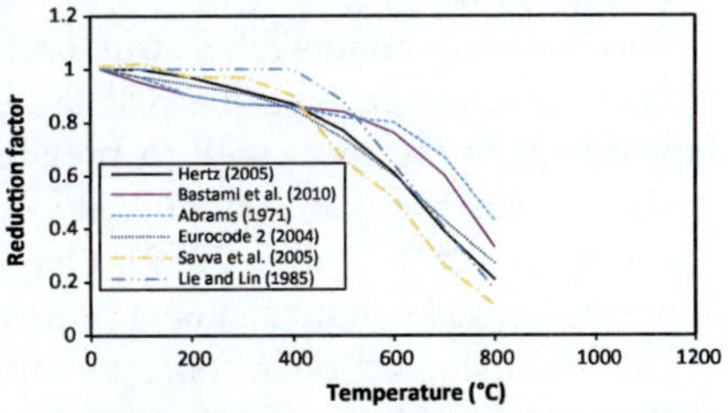


(b) Normal strength concrete – carbonate

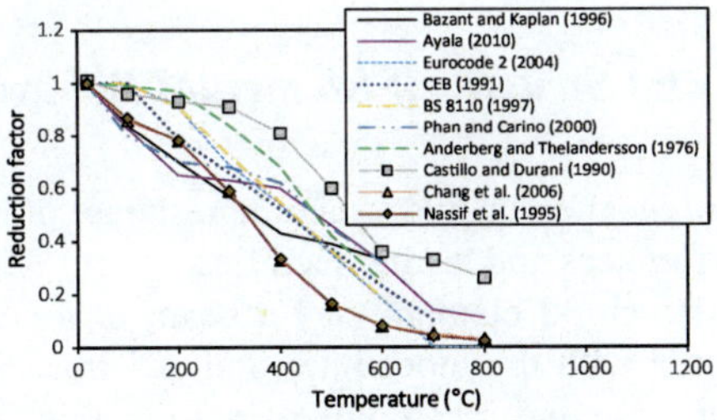


(c) Normal strength concrete

**Fig. 10.9** Variation of mechanical properties of normal weight concrete with temperature rise

The mechanical properties (e.g., strength and modulus) which govern the magnitude of load bearing capabilities are also affected by the rise in temperature. These properties are often measured in material tests carried out on cubes/cylinders. These tests can be carried out in two set-ups: *steady-state* (where a specimen is heated to reach a target temperature and then linearly loaded to failure) and *transient* (where a specimen is mechanically loaded first and then heated). It is worth noting that material property tests are sensitive to specimen size, and testing conditions and as such there exist wide discrepancies in published works (Kodur 2014). Figure 10.9 shows such discrepancies and highlights how the mechanical properties of concrete degrade with rise in temperature. It is commonly accepted that this degradation is influenced by the type of aggregates, water–cement ratio, heating rate, etc. (Arioz 2007).

A comparison between Figs. 10.8 and 10.9 shows the discrepancies in measured properties of concrete.

Such discrepancies are not only existing between test data points reported 1950s–70s and late 1990s–2000s, but such discrepancies are also noticeable in data points obtained within the same era. This variation arises from the fact that concretes significantly vary in composition due to the use of raws of different origins, as well as to the different testing regimes/equipment used in those studies (Neville 2012). These observations deduce that while there exists a good amount of high-temperature material tests on thermal and mechanical properties of concrete, the outcome of these tests significantly varies.

This not only complicates structural design for scenarios where structures are expected to undergo high thermal cycles/gradients or fire events but may hinder standardization efforts as well. A previous study has pointed out that discrepancies between material models could lead to under- or over-estimating structural performance by 20–25% (Naser 2018). One solution to this predicament is to develop unbiased thermal and mechanical property models that have the capability to properly represent fluctuation in concrete. Such models can be developed using intelligent systems such as ANN and GA.

### *10.4.2 Thermal and Structural Elemental Behavior*

Researchers often examine thermal and structural response of load bearing elements in specially designed furnaces and testing facilities.

In this process, a structural element (say a beam or a column) is installed in a furnace, supplemented with thermocouples and deformation measuring devices to monitor temperature rise and magnitude of deformation, subjected to constant gravity loading, and exposed to a *standard* temperature–time curve (ASTM 2016). As one can imagine, fire testing is a complex process that requires the availability of specialized equipment with the capability to provide a consistent heat output in excess of 1000 °C for couple of hours. As such, fire testing is both costly and tedious. Due to this growing complexity, researchers sought other means to trace the response of structures under fire. One such mean is numerical techniques (i.e., finite difference analysis, finite element simulation) (Kodur et al. 2017a; Kodur and Bhatt 2018). However, these techniques still lack proper validation, require availability of various input parameters (e.g., unbiased thermal and mechanical properties, heat transfer coefficients and special software with demanding resources) (Naser 2016).

To overcome these issues, this chapter seeks the development of an AI-driven approach using GA to trace response of RC beams under extreme temperatures. This approach aims to derive simple expressions that can evaluate temperature rise and deformation in RC beams; at a specific point in time, or throughout the complete fire exposure time. To maximize the efficiency of these expressions, these were derived to account for geometry aspects, aggregate type, steel reinforcement ratio, magnitude of gravity loading, concrete cover thickness, fire characteristics as well as strength of concrete and reinforcing steel (see Table 10.1). Those expressions were specifically derived to implicitly account for temperature-dependent material properties of constituent materials, together with associated phenomena, i.e., creep, and hence do not require the input of these parameters (Naser 2019d).

The above listed parameters were chosen as a result of a cross-examination of published reports that identified such parameters to be of critical nature based on researchers recommendations and observations from tests (Ellingwood and Lin 2007; Kodur and Dwaikat 2007; Dwaikat and Kodur 2009; Choi and Shin 2011; Palmieri et al. 2012; Zhu et al. 2014; Jiangtao et al. 2017; Albuquerque et al. 2018; Carlos et al. 2018).

For example, the temperature rise in steel rebars during fire is said to be governed by the duration of burning, $t$, as well as thickness of concrete cover to rebars, $C_b$ and $C_s$.[6] Through this rationale, the developed GA model relates the phenomenon of temperature rise in steel rebars to these three identified parameters through an expression that implicitly incorporates thermal properties of concrete. In a similar rationale, the geometric and mechanical properties of RC beams including steel reinforcement ratio, concrete cover, compressive strength of concrete and yield strength of steel as well as applied load level at ambient conditions were also collected, and these were analyzed as part of the structural-based GA model (Naser 2019d). Overall, all above parameters were collected for various beams at 1–5 min time intervals.

### 10.4.3 Systemal Analysis

As in the case of other two phenomena examined in this chapter, a third database was compiled through a literature survey aimed to identify well-documented infrastructure failures that occurred as a result of an extreme event.

For illustration purposes, the selected type of infrastructure was bridges. These were selected since they present a good representation of infrastructure that is

[6] Keeping in mind that strength of concrete or presence of reinforcing steel does not contribute to temperature rise and hence was neglected.

**Table 10.1** Selected input parameters thermal and structural elemental behavior of RC beams

| Parameter and case | Inputs | | | | | | | Outputs | |
|---|---|---|---|---|---|---|---|---|---|
| | Fire exposure time ($t$) | Compressive strength of concrete ($f_c$) | Yield strength of steel ($f_y$) | Steel reinforcement ratio ($\rho$) | Load level ($P$) | Bottom cover to steel reinforcement ($C_b$) | Side cover to steel reinforcement ($C_s$) | Temperature (°C) | Mid-spam deflection (mm) |
| Temperature rise in rebars (°C) | ✓ | – | – | – | – | ✓ | ✓ | ✓ | – |
| Mid-span deflection (mm) | ✓ | ✓ | ✓ | ✓ | ✓ | ✓ | – | – | ✓ |

designed to endure a multitude of loading conditions and are expected to serve for a number of decades and hence must withstand a continuous increase in demand, exposure to harsh seasonal weathering as well as possible extreme loading events (Garlock et al. 2012).[7]

Three different extreme events were considered here: fire, high wind and flooding (Smith 1976; Wikipedia 2019).[8] The outcome of this extensive survey led to collecting information on 100 international bridges. This information covers various aspects on characteristics of bridges, associated traffic demands as well as event type and damage magnitude. Out of the immense amount of information collected in this survey, a challenge is to link each bridge's characteristics with expected magnitude of damage during an extreme event. From construction and structural engineering points of view, the response of a given bridge to an extreme event is a function of the *structural characteristics*, *traffic features* and *type of event* the bridge is exposed to. The validity of these characteristics and features was also cross-checked by examining recent works, and the rationale behind the selected characteristics and features is further outlined below (Garlock et al. 2012; Kodur and Naser 2013; Naser and Kodur 2015; Peris-Sayol and Payá-Zaforteza 2017; Peris-Sayol et al. 2017).

The structural characteristics that make up the load bearing elements a bridge include: type of structural system and construction material. Typical *load bearing systems, S,* in bridges are assembled under truss, arch, girder-type, cable-stayed and suspension. The last two are infamous for their complex design and susceptibility to lateral and sudden loadings (due to their slenderness). This is unlike that of other systems which are designed to be redundant and compact. While most *construction materials, M*, seem to perform well under ambient environments, some may have an edge under certain conditions. For instance, bridges made of steel or timber are susceptible to corrosion, rotting and fire damage, unlike concrete bridges. Concrete bridges can be classified under two groups, i.e., reinforced concrete, or prestressed concrete (Scheer 2010; Peris-Sayol et al. 2017). A hybrid combination between steel and concrete bridges is that comprises of steel–concrete composite construction. These three types of bridges are considered herein.

As large span bridges serve heavy traffic, the *span, P*, of a bridge indirectly governs its structural capacity. Thus, an assumption is made that the vulnerability of a bridge directly correlates to its span. Since bridges are continuously in service, their load bearing components undergo environmental deterioration (e.g. corrosion of reinforcement, excessive cracking in concrete) and this reduces the structural capacity. To account for these effects, the *age, A*, of a bridge is considered of importance.

Traffic demand is another factor that governs the structural response of a bridge under an extreme event. This factor comprises two parameters, *location/geographical significance, L* and *number of lanes, N*. The first factor is used to distinguish the

---

[7]A similar methodology to that described herein can also be applied toward other infrastructure such as airports, hospitals.

[8]Unlike other works (Harik et al. 1990; Wardhana and Hadipriono 2003; Fu et al. 2012; Cook et al. 2015; Xu et al. 2016; Peris-Sayol and Payá-Zaforteza 2017), bridges that failed during construction/demolition, or due to overloading, fatigue and other similar causes were not considered herein.

geographical significance of bridges (i.e., those located in rural areas/with low traffic are considered as common, while those located in suburbans are considered as landmarks/prestigious). Similarly, the number of lanes in a bridge indirectly reflects the average number of vehicles traveled per day (and overall bridge's load carrying capacity) and hence is of significance.

While the magnitude of an event (i.e., bridge fire) may not share exact features to all other fire events collected in the compiled database, the adverse effects of a fire (by imposing thermal and mechanical loadings) on the bridge are more or less similar. The same also goes for other types of events (e.g., wind: generates dynamic loadings, flooding: leads to scour and affects stability). Thus, this chapter hypothesizes that the type of an event, *T*, is of importance.[9] As such, the only incidents included are those of an extreme nature (i.e., an incident wherein a bridge undergoes a fire due to a motorcycle collision or a below average precipitation was not included). To better illustrate the magnitude of each type of events used in this study, Table 10.2 describes each event and lists some of the notable bridge failures used in this study.

It could be argued that other structural characteristics and traffic features can also be added into the compiled database such as average daily traffic, etc. Some of these factors were not explicitly included herein due to the lack of information on such factors (especially for older bridges and those of international origin). In any case, the compiled database can be extended upon the availability of new incidents by collecting related information on a new factor (i.e., average daily traffic) and add it to all bridges.

In the aftermath of an extreme event, bridge's integrity can be jeopardized depending on the damage magnitude the bridge experiences. Hence, an identification of this *magnitude of damage, D*, is required. It can be seen that this falls into a classification problem. As such, this systemal analysis was pursued using GEP as the main AI technique in which logistic regression was applied to arrive at proper classification of bridges. Three classes were identified, "no/minor" (that does not require shutting down a bridge), "major" (where a bridge would need to undergo major repairs) and "collapse" (where a bridge partially or fully collapses).

## 10.5 Development and Validation of Artificial Intelligence Tool

Now that the databases on the above three cases are compiled, the AI analysis using ANN, GA and GEP is to be carried out.

[9]A keynote to remember is that due to lack of proper documentations on incident magnitudes, an accurate and quantitative estimation of the size/intensity of a particular event may not be obtained (i.e., media do not usually provide the exact magnitude of a fire, speed of collision, flooding volume). This aspect, together with others, is further discussed in a subsequent section toward the end of this chapter.

**Table 10.2** Description of magnitude of extreme events and notable bridge incidents

| Event type | Description of considered events |
|---|---|
| Fire | Resulting from collision of vehicles, fuel tankers, barges and full-sized fires due to arson or wildfires/lightning |
| High wind | Effects arising from severe storms, tsunamis, level 3–5 hurricanes and exceptional precipitation |
| Scour/flood | |

*Notable bridge incidents*

| Bridge | Year of incident | City/Country | Location | System | Material | Span (m) | Age | Lanes | Type |
|---|---|---|---|---|---|---|---|---|---|
| Beaver Dam Bridge | 1963 | Canada | Rural | Cable-stayed | Reinforced concrete | 37 | *30* | 2 | Flood |
| Puente Río Abajo | 2017 | USA | Rural | Cable-stayed | Reinforced concrete | 22 | 43 | 2 | Flood |
| Cần Thơ Bridge | 2007 | Vietnam | Sub-urban | Cable-stayed | Composite | 90 | 1 | 4 | Flood |
| Hoover Dam Bypass | 2006 | USA | Sub-urban | Truss/Arch | Prestressed concrete | 323 | 1 | 2 | Wind |
| River Rega Bridge | 1913 | Germany | Sub-urban | Truss/Arch | Reinforced concrete | 65 | *30* | 1 | Flood |
| Haderslis Bridge | 1987 | Switzerland | Sub-urban | Truss/Arch | Reinforced concrete | *25* | 18 | 2 | Flood |
| I-85 Bridge | 2017 | USA | Sub-urban | I-girder | Prestressed concrete | 30 | 64 | 6 | Fire |
| Highway 310 and U.S. 175 | 2008 | Dallas, TX, USA | Sub-urban | I-girder | Prestressed concrete | *20* | 21 | 4 | Fire |
| Hatchie River Bridge | 1989 | USA | Sub-urban | I-girder | Prestressed concrete | 31 | 15 | 2 | Flood |
| Howard Avenue | 2004 | Bridgeport, CT, USA | Sub-urban | I-girder | Composite | 23 | 10 | 10 | Fire |

In all cases, the databases were first randomly assembled as to avoid influencing the biasness of the AI technique. In all cases, 70–80% of a database is used to train the AI models, while the remaining 20–30% was used to validate/test the outcome of the analysis.

This section highlights the outcome of this analysis.

### *10.5.1 Using ANN and GA to Derive Material Properties*

An ANN is first developed and trained to understand the how thermal and mechanical properties fluctuate under elevated temperature to arrive at a pattern that exemplifies such fluctuations.

Thus, data was collected on measured properties, and those adopted on building codes (i.e., Eurocode 2, AS 4100, AISC), (i.e., each property was collected at target temperatures, that is, 25, 100 °C from all collected studies) as plotted in Figs. 10.8 and 10.9, and this data was input into an ANN. This data is then processed through neurons and hidden layers to obtain AI-based values for thermal and mechanical material properties. These values are then analyzed using GA to derive simple expressions.

Figure 10.10 shows plots of ANN-predicted values and also shows how these values lay within the measured material tests carried out by other researchers. This precision can also be confirmed by examining coefficient of determination ($R^2$) utilized herein and listed in Table 10.3. This metric demonstrates that the hybrid combination of ANN and GA was able to properly trace physio-chemical changes in thermal and mechanical properties of concrete. It is then safe to assume that the ANN/GA predictions can be used to develop expressions for temperature-dependent material models of other materials (Naser 2019b, c).

### *10.5.2 Using GA to Trace Thermal and Structural Elemental Behavior*

The database compiled to attain elemental behavior of RC beams under elevated temperatures was input into the GA model proposed by Searson (2009).

The outcome of this GA analysis is listed in Table 10.4 in the form of two derived expressions.

These expressions were validated using three fitness functions, i.e., coefficient of determination ($R^2$) and mean average error (MAE) to cross-check the predicted outcomes with that recorded in experiments (i.e., measured temperature in rebars). Furthermore, Fig. 10.11 plots predictions of those expressions against that measured in two independent studies. It should be noted that similar expressions to those listed in Table 10.4 can also be extended to other types of structural members (Naser 2019d).

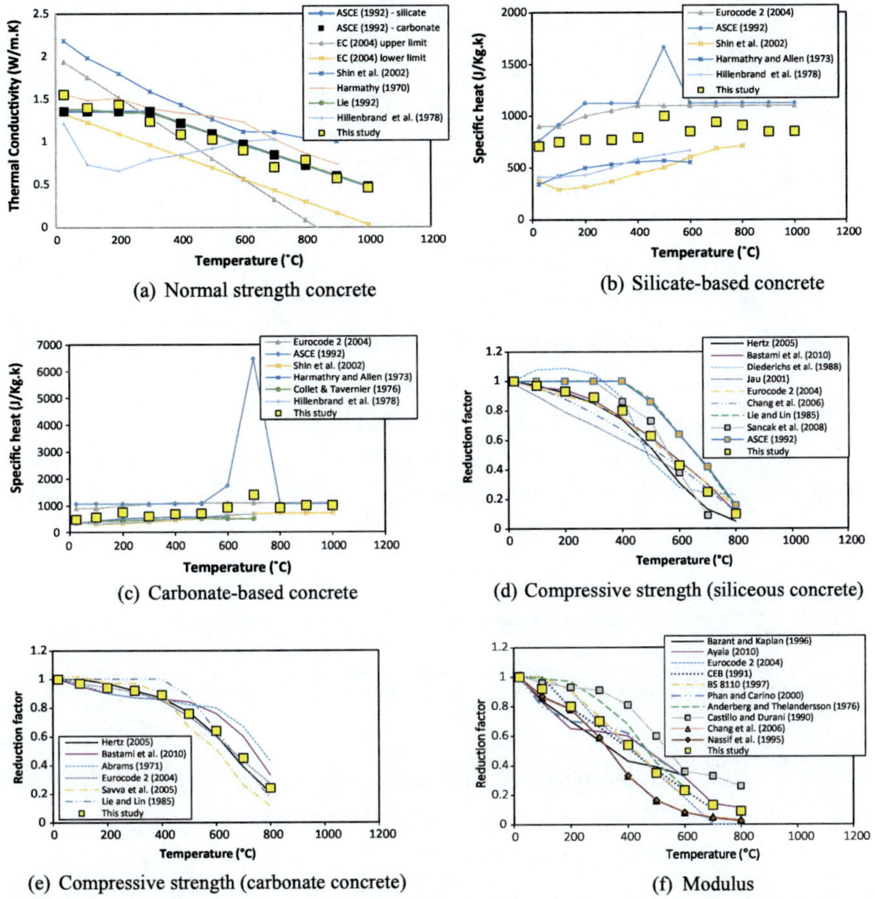


**Fig. 10.10** Material models from ANN/GA analysis as compared to those in fire codes and published studies

## *10.5.3 Classifying Infrastructure Systems with GEP*

The performance of outcome from the GEP carried out analysis is listed in Table 10.5 and Fig. 10.12.

Evidently, GEP algorithm managed to achieve a high accuracy exceeding 85% for identifying all bridges with different damage magnitudes in the training phase. The lowest accuracy in these models was 65%, and this was for the case of bridges expected to undergo major damage (part of the testing phase). Other metrics were also calculated. These include receiver operating characteristic (ROC), sensitivity (proportion of actual positives that are correctly identified), specificity (proportion

**Table 10.3** GA-based expressions for thermal and mechanical material properties of concrete

| Property | Expressions | $R^2$ |
|---|---|---|
| Thermal conductivity | $k = 1.636 + 3.682e^{-12}T^4 + 4.553e^{-7}T^2\cos(T) - 0.001404T - 0.09206\cos(T) - 3.719e^{-15}T^5$ | 99.4 |
| Specific heat | $C_{(\text{siliceous})} = 712 + 0.2185T + 0.4447\tan(T^2)^3 + 0.1053T\sin(5.272e^{-9}T^3) - 0.013\sinh(\tan(T^2)) - 2.242e^{-11}T^4\tan(T^2)$ | 95.6 |
| | $C_{(\text{carbonate})} = 452.9 + 0.8453T + 40.55\tan(564.T) - 0.000291T^2 - 8.783\tan(564.9T) - 0.02121T\tan(564.9T)$ | 99.7 |
| Compressive strength | $f_{c(\text{siliceous})} = 1.016 + 4.918e^{-6}T^2 + 9.411e^{-12}T^4 - 0.000885T - 1.408e^{-8}T^3$ | 99.5 |
| | $f_{c(carbonate)} = 1.013 + 3.337e^{-6}T^2 + 3.693e^{-12}T^4 - 0.0007362T - 7.486e^{-9}T^3$ | 99.8 |
| Modulus | $E = 1.019 + 3.413e^{-12}T^4 - 0.0009538T - 3.054e^{-9}T^3$ | 99.8 |

*T* Temperature in °C

**Table 10.4** GA-based expressions to trace response of RC beams under fire

| Case | Expressions | $R^2$ | MAE |
|---|---|---|---|
| Temperature rise in rebars (°C) | $T = 0.0169tC_b + \frac{182.64t}{C_s} + t\sin(5.21C_b) - 6.43 - 0.0098t^2$ | 95.1 | 24.6 °C |
| Mid-span deflection (mm)* | $\Delta = 36.2\exp(0.023t)\cos(\sin(23040P))\cos(\sin(2.28\times10^{-8}P)) - 0.206C_b - 9.28\sin(f_y) - 12.59\exp(0.0236t) + 4.5 + 0.105t + 2.299\times10^{-6}t^{f_c\rho P}$ | 95.6 | 4.8 mm |

*In the case this expression gives high values for initial deflections, all that is needed is to normalize this deflection

of actual negatives that are correctly identified). These metrics are also in good standing. A referral to Fig. 10.12 shows the developed trees in the outcome of GEP analysis.

## 10.6 Concluding Remarks

This chapter advocates the integration of intelligent data analytics based on AI and machine learning techniques into structural and construction engineering as tools to aid designers and engineers in decision-making.

Future generation of AI are expected to account for complex structural systems and a variety of loading conditions. Thus, these models can accelerate development of

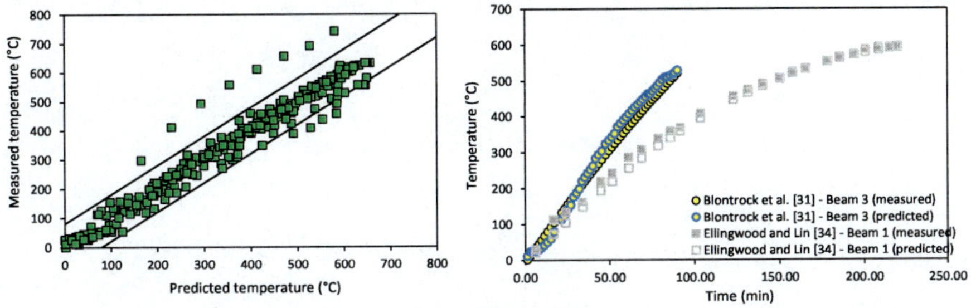


(a) Temperature in rebars

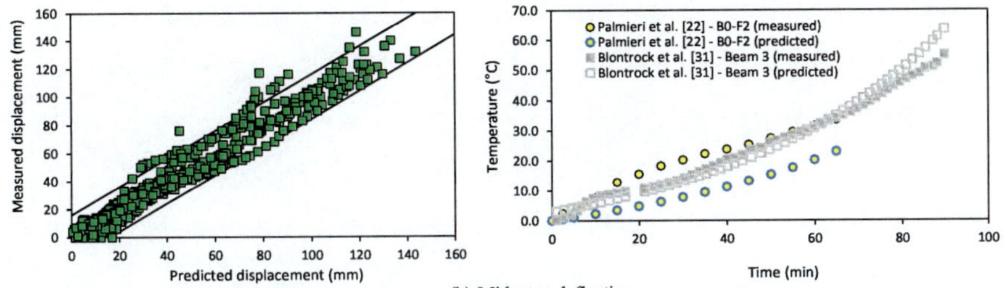


(b) Mid-span deflection

**Fig. 10.11** Performance of GA-derived expressions

**Table 10.5** Performance of GEP model in identifying damage magnitude in bridges

| | Training (%) | Testing (%) | ROC (%) | Sensitivity (%) | Specificity (%) |
|---|---|---|---|---|---|
| No or minor | 85.00 | 80.00 | 83.3 | 85.29 | 84.47 |
| Major | 90.00 | 65.00 | 89.9 | 86.21 | 92.16 |
| Collapse | 93.75 | 70.00 | 88.2 | 76.47 | 98.41 |

future infrastructure with cognitive abilities (Levitt et al. 2008). However, before such integration is fully adopted, the reader must realize that there remain few challenges that hinder utilizing AI into practical applications.

A burning issue is that related to the fact that the accuracy of a given AI model is largely dependent upon the number of inputs and training data points. From this view, the available information on extreme events is not only limited but also varies significantly. As a rule of thumb, having limited data points could potentially hinder the development of properly trained AI models.

On the positive side, this could be overcome by collaborative collection of data points (between researchers across the globe) as well as through generating new data points (via numerical simulations on validated structural failures and extreme events (Kodur et al. 2017b). The author hopes that upcoming works will lead into developing much intelligent AI platforms.

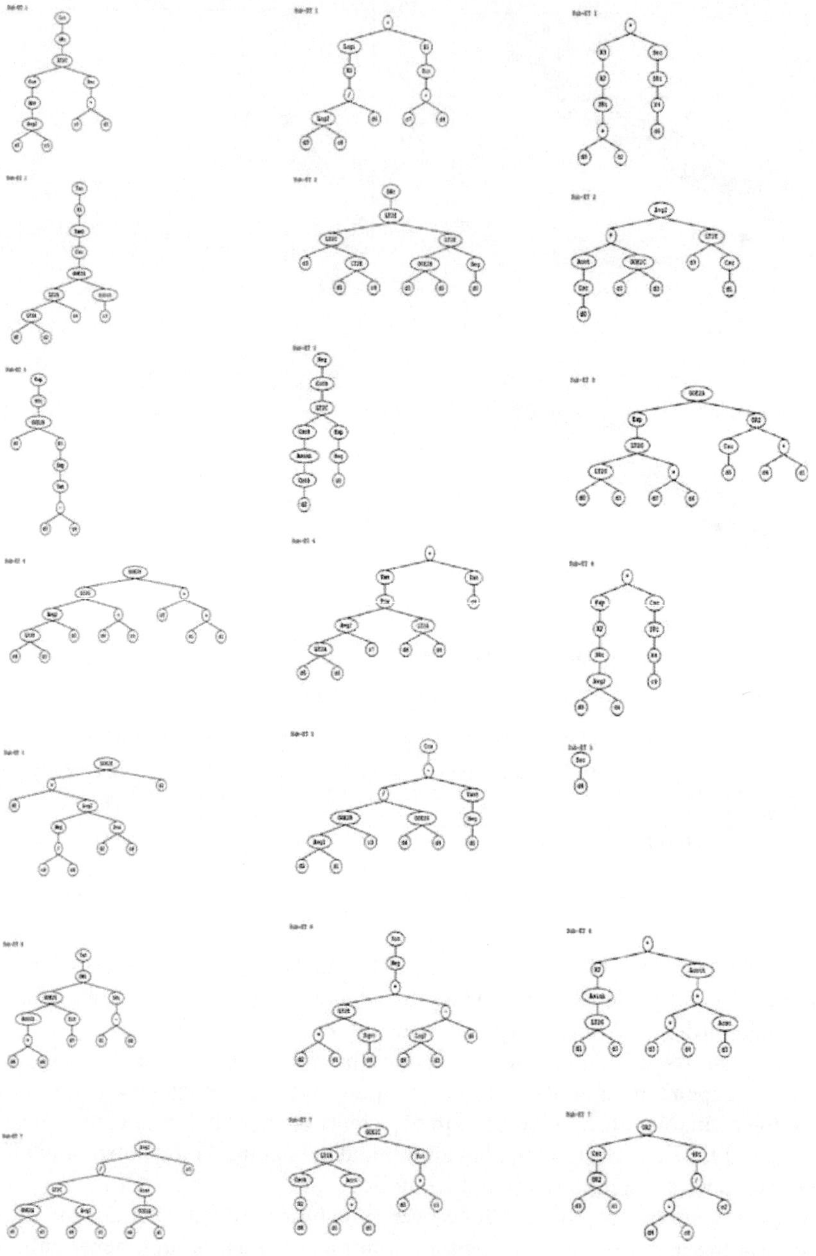

(a) No damage (b) Major damage (c) Collapse

**Fig. 10.12** GEP trees for classifying bridge damage

This chapter fosters AI as a modern analysis tool that could revolutionize the classical fields of structural and construction engineering for extreme conditions. This chapter highlights how various AI techniques, i.e., ANN, GA and GEP could be used to develop an intelligent platform to trace the response of construction materials and civil infrastructure at elemental and systemal levels.

The following three notes could also be deduced from the outcome of this chapter:

- There is an urgent demand to modernize and automate structural and construction engineering. The current advancements in AI technologies could pave the way for a new modern phase of engineering.
- The developed AI models seem to fully comprehend the complex nature of material behavior and thermal and structural response of structures under extreme loading events.
- A number of challenges remain in the way of limiting the integration of AI technologies, and these challenges can be overcame through collaborative works within the next few years.

## References


1 year since Atlanta's infamous I-85 bridge collapse |WSB-TV. https://www.wsbtv.com/news/local/1-year-since-atlantas-infamous-i-85-bridge-collapse/723090006. Accessed 10 Jun 2019

AASHTO LRFD bridge design specifications, 8th Edition (2017)

Abdalla JA, Hawileh R (2011) Modeling and simulation of low-cycle fatigue life of steel reinforcing bars using artificial neural network. J Franklin Institute

Alavi AH, Gandomi AH, Sahab MG, Gandomi M (2010) Multi expression programming: a new approach to formulation of soil classification. Eng Comput 26:111–118. https://doi.org/10.1007/s00366-009-0140-7

Albuquerque GL, Silva AB, Rodrigues JPC, Silva VP (2018) Behavior of thermally restrained RC beams in case of fire. Eng Struct 174:407–417. https://doi.org/10.1016/j.engstruct.2018.07.075

Arioz O (2007) Effects of elevated temperatures on properties of concrete. Fire Saf J. https://doi.org/10.1016/j.firesaf.2007.01.003

ASCE (2017) ASCE infrastructure report card. https://www.infrastructurereportcard.org/cat-item/bridges/. Accessed 23 May 2019

ASTM (2016) E119-16—Standard test methods for fire tests of building construction and materials. Am Soc Test Mater

Biezma MV, Schanack F Collapse of steel bridges. https://doi.org/10.1061/ASCE0887-3828200721:5398

Bisby LA (2016) Structural mechanics. In: SFPE handbook of fire protection engineering, Fifth Edition

Bruneau M, Reinhorn A (2006) Overview of the resilience concept. In: 8th US national conference on earthquake engineering

Buchanan AH, Abu AK (2016) Fire safety in buildings. In: Structural design for fire safety

Carlos TB, Rodrigues JPC, de Lima RCA, Dhima D (2018) Experimental analysis on flexural behaviour of RC beams strengthened with CFRP laminates and under fire conditions. Compos Struct 189:516–528. https://doi.org/10.1016/j.compstruct.2018.01.094

Chen SM, Tan JM (1994) Handling multicriteria fuzzy decision-making problems based on vague set theory. Fuzzy Sets Syst. https://doi.org/10.1016/0165-0114(94)90084-1

Choi EGG, Shin YSS (2011) The structural behavior and simplified thermal analysis of normal-strength and high-strength concrete beams under fire. Eng Struct 33:1123–1132. https://doi.org/10.1016/J.ENGSTRUCT.2010.12.030

Cook W, Barr PJ, Halling MW (2015) Bridge failure rate. J Perform Constr Facil 29:04014080. https://doi.org/10.1061/(ASCE)CF.1943-5509.0000571

Décugis J, Gautronneau V, Jérémie P (2019) Six questions sur l'incendie de Notre-Dame de Paris—Le Parisien. http://www.leparisien.fr/faits-divers/six-questions-sur-l-incendie-de-notre-dame-de-paris-15-04-2019-8054094.php. Accessed 27 Jun 2019

DesRoches R (2006) Hurricane Katrina : performance of transportation systems. American Society of Civil Engineers

Ding L, Rangaraju P, Poursaee A (2019) Application of generalized regression neural network method for corrosion modeling of steel embedded in soil. Soils Found. https://doi.org/10.1016/j.sandf.2018.12.016

Dwaikat MB, Kodur VKR (2009) Response of restrained concrete beams under design fire exposure. J Struct Eng. https://doi.org/10.1061/(asce)st.1943-541x.0000058

Eagar TW, Musso C (2001) Why did the World Trade Center collapse? Science, engineering, and speculation. JOM. https://doi.org/10.1007/s11837-001-0003-1

Ellingwood B, Lin TD (2007) Flexure and shear behavior of concrete beams during fires. J Struct Eng 117:440–458. https://doi.org/10.1061/(asce)0733-9445(1991)117:2(440)

Ferreira C (2001) Gene expression programming: a new adaptive algorithm for solving problems. Complex Syst 13

Fu Z, Ji B, Cheng M, Maeno H (2012) Statistical analysis of the causes of bridge collapse in China. Forensic Engineering 2012. American Society of Civil Engineers, Reston, VA, pp 75–83

Garlock M, Paya-Zaforteza I, Kodur V, Gu L (2012) Fire hazard in bridges: review, assessment and repair strategies. Eng Struct. https://doi.org/10.1016/j.engstruct.2011.11.002

Goldberg DE, Holland JH (1988) Genetic algorithms and machine learning. Mach, Learn

Harik IE, Shaaban AM, Gesund H et al (1990) United states bridge failures, 1951–1988. J Perform Constr Facil 4:272–277. https://doi.org/10.1061/(ASCE)0887-3828(1990)4:4(272)

Hawileh RA, Naser M, Rasheed HA (2011) Thermal-stress finite element analysis of CFRP strengthened concrete beam exposed to top surface fire loading. Mech Adv Mater Struct 18:172–180. https://doi.org/10.1080/15376494.2010.499019

Holling CS (2003) Resilience and stability of ecological systems. Annu Rev Ecol Syst 4:1–23. https://doi.org/10.1146/annurev.es.04.110173.000245

Hudson S, Cormie D, Tufton E, Inglis S (2012) Engineering resilient infrastructure. Proc Inst Civ Eng—Civ Eng. https://doi.org/10.1680/cien.11.00065

Jafari S, Mahini SS (2017) Lightweight concrete design using gene expression programing. Constr Build Mater. https://doi.org/10.1016/j.conbuildmat.2017.01.120

Jahangiri A, Rakha HA (2015) Applying machine learning techniques to transportation mode recognition using mobile phone sensor data. IEEE Trans Intell Transp Syst 16:2406–2417. https://doi.org/10.1109/TITS.2015.2405759

Jiangtao Y, Yichao W, Kexu H et al (2017) The performance of near-surface mounted CFRP strengthened RC beam in fire. Fire Saf J 90:86–94. https://doi.org/10.1016/j.firesaf.2017.04.031

Khorasani NE, Garlock M, Gardoni P (2016) Probabilistic performance-based evaluation of a tall steel moment resisting frame under post-earthquake fres. J Struct Fire Eng. https://doi.org/10.1108/JSFE-09-2016-014

Kim Y, Chen YS, Linderman K (2015) Supply network disruption and resilience: a network structural perspective. J Oper Manag. https://doi.org/10.1016/j.jom.2014.10.006

Kiremidjian A, Moore J, Fan YY et al (2007) Seismic risk assessment of transportation network systems. J Earthq Eng. https://doi.org/10.1080/13632460701285277

Kisi Ö, Çobaner M (2009) Modeling river stage-discharge relationships using different neural network computing techniques. CLEAN—Soil, Air, Water 37:160–169. https://doi.org/10.1002/clen.200800010

Kleindorfer PR, Saad GH (2009) Managing disruption risks in supply chains. Prod Oper Manag 14:53–68. https://doi.org/10.1111/j.1937-5956.2005.tb00009.x
Kodur V (2014) Properties of concrete at elevated temperatures. ISRN Civ, Eng
Kodur VKR, Bhatt PP (2018) A numerical approach for modeling response of fiber reinforced polymer strengthened concrete slabs exposed to fire. Compos Struct 187:226–240. https://doi.org/10.1016/J.COMPSTRUCT.2017.12.051
Kodur VKR, Dwaikat M (2007) Performance-based fire safety design of reinforced concrete beams. J Fire Prot Eng. https://doi.org/10.1177/1042391507077198
Kodur VKR, Naser MZ (2013) Importance factor for design of bridges against fire hazard. Eng Struct 54:207–220. https://doi.org/10.1016/j.engstruct.2013.03.048
Kodur VKR, Garlock M, Iwankiw N (2012) Structures in fire: state-of-the-art, research and training needs. Fire Technol 48:825–839. https://doi.org/10.1007/s10694-011-0247-4
Kodur V, Hibner D, Agrawal A (2017a) Residual response of reinforced concrete columns exposed to design fires. Procedia Eng 210:574–581. https://doi.org/10.1016/J.PROENG.2017.11.116
Kodur VK, Aziz EM, Naser MZ (2017b) Strategies for enhancing fire performance of steel bridges. Eng Struct 131:446. https://doi.org/10.1016/j.engstruct.2016.10.040
Koza JR (1992) A genetic approach to finding a controller to back up a tractor-trailer truck. In: Proceedings of the 1992 American Control Conference
Kushida M, Miyamoto A, Kinoshita K (1997) Development of concrete bridge rating prototype expert system with machine learning. J Comput Civ Eng 11:238–247. https://doi.org/10.1061/(ASCE)0887-3801(1997)11:4(238)
Ladkin P (2012) The Fukushima accident. In: Achieving systems safety—proceedings of the 20th safety-critical systems symposium, SSS 2012
Landrigan PJ, Lioy PJ, Thurston G et al (2004) Health and environmental consequences of the World Trade Center disaster. Environ Health Perspect. https://doi.org/10.1289/ehp.6702
LeLaisserPasserA38 (2019) Français: Feu dans la charpente de Notre Dame. https://commons.wikimedia.org/wiki/File:Incendie_Notre_Dame_de_Paris_cropped.jpg
Levitt RE, Kartam NA, Kunz JC (2008) Artificial intelligence techniques for generating construction project plans. J Constr Eng Manag. https://doi.org/10.1061/(asce)0733-9364(1988)114:3(329)
Levy M, Salvadori M (2002) Why buildings fall down: how structures fail
Lounis Z, McAllister TP (2016) Risk-based decision making for sustainable and resilient infrastructure systems. J Struct Eng. https://doi.org/10.1061/(ASCE)ST.1943-541X.0001545
Lu SC-Y (1991) Building layered models to support engineering decision making: a machine learning approach. J Manuf Sci Eng. doi 10(1115/1):2899617
Michael Grayson J, Pang W, Schiff S (2013) Building envelope failure assessment framework for residential communities subjected to hurricanes. Eng Struct. https://doi.org/10.1016/j.engstruct.2013.01.027
Mohan S (1990) Expert systems applications in construction management and engineering. J Constr Eng Manag 116:87–99. https://doi.org/10.1061/(ASCE)0733-9364(1990)116:1(87)
Mousavi SM, Aminian P, Gandomi AH et al (2012) A new predictive model for compressive strength of HPC using gene expression programming. Adv Eng Softw. https://doi.org/10.1016/j.advengsoft.2011.09.014
Naik TR (2008) Sustainability of concrete construction. Pract Period Struct Des Constr 13(2):98. https://doi.org/10.1061/(ASCE)1084-0680(2008)
Naser M (2016) Response of steel and composite beams subjected to combined shear and fire loading. Michigan State University
Naser MZ (2018) Deriving temperature-dependent material models for structural steel through artificial intelligence. Constr Build Mater 191:56–68. https://doi.org/10.1016/J.CONBUILDMAT.2018.09.186
Naser MZ (2019a) Can past failures help identify vulnerable bridges to extreme events? A biomimetical machine learning approach. Eng Comput https://doi.org/10.1007/s00366-019-00874-2

Naser MZ (2019b) Properties and material models for common construction materials at elevated temperatures. Constr Build Mater 10:192–206. https://doi.org/10.1016/j.conbuildmat.2019.04.182

Naser MZ (2019c) Properties and material models for modern construction materials at elevated temperatures. Comput Mater Sci 160:16–29. https://doi.org/10.1016/J.COMMATSCI.2018.12.055

Naser MZ (2019d) AI-based cognitive framework for evaluating response of concrete structures in extreme conditions. Eng Appl Artif Intell 81:437–449. https://doi.org/10.1016/J.ENGAPPAI.2019.03.004

Naser MZ, Kodur VKR (2015) A probabilistic assessment for classification of bridges against fire hazard. Fire Saf J 76:65–73. https://doi.org/10.1016/j.firesaf.2015.06.001

Naser MZ, Kodur VKR (2018) Cognitive infrastructure—a modern concept for resilient performance under extreme events. Autom Constr 90:253. https://doi.org/10.1016/j.autcon.2018.03.004

Neville A (2012) Properties of concrete, 5th edn. Prentice Hall

NYDOT (2008) Bridge fire incidents in New York State. New York State

O'Rourke TD (2007) Critical Infrastructure, Interdependencies, and Resilience. Bridhe Link Eng Soc. https://doi.org/10.1109/TIGA.1967.4180765

Palmieri A, Matthys S, Taerwe L (2012) Experimental investigation on fire endurance of insulated concrete beams strengthened with near surface mounted FRP bar reinforcement. Compos Part B Eng 43:885–895. https://doi.org/10.1016/j.compositesb.2011.11.061

Peris-Sayol G, Payá-Zaforteza I (2017) Bridge fires database

Peris-Sayol G, Paya-Zaforteza I, Balasch-Parisi S, Alós-Moya J (2017) Detailed analysis of the causes of bridge fires and their associated damage levels. J Perform Constr Facil. https://doi.org/10.1061/(ASCE)CF.1943-5509.0000977

Salehi H, Burgueño R (2018) Emerging artificial intelligence methods in structural engineering. Elsevier

Scheer J (2010) Failed bridges: case studies, causes and consequences

Searson D (2009) GPTIPS Genetic programming and symbolic regression for MATLAB User Guide

Seitllari A (2014) Traffic flow simulation by neuro-fuzzy approach. In: Second international conference on traffic. Belgrade, pp 97–102

Smith D (1976) Bridge failures. Proc Inst Civ Eng 60:367–382. https://doi.org/10.1680/iicep.1976.3389

Steinhauser G, Brandl A, Johnson TE (2014) Comparison of the Chernobyl and Fukushima nuclear accidents: a review of the environmental impacts. Sci Total Environ 470:800

Vdot Bridge inspection Definitions

Wang Y, Burgess I, Wald F, Gillie M (2012) Performance-based fire engineering of structures

Wardhana K, Hadipriono FC (2003) Analysis of recent bridge failures in the United States. J Perform Constr Facil 17:144–150. https://doi.org/10.1061/(ASCE)0887-3828(2003)17:3(144)

Watson D, Adams M (2012) Design for flooding: architecture, landscape, and urban design for resilience to flooding and climate change

Wikipedia (2019) Ponte Morandi. https://upload.wikimedia.org/wikipedia/commons/0/06/Ponte_morandi_crollato.jpg

Xu FY, Zhang MJ, Wang L, Zhang JR (2016) Recent highway bridge collapses in China: review and discussion. J Perform Constr Facil 30:04016030. https://doi.org/10.1061/(ASCE)CF.1943-5509.0000884

Zhu H, Wu G, Zhang L et al (2014) Experimental study on the fire resistance of RC beams strengthened with near-surface-mounted high-Tg BFRP bars. Compos Part B Eng 60:680–687. https://doi.org/10.1016/j.compositesb.2014.01.011

# Chapter 11
# Machine Learning to Derive Unified Material Models for Steel Under Fire Conditions

**M. Z. Naser and Huanting Zhou**

## 11.1 Introduction

Structural steel is an attractive construction material due to its good mechanical properties and ease of erection and sustainability.

Owing to its metallic nature and despite its advantages, steel has an inherently high thermal conductivity and relatively low heat capacity. Given that the majority of structural steel-shaped members comprise thin (slender) plates, temperature rise in such members can be rapid (Naser and Kodur 2017). This often translates into the notion that steel structures exhibit lower fire resistance as compared to those made of concrete or masonry (Kodur and Harmathy 2016). On a positive note, insulated steel structures have been shown to be resilient to the adverse effects of fire in a number of incidents/scenarios (Elhami Khorasani et al. 2019).

From this chapter's perspective, fire-induced material degradations in structural steel can be represented by a set of material models. These models can be obtained by measuring properties from small scale fire tests by means of charts or reduction factors at target temperatures, i.e., 25, 100, 200 … 800 °C. In general, two sets of such temperature-dependent material models are commonly available, "thermal" and "mechanical." The first set of models governs temperature rise/distribution as this can be traced by how density, thermal conductivity, and specific heat of steel changes with temperature rise. The second set of models contains yield strength, Young's

M. Z. Naser (✉)
Glenn Department of Civil Engineering, Clemson University, Clemson, SC 29634, USA
e-mail: mznaser@clemson.edu; m@mznaser.com
URL: https://www.mznaser.com

H. Zhou
School of Civil Engineering and Architecture, Wuhan University of Technology, Wuhan 430072, China
e-mail: zhouht@whut.edu.cn

© Springer Nature Singapore Pte Ltd. 2021

R. C. Deo et al. (eds.), *Intelligent Data Analytics for Decision-Support Systems in Hazard Mitigation*, Springer Transactions in Civil and Environmental Engineering,
https://doi.org/10.1007/978-981-15-5772-9_11

modulus, etc., and hence determines structural (mechanical/deflection) behavior of steel members under elevated temperatures.

The open literature seems to contain a good documentation of various experimental works that aimed to develop temperature-dependent material models for structural steel (Outinen 2007; Naser 2019a; Zhou et al. 2019). One must note that these models vary due to the absence of standardized testing (especially during the last 30 years). This has led to arising discrepancies in test methods, specimen sizes, and processing techniques, etc. On a parallel note, advancements in material sciences and fabrications have also led to other variations arising from differences in metallurgical composition of structural steel produced by fabricators around the globe.

Despite these variations, the structural fire engineering community has mainly adopted two temperature-dependent material models. These models are suggested by ASCE (Lie 1992) and Eurocode 3 (2005) and were widely used in practice and research with good success. However, using these models implies that the microstructure of structural steels is independent of its origin, composition/fabrication, as well as surrounding heating conditions (as the same models were developed for specific heating conditions). The aforementioned models also have never been updated ever since their adoption. A key point to remember is that these two models are not identical and do not represent degradation in steel in the same manner which presents a major challenge when conducting fire resistance analysis. For example, temperature-dependent reduction factor at 500 °C ($k_{500\ °C}$) in ASCE and Eurocode 3 models comes to 0.60 and 0.80, respectively. This implies that the estimated flexural capacity could potentially vary by about 20%. This variation may translate to over/under estimating flexural capacity and in return fire resistance as well especially when checking for complex load actions (i.e., buckling, etc.) or selecting fire insulation type/thickness.

A solution to the above dilemma is to derive a unified material model that considers for temperature-induced variation in properties of structural steel. Such a model also needs to be representative of modern types of steel as well as be of high fidelity and general usability/applicability. From this view, this chapter presents an intelligent approach to arrive at temperature-dependent mechanical material properties of structural steel by leveraging artificial intelligence (AI). For this purpose, property data collected from notable fire codes and standards, in addition to open literature was analyzed using artificial neural networks (ANNs) and genetic algorithms (GA). The presented work hypothesizes that using AI could potentially be successfully used to derive unified and unbiased material models and this could in turn pave the way toward developing standardized fire resistance analysis and design.

## 11.2 Mechanical Properties of Structural Steel at Elevated Temperatures

The main two mechanical properties of structural steel that are needed for fire analysis and design are yield strength, $f_y$, and stiffness, *E*; both of which are measured from outcome of small scale tensile-based tests conducted on steel coupons. Such tests can be carried out in two domains; steady-state and transient. In the first domain, a steel coupon is heated to an elevated temperature, and once such a particular temperature is attained, the coupon is then loaded with tension, and the stress–strain response of the heated coupon is measured. In the second domain, the steel coupon is mechanically loaded while being heated, and its response is also measured.

One can see that it is due to key variances in aforementioned testing that there exist large differences in published material constitutive models. Furthermore, due to the lack of a standard testing procedure, the open literature shows a large amount of test data without reporting specific information such as those related to heating rate, strain rate, and coupon size/shape. Luckily, ASCE and Eurocode 3 provide the following expressions for temperature-dependent reduction factors for mechanical properties of structural steel as:

**Yield Strength**

*ASCE*:

$$\frac{f_{y,T}}{f_y} = \begin{cases} 1.0 + \frac{T}{900\ln\left(\frac{T}{1750}\right)} & \text{for } T \leq 600\,^\circ\text{C} \\ \frac{340-0.34T}{T-240} & \text{for } T > 600\,^\circ\text{C} \end{cases} \tag{11.1}$$

*Eurocode 3*:

$$\frac{f_{y,T}}{f_y} = \begin{Bmatrix} 1.0 & T < 100\,^\circ\text{C} \\ -1.933 \times 10^{-3}T + 1.193 & \text{for } 100 \leq T < 400\,^\circ\text{C} \\ -0.6 \times 10^{-3}T + 0.66 & \text{for } 400 \leq T < 500\,^\circ\text{C} \\ -1.8 \times 10^{-3}T + 1.26 & \text{for } 500 \leq T < 600\,^\circ\text{C} \\ -1.05 \times 10^{-3}T + 0.81 & \text{for } 600 \leq T < 700\,^\circ\text{C} \\ -2.5 \times 10^{-3}T + 0.25 & \text{for } 700 \leq T < 800\,^\circ\text{C} \\ -1.25 \times 10^{-3}T + 0.15 & \text{for } 800 \leq T < 1200\,^\circ\text{C} \end{Bmatrix} \tag{11.2}$$

**Modulus of Elasticity**

*ASCE*:

$$\frac{E_{y,T}}{E_y} = \begin{cases} 1.0 + \frac{T}{2000\ln\left(\frac{T}{1100}\right)} & \text{for } T \leq 600\,^\circ\text{C} \\ \frac{690-0.69T}{T-53.5} & \text{for } T > 600\,^\circ\text{C} \end{cases} \tag{11.3}$$

*Eurocode 3*:

$$\frac{E_{y,T}}{E_y} = \begin{Bmatrix} 1.0 & T < 100\,^{\circ}\mathrm{C} \\ -1.0 \times 10^{-3}T + 1.1 & \text{for } 100 \le T < 500\,^{\circ}\mathrm{C} \\ -2.9 \times 10^{-3}T + 2.05 & \text{for } 500 \le T < 600\,^{\circ}\mathrm{C} \\ -1.8 \times 10^{-3}T + 1.39 & \text{for } 600 \le T < 700\,^{\circ}\mathrm{C} \\ -4.0 \times 10^{-3}T + 0.41 & \text{for } 700 \le T < 800\,^{\circ}\mathrm{C} \\ -2.25 \times 10^{-3}T + 0.27 & \text{for } 800 \le T < 1200\,^{\circ}\mathrm{C} \end{Bmatrix} \tag{11.4}$$

Figure 11.1a–b compares ASCE and Eurocode 3 listed expressions for yield strength and modulus of elasticity, together with other material models obtained from various researchers (Mäkeläinen et al. 1998; Poh 2001; Outinen 2007; Qiang et al. 2012; Huang and Young 2017; He et al. 2019).

These plots show the wide scatter reported on mechanical properties of structural steel; possibly due to variations in steel composition, testing regime, applied heating/loading rates, among others. A look into Fig. 11.1a shows how Eurocode 3 assumes a relatively slower degradation in yield strength as compared to that in ASCE's model. More specifically, Eurocode 3 also expressed that yield strength remains unaffected by rise in temperature up to an exposure to 400 °C. Conversely, the ASCE model assumes a relatively large loss (equal to 30%) at the same temperature. A look into Fig. 11.2b shows that both models presume a similar degradation in stiffness which starts to take place at 150 °C. This early loss in modulus property occurs as only a slight increase in temperature is required to weaken interatomic bonds and initiates plane dislocations within structural steel's microstructure. It should be noted that plots shown as part of Fig. 11.1 represent the database used in this chapter, wherein a much larger database was developed earlier (Mäkeläinen et al. 1998; Poh 2001; Outinen 2007; Qiang et al. 2012; Huang and Young 2017; He et al. 2019; Naser 2019a, b).

## 11.3 Development of an Intelligent Data Analytics (AI) Framework

The earliest modern discussion of AI as a methodology to solve complex problems by mimicking how the brain solves problems was documented in mid-1950s.

Since then, AI has evolved into an exciting interdisciplinary field. It is interesting to note that AI does not follow traditional solution procedures that rely on allocating assumptions. Rather, AI attempts to follow a systematic adaptive learning procedure to analyze a given problem via novel methods and algorithms to identify key factors, examine the effects of these factors and how they interact and then arrive at solutions that can best represent the phenomenon on hand. While a number of AI algorithms and techniques were developed, two main techniques have been primarily used in applications similar to that to be addressed in this chapter. As discussed above, these

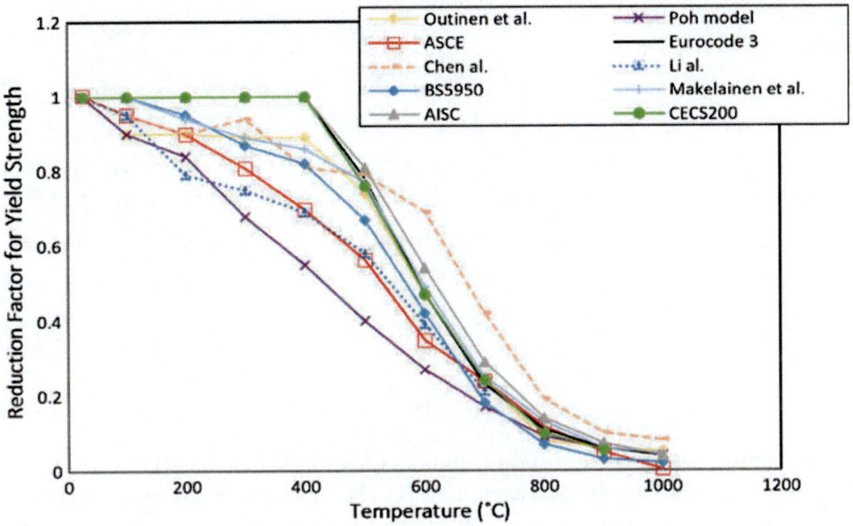


(a) Yield strength

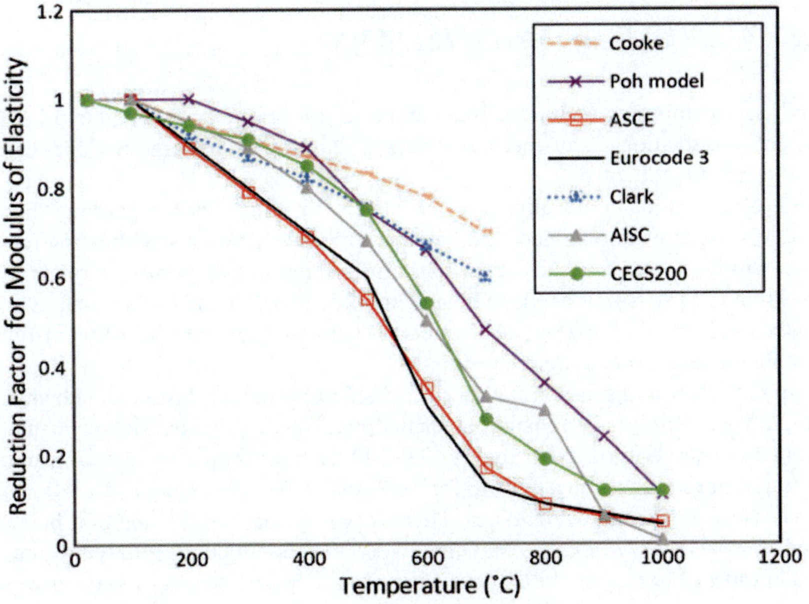


(b) Modulus of elasticity

**Fig. 11.1** Mechanical properties of structural steel as prescribed in the literature

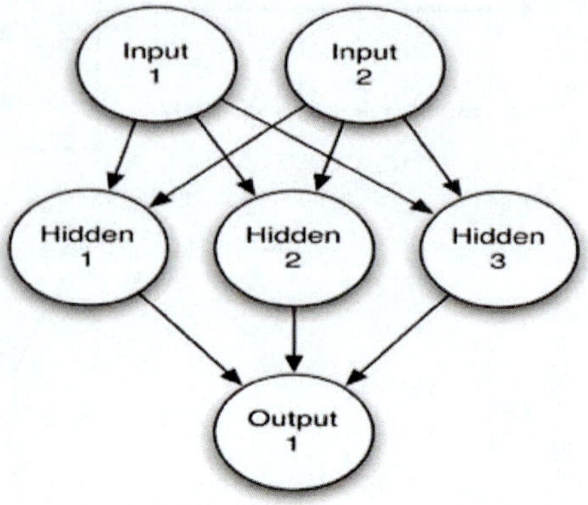


**Fig. 11.2** Typical ANN structure

two techniques are artificial neural networks (ANN), genetic algorithms (GA), etc. (Goldberg and Holland 1988; Ferreira 2001). These two techniques are discussed herein, and a more in-depth discussion can be found elsewhere (Seitllari and Kutay 2018; Naser 2019c).

### *11.3.1 Artificial Neural Networks (ANN)*

An ANN is a computing technique that can be useful in scenarios wherein a large amount of observations (say measured data points from temperature-dependent material tests) are collected.

These observations could also be associated with information regarding "how these observations were obtained," i.e., temperature rise, coupon size, heating rate, exposure duration, etc. An ANN learns patterns hidden in data points. In order for an ANN to be successful, it needs to be able to understand the reason(s) connecting associated random variables (i.e., mechanical properties of structural steel) governing a given phenomenon (i.e., temperature rise).

A typical ANN comprises a number of visible and hidden layers as shown in Fig. 11.2. The first layer is a visible layer that receives the inputs. This layer then sends out these inputs into hidden layer(s). The hidden layers contain processing units called (neurons) which are arranged to form a network for processing and analyzing the inputs through specific weightages. This process of forward flowing of inputs is referred to as *feed-forward network*. This process may continue till satisfying a pre-defined number of iterations and/or once a pre-identified performance metric error is achieved between experimental and ANN-predicted output (Kisi and Çobaner 2009). One of the most commonly used optimization methods is *Leveberg-Marquard,* which evaluates the error in terms of Mean Squared Error (MSE). In this method, if $z$ is the experimental dataset, then MSE can be calculated using Eq. 11.5.

$$\text{MSE} = \frac{1}{z}\sum_{i=0}^{z}(e_i)^2 = \frac{1}{z}\sum_{i=0}^{z}(m_i - p_i)^2 \tag{11.5}$$

where $z$ = the total number of datasets, $e_i$ = the error for each input set, $m_i$ = the measured output, and $p_i$ = the estimated output.

The best fitting model can also be statistically evaluated in terms of MSE as well as coefficient of determination ($R^2$) or mean absolute relative error (MARE) (see equations below).

$$R^2 = \frac{\Sigma(m - p)^2}{\Sigma\left(p - p_{\text{avg}}\right)^2} \tag{11.6}$$

$$\text{MARE} = \frac{1}{z}\sum_{i=1}^{z}\left|\frac{m - p}{m}\right| \times 100 \tag{11.7}$$

where $z$ = the total number of datasets, $e_i$ = the error for each input set, $m_i$ = the measured output, and $p_i$ = the estimated output and, $p_{\text{avg}}$ = the average estimate output.

It is due to this process that the network can learn about relevant data patterns in order to arrive at a linear/nonlinear transformation function that can link the inputs to the output(s).

### *11.3.2 Genetic Algorithm (GA)*

GA is quite different from ANN. GA is an evolutionary search technique that was introduced by Koza (Koza 1992) to attempt to solve problems via mimicking the Darwinian natural selection process. The GA analysis starts by a random population of candidate solutions. These solutions are structured in strings comprising of mathematical functions, symbols, and terminals. For example, a function may contain basic algebraic operations such as addition or multiplication and may also contain power or logic functions (exponential, NOR, etc.). Conversely, a terminal ($T$) may comprise arguments and/or numerical constants/variables. All in, a developed GA model has a tree-like configuration in which branches can extend from a function and end in a terminal (see Fig. 11.3). Once a candidate expression (expected solution) is arrived at, the GA evaluates the fitness of such expression. If the fitness deemed poor, then the candidate expression is manipulated through genetic operations, i.e., reproduction, crossover and mutation to improve its fitness (Koza 1992).

Is manipulated through genetic operations, i.e., reproduction, crossover and mutation to improve its fitness (Koza 1992).

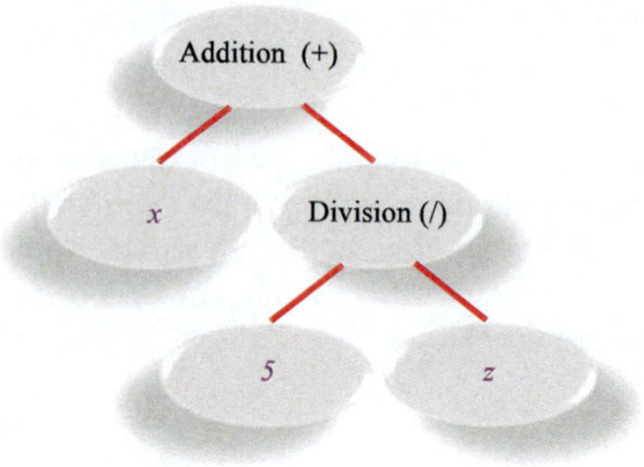


**Fig. 11.3** Typical candidate expression: $x - \frac{5}{z}$ in GA

## 11.4 AI Application on Structural Steel Material Properties

Before starting AI analysis, the database plotted in Fig. 11.1 is randomly arranged as to mitigate any arising biasness of a particular test data or test setup. In all cases, 70% of a database is used to train the AI models, while the remaining 20% was used to validate/test the outcome of the analysis. In designed analysis, an ANN is first developed and trained to understand how temperature-dependent mechanical properties degrade once exposed to rise in temperature. The goal of this ANN is to arrive at a pattern that exemplifies material degradation and generate reduction factors at target temperatures, i.e., 25, 100 °C, etc., as plotted in Fig. 11.1. These values are plotted in Fig. 11.4 and analyzed using GA to derive simple expressions.

Figure 11.4 shows plots of ANN-predicted values and also shows how these values lay within the measured material tests available in fire codes and literature. A similar conclusion can also be seen by examining coefficient of determination ($R^2$) and listed in Table 11.1. This metric demonstrates that the hybrid combination of ANN and GA was able to properly trace temperature-induced degradation in yield strength and modulus properties in structural steel (Naser 2019a, b).

## 11.5 Development of Finite Element Model

This section establishes the validity and prediction capabilities of the proposed ML temperature-based material models. For this, the ML models were used as input into a highly nonlinear three-dimensional finite element (FE) model to trace mechanical response of steel beams. This particular FE model was generated in ANSYS software package to take into account geometric and material nonlinearities and temperature-dependent material properties of steel beams. The steel beam, tested by Wainman and Kirby (Wainman and Kirby 1988), was made of Grade 400 MPa steel and a shape of UB305 × 165. Figure 11.5 shows restraints and loading applied to this beam.

The above beam was discretized with thermal elements (SURF152 and SHELL131) to investigate its thermal response. SURF152 is first overlaid on

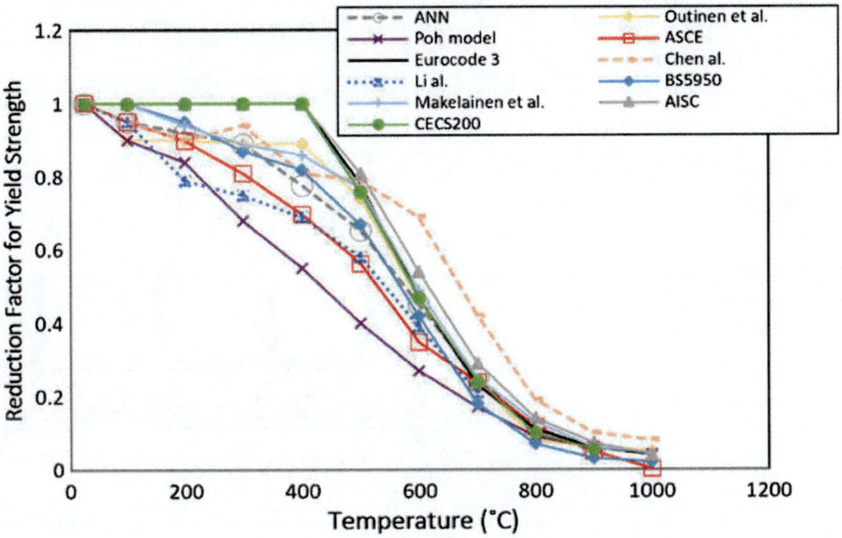


(a) Yield strength

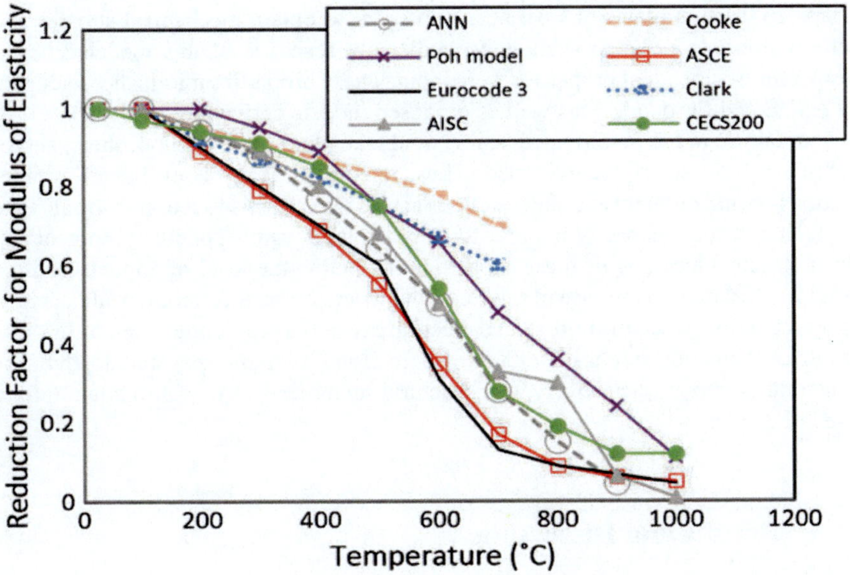


(b) Modulus of elasticity

**Fig. 11.4** Mechanical properties of structural steel as a function of temperature

**Table 11.1** GA-based expressions for mechanical material properties of structural steel

| Property | Expressions |
|---|---|
| Yield strength | $f_y = 1.022 + 4.382\mathrm{e}^{-6}T^2 + 8.204\mathrm{e}^{-18}T^6 - 0.001T - 4.024\mathrm{e}^{-24}T^8 - 8.524\mathrm{e}^{-9}T^3$ |
| Modulus | $E = 1.007 + 4.838\mathrm{e} - 25T^8 - 1.419\mathrm{e}^{-6}T^2 - 1.239e^{-7}T^2\sin(0.962T)^2$ |

*T* Temperature in °C

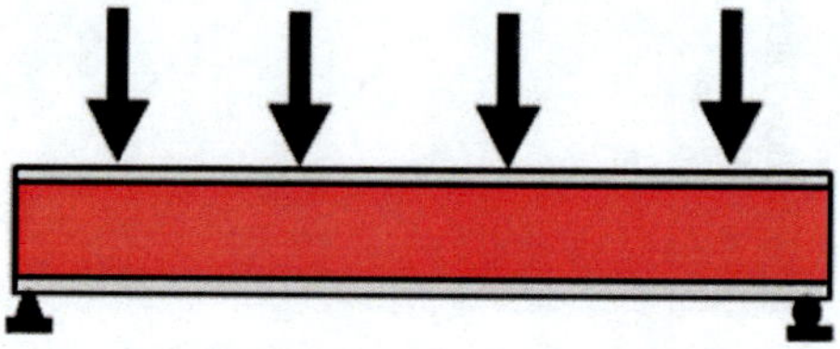

**Fig. 11.5** Steel beams (tested by Wainman and Kirby (1988))

SHELL131 elements as surface elements can model convective and radiative heat transfer from flames to the beam. Upon the completion of the thermal analysis, nodal temperatures are generated, and these are then input as another parameter into the second stage of analysis which comprises a structural model. In the second model, SHELL 131 was replaced with SHELL181 as SHELL181 is formulated to discretize the thin plates of steel beams in order to obtain mechanical stresses and deformations. In order to accurately predict fire response of this modeled beam, temperature-dependent properties were input. These properties mimic those adopted by ASCE and Eurocode 3, as well as proposed models derived by ML.

Fire failure in the investigated beam could take place once flexural, shear, and/or deflection limit states are exceeded. Thus, internal capacity is evaluated by integrating internal moment and shear stresses across the mid-span section as to calculate moment and shear capacity in order to compare them against bending moment and shear effects. Once any of these capacities fall below the level of applied loading, failure is declared. In the case of failure through deflection, this failure could occur at the point in time when the mid-span deflection exceeds a pre-defined limit of $L^2/400d$ or rate of deflection reaches $L^2/9000d$, where $L$ and $d$ are the span and depth of the structural member, respectively, the structural member is also said to attain failure (ASTM 2016).

## 11.6 Results and Discussion

The FE analysis results are plotted in Fig. 11.6. This figure shows the time-history of mid-span deflection during fire exposure time. The same figure also illustrates that the predictions from ASCE and Eurocode 3 models do not agree as well with that measured in fire test as predictions obtained from the proposed ML-derived models. This can be attributed to the fundamental differences in constitutive material models.

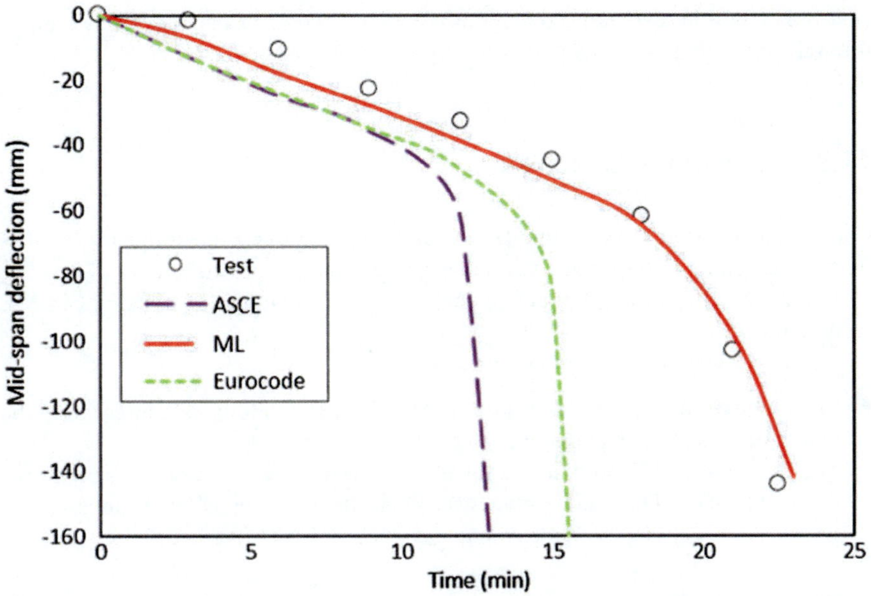


**Fig. 11.6** Prediction of mid-span deflection in UB305 × 165 using different material models

In summary, it can be inferred that using codal provisions in this particular case may not perform as well as ML models.

## 11.7 Future Works

In the practice of fire engineering, academics and design engineers seek to satisfy fire-based requirements. Since behavior of a steel structure might not be known beforehand in most practical scenarios, then predictions from fire calculations/simulations are to be perceived with caution. As material properties are a key input parameter and a first step in any fire analysis, these practitioners face difficulties in making selecting appropriate properties due to the lack of guidance in this area.

This chapter advocates the integration of data intelligent (AI and machine learning) techniques in structural and construction engineering as tools to aid engineers and designers. While this chapter proposes an intelligent approach to derive material models for structural steel using ML techniques, future studies are invited to refine the presented approach for other construction materials as well as complex structural systems and a variety of loading conditions. A burning issue is one that is related to the fact that the accuracy of a given AI model is largely dependent upon the number of available data points. As a rule of thumb, having limited data points could potentially hinder the development of properly trained AI models. On the positive side, this could be overcome by collaborative collection of data points as well as through generating new data points from tests (Kodur et al. 2017).

## 11.8 Concluding Remarks

This chapter explored how AI could revolutionize fire analysis and design by deriving unified material property models suitable for elevated temperatures.

In this, the chapter highlights how such models can be arrived at using two ML techniques, i.e., ANN and GA.

Other conclusions can also be found herein:

- The proposed material models are capable of understanding and representing the nature of structural steel under fire conditions.
- The adoption of ML into structural fire engineering is expected to grow in the following years. This can be best ensured via encouraging collaborations.

## References

ASTM (2016) E119-16—Standard test methods for fire tests of building construction and materials. Am Soc Test Mater

Elhami Khorasani N, Gernay T, Fang C (2019) Parametric study for performance-based fire design of US prototype composite floor systems. J Struct Eng (United States). https://doi.org/10.1061/(ASCE)ST.1943-541X.0002315

Ferreira C (2001) Gene expression programming: a new adaptive algorithm for solving problems. Complex Syst 13

Goldberg DE, Holland JH (1988) Genetic algorithms and machine learning

He A, Liang Y, Zhao O (2019) Experimental and numerical studies of austenitic stainless steel CHS stub columns after exposed to elevated temperatures. J Constr Steel Res 154:293–305. https://doi.org/10.1016/j.jcsr.2018.12.005

Huang Y, Young B (2017) Post-fire behaviour of ferritic stainless steel material. Constr Build Mater. https://doi.org/10.1016/j.conbuildmat.2017.09.082

Kisi Ö, Çobaner M (2009) Modeling river stage-discharge relationships using different neural network computing techniques. CLEAN Soil Air Water 37:160–169. https://doi.org/10.1002/clen.200800010

Kodur VK, Aziz EM, Naser MZ (2017) Strategies for enhancing fire performance of steel bridges. Eng Struct 131. https://doi.org/10.1016/j.engstruct.2016.10.040

Kodur VKR, Harmathy TZ (2016) Properties of building materials. SFPE handbook of fire protection engineering. Springer, New York, pp 277–324

Koza JR (1992) A genetic approach to finding a controller to back up a tractor-trailer truck. In: Proceedings of the 1992 American control conference

Lie TT (ed) (1992) Structural fire protection. American Society of Civil Engineers, New York

Mäkeläinen P, Outinen J, Kesti J (1998) Fire design model for structural steel S420M based upon transient-state tensile test results. J Constr Steel Res. https://doi.org/10.1016/S0143-974X(98)00005-4

Naser MZ (2019a) Properties and material models for common construction materials at elevated temperatures. Constr Build Mater 10:192–206. https://doi.org/10.1016/j.conbuildmat.2019.04.182

Naser MZ (2019b) Properties and material models for modern construction materials at elevated temperatures. Comput Mater Sci 160:16–29. https://doi.org/10.1016/J.COMMATSCI.2018.12.055

Naser MZ (2019c) Heuristic machine cognition to predict fire-induced spalling and fire resistance of concrete structures. Autom Constr 106:102916. https://doi.org/10.1016/J.AUTCON.2019.102916

Naser MZ, Kodur VKRKR (2017) Comparative fire behavior of composite girders under flexural and shear loading. Thin-Walled Struct 116. https://doi.org/10.1016/j.tws.2017.03.003

Outinen J (2007) Mechanical properties of structural steel at elevated temperatures and after cooling down. In: 10th international conference—fire and materials 2007

Poh KW (2001) Stress-strain-temperature relationship for structural steel. J Mater Civ Eng 13(5):371. https://doi.org/10.1061/(ASCE)0899-1561

Qiang X, Bijlaard FSK, Kolstein H (2012) Post-fire mechanical properties of high strength structural steels S460 and S690. Eng Struct. https://doi.org/10.1016/j.engstruct.2011.11.005

Seitllari A, Kutay ME (2018) Soft computing tools to predict progression of percent embedment of aggregates in chip seals. Transp Res Rec 036119811875686

Wainman D, Kirby B (1988) Compendium of UK standard fire test data: unprotected structural steel. British Steel Corporation

Zhou H, Wang W, Wang K, Xu L (2019) Mechanical properties deterioration of high strength steels after high temperature exposure. Constr Build Mater 199:664–675. https://doi.org/10.1016/j.conbuildmat.2018.12.058

# Chapter 12
# Energy Dissipation in Rough Chute: Experimental Approach Versus Artificial Intelligence Modeling

**Sungwon Kim, Farzin Salmasi, Mohammad Ali Ghorbani, Vahid Karimi, Anurag Malik, and Ercan Kahya**

## Abbreviations

$B$ flume width;
$D_{50}$ mean diameter of gravel which used as roughness in chute surface;
$g$ acceleration due to gravity;
$L_1$ upstream length of chute for establishment of uniform flow;
$L_2$ horizontal length of chute;

S. Kim
Department of Railroad Construction and Safety Engineering, Dongyang University, Yeongju 36040, Republic of Korea
e-mail: swkim1968@dyu.ac.kr

F. Salmasi (✉) · M. A. Ghorbani · V. Karimi
Department of Water Engineering, Faculty of Agriculture, University of Tabriz, Tabriz, Iran
e-mail: Salmasi@tabrizu.ac.ir

M. A. Ghorbani
e-mail: Ghorbani@tabrizu.ac.ir; mohammadalighorbani@tdtu.edu.vn

V. Karimi
e-mail: vahid.karimi22@yahoo.com

M. A. Ghorbani
Sustainable Management of Natural Resources and Environment Research Group, Faculty of Environment and Labour Safety, Ton Duc Thang University, Ho Chi Minh City, Vietnam

A. Malik
Department of Soil and Water Conservation Engineering, College of Technology, G.B. Pant, University of Agriculture and Technology, Pantnagar, Uttarakhand 263145, India
e-mail: anuragmalik_swce2014@rediffmail.com

E. Kahya
Department of Civil Engineering, Istanbul Technical University, Istanbul, Turkey
e-mail: kahyae@itu.edu.tr

© Springer Nature Singapore Pte Ltd. 2021

R. C. Deo et al. (eds.), *Intelligent Data Analytics for Decision-Support Systems in Hazard Mitigation*, Springer Transactions in Civil and Environmental Engineering,
https://doi.org/10.1007/978-981-15-5772-9_12

$P$ chute height;
$R_e$ Reynolds number in upstream of chute ($R_e = \frac{\rho V(P+y_0)}{\mu}$);
$S$ slope of the chute;
$V_1$ mean water velocity in downstream of chute before hydraulic jump $\left(V_1 = \frac{Q}{By_1}\right)$;
$V_0$ mean water velocity in upstream of chute $\left(V_0 = \frac{Q}{B(P+y_0)}\right)$;
$W_e$ Weber number($W_e = \frac{\rho L V^2}{\sigma}$);
$y_c$ critical depth ($y_c = \left(q^2/g\right)^{1/3}$);
$y_0$ water depth upstream of chute crest;
$y_1$ downstream depth of water before hydraulic jump;
$y_2$ downstream depth of water after hydraulic jump;
$\mu$ dynamic viscosity of water;
$\rho$ water density and
$\sigma$ surface tension of water

## 12.1 Introduction

Control of energy in high-velocity discharges is one of the challenges in the design and operation of hydraulic structures. Such high-speed flows may occur, for example, over the spillway structure of dams in the irrigation canals, drainage systems of urban areas and steep mountain rivers. The excessive energy of these flows can cause serious damage to the surface of the structure itself and the downstream ends by scouring. Energy dissipaters for dam spillways include the conventional stilling basins and flip buckets (Chinnarasri and Wongwises 2006; Gonzalez et al. 2008).

In irrigation and drainage projects, chutes are used to convey water from a higher to a lower elevation. Chutes are similar to drops except that they carry the water over longer distances and flatter slopes through greater changes in grade (USBR 1978). Rock chutes are also natural river training structures and efficient energy dissipaters. From the hydraulic and environmental point of view, rock chutes have become the important hydraulic structures in the natural river morphology (Pagliara et al. 2015).

Recent studies on particle stability and energy dissipation over rough beds have been focused on the effects of chute slope on energy dissipation. The researches on the effects of slope and energy dissipation have been accomplished continuously (Pagliara and Chiavaccini 2006). They investigated the energy dissipation on block ramps using build models with different slopes (1:4, 1:6, 1:8, 1:10, 1:12) and using stones with different sizes. Results indicated that as increasing the slope, energy dissipation is reduced. They also provided a relationship to predict energy dissipation.

A block ramp offers high resistance to the flow for dissipating more energy compared with other traditional drop. In practical applications, block ramps have a height of several meters and are made of rocks which mean diameter varies between 0.3 and 1.5 m (Pagliara et al. 2008).

Many works are presented in the literature which allows the evaluation of the energy dissipation on a stepped channel including hydraulic behavior in the skimming flow. Researches show that energy dissipation in stepped spillways can be more efficient compared with spillways with smooth bed (without steps) of the same size (Salmasi and Özger 2014). Most researchers have pointed out two different flow regimes including nappe and skimming. The first type occurs in low discharge and high step height, while the second type occurs in spillways with high discharge and low step height (Chanson 1994). The division of flows over stepped spillways into nappe and skimming flow regimes goes back to the 1970–80s (Sorensen 1985), and this type of research has been performed further by Rajaratnam 1990; Peyras et al. 1992; Pegram et al. 1999. On the basis of transformation of nappe to skimming flow regimes, energy dissipation is reduced, and the size of stilling basin is increased (Christodoulou 1993).

Different research experiments on the chute modeling of different types (horizontal, inclined and steps with terminal appendages) and numbers of steps have been carried out. Results provided that the step with terminal appendages yielded more effective than other types of steps on energy dissipation (Chinnarasri and Wongwises 2006). In a different research, a stepped spillway model of *Salado Creek* dam in Texas State was built.

Christodoulou (1993) built the slope stepped spillway with slope of 55° suggested by Rice and Kadavy (1996). This research explained that energy dissipation should be more effective in steep slopes compared with mild slopes conditions. The experimental results indicated that the value of energy dissipation yielded the positive. This result is in agreement with Christodoulou (1993).

Chamani and Rajaratnam (1999) carried out the experiments on stepped spillway with slope of 59°. They presented that the energy dissipation for skimming flow regime was changed between 48 and 63%.

Experiment results suggested that design guidelines for smooth (concrete) stepped spillway may not be suitable to rough stepped chutes including gabion stepped weirs and older stepped chutes with damaged steps. In the aerated flow region, the velocities on rough step chutes were larger than those of smooth chute flows for a given flow rate and dimensionless location from the inception point of free-surface aeration (Gonzalez et al. 2008). In detection of air/water surface roughness in self-aerated chute flows, it was found that the amount of entrapped air (e.g., at the water level where the air concentration is 90%) was reduced when a stepped spillway was considered. In case of decrease in step height on smooth invert chutes, entrapped air became more relevant (Bung 2013). Takahashi and Ohtsu (2014) experimented the aerated flow characteristics of skimming flow regime over stepped chutes. They found that the specific energy was determined based on the aerated flow characteristics, revealing the effects of chute angle and step height. The ratio of specific energies from stepped to smooth chutes ($E_{step}/E_{smooth}$) varied between 0.21 and 0.65.

An experimental study of energy dissipation over stepped gabion spillways with low heights was conducted by Salmasi et al. (2012) in University of Tabriz, Iran. In addition, gabion weirs can be helpful for measurement of discharge in irrigation canals. In this case, determination of discharge coefficient as a function of weir geometry and its hydraulic behavior is important. For example, Salmasi and Sattari (2017) carried out a study for predicting discharge coefficient of rectangular broad-crested gabion weir using M5 tree model. Salmasi and Samadi (2018) conducted an experimental and numerical simulation of flow over stepped spillways. They used computational fluid dynamic (CFD) to verify experimental results. In that study, fluctuation of velocity vectors, shear stress and pressure during the flow on each step is compared. For this purpose, a physical model of the stepped spillway was built with slope at a ratio of 2:1 (horizontal: vertical) and experiments were performed with 10 different flow rates. The numerical simulation was based on RNG k-ε turbulence model. The different application of artificial intelligence (AI) for the energy dissipation in hydraulic structures has been investigated as follows.

Khatibi et al. (2014) experimented the energy dissipation over stepped gabion weirs by carrying out a series of laboratory experiments, building models to explain the experimental data and testing their robustness. They provided the multiple regression equations based on dimensional analysis theory, ANNs and GEP for modeling the energy dissipation over stepped gabion weirs. Regression approach can be also used for modeling of the energy dissipation. Other applications of AI techniques in water engineering topics are as: Raheli et al. (2017) for prediction of biochemical oxygen demand and dissolved oxygen; Khatibi et al. (2011) for discharge routing; Ghorbani et al. (2016) for modeling river flow time series and Ghorbani et al. (2013) for wind speed prediction.

This chapter is focused on experimental approaches and intelligent data analytics (AI) modeling for the relative energy dissipation over low-height chutes. The investigation on energy dissipation is based on a set of experimental data conducted on different physical models. These models comprise six slopes for chutes with three uniform roughness heights on them. To test and evaluate the capability of AI methods for the prediction of energy dissipation, ten methodologies were selected and applied for comparison. These methodologies are multi-layer perceptron artificial neural network (ANN), radial basis function (RBF), gene expression programming (GEP), multiple linear regression (MLR), support vector machines (SVM), decision tree (DT), self-organizing feature map (SOFM), random forest (RF), nearest neighbor (NN) and cascade correlation neural networks (CCNN).

## 12.2 Materials and Method

### *12.2.1 Experimental Setup*

The experimental tests were carried out at the Hydraulic Laboratory of Water Engineering Department, Faculty of Agriculture at the University of Tabriz, Iran. The experimental models were installed in a flume, 0.25 m wide, 10 m long and 0.50 m high. An adjustable sluice gate at the end of the flume allows the creation of required downstream water level. A total of 240 runs have been carried out. Variations of water discharge were between 0.004 and 0.045 $m^3s^{-1}$. In the present study, to investigate the effect of chute slope on energy dissipation, after identification of parameters, 54 physical models were set using six slopes (16.4, 20.6, 22.6, 25.0, 28.7 and 35.0°), three heights (20, 25 and 30 cm) for chutes and three different uniform roughness heights (3.4, 12.7 and 38.1 mm). To prepare the roughness, the stones were used with uniform forms. In addition, stones were attached with adhesive on the metal plates. The relevant plate was installed on chute to use each roughness. Should be noted that the roughness steadily was attached on the metal plates thoroughly, and no distance between stones was considered.

Metal piers were used under the chute to prevent the arc creation by the water weight. A physical model of chute with rough and smooth bed is shown in Fig. 12.1.

To measure the depth of water in the upstream and downstream of chute, ultrasonic sensors were used. Sensors were installed on the upstream and downstream of the flume and able to measure five depths per second. In this experiment, three hundred depths were measured in one minute, and then, the average depth of water was calculated using them.

Physical model of chutes was made of steel plates with 1 mm thickness. They were made of two parts, respectively; first can be explained as a broad-crested weir with length of 2 m for established uniform flow before reaching to the chute crest. Second can be categorized as chutes with six different slopes. Figure 12.2 shows the longitudinal profile of chute and its components. The characteristics of physical models are presented in Table 12.1. For example, "S16.4p20" in Table 12.1 represents

**Fig. 12.1** Experimental physical model of chute with 38.1 mm in diameter gravel on bed (left side) and smooth chute type (right side)

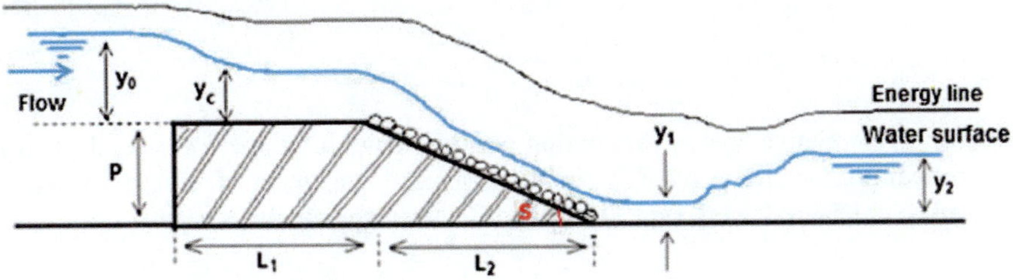


**Fig. 12.2** Longitudinal profile of chute

**Table 12.1** Characteristics of physical models in this study

| Model No. | $B$ (cm) | $P$ (cm) | $L_1$ (cm) | $L_2$ (cm) | Slope (degrees) | Models symbol |
|---|---|---|---|---|---|---|
| 1 | 25 | 20 | 200 | 68.0 | 16.4 | S16.4 p20 |
| 2 | 25 | 25 | 200 | 66.4 | 20.6 | S20.6 p25 |
| 3 | 25 | 20 | 200 | 48.0 | 22.6 | S22.6 p20 |
| 4 | 25 | 30 | 200 | 64.0 | 25.0 | S25.0 p30 |
| 5 | 25 | 25 | 200 | 45.6 | 28.7 | S28.7 p25 |
| 6 | 25 | 30 | 200 | 42.0 | 35.0 | S35.0 p30 |

a chute with slope 16.4° and $P = 20$ cm. In Fig. 12.2 and Table 12.1, $P$ = chute height, $B$ = flume width ($B = 0.25$ cm), $L_1$ = upstream length of chute for establishment of uniform flow, $L_2$ = horizontal length of chute, $y_c$ = critical depth, $y_0$ = water depth upstream of chute crest, $y_1$ = downstream depth of water before hydraulic jump, $y_2$ = downstream depth of water after hydraulic jump and $S$ is the slope of chute, respectively.

Corresponding discharges were measured with using a calibrated triangular sharp crested weir at the outlet of flume.

### 12.2.2 *Dimensional Analysis*

The energy dissipation per unit length of chute ($\Delta E/L$) can be expressed using the following explicit equation:

$$\frac{\Delta E}{L} = f\left(\frac{D_{50}}{y_c}, \frac{y_c}{P}, S\right) \tag{12.1}$$

where $f$ = functional symbol; $\Delta E/L$ = relative energy dissipation (head loss) per unit length of spillway; $D_{50}$ = representative diameter of the material constituting the chute; $P$ = chute height; $S$ = chute slope and $y_c$ = the critical depth in rectangular flume. Energy dissipation per unit length of spillway can be calculated using Eq. (12.2):

$$\frac{\Delta E}{L} = \frac{E_0 - E_1}{L} \tag{12.2}$$

where $\frac{\Delta E}{L}$ = relative energy dissipation per unit length of chute, $E_0$ = total energy in upstream of the chute and $E_1$ = total energy in downstream of the chute. $E_0$ and $E_1$ can be calculated using Eqs. (12.3) and (12.4), respectively.

$$E_0 = P + y_0 + \frac{V_0^2}{2g} = P + y_0 + \frac{q^2}{2g(p + y_0)^2} \tag{12.3}$$

**Table 12.2** Range of parameters variations

| Range of variations | $Q$ ls$^{-1}$ | $D_{50}$ (cm) | $S$ (degrees) | $D_{50}/y_c$ | $y_c/P$ | Water depth on the crest (cm) | $R_e$ |
|---|---|---|---|---|---|---|---|
| Minimum | 4 | 3.38 | 16.4 | 0.00 | 0.09 | 5 | 3406.3 |
| Maximum | 45 | 38.1 | 35.0 | 1.26 | 0.80 | 19 | 42,876.1 |

$$E_1 = y_1 + \frac{V_1^2}{2g} = y_1 + \frac{q^2}{2gy_1^2} \tag{12.4}$$

In this study, mean flow velocity was calculated from continuity equation $\left(V_0 = \frac{Q}{B(P+y_0)}\right)$ and $\left(V_1 = \frac{Q}{By_1}\right)$ where $V_0$ and $V_1$ = mean water velocity in upstream and downstream, respectively. $B$ = flume width ($B$ = 25 cm), $y_0$ and $y_1$ = the depths of water in the upstream and downstream, respectively. In Eq. (12.1), $y_c$ = the critical depth in rectangular flume that can be obtained using ($y_c = \left(q^2/g\right)^{1/3}$).

Pagliara and Chiavaccini (2006) defined the relative energy dissipation as $\Delta E/E_0$. Because this study used "$L$" instead of "$E_0$" in the function, the relative energy dissipation is quite different from identified by Pagliara and Chiavaccini (2006). Therefore, this study uses $\Delta E/L$ for considering the chute length effect in head loss. Table 12.2 shows the range of parameter variations in the present study.

### 12.2.3 *Artificial Neural Network (ANN)*

Multi-layer perceptron (MLP) is a branch of artificial neural network (ANN) which was first introduced by Haykin (1999). The multi-layer perceptron is called a feed-forward network (Cigizoglu and Kisi 2005). Generally, MLP network consists of one input, one hidden and one output layer. The input layer passes the signals to the hidden layer unprocessed. The values are distributed to every node in the middle layer depending on the connection weights between the input and middle (hidden) layers. The weights are determined for all connections. Biases and activation functions are suggested for each of the middle and output nodes.

In this chapter, the number of neurons in the hidden layer was found using trial-and-error method. The MLP training was stopped after 1000 epochs with a threshold value of 0.001. Each neuron in middle and output layers receives the weighted sum of outputs from the previous layer as input. The *NET* output for layer *j* is given as:

$$\text{NET}_h = \sum_{i=1}^{n} W_{ih} O_{pi} + b_h \tag{12.5}$$

where $b_h$ is the neuron threshold value for $h$; $O_{pi}$ is the $i$th output of the previous layer and $W_{ih}$ is the weight between the layers $i$ and $h$. The effective incoming signal,

$NET_h$, is passed through a nonlinear activation function to produce output from each neuron in layers h and O. The architecture of MLP in this study was decided by trial-and-error method in MATLAB version 2016. In the recent time, few applications of ANN have been found in modeling energy dissipation by (Khatibi et al. 2014).

### 12.2.4 Radial Basis Functions

The radial basis function (RBF) is the most widely used architecture in ANN and simpler than MLP neural network. The RBF has an input, hidden and output layer. The hidden layer consists of RBF activation function or $h(x)$ as networks neuron. There are different types of radial basis functions, but the most widely used type is the Gaussian function. The basic functions in the hidden layer produce a localized response to the input. That is, each hidden neuron has a localized receptive field (Ghorbani et al. 2016).

### 12.2.5 Genetic Expression Programming

The concept of genetic expression programming (GEP) was developed by Ferreira (2001) using the fundamental principles of the genetic algorithms (GA) and genetic programming (GP). GEP is a procedure that mimics biological evolution to create a computer program to model some phenomenon. The problems are encoded in linear chromosomes of fixed length as a computer program. GEP performs the symbolic regression using most of the genetic operators of genetic algorithm (GA). GEP encoded as simple strings of fixed length (chromosomes) which are subsequently expressed as expression trees of different sizes and shapes (Sonebi and Cevik 2009). GEP algorithm begins by selecting the five elements such as the function set, terminal set, fitness function, control parameters and stop condition. In this study, GeneXpro tool was applied to perform GEP for energy dissipation in rough chute. The detailed information and applications of GEP can be found in modeling energy dissipation (Khatibi et al. 2014) and suspended sediment modeling (Kisi et al. 2006).

### 12.2.6 Multiple Linear Regression

Multiple linear regression (MLR) is a multivariate statistical technique used to model the linear correlation between a single dependent variable and two or more independent variables (Malik and Kumar 2015; Malik et al. 2017). The regression equation can be written as:

$$Y = \beta_0 + \beta_1 X_1 + \beta_2 X_2 + \cdots + \beta_k X_n \quad (12.6)$$

where $Y$ is the dependent variable; $X_1, X_2, \ldots, X_n$ are the independent variables and $\beta_0, \beta_1, \beta_2, \ldots, \beta_k$ are the regression coefficients. In this study, the values of these coefficients were determined using SPSS statistics software.

### *12.2.7 Support Vector Machine*

The concept of support vector machine (SVM) was first given by Vapnik (1995), which is a supervised learning algorithm, and used for classification and regression. The formulation of SVM includes the principle of structural risk minimization (SRM) on traditional empirical risk minimization (ERM), employed by conventional neural networks. In SVM, the kernel function (polynomial, gaussian radial basis, linear and sigmoid) is used to perform all the operations in the input space not in the potentially high-dimensional feature space. The detailed information and applications of SVM can be found in lake-level fluctuation estimation (Çimen and Kisi 2009) and pan evaporation simulation (Goyal et al. 2014).

### *12.2.8 Decision Tree*

Decision tree (DT) model is employed to classify data and find the regression model for data by extracting logical values involved in the dataset. In general, decision tree model continuously breaks down the dataset into smaller subsets according to the priority of features till to reach an appropriate breakdown level. A decision tree model has three types of nodes viz: (i) a root node; (ii) internal nodes and (iii) leaf or terminal nodes. The results come out in the form of a tree with decision and leaf nodes which can be easily interpreted. In decision tree, each leaf node is assigned a class label. The topmost decision node of a tree as the root node corresponds to the best predictor from which the classification process is begun and comes down to make sub-branches of leaf nodes.

### *12.2.9 Self-organizing Feature Map (SOFM)*

Kohonen (1982) introduced the concept of self-organizing feature map (SOFM) algorithm, which is effective and powerful clustering method. The structure of the SOFM is composed of input layer, clustering layer and output layer. The size of SOFM, i.e., rows $\times$ column metric, neighborhood shape, hidden layer, processing elements, activation function and learning algorithm, was decided using error-and-trial procedure, to train and test the obtained networks based on training and testing datasets, respectively. In SOFM, neighborhood function is used to preserve the topological features of the input space. SOFM projects high-dimensional data onto smaller dimension while preserving neighborhoods, thus facilitating the detection of the inherent structure and the interrelationship of data. The detailed information about SOFM was given by Chang et al. (2007) for flood forecasting and Chang et al. (2010) in estimation of daily pan evaporation.

### *12.2.10 Random Forest (RF)*

Random forest (RF) algorithm is a supervised classification algorithm and used for classification and the regression kind of problems. As the name suggest, this algorithm creates the forest with a number of trees. Each tree is fit using a small, randomly selected subset of predictor variables, resulting in reduced correlation between trees (Breiman 2001). The RF provides reliable error estimates by using the so-called out-of-bag (OOB) data. RF estimates covariate importance by changing the order of arrangement/arranging in all possible ways the values of each covariate in the OOB sample and predicting OOB samples using the permuted variable. The change in OOB error is then an indication of importance of that covariate. In general, the RF is created using the following procedure: (i) randomly select K features from total m features; (ii) calculate the node d using the best split point; (iii) split the node into daughter nodes using the best split; (iv) repeat the steps 1–3 until one number of nodes has been reached; (iv) build forest by repeating steps 1–4 for *n* number times to create *n* number of trees (Hengl et al. 2015; Were et al. 2015).

### *12.2.11 Nearest Neighbor (NN)*

Nearest neighbor (NN) approach is based on the principle of similarity and proximity of data.

The reference dataset contains a wide variety of information and explored it to the target based on the selected input attributes. The similarity distance between the targets is measured in terms of Euclidean distance after normalization and rescaling of the attribute data in the reference dataset.

This ensures that different input attributes will receive equal weight (Nemes et al. 2006).

### *12.2.12 Cascade Correlation Neural Network (CCNN)*

Fahlman and Lebiere (1990) developed the cascade correlation neural networks (CCNN), which combine the idea of incremental structure and learning during its training.

In CCNN, training starts with a minimal network consisting of an input and output layer without hidden layer. Training phase stopped when no longer reduces the residual error of this phase and enters the next phase for the training of the potential hidden node. The potential hidden node is associated with connection weights from the input layer and all pre-existing hidden nodes but not toward the output layer. The connection weights associated with the potential hidden nodes are optimized by the gradient ascent method to maximize the correlation between its output and the residual error of CCNN. When a potential hidden node is trained, connection weights associated with the output layer are kept unchanged. When a potential hidden node is added to CCNN structure, it becomes a new hidden node, and its incoming connection weights are fixed for the remainder of training. After installing a hidden node successfully, the training updates all of the connection weights, which directly feed the output layer. CCNN automatically constructs a suitable structure for a given problem. In the recent decade, limited application of CCNN found in river flow forecasting (Karunanithi et al. 1994), river stage forecasting (Thirumalaiah and Deo 1998), stream flow modeling (Kişi 2007) and pan evaporation modeling (Kim et al. 2014).

### *12.2.13 Performance Evaluation*

Performance measures are assessed by comparing modeled values with their corresponding observed values using the following criteria:

I. Root-mean-square error (RMSE) (Willmott and Matsuura 2005)

$$\text{RMSE} = \sqrt{\frac{\sum_{i=1}^{N}\left(y_p - y_0\right)^2}{N}}, \quad 0 < \text{RMSE} < \infty \tag{12.7}$$

II. Nash–Sutcliffe coefficient (NS) (Nash and Sutcliffe 1970)

$$\text{NS} = 1 - \left[\frac{\sum_{i=1}^{N}\left(y_0 - y_p\right)^2}{\sum_{i=1}^{N}\left(y_0 - \overline{y}_0\right)^2}\right], \quad -\infty < \text{NS} \leq 1 \tag{12.8}$$

III. Willmott's index of agreement (WI) (Willmott et al. 2012)

$$\text{WI} = 1 - \left[\frac{\sum_{i=1}^{N}\left(y_0 - y_p\right)^2}{\sum_{i=1}^{N}\left(\left|y_p - \overline{y}_0\right| + \left|y_0 - \overline{y}_0\right|\right)^2}\right], \quad 0 < \text{WI} \leq 1 \tag{12.9}$$

IV. Relative root-mean-square error (RRMSE) (Notton et al. 2013)

$$\text{RRMSE} = \frac{\text{RMSE}}{\overline{y}_0} \tag{12.10}$$

where $N$ = number of observations; $y_0$ = observed data; $y_p$ =predicted data; $\overline{y}_0$ = mean of the observed data and $\overline{y}_p$ = mean of the predicted data.

Other than statistical metrics described in Eqs. (12.7–12.10), the authors also utilized Taylor diagram (Taylor 2001) to demonstrate their importance in hydraulic engineering. Essentially, a Taylor diagram is a visual representation of the observed and modeled data in a climate-related assessment (IPCC 2007). It is noteworthy that, Taylor introduced a single diagram to summarize multiple aspects of the model and observed parameters, including the RMSE and the correlation information that is contained simultaneously in the evaluation of the respective model. In order to provide a complete picture of model assessment, the ratio of the variance is computed to indicate the relative amplitude of the forecasted and observed variations, whereas the correlation in the plot can indicate whether the fields have similar patterns of variation, regardless of amplitude. Moreover, the normalized RMSE can be resolved into a part due to differences in the overall means and a part due to errors in the pattern of variations (Taylor 2001; IPCC 2007; Gleckler et al. 2008).

## 12.3 Results and Discussion

As mentioned in introduction, the main objective of this study is to investigate the relative energy dissipation using the experimental data which were accomplished in different physical models. A regression-based and nine AI-based models were applied to predict the relative energy dissipation. The capability and efficiency of developed models were assessed using the performance measures, including RMSE, NS, WI and RRMSE, respectively.

Table 12.3 presents the input combinations for the structure of developed models in this study. The input combinations consist of one ($D_{50}/y_c$), two ($D_{50}/y_c$ and $S$) and three ($D_{50}/y_c$, $S$ and $y_c/p$) inputs, respectively. In addition, output is only made of the relative energy dissipation per unit length of spillway ($\Delta E/L$). It can be judged from Table 12.3 that the individual model has the corresponding three sub-models, respectively. Therefore, all developed models can be classified as 30 models in this study.

Table 12.4 summarizes the results of performance measures for the capability and efficiency of developed models.

Evidently, any model producing higher NS and WI values and lower RMSE and RRMSE values so it offers a better efficiency than the others. When the input combination of developed models was compared for the prediction of relative energy dissipation in the training and testing phases, the comparison explained that three input combinations provided better results than other combinations except for RBF2 and SOFM2 models among all developed models, respectively. It can be also suggested

**Table 12.3** Input combinations for the structure of developed models

| No. | Input combinations | Output | ANN | RBF | SVM | GEP | MLR | DT | SOFM | RF | NN | CCNN |
|---|---|---|---|---|---|---|---|---|---|---|---|---|
| 1 | $D_{50}/y_c$ | $\Delta E/L$ | ANN1 | RBF1 | SVM1 | GEP1 | MLR1 | DT1 | SOFM1 | RF1 | NN1 | CCNN1 |
| 2 | $D_{50}/y_c$, $S$ | $\Delta E/L$ | ANN2 | RBF2 | SVM2 | GEP2 | MLR2 | DT2 | SOFM2 | RF2 | NN2 | CCNN2 |
| 3 | $D_{50}/y_c$, $S$, $y_c/P$ | $\Delta E/L$ | ANN3 | RBF3 | SVM3 | GEP3 | MLR3 | DT3 | SOFM3 | RF3 | NN3 | CCNN3 |

**Table 12.4** Results of performance measures for efficiency of developed models

| No. | Model | Model structure[a] | Training period | | | | Testing period | | | |
|---|---|---|---|---|---|---|---|---|---|---|
| | | | RMSE | NS | WI | RRMSE[b] (%) | RMSE | NS | WI | RRMSE[b] (%) |
| 1 | ANN1 | 1-8-1 | 0.086 | 0.533 | 0.808 | 25.50 | 0.120 | 0.318 | 0.764 | 26.77 |
| 2 | ANN2 | 2-3-1 | 0.039 | 0.906 | 0.974 | 11.42 | 0.083 | 0.672 | 0.909 | 18.55 |
| 3 | **ANN3** | **3-5-1** | **0.022** | **0.971** | **0.992** | **6.39** | **0.081** | **0.687** | **0.924** | **18.14** |
| 4 | RBF1 | HN: 48; *r*: 72; BF: Gaussian | 0.086 | 0.533 | 0.828 | 25.49 | 0.138 | 0.094 | 0.759 | 30.85 |
| 5 | **RBF2** | **HN: 39; r: 26; BF: Gaussian** | **0.043** | **0.883** | **0.968** | **12.78** | **0.076** | **0.724** | **0.917** | **17.02** |
| 6 | RBF3 | HN: 50; *r*: 78; BF: Gaussian | 0.028 | 0.952 | 0.988 | 8.17 | 0.082 | 0.681 | 0.924 | 18.29 |
| 7 | SVM1 | $C$: 3.42; $\varepsilon$: 0.085; $\gamma$: 0.67 | 0.085 | 0.541 | 0.828 | 25.28 | 0.121 | 0.298 | 0.755 | 27.14 |
| 8 | SVM2 | $C$: 11,924.83; $\varepsilon$: 0.017; $\gamma$: 1.96 | 0.036 | 0.916 | 0.978 | 10.80 | 0.089 | 0.621 | 0.909 | 19.95 |
| 9 | **SVM3** | **C: 14,179.29; ε: 0.004; γ: 2.04** | **0.015** | **0.986** | **0.996** | **4.45** | **0.087** | **0.640** | **0.919** | **19.45** |
| 10 | GEP1 | + − × / | 0.086 | 0.538 | 0.821 | 25.37 | 0.101 | 0.513 | 0.810 | 22.61 |
| 11 | GEP2 | + − × / | 0.041 | 0.892 | 0.970 | 12.27 | 0.050 | 0.879 | 0.965 | 11.27 |
| 12 | **GEP3** | **+ − × /** | **0.038** | **0.909** | **0.973** | **11.24** | **0.045** | **0.902** | **0.973** | **10.15** |
| 13 | MLR1 | $\Delta E/L = 0.279 + 0.436 * (D_{50}/y_c)$ | 0.095 | 0.425 | 0.759 | 28.29 | 0.107 | 0.455 | 0.837 | 23.91 |
| 14 | MLR2 | $\Delta E/L = 0.587 - 0.012 * S + 0.428 * (D_{50}/y_c)$ | 0.060 | 0.771 | 0.931 | 17.87 | 0.062 | 0.815 | 0.954 | 13.93 |
| 15 | **MLR3** | $\boldsymbol{\Delta E/L = 0.639 - 0.013 * S - 0.095 * (y_c/P) + 0.422 * (D_{50}/y_c)}$ | **0.058** | **0.787** | **0.937** | **17.22** | **0.060** | **0.831** | **0.960** | **13.32** |
| 16 | DT1 | – | 0.089 | 0.502 | 0.813 | 26.32 | 0.117 | 0.346 | 0.745 | 26.20 |
| 17 | DT2 | – | 0.034 | 0.926 | 0.980 | 10.15 | 0.070 | 0.769 | 0.932 | 15.58 |
| 18 | **DT3** | – | **0.026** | **0.958** | **0.989** | **7.65** | **0.059** | **0.831** | **0.952** | **13.30** |
| 19 | SOFM1 | Architecture of model = 5 × 5; HN = 4 | 0.095 | 0.428 | 0.755 | 28.23 | 0.113 | 0.395 | 0.685 | 25.20 |
| 20 | **SOFM2** | **Architecture of model = 5 × 5; HN = 6** | **0.049** | **0.849** | **0.958** | **14.49** | **0.061** | **0.822** | **0.949** | **13.68** |

(continued)

**Table 12.4** (continued)

| No. | Model | Model structure[a] | Training period | | | | Testing period | | | |
|---|---|---|---|---|---|---|---|---|---|---|
| | | | RMSE | NS | WI | RRMSE[b] (%) | RMSE | NS | WI | RRMSE[b] (%) |
| 21 | SOFM3 | Architecture of model = 5 × 5; HN = 5 | 0.046 | 0.866 | 0.963 | 13.64 | 0.063 | 0.811 | 0.947 | 14.08 |
| 22 | RF1 | No. of trees: 200; leaf size: 5 | 0.060 | 0.775 | 0.926 | 17.71 | 0.121 | 0.301 | 0.704 | 27.09 |
| 23 | RF2 | No. of trees: 200; leaf size: 5 | 0.026 | 0.958 | 0.989 | 7.65 | 0.061 | 0.822 | 0.945 | 13.66 |
| 24 | **RF3** | **No. of trees: 200**; leaf size: 5 | **0.014** | **0.987** | **0.997** | **4.18** | **0.048** | **0.888** | **0.965** | **10.84** |
| 25 | NN1 | DF: Euclidean distance; No. of neighbors: 10 | 0.088 | 0.512 | 0.802 | 26.07 | 0.113 | 0.390 | 0.713 | 25.31 |
| 26 | NN2 | DF: Euclidean distance; No. of neighbors: 5 | 0.038 | 0.911 | 0.975 | 11.16 | 0.068 | 0.780 | 0.926 | 15.20 |
| 27 | **NN3** | **DF: Euclidean distance; No. of neighbors: 2** | **0.027** | **0.955** | **0.988** | **7.95** | **0.053** | **0.866** | **0.964** | **11.88** |
| 28 | CCNN1 | 1-2-1 | 0.084 | 0.544 | 0.836 | 25.19 | 0.101 | 0.508 | 0.809 | 22.72 |
| 29 | CCNN2 | 2-3-1 | 0.039 | 0.903 | 0.975 | 11.62 | 0.021 | 0.911 | 0.975 | 4.84 |
| 30 | **CCNN3** | **3-6-1** | **0.020** | **0.974** | **0.993** | **6.03** | **0.020** | **0.982** | **0.995** | **4.40** |

[a]A model precision based on different ranges of RRMSE (Jamieson et al. 1991); Excellent (RRMSE < 10%); Good (10% < RRMSE < 20%); Fair (20% < RRMSE < 30%); Poor (RRMSE > 30%)
[b]*HN* hidden neurons, *r* spread value, *BF* basic function
*C* magnitude of penalty term, $\varepsilon$ width/deviation of the error margin, $\gamma$ Gaussian radial basis function parameter
*DF* distance function

from Table 12.4 that SVM3 model produced the best results, whereas MLR1 and SOFM1 models provided the worst results among all developed models in the training phase.

In the testing phase, CCNN3 model offered the best efficiency, and RBF1 model gave the worst efficiency, respectively. It can be imagined from Table 12.4 that CCNN3 model (RMSE = 0.020, NS = 0.982, WI = 0.995 and RRMSE = 4.40(%)) is the superior to other developed models for predicting the relative energy dissipation in this study. One advantage of GEP models compared with other AI approaches is classified that they can provide the predictive equations for the addressed problem. Table 12.5 proposed the general GEP expressions for one, two and three input combinations, respectively. The generalization ability of GEP models can estimate the relative energy dissipation using three GEP equations in this study.

The graphical comparisons, including error histograms and scatter plots, were also applied to compare the accuracy of developed models for predicting the relative energy dissipation. Figure 12.3 shows the histogram distributions for the performance measures (RMSE, NS, WI and RRMSE) of outstanding sub-models (ANN3, RBF2, SVM3, GEP3, MLR3, DT3, SOFM2, RF3, NN3 and CCNN3), respectively. Figure 12.3 also explained that CCNN3 model provided the best accuracy from the viewpoint of optical assistances including training and testing phases.

Figure 12.4 displays the observed and predicted relative energy dissipation values for outstanding sub-models, respectively. The centered dash line (dark red) in Fig. 12.4 shows the best-fit line equation. The best equation for $y = ax + b$ pattern can be explained that "$a$" offers near to one and "$b$" approaches near to zero (Seo et al. 2015). It was clearly seen from the best-fit line equations and $R^2$ values that CCNN3 could predict the relative energy dissipation than other models. Therefore, CCNN3 model can be classified as the best one to predict the relative energy dissipation using the performance measures and graphical comparisons in this study. In addition, different experimental data are required to confirm the reliable model choice for predicting the relative energy dissipation.

**Table 12.5** GEP expressions for one, two and three input combinations

| Model | Model structure |
|---|---|
| GEP1 | $\Delta E/L = ((D_{50}/y_c)/(1.704 + (D_{50}/y_c))) + \left((D_{50}/y_c)/\left(\left(74.329 * (D_{50}/y_c)^2 + 0.868\right)\right)\right) + \left(2.929/\left(12.057 + (D_{50}/y_c)^2\right)\right)$ |
| GEP2 | $\Delta E/L = -(0.027) + (6.022/(S - (7.022 * (D_{50}/y_c))) + (D_{50}/y_c)/(0.384 + (4 * (D_{50}/y_c)))$ |
| GEP3 | $\Delta E/L = (-2.266 * (((y_c/P) - 3.08)/(S - (D_{50}/y_c)))) + \left(\left((D_{50}/y_c)/\left((y_c/P) * S^2\right) * (17.202 + (y_c/P) - (D_{50}/y_c))\right)\right) + \left(((D_{50}/y_c)/(2.01 - 1.418 * (D_{50}/y_c))) * \left((-1.418 + (D_{50}/y_c))^2\right)\right)$ |

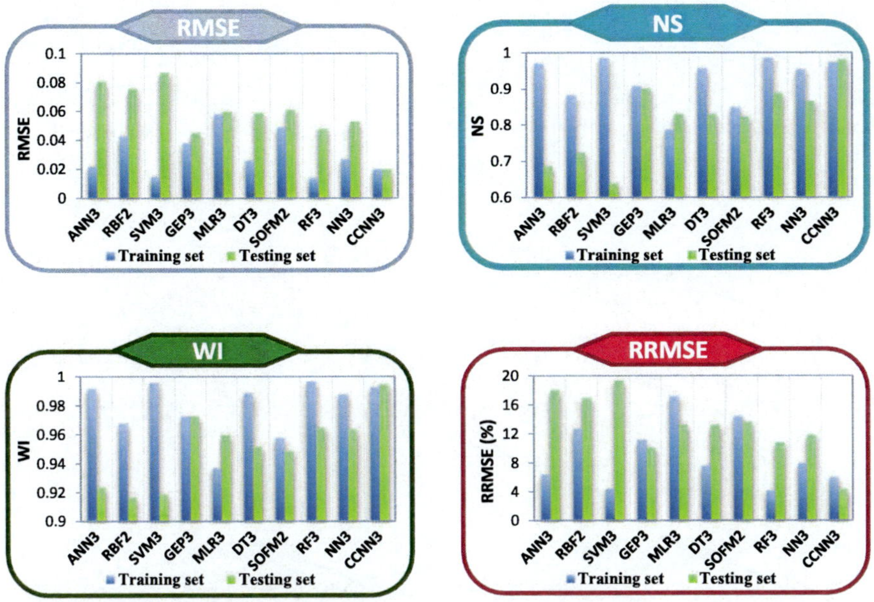


**Fig. 12.3** Histogram of the performance measures for ten techniques

As a further diagnostic tool, a Taylor plot is used in Fig. 12.5 to collectively assess several approximate representations of the developed models for energy dissipation estimation. Taylor diagram gives a method for graphically condensing how intently an example (or an arrangement of examples) matches observations. In its general sense, the Taylor diagram is used to quantify the degree of correspondence between modeled and observed behavior of energy dissipation in the tested dataset in terms of three statistics: the Pearson correlation coefficient, the root-mean-square error and the standard deviation for ten models. The distance from the reference point (observed) is a measure of the centered RMSE (Taylor 2001).

Accordingly, a perfect model (being in full concurrence with the observations) is set apart by the reference point with the correlation coefficient equivalent to 1, and a similar abundancy of varieties contrasted with the observations (Heo et al. 2014). According to the visualization of the results, the CCNN3 results were closer to the observation points than other models. Figure 12.6 shows performance of the proposed models using violin and point density plots. The point density plot effectively reveals the positions of the points, and the violin plot shows the shape of the density mass function or probability density function (PDF) for a dataset (Choubin and Malekian 2017; Choubin et al. 2017). Given the obtained results in Fig. 12.6, it is indicated CCNN3 model density mass is close fit to observed data and has an appropriate performance in relative energy dissipation prediction.

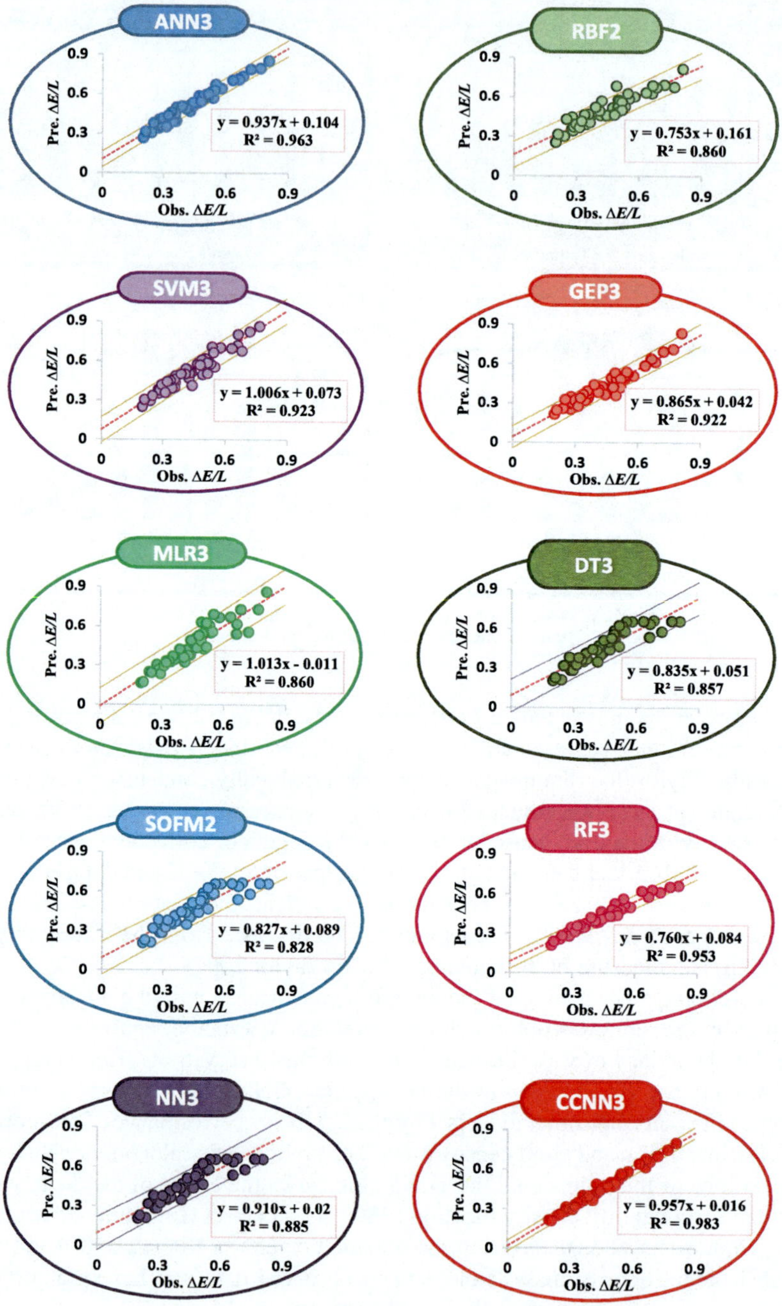


**Fig. 12.4** Comparison for the observed and predicted relative energy dissipation

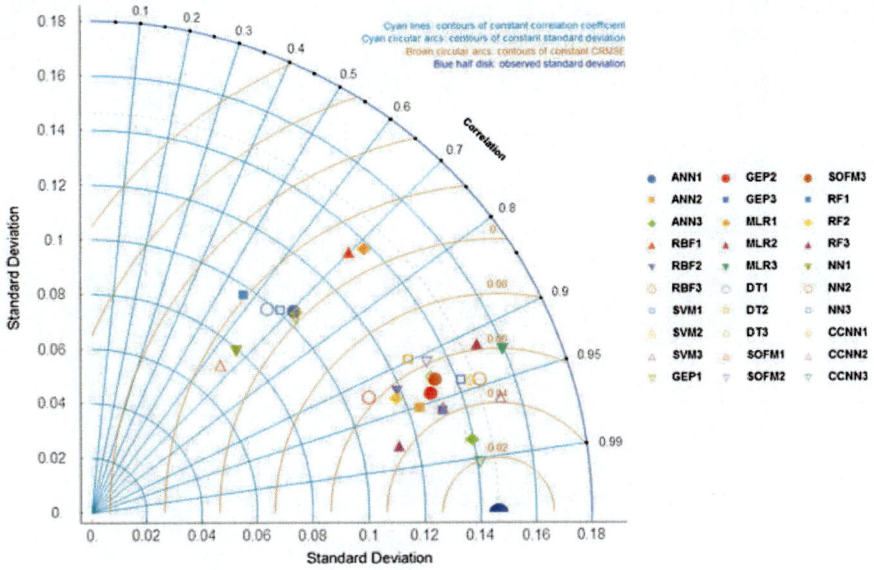


**Fig. 12.5** Taylor's plot for the proposed models

## 12.4 Conclusions

Chutes are hydraulic structures that are used in irrigation canals and storage dams.

These structures convey water from high to the lower levels. Usually, downstream of chute is provided with a stilling basin for energy dissipation. The length and height of stilling basin wall made from reinforced concrete increase cost of construction. Reduction in dimension of stilling basin causes reduction in construction costs. For this purpose, in this study, chute surface is roughened with stone arrangement for creation of more energy loss. Experimental data are collected from 54 physical models of chutes with variable chute slope, chute height and chute surface roughness. Datasets are used to develop both a regression model and nine artificial intelligence (AI) models.

The most proper models for predicting of energy loss over the chute are determined as CCNN3, GEP3 and RF3, respectively. The ratio of $D_{50}/y_c$ has the most impact on energy dissipation of flow over chute, whereas chute slope (S) has lowest impact on energy dissipation.

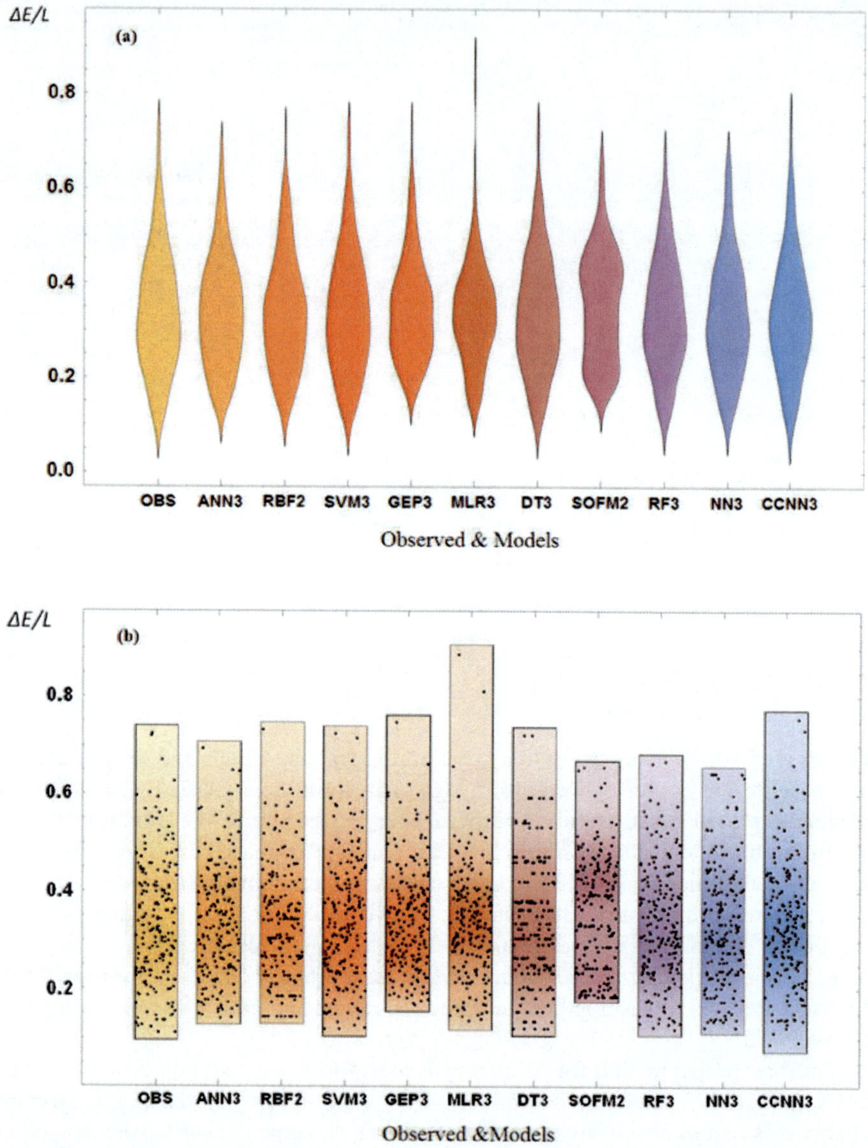


**Fig. 12.6** Comparison of the proposed models performance: **a** violin plot, **b** point density plot

# References


Breiman L (2001) Random forests. Mach Learn 45:5–32. https://doi.org/10.1023/A:1010933404324

Bung DB (2013) Non-intrusive detection of air-water surface roughness in self-aerated chute flows. J Hydraul Res 51:322–329. https://doi.org/10.1080/00221686.2013.777373

Chamani MR, Rajaratnam N (1999) Characteristics of skimming flow over stepped spillways. J Hydraul Eng 125:361–368. https://doi.org/10.1061/(ASCE)0733-9429(1999)125:4(361)

Chang F-J, Chang L-C, Wang Y-S (2007) Enforced self-organizing map neural networks for river flood forecasting. Hydrol Process 21:741–749. https://doi.org/10.1002/hyp.6262

Chang FJ, Chang LC, Kao HS, Wu GR (2010) Assessing the effort of meteorological variables for evaporation estimation by self-organizing map neural network. J Hydrol 384:118–129. https://doi.org/10.1016/j.jhydrol.2010.01.016

Chanson H (1994) Comparison of energy dissipation in nappe and skimming flow regimes on stepped chutes. J Hydraul Res 32:213–218

Chinnarasri C, Wongwises S (2006) Flow patterns and energy dissipation over various stepped chutes. J Irrig Drain Eng 132:70–76. https://doi.org/10.1061/(ASCE)0733-9437

Choubin B, Malekian A (2017) Combined gamma and M-test-based ANN and ARIMA models for groundwater fluctuation forecasting in semiarid regions. Environ Earth Sci. https://doi.org/10.1007/s12665-017-6870-8

Choubin B, Malekian A, Samadi S et al (2017) An ensemble forecast of semi-arid rainfall using large-scale climate predictors. Meteorol Appl 24:376–386. https://doi.org/10.1002/met.1635

Christodoulou GC (1993) Energy-dissipation on stepped spillways. J Hydraul Eng 119:644–650

Cigizoglu HK, Kisi O (2005) Flow prediction by three back propagation techniques using k-fold partitioning of neural network training data. Nord Hydrol 36:49–64

Çimen M, Kisi O (2009) Comparison of two different data-driven techniques in modeling lake level fluctuations in Turkey. J Hydrol 378:253–262. https://doi.org/10.1016/j.jhydrol.2009.09.029

Fahlman SE, Lebiere C (1990) The cascade-correlation learning architecture. Adv Neural Inf Process Syst 524–532. https://doi.org/10.1190/1.1821929

Ferreira C (2001) Gene expression programming: a new adaptive algorithm for solving problems. Complex Syst 13:1–22

Ghorbani MA, Khatibi R, Hosseini B, Bilgili M (2013) Relative importance of parameters affecting wind speed prediction using artificial neural networks. Theor Appl Climatol 114:107–114

Ghorbani MA, Zadeh HA, Isazadeh M, Terzi O (2016) A comparative study of artificial neural network (MLP, RBF) and support vector machine models for river flow prediction. Environ Earth Sci. https://doi.org/10.1007/s12665-015-5096-x

Gleckler PJ, Taylor KE, Doutriaux C (2008) Performance metrics for climate models. J Geophys Res Atmos. https://doi.org/10.1029/2007JD008972

Gonzalez CA, Takahashi M, Chanson H (2008) An experimental study of effects of step roughness in skimming flows on stepped chutes. J Hydraul Res 46:24–35. https://doi.org/10.1080/00221686.2008.9521937

Goyal MK, Bharti B, Quilty J et al (2014) Modeling of daily pan evaporation in sub tropical climates using ANN, LS-SVR, fuzzy logic, and ANFIS. Expert Syst Appl 41:5267–5276. https://doi.org/10.1016/j.eswa.2014.02.047

Haykin S (1999) Neural networks, a comprehensive foundation, Second. Prentice Hall, Upper Saddle River, NJ, USA

Hengl T, Heuvelink GBM, Kempen B et al (2015) Mapping soil properties of Africa at 250 m resolution: random forests significantly improve current predictions. PLoS ONE. https://doi.org/10.1371/journal.pone.0125814

Heo K, Ha K, Yun K et al (2014) Methods for uncertainty assessment of climate models and model predictions over East Asia. Int J Climatol 34:377–390

IPCC (2007) Climate change 2007: the physical science basis. Intergov Panel Clim Chang 446:727–728. https://doi.org/10.1038/446727a

Jamieson PD, Porter JR, Wilson DR (1991) A test of the computer simulation model ARCWHEAT1 on wheat crops grown in New Zealand. F Crop Res 27:337–350. https://doi.org/10.1016/0378-4290(91)90040-3

Karunanithi N, Grenney WJ, Whitley D, Bovee K (1994) Neural networks for river flow prediction. J Comput Civ Eng 8:201–220. https://doi.org/10.1061/(ASCE)0887-3801(1994)8:2(201)

Khatibi R, Ghorbani MA, Kashani MH, Kisi O (2011) Comparison of three artificial intelligence techniques for discharge routing. J Hydrol 403:201–212. https://doi.org/10.1016/j.jhydrol.2011.03.007

Khatibi R, Salmasi F, Ghorbani MA, Asadi H (2014) Modelling energy dissipation over stepped-gabion weirs by artificial intelligence. Water Resour Manage 28:1807–1821. https://doi.org/10.1007/s11269-014-0545-y

Kim S, Singh VP, Seo Y (2014) Evaluation of pan evaporation modeling with two different neural networks and weather station data. Theor Appl Climatol 117:1–13. https://doi.org/10.1007/s00704-013-0985-y

Kişi Ö (2007) Streamflow forecasting using different artificial neural network algorithms. J Hydrol Eng 12:532–539. https://doi.org/10.1061/(ASCE)1084-0699(2007)12:5(532)

Kisi O, Karahan ME, Şen Z (2006) River suspended sediment modelling using a fuzzy logic approach. Hydrol Process 20:4351–4362. https://doi.org/10.1002/hyp.6166

Kohonen T (1982) Self-organized formation of topologically correct feature maps. Biol Cybern 43:59–69. https://doi.org/10.1007/BF00337288

Malik A, Kumar A (2015) Pan evaporation simulation based on daily meteorological data using soft computing techniques and multiple linear regression. Water Resour Manage 29:1859–1872. https://doi.org/10.1007/s11269-015-0915-0

Malik A, Kumar A, Piri J (2017) Daily suspended sediment concentration simulation using hydrological data of Pranhita River Basin, India. Comput Electron Agric 138:20–28. https://doi.org/10.1016/j.compag.2017.04.005

Nash JE, Sutcliffe JV (1970) River flow forecasting through conceptual models part 1—a discussion of principles. J Hydrol 10:282–290. https://doi.org/10.1016/0022-1694(70)90255-6

Nemes A, Rawls WJ, Pachepsky YA (2006) Use of the nonparametric nearest neighbor approach to estimate soil hydraulic properties. Soil Sci Soc Am J 70:327–336. https://doi.org/10.2136/sssaj2005.0128

Notton G, Paoli C, Ivanova L et al (2013) Neural network approach to estimate 10-min solar global irradiation values on tilted planes. Renew Energy 50:576–584. https://doi.org/10.1016/j.renene.2012.07.035

Pagliara S, Chiavaccini P (2006) Energy dissipation on block ramps. J Hydraul Eng 132:41–48. https://doi.org/10.1061/(ASCE)0733-9429(2006)132:1(41)

Pagliara S, Das R, Palermo M (2008) Energy dissipation on submerged block ramps. J Irrig Drain Eng 134:527–532. https://doi.org/10.1061/(ASCE)0733-9437(2008)134:4(527)

Pagliara S, Roshni T, Palermo M (2015) Energy dissipation over large-scale roughness for both transition and uniform flow conditions. Int J Civ Eng 13:341–346

Pegram GGS, Officer AK, Mottram SR (1999) Hydraulics of skimming flow on modeled stepped spillways. J Hydraul Eng 125:500–510. https://doi.org/10.1061/(ASCE)0733-9429

Peyras L, Royet P, Degoutte G (1992) Flow and energy dissipation over stepped gabion weirs. J Hydraul Eng ASCE 118:707–717. https://doi.org/10.1061/(ASCE)0733-9429(1992)118:5(707)

Raheli B, Aalami MT, El-Shafie A, Ghorbani MA, Deo RC (2017) Uncertainty assessment of the multilayer perceptron (MLP) neural network model with implementation of the novel hybrid MLP-FFA method for prediction of biochemical oxygen demand and dissolved oxygen: a case study of Langat River. Environ Earth Sci 76(503). https://doi.org/10.1007/s12665-017-6842-z

Rajaratnam N (1990) Skimming flow in stepped spillways. J Hydraul Eng ASCE 116:587–591. https://doi.org/10.1061/(ASCE)0733-9429(1990)116:4(587)

Rice CE, Kadavy KC (1996) Model study of a roller compacted concrete stepped spillway. J Hydraul Eng 122:292–297. https://doi.org/10.1061/(ASCE)0733-9429(1996)122:6(292)

Salmasi F, Özger M (2014) Neuro-fuzzy approach for estimating energy dissipation in skimming flow over stepped spillways. Arabian J Sci Eng 39:6099–6108. https://doi.org/10.1007/s13369-014-1240-2

Salmasi F, Sattari MT (2017) Predicting discharge coefficient of rectangular broad-crested gabion weir using M5 tree model. Iranian J Sci Technol Trans Civil Eng Shiraz Univ 41(2): 205–212. https://doi.org/10.1007/s40996-017-0052-5

Salmasi F, Cahamani MR, Farsadi Zadeh D (2012) Experimental study of energy dissipation over stepped gabion spillways with low heights. Iranian J Sci Technol IJST Trans B Civil Eng Shiraz Univ 36(C2): 253–264. https://doi.org/10.22099/IJSTC.2012.640

Salmasi F, Samadi A (2018) Experimental and numerical simulation of flow over stepped spillways. Appl Water Sci 8(229):1–11. https://doi.org/10.1007/s13201-018-0877-5

Seo Y, Kim S, Kisi O, Singh VP (2015) Daily water level forecasting using wavelet decomposition and artificial intelligence techniques. J Hydrol 520:224–243. https://doi.org/10.1016/j.jhydrol.2014.11.050

Sonebi M, Cevik A (2009) Genetic programming based formulation for fresh and hardened properties of self-compacting concrete containing pulverised fuel ash. Constr Build Mater 23:2614–2622. https://doi.org/10.1016/j.conbuildmat.2009.02.012

Sorensen RM (1985) Stepped spillway hydraulic model investigation. J Hydraul Eng 111:1461–1472. https://doi.org/10.1061/(ASCE)0733-9429

Takahashi M, Ohtsu I (2014) Analysis of nonuniform aerated skimming flows on stepped channels. In: ISHS 2014-hydraulic structures and society-engineering challenges and extremes: proceedings of the 5th IAHR international symposium on hydraulic structures. The University of Queensland, pp 1–9

Taylor KE (2001) Summarizing multiple aspects of model performance in a single diagram. J Geophys Res Atmos 106:7183–7192. https://doi.org/10.1029/2000JD900719

Thirumalaiah K, Deo MC (1998) River stage forecasting using artificial neural networks. J Hydrol Eng 3:26–32. https://doi.org/10.1061/(ASCE)1084-0699(1998)3:1(26)

USBR (1978) Design of small canal structures

Vapnik VN (1995) The nature of statistical learning theory. Springer 8:188

Were K, Bui DT, Dick OB, Singh BR (2015) A comparative assessment of support vector regression, artificial neural networks, and random forests for predicting and mapping soil organic carbon stocks across an Afromontane landscape. Ecol Indic 52:394–403. https://doi.org/10.1016/j.ecolind.2014.12.028

Willmott CJ, Matsuura K (2005) Advantages of the mean absolute error (MAE) over the root mean square error (RMSE) in assessing average model performance. Clim Res 30:79–82. https://doi.org/10.3354/cr030079

Willmott CJ, Robeson SM, Matsuura K (2012) A refined index of model performance. Int J Climatol 32:2088–2094. https://doi.org/10.1002/joc.2419

# Chapter 13
# Morphological Changes of Floodplain Reach of Jhelum River, India, from 1984 to 2018

**Thendiyath Roshni, Dar Himayoun, and Mohammad Danish Azim**

## 13.1 Introduction

Changes in river morphology and its relation with natural and anthropogenic activities are considered an important tool for sustainable river restoration and integrated watershed management. Morphology is the study about the changing shapes of river channel. It is related with the forms of structure of river, which includes channel geometry, channel configuration, its profile characteristics and the bed forms.

The riverbank line change is related to a number of processes and environmental conditions like flood erosion, sedimentation loads and man-made activities. The erosion takes place from the consistency of current, which affects the formation of river's path and sinuosity. The availability of sediments, its size and composition of the sediments have an important role in the bank line shift and sinuosity index variation (Hemmelder et al. 2018). Running water has its velocity and possesses kinetic energy which causes two types of processes such as erosion and corrosion. Erosion has been derived from the energy involved in running water, and it is said to be a hydraulic action of the water waves. Gravel solid is being brought by the running water that scours the channel and moves sediment from the river bed. Erosion process makes the channel deeper and broader.

Corrosion is the process where the stream water reacts chemically with the rocks and gets dissolved them. (Mao, 2018; Monegaglia et al. 2018; Pal and Pani 2018). The three procedures which are primarily engaged with transporting residue load are suspension, corrosion and traction. Suspension is the procedure, wherein fine materials, for example, earth, residue, fine sand and materials which are lighter than water get transported in the water or on the outside of water without interacting with the bed of the stream. Corrosion is the different procedure, wherein stream water

T. Roshni (✉) · D. Himayoun · M. D. Azim
Department of Civil Engineering, National Institute of Technology Patna, Patna 800005, India
e-mail: roshni@nitp.ac.in

© Springer Nature Singapore Pte Ltd. 2021

R. C. Deo et al. (eds.), *Intelligent Data Analytics for Decision-Support Systems in Hazard Mitigation*, Springer Transactions in Civil and Environmental Engineering,
https://doi.org/10.1007/978-981-15-5772-9_13

having greater velocity erodes, shakes and brings them into solution. (Cenderelli and Wohl 2003; Clerici et al. 2015; Norbiato et al. 2007).

Rivers are a unique feature in the geography of the earth and have a major role in shaping the landscape by erosion, deposition and transportation of the sediment. The nature and functioning of the rivers depend on many variables like water quality and discharge. Rivers provide services like inland navigation, ecological habitats, recreation, irrigation water and drinking water. The largest rivers of the world are characterized by multiple channels that are often highly sinuous and consist of a branching channel patterns (Arróspide et al. 2018; Hekal 2018; Hemmelder et al. 2018; Strick et al. 2018). Rivers observe different flow patterns like meandering and braiding depending upon hydrodynamic forces, floodplain properties and discharge regime.

The morphological changes in rivers are influenced by variables like sediment load, sediment size, channel width, depth, discharge, velocity, roughness of channel materials and channel slopes. A slight change in one variable leads to a series of channel adjustments and consequently the changes in the other variables as well (Alayande and Ogunwamba 2010; Camporeale and Ridolfi 2010; Guan and Liang 2017; Ray et al. 2015; Strick et al. 2018). These morphological changes can be because of natural events like floods and volcanic eruptions or human interventions like land use change, sediment mining and dams (Dean and Schmidt 2013; Guan and Liang 2017; Marchese et al. 2017; Romshoo et al. 2018; Strick et al. 2018; Yousefi et al. 2018).

A stream or waterway in an alluvial territory is viewed as in the condition of equilibrium if the sediment load, discharge and sediment size are adjusted over a long span of time in a particular reach, and no change in the elevation takes place. Any adjustment in these controlling factors or the inconvenience of the counterfeit change over the waterway may disturb its balance, which results in aggradation or degradation of the stream, which can proceed for quite a while until another equilibrium is accomplished (Bechter et al. 2018; Billi and Fazzini 2017; Erskine and Saynor 1996; Hagstrom et al. 2018; Pal and Pani 2018). Prediction of when and where future erosion will occur and its extent is uncertain because of many factors involved. Proper understanding of channel pattern changes, and meander development of river is very important.

The awareness of the human activity in the vicinity of channels has unfortunately preceded in the ignorance of the pattern change (Strick et al. 2018). Alluvial stream patterns can be divided as straight channels, meandering channels and braided rivers which are according to the static properties and dynamic characteristics. Straight channels are having less sinuosity (<1.5) at the bank. Generally, rivers are simply straight open channels which have an existence over short reaches; however long, straight rivers rarely occur in nature (Dar et al. 2019; Fuller 2008; Srivastava et al. 2012).

Meandering rivers are having sinuosity greater than 1.5 that consists of a number of turns having alternate curvatures which is connected at the inflection points or it can be a short straight crossings. It is having relatively less gradient. The natural meandering occurred at the rivers is quite unstable because of the predomination of the bank erosion at the downstream of the concave bank. The bends have deeper flows and higher velocities along the outer side of the concave banks. The flow depths at the crossings are relatively low depth as compared to that of turns or bends. Meandering rivers have the tendency to migrate gradually, and hence, the sinuosity has the tendency to increase (Dean and Schmidt 2013; Monegaglia et al. 2018; Pradhan et al. 2018). Sometimes, the channel is just in the shape of a closed loop, and so, the meander has the tendency to cut off during the flood. Therefore, the meandering channel is the result of streambed instability, particularly when instability acts on the banks. Sinuosity having less than 1.1 comes under straight, between 1.1 and 1.5 are sinuous, and having greater than 1.5 comes under meandering. Therefore, sinuous rivers can be said as the transition between straight and meandering rivers (Himayoun and Roshni 2020). Remote sensing techniques provide information through time and space, which can never be appreciated from ground. Aerial photographs and satellite sensor images provide supreme way to observe the changes in the cross section of the river course. Advantages of the information acquired by satellite remote sensing are of synoptic coverage, and the dynamic changes in surface can be monitored effectively. Various satellites having sensors work in different spectrum at different spatial resolution, so can be used for obtaining various information on planform characteristic of river courses.

The aim of this chapter is to evaluate the changes in channel morphology of the Jhelum River, India, from 1994 to 2018 based on the bank shift and sinuosity index.

## 13.2 Study Area and Data

### *13.2.1 Study Area*

Jhelum River (a tributary of Indus River) flows through Kashmir Valley in Jammu and Kashmir, India.

The Jhelum catchment, India, is located in vicinity of great western Himalayan mountain ranges and is bounded from east to west by Pakistan, China and Afghanistan. It is spread between 33.50–35° N and 73.50–75.50° E having sharp rise in elevation from 1000 to 28,250 feet above mean sea level within the three degree of latitude. Further details of the Jhelum River and basin are found in (Himayoun and Roshni 2019; Himayoun et al. 2019). Jhelum River drains a total geographic area 17,622 km$^2$ up to Indian boarder. It originates from Karstic mountainous terrain, flows through the Kashmir Valley before entering the Pakistan, where it joins Indus River.

Figure 13.1 shows the location map of the study area, and it consists of 150 km long stretch of Jhelum River. It includes the Srinagar city, which is the summer capital of Jammu and Kashmir. The upstream part of the reach is located near Sangam, which is the main runoff gauging station of Jhelum River. The downstream runoff gauging station of Jhelum River is located near Baramulla. A buffer of 2–5 km was

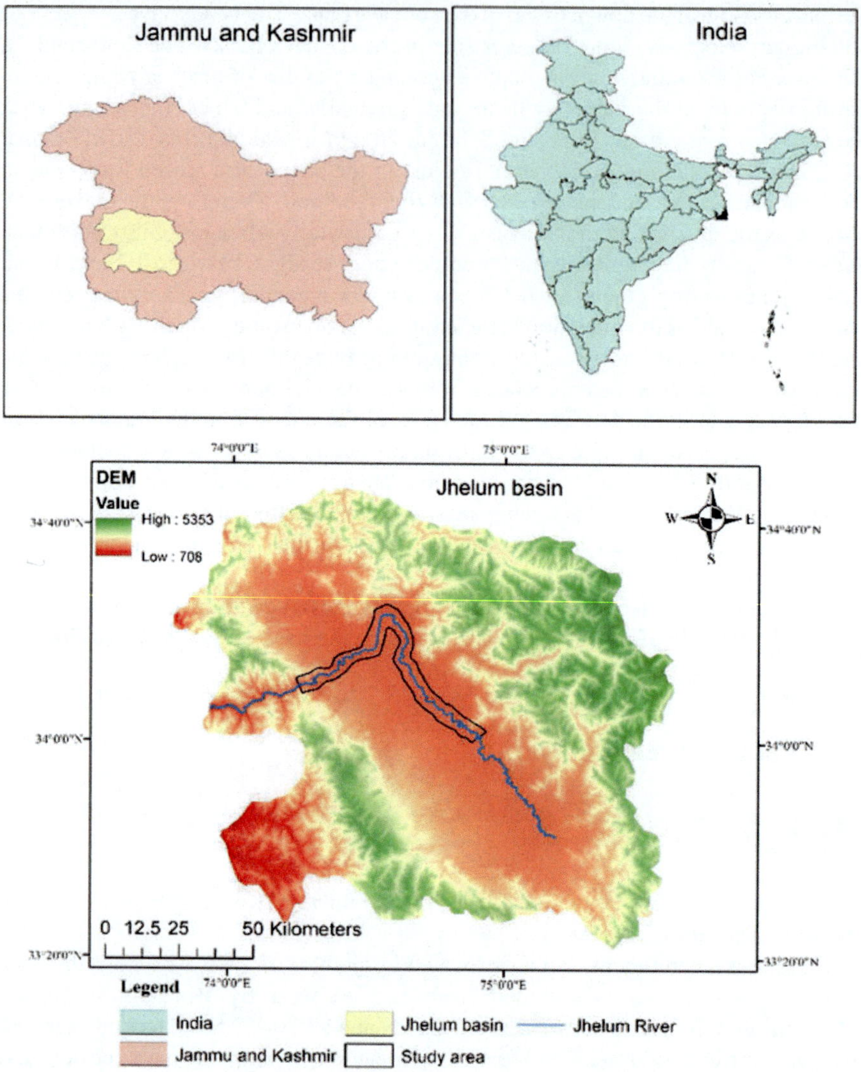


**Fig. 13.1** Digital elevation model of the basin, showing location of the study area

taken along the reach for both sides of the river to assess the morphological changes. Figure 13.2 shows the location of segments on the river stretch. Thirteen segments (Segments *A*–*M*) have been selected for the evaluation of morphological changes.

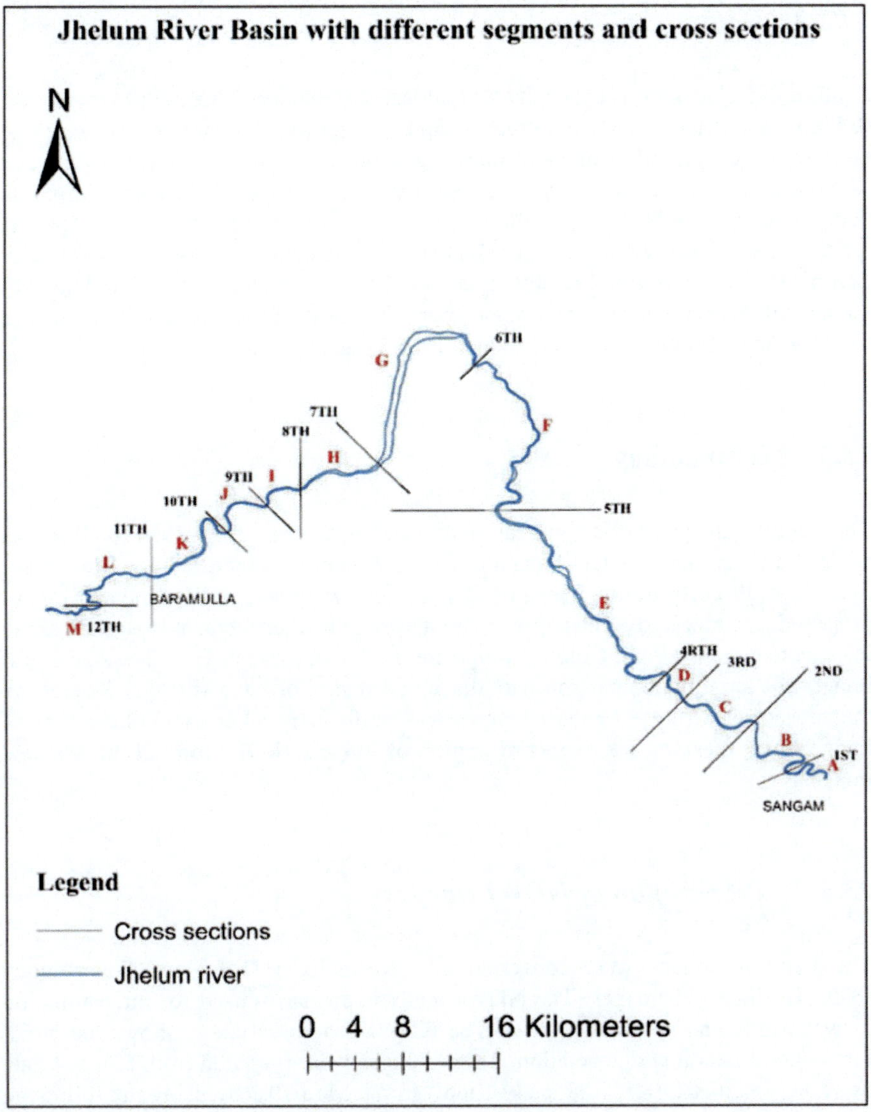


**Fig. 13.2** River stretch with the location of different segments

### 13.2.2 Data

Landsat 4-5 Thematic Mapper (TM), Landsat 7 Enhanced Thematic Mapper Plus (ETM+) and Landsat 8 Operational Land Imager (OLI) satellite images were employed for evaluation of bank line migration of the Jhelum River, India. The satellite images covering study area were obtained from earth explorer (http://earthexplorer.usgs.gov/in). The satellite takes the image of the earth every 16th day repeat cycle. These satellite images consist of seven spectral bands with 30 m resolution. However, Landsat 7 satellite images consist of an additional spectral band of 15 m resolution known as panchromatic band. Landsat 8 provides redefined heritage bands, along with deep blue band which is for coastal studies.

## 13.3 Methodology

The satellite imagery collected at an interval of eight years from 1994 to 2018 was subjected to dark spot reduction and reflectance, geometric and radiometric corrections. The 30 m resolution bands of this imagery were selected for the analysis of proposed research study. The corrected images were further subjected to layer staking in order to generate False Colour Composite (FCC) for these images. The size of full image was large in comparison with the actual region of interest (ROI). Therefore, the subsetting of these Landsat images was performed either by clipping it from image or by masking for extracting region of interest (ROI) from all the selected images.

### 13.3.1 Preparation of NDWI Images

The Landsat images were converted into Normalized Difference Water Index (NDWI) processed images. The NDWI index is primarily used for the estimation of water area and vegetation cover. The NDWI processes image shows the moisture content in soil and vegetation. These images were prepared from Landsat data by raster calculator (a spatial analyst tool) in ArcMap 10.3 by using the following formula (McFeeters 1996):

$$\mathrm{NDWI} = \frac{(\mathrm{Green} - \mathrm{NIR})}{(\mathrm{Green} + \mathrm{NIR})} \tag{13.1}$$

where the NRI covers the reflected near-infrared radiation and Green involves the reflected green light. These two wavelengths are selected because green light maximizes the reflectance of water feature and NIR maximizes the reflectance by vegetation and soil cover but not by water content.

### *13.3.2 Delineation of RiverBank Line*

The entire river stretch from Sangam to Baramulla has been divided into 13 segments.

The four sets of NDWI processed images prepared at an interval eight years from 1994 to 2018 were used to delineate active channels. Fluvial corridor and visual digitizing of the channel planforms were used for the assessment of morphologic changes. The digitization of bank lines has been performed using HEC-GeoRAS tool in ArcGIS. The NDWI images proved to be very useful in detecting the active riverbank lines. The NDWI is most appropriate index for water body mapping.

The water body has strong absorbability and low radiation in the range from visible to infrared wavelengths. Shallow water channel is considered as the part of river, whereas old and new soil deposits at riverbanks create ambiguity in interpretation. Sand patches far from active water channel have dark tone in satellite image indicating higher moisture. However, soil patches close to active water channel show bright tone signifying low moisture. Considering the data set of 1994 as base year, the changes in the channel morphology have been calculated at an interval of eight years.

### *13.3.3 Computed Parameters*

The evaluation of morphological changes from satellite imagery requires the manual digitization of the active channel banks. The active channel banks represent the top of the scarp delimiting the channel at bank-full stage. A number of morphological indices and parameters have been proposed by the authors across globe for the effective and quantitative analysis of historical changes in channel morphology. In this research study, the following parameters and indices have been calculated for the analysis of detecting the changes in channel morphology.

#### **13.3.3.1 Sinuosity**

This is usually defined as the ratio of centreline channel length of the river to valley length of the river (Pradhan et al. 2018; Clerici et al. 2015). In this study, the entire river stretch has been divided into 12 rectilinear segments. The sinuosity of each segment is calculated by measuring the centreline length encompassed by the channel segment and dividing it by length of the straight channel segment.

$$\text{Sinuosity}\,(C) = \frac{S}{L} \tag{13.2}$$

where ($S$) is the centerline flow length of the segment and ($L$) is the straight channel length of the segment.

#### 13.3.3.2 Bank Line Shifting

The shifting of channel between two dates is calculated using measuring tool in ArcGIS.

In this procedure, two Landsat images at an interval of eight years are patched in the same map, and the active channels corresponding to these two different dates are overlapped. The average bank shift between two dates is computed by measuring the average area between two bank lines on the same side of the river channel and dividing it by the length of the segment.

This procedure was implemented for all the segments of both banks for the computation of bank shift.

## 13.4 Results and Discussion

In this section, the planimetric changes based on channel shift and sinuosity were analysed at an interval of eight years (from 1994 to 2018) for the selected study reach.

### *13.4.1 Channel Shift*

The active river channels extracted from NDWI images at an interval of 8 years were employed for the calculation of bank shift. The selected river reach which is approximately 150 km is stretched between Sangam and Baramulla. A total of twelve cross sections which consist of 13 segments has been taken for the analysis. Figures 13.3, 13.4 and 13.5 demonstrated the shifting patterns of study stretch in all the 13 segments from 1994 to 2018. The total shift from 1994 at an interval of eight years up to 2018 was also calculated for both the banks and is presented in Table 13.1.

In most of the segments, the processes of erosion and deposition take place simultaneously. The negative shift representing the erosion occurs in one bank, and the deposition as represented by positive shift takes place on the other bank. The maximum positive of 100 m has been observed in segment *G,* and maximum negative shift −170 m has also been observed in segment *G*. A shift of −170 m occurred in 2002 and persisted up to 2010, then started decreasing up to 2018. Similarly, a shift of 100 m was observed in segment *M* during 2010–2108.

The shifting pattern in all the segments at an interval of eight years for the selected study reach is presented in Fig. 13.4.

Evidently, from Fig. 13.6a–c, the maximum shift occurred in segment 7 (*M*) throughout the study period followed by segment 4 (*D*) during 2010 to 2018

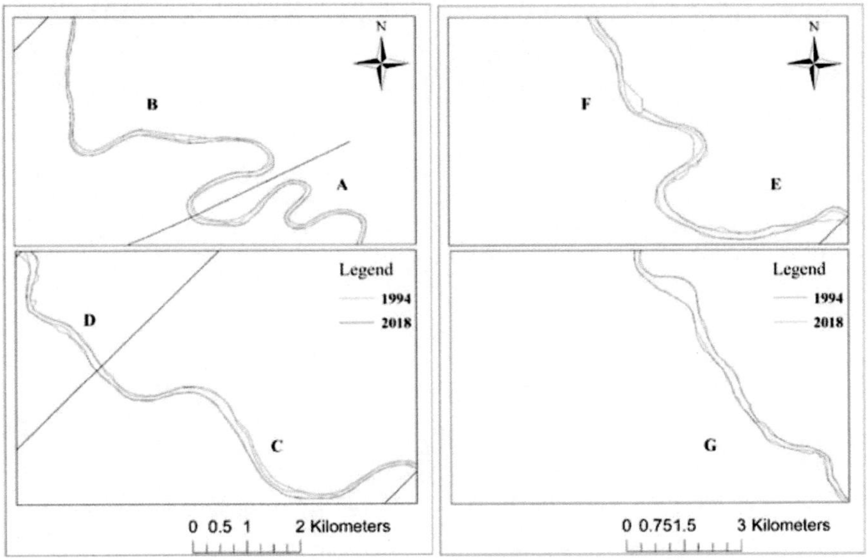


**Fig. 13.3** Shifting view of segments *A–G* from 1994 to 2018

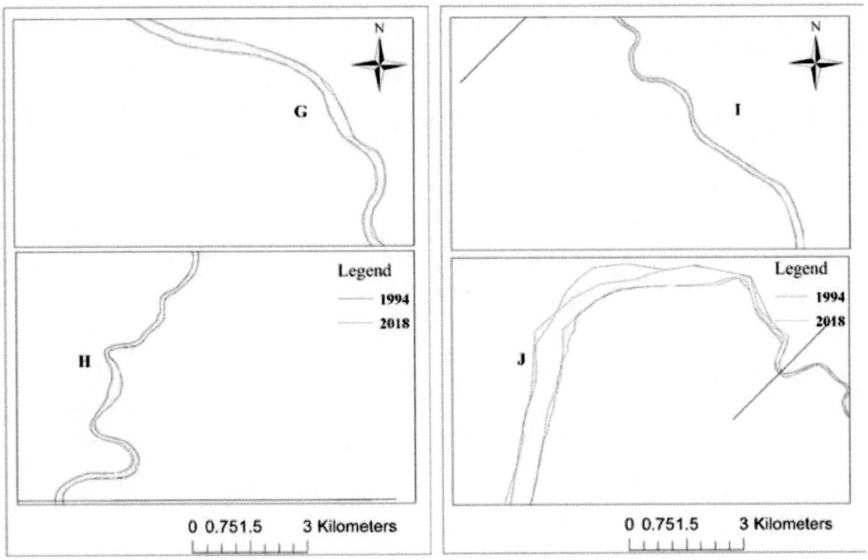


**Fig. 13.4** Shifting view of segments *G–J* from 1994 to 2018

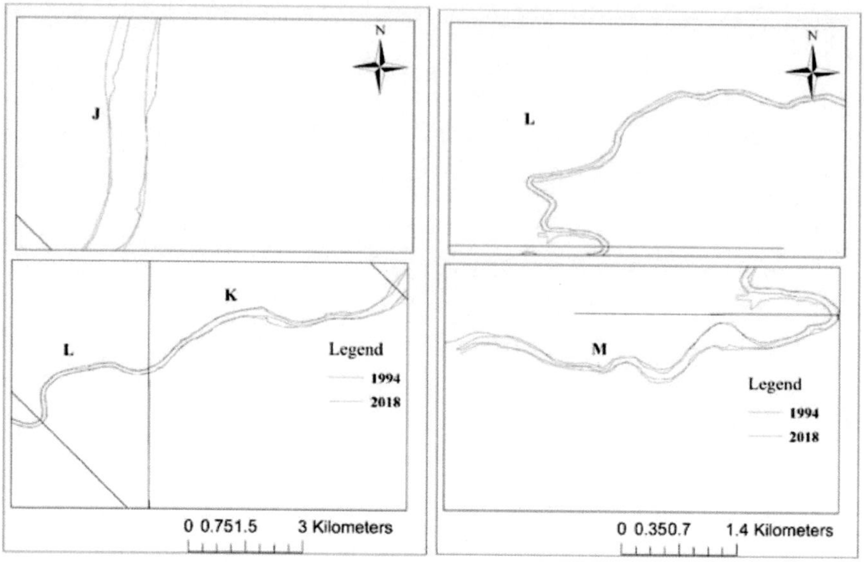


**Fig. 13.5** Shifting view of segments *J–M* from 1994 to 2018

**Table 13.1** Bankline shift (*m*) with respect to 1994, negative and positive signs indicate erosion and deposition, respectively

| Segments | 1994–1999 | | 1994–2008 | | 1994–2018 | |
|---|---|---|---|---|---|---|
| | Left bank | Right bank | Left bank | Right bank | Left bank | Right bank |
| *A* | 20 | 30 | −10 | −10 | 20 | 40 |
| *B* | −10 | 40 | −10 | 20 | 10 | 20 |
| *C* | −10 | 20 | 0 | 10 | 10 | 30 |
| *D* | −70 | −10 | −40 | −20 | −120 | −10 |
| *E* | −20 | −10 | −20 | −70 | −20 | −10 |
| *F* | −10 | 0 | −20 | 0 | −10 | −10 |
| *G* | 80 | −170 | 80 | −170 | 100 | −120 |
| *H* | −20 | 30 | −20 | 10 | −20 | 40 |
| *I* | 10 | 20 | −10 | −20 | 10 | 20 |
| *J* | 20 | 30 | −70 | −30 | 20 | −20 |
| *K* | 10 | 30 | −10 | −30 | 10 | 10 |
| *L* | −20 | 10 | −10 | −40 | −10 | −20 |
| *M* | −30 | 20 | −20 | −40 | −10 | −20 |

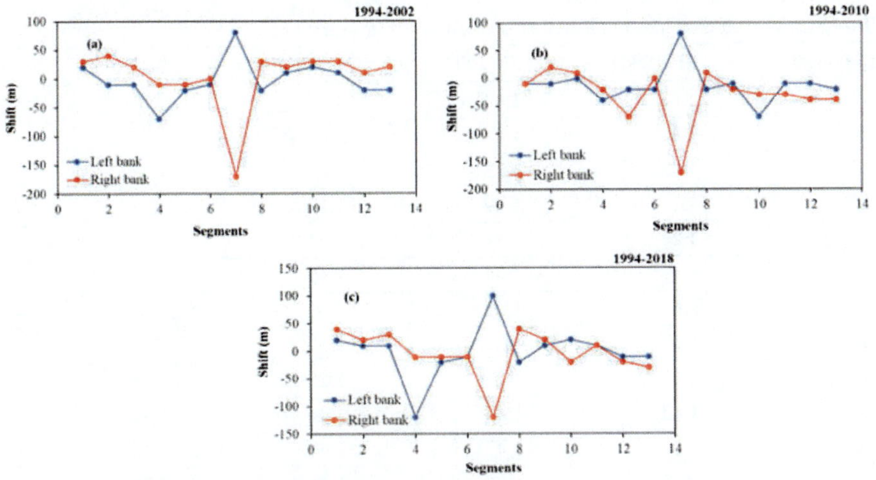


**Fig. 13.6** Shifting view of riverbank in 1999

Fig. 13.6c. This is an indication of significant erosion and deposition and could be attributed to the lack of bank protection works. The land use/land cover (LULC) have significant role in the processes of erosion and deposition and consequently on the channel shift. Yousefi et al. (2018) reported the effects on channel morphology (155 km Karoon River reach) due to an extreme flood event. They found that the river channel was less effected in residential areas as compared to the other LULC types.

## *13.4.2 Sinuosity Analysis*

In addition to this, the sinuosity index was also calculated for all the segments at an interval of eight years. Table 13.2 presents the sinuosity values for all the segments. The values of sinuosity index in most of the segments are in between 1.5 and 2 that comes under "meandering" type segment. At segment *H*, the values are in between 1.05 and 1.5, which comes under "sinuous" type segment. As can be seen from the table, the maximum sinuosity index of 1.72, 1.72, 1.62 and 1.72 was observed in *A*, *B*, *G* and *L* segments, respectively. Similarly, the minimum sinuosity values of 1.1, 1.21, 1.07 and 1.17 were observed in *D*, *E*, *H* and *M* segments, respectively.

Figures 13.7 and 13.8 show the variation in sinuosity for all the segments selected for the study during the observation period. As evident from the figure that the maximum change in sinuosity index occurred in *B*, *H*, *L* and *M* segments and the minimum change in sinuosity has been observed in *D* and *E* segments. However, most of the segments such as *C*, *F*, *G* and *K* evidenced no change in sinuosity values

**Table 13.2** Sinuosity index at different segments in four years

| Segment | SI (1994) | SI (2002) | SI (2010) | SI (2018) |
|---|---|---|---|---|
| *A* | 1.82 | 1.83 | 1.83 | 1.8 |
| *B* | 1.82 | 1.8 | 1.8 | 1.79 |
| *C* | 1.23 | 1.23 | 1.23 | 1.23 |
| *D* | 1.1 | 1.09 | 1.1 | 1.11 |
| *E* | 1.21 | 1.21 | 1.21 | 1.2 |
| *F* | 1.44 | 1.44 | 1.45 | 1.44 |
| *G* | 1.62 | 1.62 | 1.62 | 1.62 |
| *H* | 1.07 | 1.09 | 1.09 | 1.09 |
| *I* | 1.27 | 1.27 | 1.27 | 1.26 |
| *J* | 1.43 | 1.44 | 1.44 | 1.44 |
| *K* | 1.46 | 1.47 | 1.46 | 1.46 |
| *L* | 1.72 | 1.73 | 1.74 | 1.75 |
| *M* | 1.17 | 1.15 | 1.14 | 1.15 |

towards the end of the observation period. The change in sinuosity values could be attributed to the processes of erosion and deposition. The results of this study are in agreement with (Clerici et al. 2015). They carried out the detailed analysis of the channel planform changes of the floodplain reach of the Taro River in the last two centuries. The analysis of their results revealed that channel sinuosity decreased by 29% because of the erosion and deposition.

## 13.5 Concluding Remarks

The quantitative analysis of the planform changes of 150 km stretch from Sangam to Baramulla of the Jhelum River, India, has outlined the channel evolution during the last 24 years. The remote sensing and GIS techniques proved to be useful tools in the evaluation of morphological changes. The analysis of the results revealed that most of the segments exhibit the processes of erosion and deposition simultaneously. The negative shift representing the erosion occurs in one bank, and the deposition as represented by positive shift takes place on the other bank. The maximum positive of 100 m has been observed in segment *G*, and maximum negative shift −170 m has also been observed in segment *G*. A shift of −170 occurred in 2002 and persisted up to 2010, then started decreasing up to 2018. Similarly, a shift of 100 m was observed in segment *M* during 2010 to 2018. The values of sinuosity index in most of the segments are in between 1.5 and 2 that comes under "meandering" type segment. At segment *H*, the values are lying in between 1.05 and 1.5, which comes under

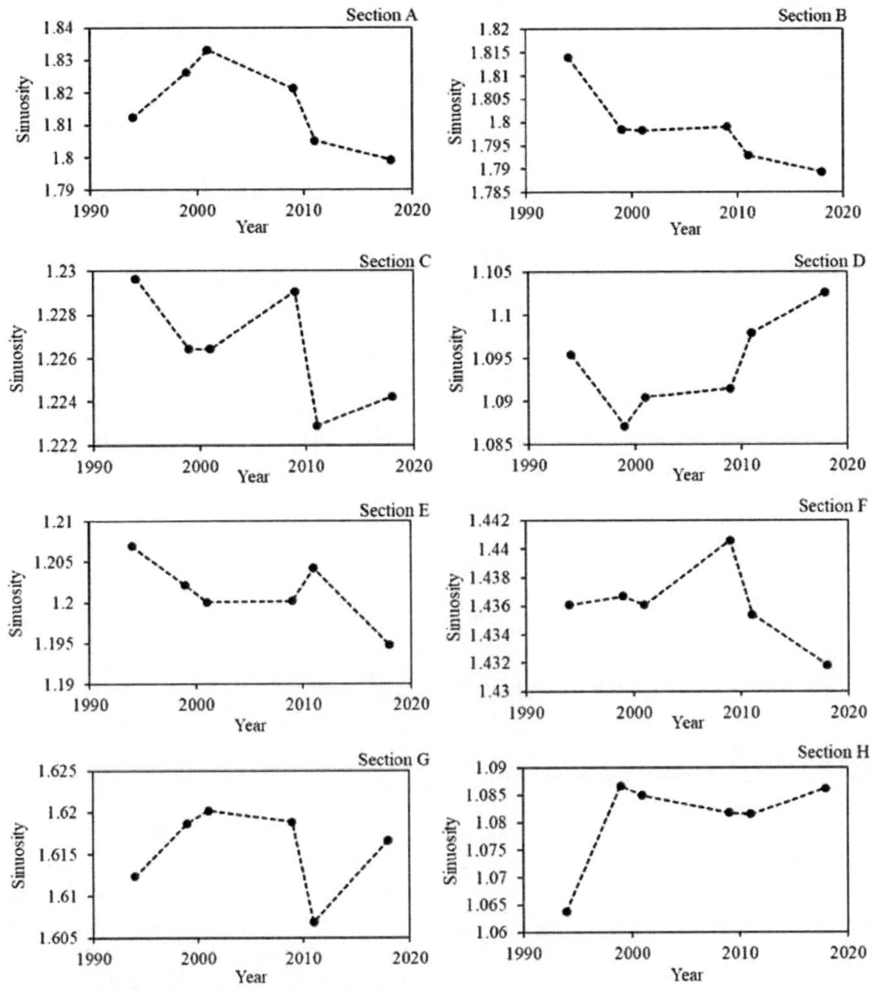


**Fig. 13.7** Sinuosity index at segment *A*–*H*

"sinuous" type segment. The maximum change in sinuosity index occurred in *B*, *H*, *L* and *M* segments, and the minimum change in sinuosity has been observed in *D* and *E* segments. However, most of the segments such as *C*, *F*, *G* and *K* evidenced no change in sinuosity values towards the end of the observation period.

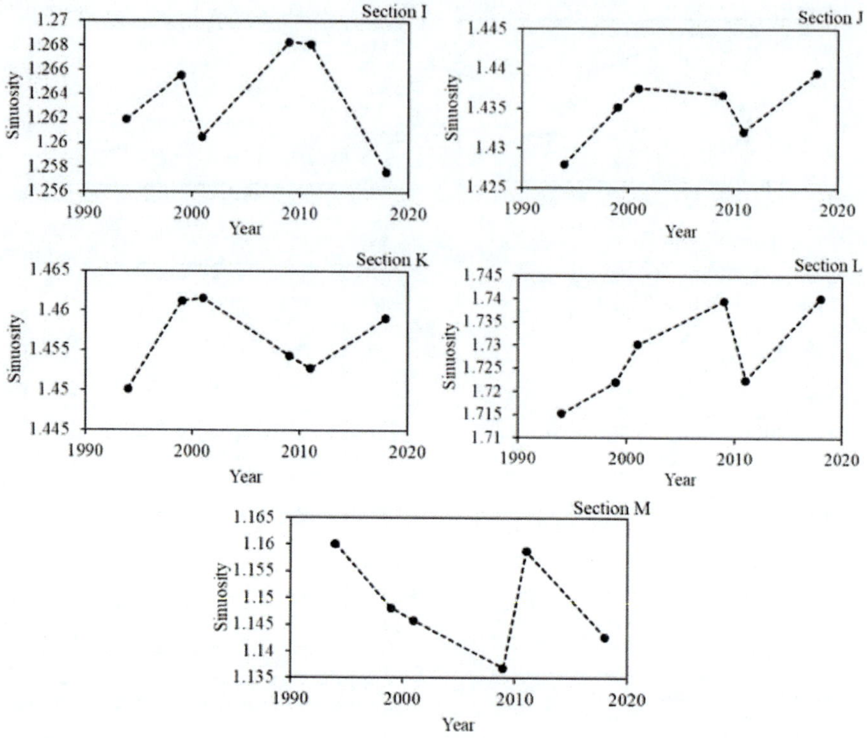


**Fig. 13.8** Sinuosity index at segments *I–M*

## References


Alayande AC, Ogunwamba JC (2010) The impacts of Urbanisation on Kaduna River Flooding. J Am Sci 6(5):28–35

Arróspide F, Mao L, Escauriaza C (2018) Morphological evolution of the Maipo River in central Chile: influence of instream gravel mining. Geomorphology 306:182–197. https://doi.org/10.1016/j.geomorph.2018.01.019

Bechter T, Baumann K, Birk S, Bolik F, Graf W, Pletterbauer F (2018) A simple and efficient GIS-based approach for large-scale morphological assessment of large European rivers. Sci Total Environ 628–629:1191–1199. https://doi.org/10.1016/j.scitotenv.2018.02.084

Billi P, Fazzini M (2017) Global change and river fl ow in Italy. Global Planet Change 155(04):234–246. https://doi.org/10.1016/j.gloplacha.2017.07.008

Camporeale C, Ridolfi L (2010) Interplay among river meandering, discharge stochasticity and riparian vegetation. J Hydrol 382(1–4):138–144. https://doi.org/10.1016/j.jhydrol.2009.12.024

Cenderelli DA, Wohl EE (2003) Flow hydraulics and geomorphic effects of glacial-lake outburst floods in the Mount Everest region. Nepal Earth Surface Process Landforms 28(4):385–407. https://doi.org/10.1002/esp.448

Clerici A, Perego S, Chelli A, Tellini C (2015) Morphological changes of the floodplain reach of the Taro River in the last two centuries. J Hydrol 527:1106–1122. https://doi.org/10.1016/j.jhydrol.2015.05.063

Dar RA, Mir SA, Romshoo SA (2019). Influence of geomorphic and anthropogenic activities on channel morphology of River Jhelum in Kashmir Valley, NW Himalayas. Quat Int. https://doi.org/10.1016/j.quaint.2018.12.014

Dean DJ, Schmidt JC (2013) The geomorphic effectiveness of a large flood on the Rio Grande in the big bend region: insights on geomorphic controls and post-flood geomorphic response. Geomorphology 201:183–198. https://doi.org/10.1016/j.geomorph.2013.06.020

Erskine WD, Saynor MJ (1996) Effects of catastrophic floods on sediment yields in southeastern Australia. Erosion Sediment Yield Glob Reg Perspect 236:381–388

Fuller IC (2008) Geomorphic impacts of a 100-year flood: Kiwitea Stream, Manawatu catchment, New Zealand. Geomorphology 98(1–2):84–95. https://doi.org/10.1016/j.geomorph.2007.02.026

Guan M, Liang Q (2017) A two-dimensional hydro-morphological model for river hydraulics and morphology with vegetation. Environ Model Softw 88:10–21. https://doi.org/10.1016/j.envsoft.2016.11.008

Hagstrom CA, Leckie DA, Smith MG (2018) Point bar sedimentation and erosion produced by an extreme flood in a sand and gravel-bed meandering river. Sed Geol. https://doi.org/10.1016/j.sedgeo.2018.09.003

Hekal N (2018) Evaluation of the equilibrium of the River Nile morphological changes throughout "Assuit-Delta Barrages " reach. Water Sci 32(2):230–240. https://doi.org/10.1016/j.wsj.2018.09.001

Hemmelder S, Marra W, Markies H, De Jong SM (2018) Monitoring river morphology and bank erosion using UAV imagery—a case study of the river Buëch, Hautes-Alpes, France. Int J Appl Earth Obs Geoinf 73(04):428–437. https://doi.org/10.1016/j.jag.2018.07.016

Himayoun D, Roshni T (2019) Spatio-temporal variation of drought characteristics, water resource availability and the relation of drought with large scale climate indices: a case study of Jhelum basin, India. Quat Int 525:140–150

Himayoun D, Roshni T (2020) Geomorphic changes in the Jhelum river due to an extreme flood event: a case study. Arab J Geosci 13–23

Himayoun D, Mohsin F, Roshni T (2019) Efficacy in simulating the peak discharge response using soft computing techniques in the Jhelum river basin, India. Int J River Basin Manage. https://doi.org/10.1080/15715124.2019.1570934

Mao L (2018) The effects of flood history on sediment transport in gravel-bed rivers. Geomorphology 322:196–205. https://doi.org/10.1016/j.geomorph.2018.08.046

Marchese E, Scorpio V, Fuller I, McColl S, Comiti F (2017) Morphological changes in Alpine rivers following the end of the little ice age. Geomorphology 295:811–826. https://doi.org/10.1016/j.geomorph.2017.07.018

McFeeters SK (1996) The use of the normalized difference water index (NDWI) in the delineation of open water features. Int J Remote Sens 17:1425–1432

Monegaglia F, Zolezzi G, Güneralp I, Henshaw AJ, Tubino M (2018) Environmental modelling and software automated extraction of meandering river morphodynamics from multitemporal remotely sensed data. Environ Model Softw 105:171–186. https://doi.org/10.1016/j.envsoft.2018.03.028

Norbiato D, Borga M, Sangati M, Zanon F (2007) Regional frequency analysis of extreme precipitation in the eastern Italian Alps and the August 29, 2003 flash flood. J Hydrol 345(3–4):149–166. https://doi.org/10.1016/j.jhydrol.2007.07.009

Pal R, Pani P (2018) The Egyptian journal of remote sensing and space sciences remote sensing and GIS-based analysis of evolving planform morphology of the middle-lower part of the Ganga River, India. Egyptian J Remote Sens Space Sci. https://doi.org/10.1016/j.ejrs.2018.01.007

Pradhan C, Chembolu V, Dutta S (2018) Impact of river interventions on alluvial channel morphology. ISH J Hydraulic Eng 5010:1–7. https://doi.org/10.1080/09715010.2018.1453878

Ray K, Bhan SC, Bandopadhyay BK (2015) The catastrophe over Jammu and Kashmir in September 2014: a meteorological observational analysis. Curr Sci 109(3):580–591

Romshoo SA, Altaf S, Rashid I, Ahmad Dar R (2018) Climatic, geomorphic and anthropogenic drivers of the 2014 extreme flooding in the Jhelum basin of Kashmir, India. Geomatics Nat Hazards Risk 9(1):224–248. https://doi.org/10.1080/19475705.2017.1417332

Srivastava PK, Han D, Rico-Ramirez MA, Bray M, Islam T (2012) Selection of classification techniques for land use/land cover change investigation. Adv Space Res 50(9):1250–1265. https://doi.org/10.1016/j.asr.2012.06.032

Strick RJP, Ashworth PJ, Awcock G, Lewin J (2018) Geomorphology morphology and spacing of river meander scrolls. Geomorphology 310:57–68. https://doi.org/10.1016/j.geomorph.2018.03.005

Yousefi S, Mirzaee S, Keesstra S, Surian N, Pourghasemi HR, Zakizadeh HR, Tabibian S (2018) Effects of an extreme flood on river morphology (case study: Karoon River, Iran). Geomorphology 304:30–39. https://doi.org/10.1016/j.geomorph.2017.12.034

# Chapter 14
# Spatial Modeling of Soil Erosion Susceptibility with Support Vector Machine

**Omid Rahmati and Abolfazl Jaafari**

## 14.1 Introduction

Different types of soil erosion often occur due to the concentration of overland flow. Soil erosion is a ubiquitous phenomenon around the world that has different environmental consequences. Impacts of soil erosion can be classified into two main groups: on-site (i.e. at the place where the soil is detached) and off-site (i.e. wherever the eroded soil ends up) effects (Cama et al. 2017). The main on-site impact of soil erosion is land and environmental degradation which result in soil quality and water-holding capacity of soils (Garcia 2018). Water erosion, as a threatening the sustainability of terrestrial/aquatic ecosystems in downstream, has some off-site effects including the movement of sediments (eroded soil) and agricultural pollutants into streams and reservoirs, and pollution of drinking water (Colazo et al. 2019). In addition, muddy floods due to extreme runoff and soil erosion damage infrastructures, houses and other buildings (Vandaele et al. 2013). Therefore, it is important to understand erosion processes to predict soil erosion and prevent all associated damages.

Various geographic information system (GIS)-based machine learning and statistical approaches such as frequency ratio, weights of evidence, logistic regression, linear regression, conditional analysis, maximum entropy, analytical hierarchy process, classification and regression trees, random forest and artificial neural networks (Dewitte et al. 2015; Pourghasemi et al. 2017; Rahmati et al. 2016, 2017)

O. Rahmati (✉)
Soil Conservation and Watershed Management Research Department, Kurdistan Agricultural and Natural Resources Research and Education Center, AREEO, Sanandaj, Iran
e-mail: o.rahmati@areeo.ac.ir; orahmati68@gmail.com

A. Jaafari
Research Institute of Forests and Rangelands, Agricultural Research, Education and Extension Organization (AREEO), Tehran, Iran

© Springer Nature Singapore Pte Ltd. 2021

R. C. Deo et al. (eds.), *Intelligent Data Analytics for Decision-Support Systems in Hazard Mitigation*, Springer Transactions in Civil and Environmental Engineering,
https://doi.org/10.1007/978-981-15-5772-9_14

have been used for soil erosion susceptibility. Recently, Rahmati et al. (2017) compared the performance of seven machine learning models (L-SVM, P-SVM, RBF-SVM, S-SVM, BP-ANN, RF and BRT) for mapping gully erosion susceptibility in the Kashkan–Poldokhtar Watershed, Iran, and demonstrated the superiority of RF and RBF-SVM over other models.

This chapter aims to design support vector machine (SVM) models: a state-of-the-art machine learning model to spatially explicit prediction of soil erosion. This model was used because of the following reasons: (1) SVM works with various types of environmental variables (e.g. continuous, categorical, binary or ordinal); (2) SVM has not been widely used for predicting soil erosion susceptibility; (3) SVM can effectively model the nonlinear relationship between the dependent and independent variables; and (4) no preliminary assumptions (normal distributions of independent variables) are required for performing SVM (Rahmati et al. 2017). Therefore, the specific objectives of this chapter are to

(1) Explore the capability of the SVM model for the prediction of soil erosion susceptibility,
(2) Produce a reliable distribution map of soil erosion susceptibility for the Golgol watershed, Ilam province, Iran.

The soil erosion susceptibility map enables decision makers and land-use planners to adequately delineate the landscapes in terms of the soil erosion susceptibility towards selecting favourable locations for the development infrastructures and for the sustainable management of soil and water resources.

## 14.2 Study Area

The Golgol watershed (Fig. 14.1) is located in Ilam province, western Iran, and lies between east longitudes of 46° 26′ to 46° 38′ and north latitudes of 33° 24′ to 33° 39′, with an altitude ranging from 1,068 to 2,580 m. This region has an average temperature of 16.9°C/year and an average rainfall of 580 mm/year. *Quercus brantii* is the dominant species in the study area. Ilam dam is a reservoir dam which supplies drinking water to Ilam city and placed at the downstream of the study area. Sediments of the upstream sub-watersheds often fill the dam and are known as the most serious threats to this area. Geographically, the study area is a part of Folded-Zagros Zone with various lithological units including shale, limestone and valley terrace deposits.

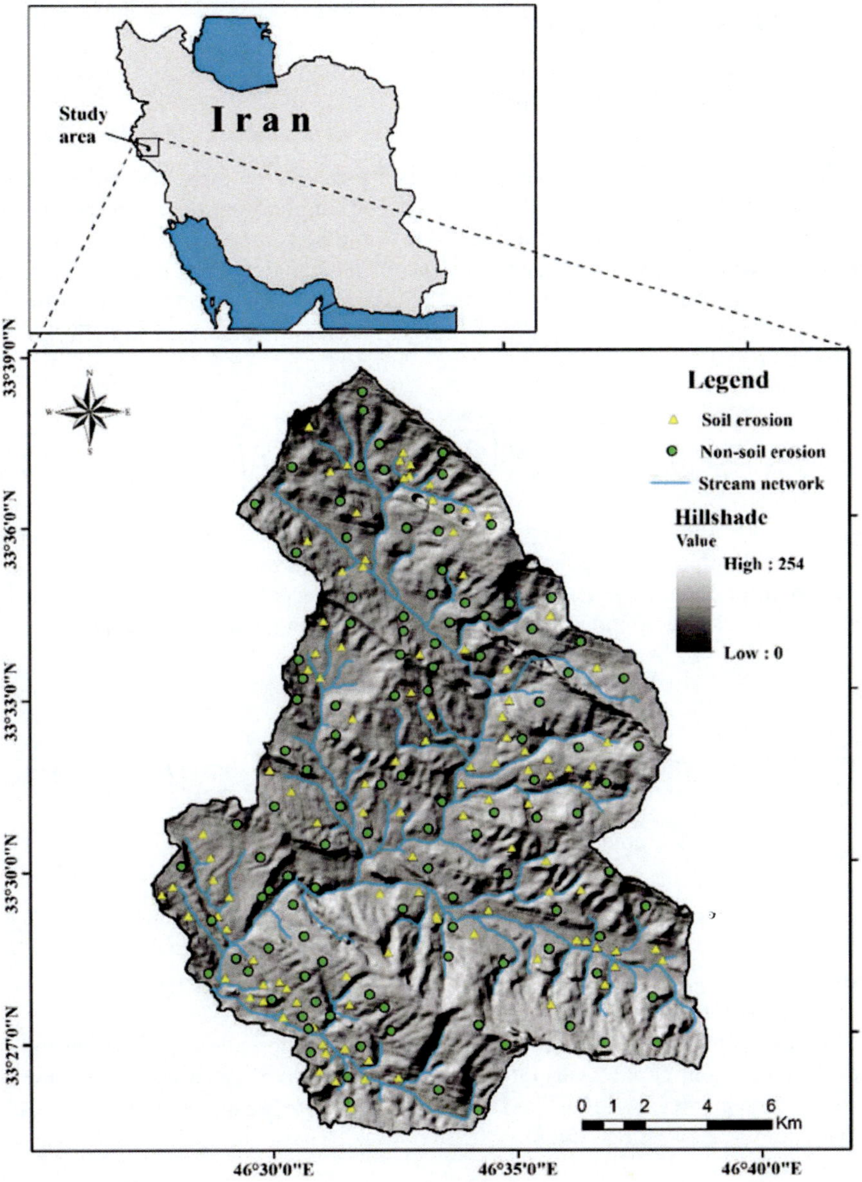


**Fig. 14.1** Location of the study area and soil erosion and non-erosion points

## 14.3 Materials and Method

### 14.3.1 Support Vector Machines

SVM, proposed by Vapnik et al. (1995), is a machine learning method that uses a learning algorithm based on the statistical learning and optimization theories to deal with classification and regression tasks. Compared to the conventional machine learning methods, SVM is popular because of its superior empirical performance and promising outputs. SVM separates the classes of a given dataset with a linear decision surface that is developed by solving a classification function (Eq. 14.1) (Vapnik et al. 1995):

$$f(v) = \mathrm{sgn}\left(\sum_{i=1}^{n} \alpha_i L_i K(v, v_1^n) + b\right) \quad (14.1)$$

where $n$ is the number of independent variables, $\alpha_i$ is the Lagrange multiplier, $v$ is the vectors of variables, $L$ is the class labels, $b$ is the constant value and $K(v, v_i)$ represents the kernel function. In general, SVM includes four kernel functions, namely linear, polynominal, sigmoid and radial basis function (RBF) that are expressed by:

$$\text{Linear: } K(v, v_1) = V^T V_1 \quad (14.2)$$

$$\text{Polynominal: } K(v, v_1) = (\gamma V^T V_j + r)^d, \quad \gamma > 0 \quad (14.3)$$

$$\text{Sigmoid: } K(v, v_1) = \tanh(\gamma V^T V_j + r), \quad \gamma > 0 \quad (14.4)$$

$$\text{RBF: } K(v, v_1) = (-\gamma \| V - V_1 \|), \quad \gamma > 0 \quad (14.5)$$

where $\gamma$ is the gamma term in the kernel function, $r$ is the bias term in the kernel function and d is the polynominal degree.

In this chapter, we used nonlinear RBF kernel function allowing efficient transformation of nonlinear classes into a linear one in high-dimensional space, giving rise to promising results for different modeling tasks (Jaafari and Pourghasemi 2019; Pourghasemi and Rahmati 2018).

**Fig. 14.2** A field photograph of soil erosion occurred in the study area

### 14.3.2 Data and Study Area

**Soil Erosion Inventory Map**

To generate a susceptibility map of soil erosion for the study area, the location of historical soil erosion was identified within the Golgol watershed through multiple field observations and surveys (Fig. 14.2).

As a result, 157 soil erosion events were recorded by a global positioning system (GPS). These locations were then grouped into two groups of training and validation with a ratio of 70:30 (Conoscenti et al. 2014). Therefore, from all the inventory locations, 110 cases were haphazardly selected for training the model (calibration step) and 47 cases for validation purpose (Fig. 14.3). The training dataset of soil erosion events was used in the SVM model as dependent variable.

### 14.3.3 Geo-environmental Factors

There is no a specific framework or a standard methodology to select conditioning factors for modeling the soil erosion susceptibility. Following a review of the literature (Angileri et al. 2016; Pournader et al. 2018), eight soil erosion conditioning factors (elevation, slope per cent, slope aspect, topographic wetness index (TWI),

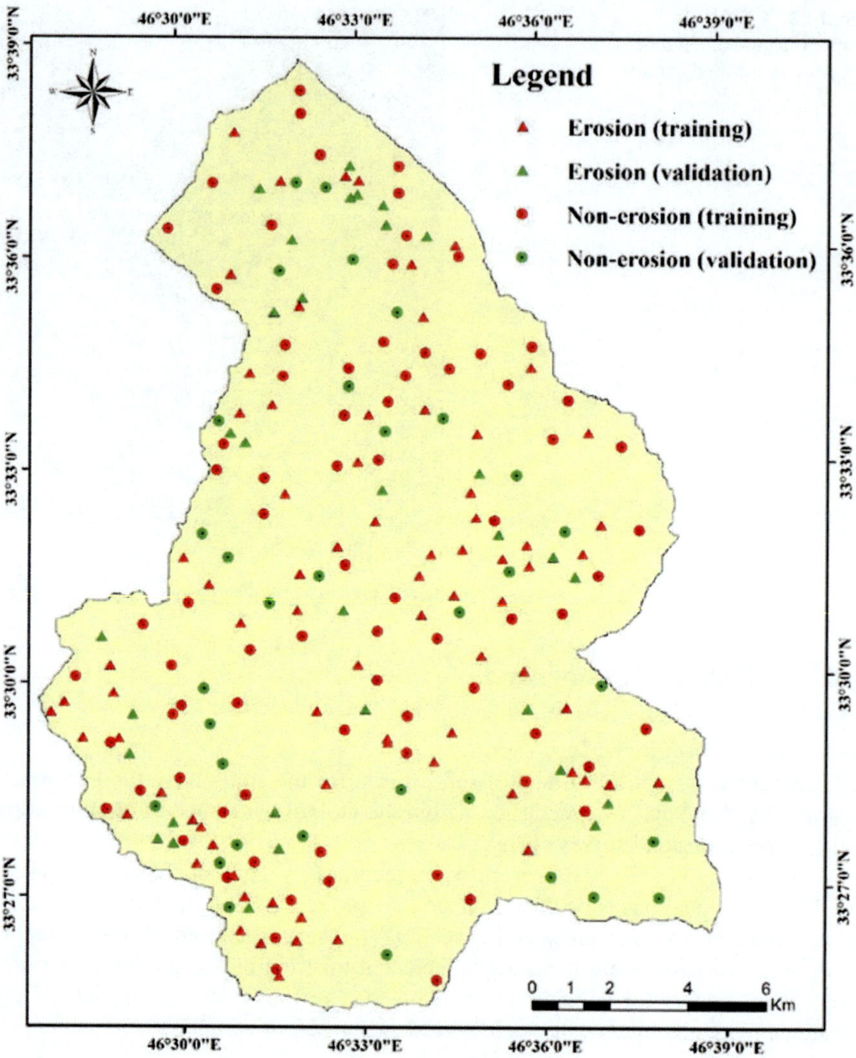


**Fig. 14.3** Training and validation groups of erosion and non-erosion

distance from the stream, plan curvature, lithology and land use) were selected for this study.

Several factors such as slope aspect, plan curvature, elevation, slope per cent (Fig. 14.4a–d) and TWI were generated based on the digital elevation model (DEM) with the spatial resolution of 30 × 30 m layer (Fig. 14.4h). Slope per cent affects the speed of water flow and subsequently, the high-graded slopes would be more susceptible to soil erosion. The range of calculated slope percentage in this study was from 0 to 74%. Slope aspect was prepared and classified into nine classes. TWI was another topo-hydrological factor that was used in this study. This factor was developed using the equation suggested by Moore et al. (1991):

$$\text{TWI} = \ln(\alpha/\tan\beta) \tag{14.6}$$

where α represents the cumulative area that flows to appoint and β is slope angle.

The other topographic factor that was used was the plan curvature. This factor was regarded as the curvature of a contour line shaped by the correspondence of a horizontal plane with the surface. Lithology was obtained from a 1: 100,000-scaled geology map of the study area (Fig. 14.4f). There are ten units of lithology in the study watershed: Pd, Gu, As, Gs, Qa, Sv, Sg, Il, Ehm and Qls(As) (Table 14.1). The Enhanced Thematic Mapper Plus (ETM +) satellite data were used to generate the land-use map of the study area that exhibited five classes of land use: agriculture, forest, orchard and rangeland (Fig. 14.4e). In addition, the distance from streams was calculated based on the Euclidian distance algorithms from the drainage network in the ArcGIS 10.2 environment (Figure 14.4g).

### *14.3.4 Validation of Susceptibility Map*

The results of erosion susceptibility mapping were quantitatively validated using the observed soil erosion dataset (validation dataset) and the receiver operating characteristic (ROC) method (Conoscenti et al. 2014). According to the literature, this method has been widely used to evaluate the performance of models in different studies (Angileri et al. 2016). The area under ROC curve (AUC) statistic, as a threshold independent evaluation of the performance of a model, ranges from 0 to 1. The AUC statistic was used to indicate the accuracy of the model. An AUC value of 0.5 indicates that the model prediction is no better than random predictions, whereas a value of 1.0 shows a perfect prediction accuracy. In this study, performance measure tool (PMT) extension was used in ArcGIS 10.2 to evaluate the accuracy of the model in the training and validation steps (Rahmati et al. 2019).

## 14.4 Results and Discussion

### *14.4.1 Soil Erosion Susceptibility Model*

Figure 14.5 shows the spatial variability of soil erosion susceptibility. Soil erosion susceptibility ranges from zero to 0.98. According to this figure, areas near drainage network where are steeper have high susceptibility.

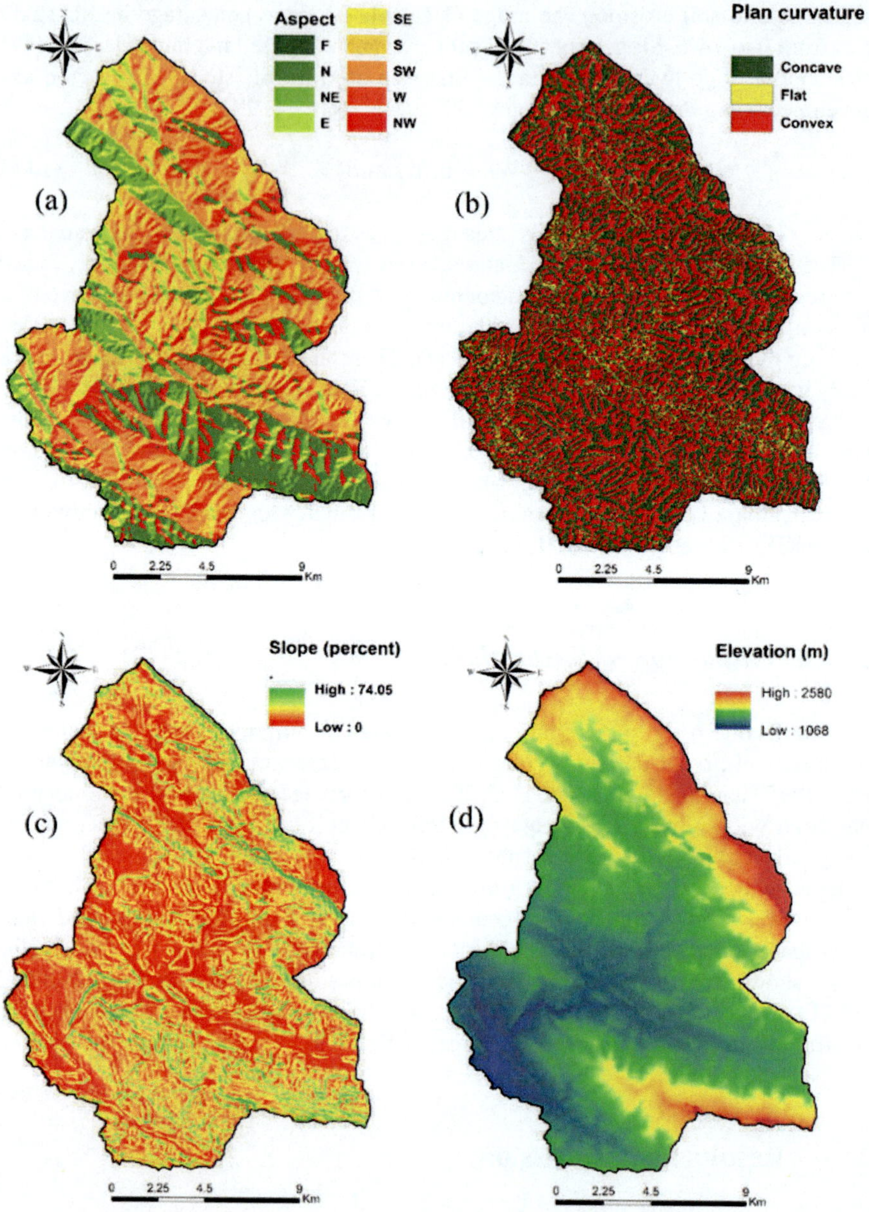


**Fig. 14.4** Soil erosion conditioning factors: **a** aspect, **b** plan curvature, **c** slope, **d** elevation, **e** land use, **f** lithology, **g** distance from stream and **h** TWI

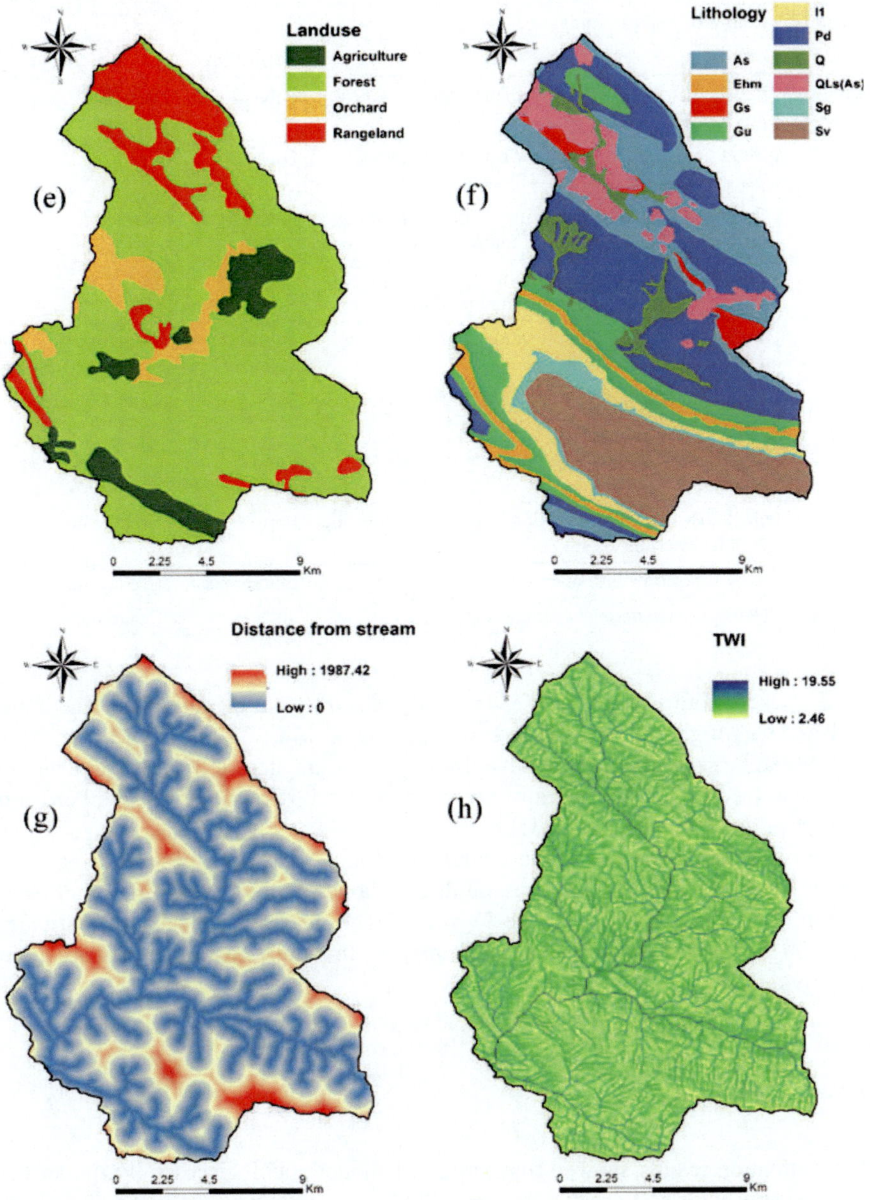


**Fig. 14.4** (continued)

**Table 14.1** Lithology of the study area

| Code | Lithology | Formation | Geological age |
|---|---|---|---|
| Pd | Red sandstone and shale with subordinate sandy limestone | Pabdeh | Permian |
| Gu | Bluish grey marl and shale with subordinate thin-bedded argillaceous—limestone | Gurpi | Cretaceous |
| As | Cream to brown—weathering, feature—forming, well-jointed limestone with intercalations of shale | Asmari | Miocene |
| Gs | Anhydrite, salt, grey and red marl alternating with anhydrite, argillaceous limestone and limestone | Gachsaran | Miocene |
| Qa | Stream channel, braided channel and flood plain deposits | Quaternary | Quaternary |
| Sv | Grey, thick-bedded to massive limestone with thin marl intercalations in upper part | Sarvak | Late. Cretaceous |
| Sg | Dark grey shale including phyllite and yellow limestone | Surgah | Late. Cretaceous |
| Il | Grey-to-white thin-to-medium layers of limestone with interactions of silt | Ilam | Cretaceous |
| Ehm | Grey limestone with interactions of grey marl | Emam hasan | Cretaceous |
| Qls(As) | *Quaternary-Asmary landslide* deposits | – | Quaternary |

The susceptibility map classified the study area into four classes of low, medium, high and very high susceptibility to soil erosion (Fig. 14.6). The distribution of these classes can be seen in this map that helps engineers and decision makers to develop soil erosion conservation programs for the study area. The area of each soil erosion susceptibility class is shown in Table 14.2.

According to the results of the model, 30.14 and 11.12% of the study area were classified as high and very high susceptibility classes, respectively. The soil erosion susceptibility map produced by the SVM model confirmed the results of Pournader et al. (2018) who have produced a similar map for this study area using the maximum entropy model.

### 14.4.2 *Validation*

The validation results showed that the SVM model could successfully model the spatial patterns of soil erosion susceptibility in the training step (goodness-of-fit) according to both evaluation metrics (AUC = 84.1% and TSS = 0.651).

However, the accuracy of the model in the training step only gives information about the degree of data fitness to the model. Therefore, the predictive performance of the model should be assessed. In the validation step, the results clearly indicated that SVM had a good predictive performance based on the AUC (81.2%) and TSS

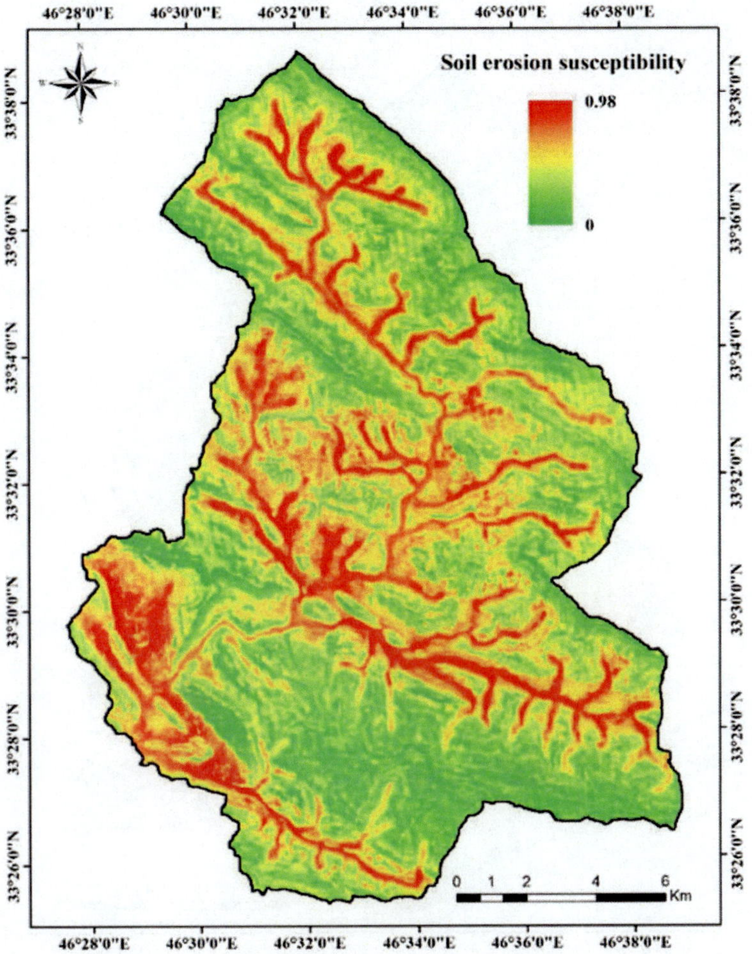


**Fig. 14.5** Soil erosion susceptibility map

(0.62) metrics. Therefore, SVM produced a valid soil erosion susceptibility map that can be considered as the main primary tool for reducing and controlling soil erosion in this data-scarce region (Kornejady et al. 2017).

In this study area, Pournader et al. (2018) have previously investigated the performance of the maximum entropy model to spatially predict soil erosion susceptibility. Their results indicated that the maximum entropy model had a good efficiency in the training step with an AUC of 0.867 and in the validation step with an AUC = 0.794. Therefore, considering the same conditions of the study area and predictive factors, the SVM model outperformed the maximum entropy model. Our results also confirmed the study of Rahmati et al. (2017) and Yunkai et al. (2010) in that SVM

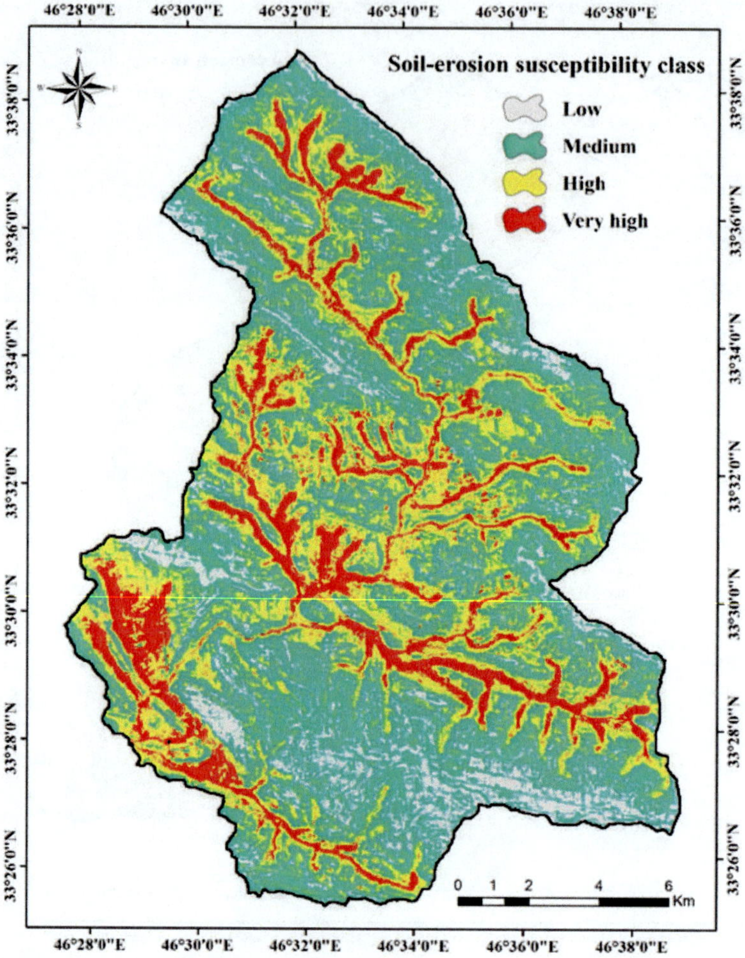


**Fig. 14.6** Soil erosion susceptibility class map

**Table 14.2** Area of soil erosion susceptibility class

| No. | Susceptibility class | Area (ha) | Area (%) |
|---|---|---|---|
| 1 | Low | 3379.03 | 13.2 |
| 2 | Medium | 11768.65 | 45.54 |
| 3 | High | 7769.44 | 30.14 |
| 4 | Very high | 2867.97 | 11.12 |

**Table 14.3** Results of the model accuracy in the training and validation steps

| Evaluation metric | Training | Validation |
|---|---|---|
| AUC (%) | 84.1 | 81.2 |
| TSS | 0.651 | 0.62 |

can successfully extract relationships between soil erosion events and erosion-related factors (predictor variables) (Table 14.3).

## 14.5 Concluding Remarks

In this chapter, the efficiency of the SVM model to model soil erosion susceptibility was investigated in both training and validation steps in the Golgol watershed, Ilam province, Iran. Results clearly demonstrated that the SVM model had good accuracy in terms of the goodness-of-fit (AUC = 84.1% and TSS = 0.651) and predictive performance (AUC = 81.2% and TSS = 0.62). Therefore, as the main conclusion, we found that the SVM model is capable to produce an accurate soil erosion susceptibility map even in data-scarce regions. This study provides a practical tool for mapping soil erosion susceptibility that helps controlling soil erosion.

**Acknowledgements** Authors thank the Iranian Department of Water Resources and the Iranian Department of Geology Survey (IDGS) for providing necessary data and maps.

## References

Angileri SE, Conoscenti C, Hochschild V, Märker M, Rotigliano E, Agnesi V (2016) Water erosion susceptibility mapping by applying stochastic gradient treeboost to the imera Meridionale river basin (Sicily, Italy). Geomorphology 262:61–76

Cama M, Lombardo L, Conoscenti C, Rotigliano E (2017) Improving transferability strategies for debris flow susceptibility assessment: application to the Saponara and Itala catchments (Messina, Italy). Geomorphology 288:52–65

Colazo JC, Carfagno P, Gvozdenovich J, Buschiazzo D (2019) Soil erosion. In: The soils of Argentina, Springer, Cham, pp 239–250

Conoscenti C, Angileri S, Cappadonia C, Rotigliano E, Agnesi V, Märker M (2014) Gully erosion susceptibility assessment by means of GIS-based logistic regression: a case of Sicily (Italy). Geomorphology 204:399–411

Dewitte O, Daoudi M, Bosco C, Van Den Eeckhaut M (2015) Predicting the susceptibility to gully initiation in data-poor regions. Geomorphology 228:101–115

Garcia RC (2018) Estimated soil loss of makatipo catchment under different climate change scenarios. Ecosyst Dev J 6(1)

Jaafari A, Pourghasemi HR (2019) Factors influencing regional-scale wildfire probability in Iran: an application of random forest and support vector machine. In: Spatial modeling in GIS and R for earth and environmental sciences, Elsevier, pp 607–619

Kornejady A, Ownegh M, Bahremand A (2017) Landslide susceptibility assessment using maximum entropy model with two different data sampling methods. CATENA 152:144–162

Moore ID, Grayson RB, Ladson AR (1991) Digital terrain modelling: a review of hydrological, geomorphological, and biological applications. Hydrol Process 5(1):3–30

Pourghasemi HR, Rahmati O (2018) Prediction of the landslide susceptibility: which algorithm, which precision? CATENA 162:177–192

Pourghasemi HR, Yousefi S, Kornejady A, Cerdà A (2017) Performance assessment of individual and ensemble data-mining techniques for gully erosion modeling. Sci Total Environ 609:764–775

Pournader M, Ahmadi H, Feiznia S, Karimi H, Peirovan HR (2018) Spatial prediction of soil erosion susceptibility: an evaluation of the maximum entropy model. Earth Sci Inform. https://doi.org/10.1007/s12145-018-0338-6

Rahmati O, Haghizadeh A, Pourghasemi HR, Noormohamadi F (2016) Gully erosion susceptibility mapping: the role of GIS-based bivariate statistical models and their comparison. Nat Hazards 82(2):1231–1258

Rahmati O, Tahmasebipour N, Haghizadeh A, Pourghasemi HR, Feizizadeh B (2017) Evaluation of different machine learning models for predicting and mapping the susceptibility of gully erosion. Geomorphology 298:118–137

Rahmati O, Kornejady A, Samadi M, Deo RC, Conoscenti C, Lombardo L, Bui DT (2019) PMT: New analytical framework for automated evaluation of geo-environmental modelling approaches. Sci Total Environ 664:296–311

Vandaele K, Lammens J, Priemen P, Evrard E (2013) How to control muddy floods from cultivated catchments. lessons from the Melsterbeek catchment in Flanders (Belgium), (On-line), Samenkering Land en water, St-Truiden, Belgium

Vapnik V, Guyon I, Hastie T (1995) Support vector machines. Mach Learn 20(3):273–297

Yunkai L, Yingjie T, Zhiyun O, Lingyan W, Tingwu X, Peiling Y, Huanxun Z (2010) Analysis of soil erosion characteristics in small watersheds with particle swarm optimization, support vector machine, and artificial neuronal networks. Environ Earth Sci 60(7):1559–1568

# Chapter 15
# Spatial Prediction of Landslide Susceptibility Using Random Forest Algorithm

**Omid Rahmati, Aiding Kornejady, and Ravinesh C. Deo**

## 15.1 Introduction

This chapter develops a random forest model for spatial prediction of landslide susceptibility. A landslide is the downward movement of slope materials mainly due to the pull of gravity as well as the interconnection between different predisposing and triggering factors (Hungr et al. 2014). Landslides are among the most frequent natural disasters that have been imposing considerable human casualties and socioeconomic losses for decades. Over the course of advancements in modeling natural phenomena, many conceptual and numerical models have been developed with different computational algorithms, and many studies have been devoted to conceptualizing the process of landsliding and, in particular, assessing landslide susceptibility as the backbone of higher hierarchical computations. General consensus categorizes these models in four different groups (Van Westen et al. 2006): (1) inventory-based approaches; (2) heuristic approaches; (3) statistical and probabilistic approaches; (4) deterministic approaches.

Intelligent data analytics, also known as data mining-based models, have received growing attention in different scientific communities particularly due to handling data scarcity, scale-invariant features, and engaging a diverse range of controlling factors of a phenomenon. Some of these models are maximum entropy, boosted regression

O. Rahmati (✉)
Soil Conservation and Watershed Management Research Department, Kurdistan Agricultural and Natural Resources Research and Education Center, AREEO, Sanandaj 6616936311, Iran
e-mail: orahmati68@gmail.com

A. Kornejady
Department of Watershed Management, Gorgan University of Agricultural Sciences and Natural Resources, Gorgan 4918943464, Iran

R. C. Deo
School of Sciences, University of Southern Queensland, Springfield Central, QLD 4300, Australia

© Springer Nature Singapore Pte Ltd. 2021 
R. C. Deo et al. (eds.), *Intelligent Data Analytics for Decision-Support Systems in Hazard Mitigation*, Springer Transactions in Civil and Environmental Engineering,
https://doi.org/10.1007/978-981-15-5772-9_15

tree, classification and regression tree, support vector machine, general linear model, and random forest (RF) (Bachmair and Weiler 2012; Catani et al. 2013; Pourghasemi and Kerle 2016; Prasad et al. 2006). Adopting such models has shed more light on the ways to cope with low performance in terms of learning and prediction aspects. In particular, the dynamics of the reciprocal cause-and-effect relationship between each causative factor and the phenomenon can be visually checked and even rendered into quantitative terms. Additionally, the hierarchy of factors importance, vast choice of model validation techniques, and pinpointing the locations with inadequate modeling inputs are considered some of the by-products of using data mining models.

North of Iran including the Mazandaran Province is home to diverse natural disasters such as floods, droughts, erosional processes, earthquakes, and landslides which together have limited the domains of human habitats and impeded development activities. The Klijanerestagh watershed is a mountainous landslide-prone interprovincial (i.e., Mazandaran and Semann provinces) basin that has witnessed many landslides through the years. It has been a pilot area for many research endeavors; however, the trail of studies is far away from the new advancements in the modeling area so far as the traditional empirical models have been the basis of landslide susceptibility assessment for a long time. These premises necessitated adopting a data mining model for landslide susceptibility in this area.

To this end, random forest, as one of the most powerful intelligent data analytic models with a successful record of applications in landslide spatial assessment, is used to map landslide susceptibility in the Klijanerestagh watershed and extract the inferences therein exist.

## 15.2 Materials and Methods

### *15.2.1 Study Area*

The Klijanerestagh watershed extends an area of 3921.7 km$^2$ located between 53° 00′ E to 54° 08′ E longitudes and 35° 08′ N to 36° 38′ N latitudes. Elevation ranges between 43 m and 3711 m.a.s.l. (Fig. 15.1). The average annual rainfall is 580 mm (Mazandaran's FRWO[1] Office). Forests account for about 66.5% of the study area, followed by agriculture (14%), rangelands (15.4%), and the remaining area is distributed among other land uses such as orchards and water bodies. From a geological viewpoint, diverse lithological settings have covered the area in which the largest area pertains to Mm, s, l class (marl, calcareous sandstone, sandy limestone, and minor conglomerate) (24.4%), followed by TRJs (dark gray shale and sandstone) (14.7%), and Pr (dark gray medium—bedded to massive limestone) (7.6%), and the remaining area is shared between other formations. Table 15.1 presents in detail some information regarding the constituent materials of the formations.

[1]Forest, Range and Watershed Management Organization.

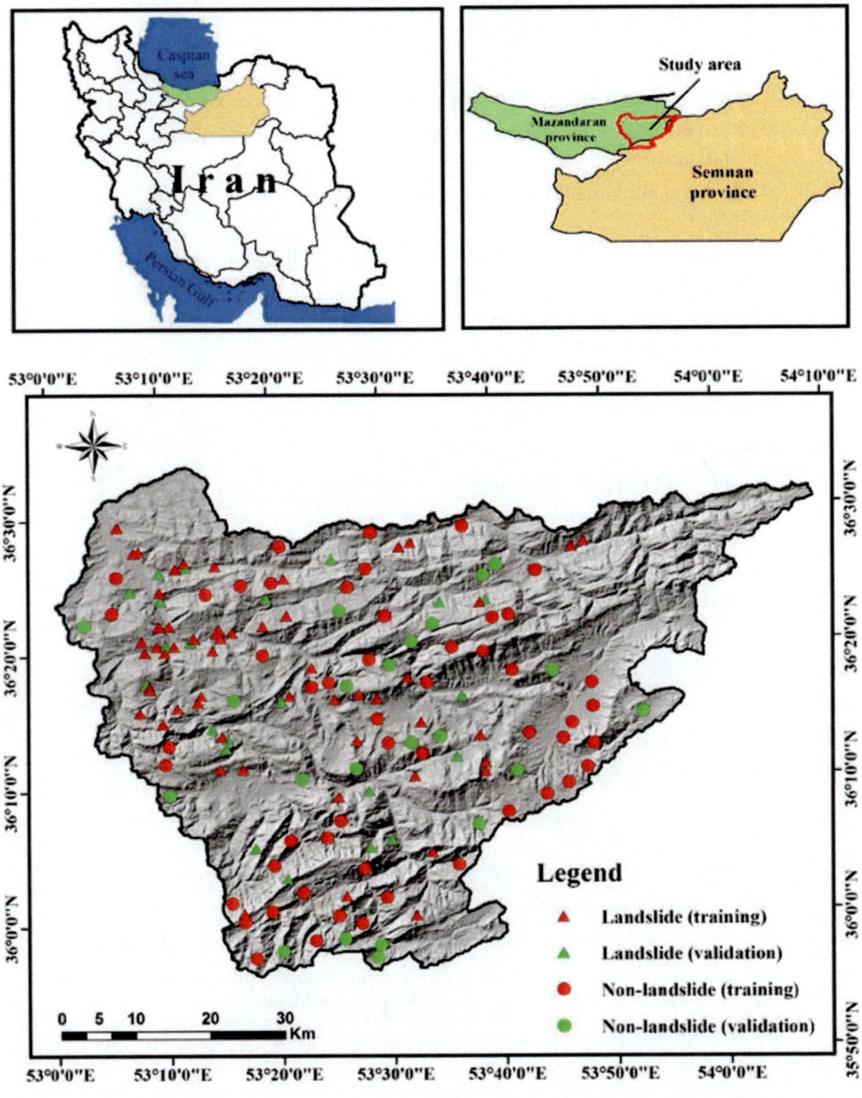


**Fig. 15.1** Location of the Klijanerestagh watershed with landslide and non-landslide points

## 15.2.2 Landslide Inventory Dataset

Landslide inventory is a pivotal part of any spatial modeling endeavor. In this study, a total of 78 landslides were recorded as point features by using the available geoinformatics (Google Earth, a handheld GPS device, archived organizational data, and local information) during extensive field surveys and its final digitized map was generated

**Table 15.1** Description of lithological units of the study area

| Code | Description |
|---|---|
| Cb | Alternation of dolomite, limestone, and variegated shale (Barut Formation) |
| Cl | Dark red medium-grained arkosic to subarkosic sandstone and micaseous siltstone (Lalun Formation) |
| Cm | Dark gray to black fossiliferous limestone with subordinate black shale (Mobarak Formation) |
| COm | Dolomite platy and flaggy limestone containing trilobite; sandstone and shale (Mila Formation) |
| Db-sh | Undifferentiated limestone, shale, and marl |
| DCkh | Yellowish, thin to thick-bedded, fossiliferous argillaceous limestone, dark gray limestone, greenish marl and shale, locally including gypsum |
| E1l | Nummulitic limestone |
| E1m | Marl, gypsiferous marl, and limestone |
| Ek | Well-bedded green tuff and tuffaceous shale (Karaj Formation) |
| Jd | Well-bedded to thin-bedded, greenish-gray argillaceous limestone with intercalations of calcareous shale (Dalichai Formation) |
| Jk | Conglomerate, sandstone, and shale with plant remains and coal seams (Kashafrud Formation) |
| Jl | Light gray, thin-bedded to massive limestone (Lar Formation) |
| K2l1 | Hyporite bearing limestone (Senonian Formation) |
| K2l2 | Thick-bedded to massive limestone (Maastrichtian Formation) |
| K2m,l | Marl, shale, and detritic limestone |
| Ktzl | Thick-bedded to massive, white to pinkish orbitolina bearing limestone (Tizkuh Formation) |
| Kupl | Globotruncana limestone |
| Mc | Red conglomerate and sandstone |
| Mm,s,l | Marl, calcareous sandstone, sandy limestone, and minor conglomerate |
| Mur | Red marl, gypsiferous marl, sandstone, and conglomerate (Upper Red Formation) |
| pCk | Dull green gray slaty shales with subordinate intercalation of quartzitic sandstone (Kahar Formation, Morad Series, and Kalmard Formation) |
| Pd | Red sandstone and shale with subordinate sandy limestone (Dorud Formation) |
| PeEm | Marl and gypsiferous marl locally gypsiferous mudstone |
| PeEz | Reef-type limestone and gypsiferous marl (Ziarat Formation) |
| Pel | Medium to thick-bedded limestone |
| Pgkc | Light red coarse-grained, polygenic conglomerate with sandstone intercalations |
| Plc | Polymictic conglomerate and sandstone |
| Pr | Dark gray medium-bedded to massive limestone (Ruteh Limestone) |
| Qcf | Clay flat |
| Qft1 | High-level piedmont fan and valley terrace deposits |

(continued)

**Table 15.1** (continued)

| Code | Description |
|---|---|
| Qft2 | Low-level piedmont fan and valley terrace deposits |
| Qm | Swamp and marsh |
| TRe | Thick-bedded gray o'olitic limestone; thin platy, yellow to pinkish shaly limestone with worm tracks and well to thick-bedded dolomite and dolomitic limestone (Elikah Formation) |
| TRe1 | Thin-bedded, yellow to pinkish argillaceous limestone with worm tracks |
| TRJs | Dark gray shale and sandstone (Shemshak Formation) |

in ArcGIS 10.3. Further, the sample points were partitioned into two sets of training (70% of samples: 55 nos.) and validation samples (30% of samples: 23 nos.). Also, an equal presence–absence balance (i.e., 78 non-landslide locations) was used and kept intact for both training and validation datasets.

### 15.2.3 Landslide Conditioning Factors

Landslides occur owing to the synergistic interaction between topological, geo-environmental, climatic, and anthropogenic factors (i.e., so-called predictors). Although there is no general consensus as to which set of predictors could contribute the most to the modeling process, field surveys and archived data paved the way to select more relevant factors. Also, it tried to select these factors from different categories mentioned above as far as the data availability issue allowed doing so (Pourghasemi and Kerle 2016). Further, to handle data multicollinearity, the variance inflation factor (VIF) was used. The VIF values above 5 are considered strong correlation and critical multicollinearity which can cause bias.

Considering these premises, twelve predictors were selected, namely elevation (as a proxy to precipitation regime, vegetation cover, and lithological settings), slope percentage (indicating gravity), slope aspect (indicating soil humidity), plan curvature (indicating flow conference/divergence), profile curvature (indicating flow velocity), distance from roads (indicating anthropogenic interference and man-made slopes), distance from streams (indicating slope undercuts), distance from faults (indicating seismic activities), lithological formations (indicating soil/rock resistance to weathering), land use (indicating environmental and, to some extent, anthropogenic configuration of the area), soil type (indicating hydrogeological processes and hydrological response zones), and topographic wetness index (TWI) (indicating runoff generation and accumulation pattern) (Fig. 15.2).

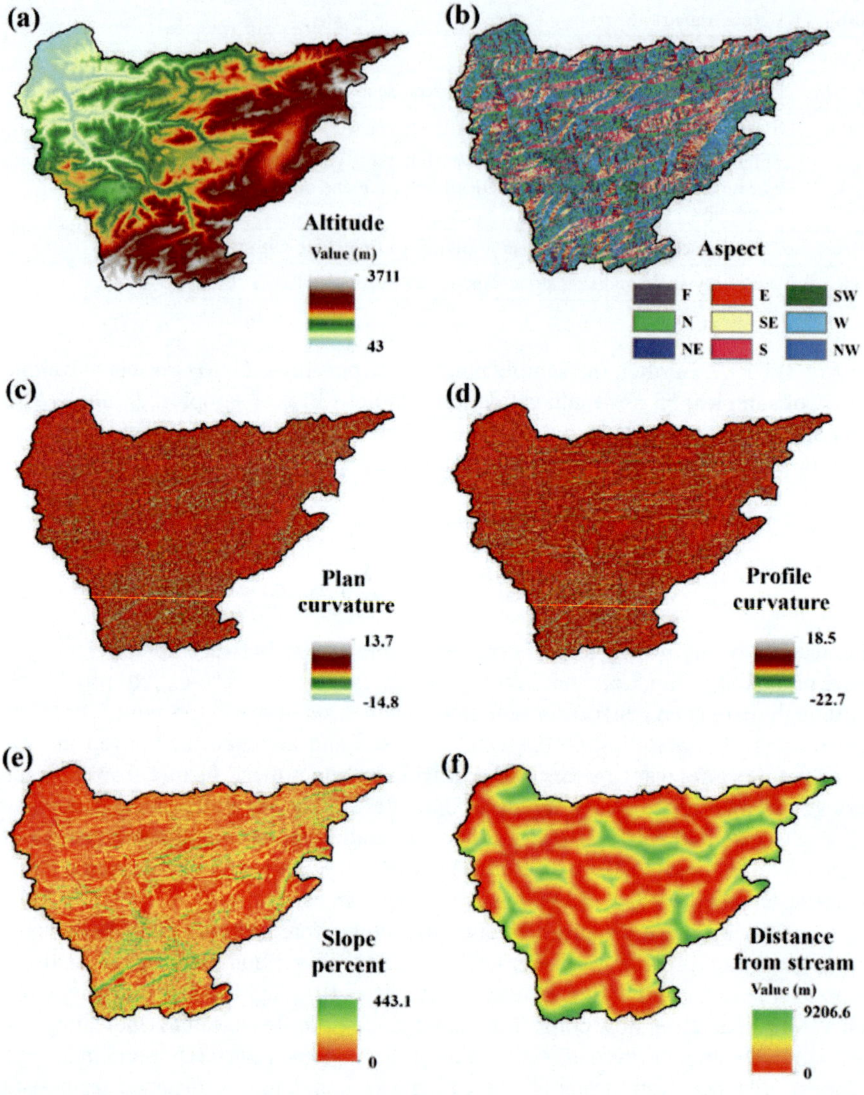


**Fig. 15.2** Landslide conditioning factors: **a** altitude, **b** aspect, **c** plan curvature, **d** profile curvature, **e** slope percent, **f** distance from stream, **g** distance from road, **h** distance from fault, **i** land use, **j** lithology, **k** soil type, and **l** TWI

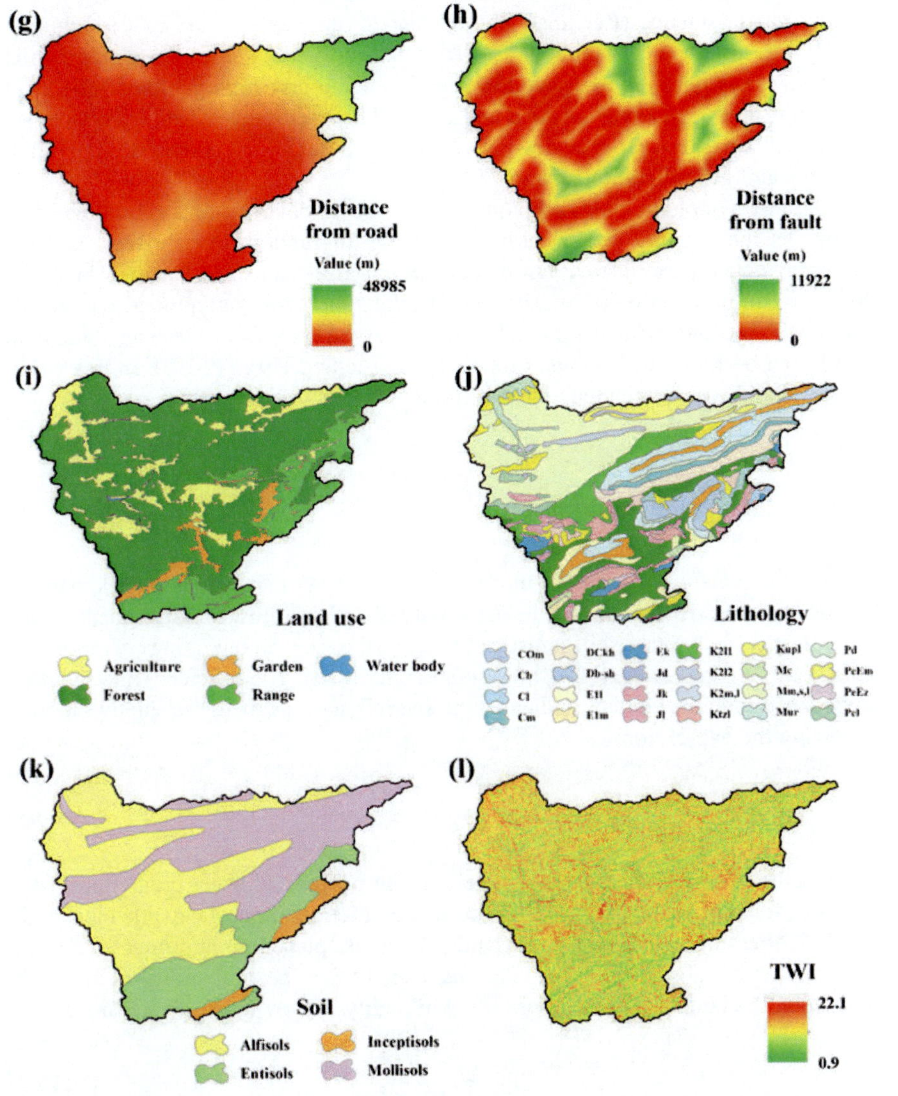


**Fig. 15.2** (continued)

## 15.3 Methodology

### *15.3.1 Design of Random Forest Model*

Random forest (RF) was first expounded by Breiman (2001). It integrates classification and regression schemes and has been widely used in various branches of environmental science (Bachmair and Weiler 2012; Prasad et al. 2006) including landslide assessment (Catani et al. 2013). Random forest is based on averaging the results of many decision trees and has been known to give high goodness-of-fit and prediction performances (Catani et al. 2013). The averaging scheme produces an outcome that has less variance, less overfitting, high flexibility, and accordingly high accuracy and generalization power.

Adopting a bootstrapping technique enables the model to select a subset of observations as the training set by taking advantage of random binary trees, and the remaining data are excluded as out-of-bag (OOB) (Breiman 2001; Catani et al. 2013; Pourghasemi and Kerle 2016). The misclassification error is computed via out-of-bag (OOB) by comparing the predicted and observed responses. More mathematical details can be found in Breiman (2001), and Calle and Urrea (2010). In this work, the RF model was implemented in the "randomForest" package.

### 15.3.2 Accuracy Assessment

Performance analysis of the RF model was carried out using five metrics, namely efficiency, true positive rate (TPR; sensitivity, recall or hit rate), false positive rate (FPR; 1-specificity), true skill statistic (TSS), and the receiver operating characteristic (ROC) curve (Frattini et al. 2010; Rahmati et al. 2019). The efficiency metric often termed as model's accuracy indicates the overall success of the predictive model, following the expression:

$$\text{Efficiency} = \frac{\text{TP} + \text{TN}}{T} \tag{15.1}$$

where TP and TN, respectively, denote the true positive (correct prediction of the landslide locations) and true negative (correct prediction of non-landslide locations), and $T$ is the sum of the positives (landslide areas; presence) and negatives (non-landslide areas; absences). The TPR index quantifies the probability of correctly predicting the landslide areas as observed in reality, following the expression:

$$\text{TPR} = \frac{\text{TP}}{\text{TP} + \text{FN}} \tag{15.2}$$

where FN denotes the incorrectly predicting landslide areas as non-landslide cases as opposed to the observations. The FPR or the fallout rate indicates the probability of incorrectly predicting a non-landslide area as landslide by the model (Eq. 15.3). It is also known as error type I as the higher values can cause economic losses due to incorrectly introducing a safe site as hazardous which then will be deprived of economic investments in the future.

$$\text{FPR} = \frac{\text{FP}}{\text{FP} + \text{TN}} \tag{15.3}$$

The TSS metric conjugates more arguments in order to give rather more determinant and inclusive connotation of the model's performance, following the expression:

$$\text{TSS} = \text{Sensitivity} + \text{Specificity} - 1 \tag{15.4}$$

The higher values of TSS indicate a model that is highly capable of distinguishing the landslide and non-landslide patterns. It is noteworthy that all the abovementioned metrics are considered cutoff-dependent as the process of their calculation requires a holdout probability value. For the latter, a 50% value was selected due to the balanced presence–absence datasets (Frattini et al. 2010). As opposed to the cutoff-dependent metrics above, the ROC curve is considered a cutoff-independent metric. It plots sensitivity or true positive (i.e., correctly predicting a landslide location as observed in nature) on the *y*-axis against the 1-specificity (i.e., FP) on the *x*-axis.

## 15.4 Results and Discussion

### *15.4.1 Landslide Susceptibility Map*

Figure 15.3 indicates the landslide susceptibility map produced by the RF model.

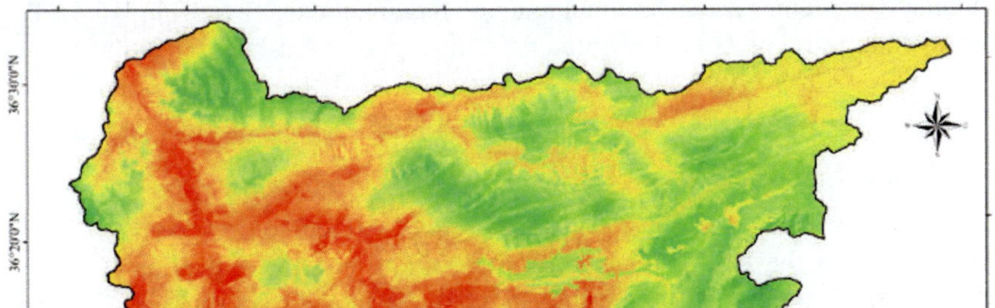


**Fig. 15.3** Landslide susceptibility map

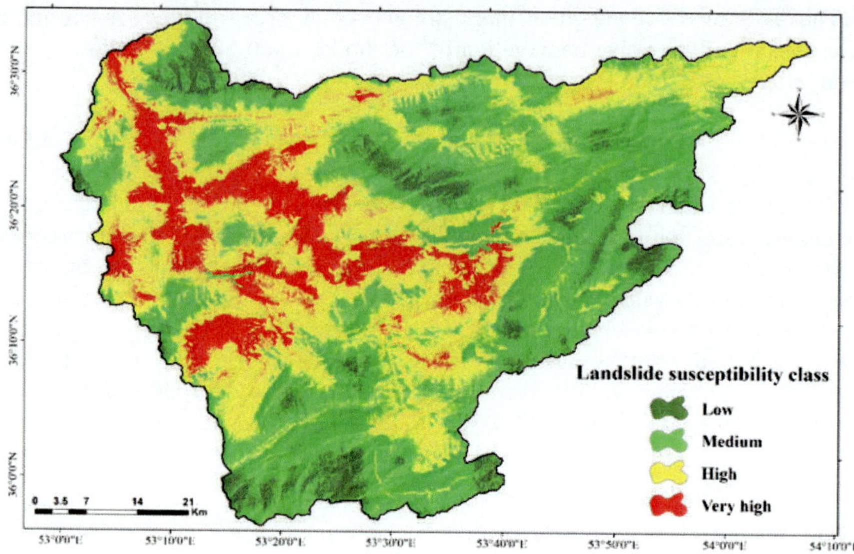


**Fig. 15.4** Landslide susceptibility class map

Spatial variability of landslide susceptibility can be seen in the study area as a quantitative index, ranging from zero to one. The landslide susceptibility map was classified into four classes of low (0–0.25), medium (0.25–0.5), high (0.5–0.75), and very high (0.75–1) by using the equal interval classification scheme (Fig. 15.4). The western and central parts of study area have the highest landslide probability.

## 15.4.2 Model Accuracy

The accuracy of the RF model in both the training and the validation steps was analyzed using different criteria (Table 15.2). In the model training step, RF model had an excellent goodness-of-fit (AUC = 0.936, $E$ = 0.887, TSS = 0.776, TPR = 0.905, FPR = 0.129). However, as the training dataset was used to calibrate the model, it could not be used to investigate their prediction capability. In the validation

**Table 15.2** Accuracy of the RF model using cutoff-dependent evaluation criteria

| Evaluation criteria | Training | Validation |
|---|---|---|
| Efficiency ($E$) | 0.887 | 0.777 |
| True skill statistic (TSS) | 0.776 | 0.559 |
| True positive rate (TPR; sensitivity) | 0.905 | 0.809 |
| False positive rate (FPR; fallout; 1-specificity) | 0.129 | 0.250 |

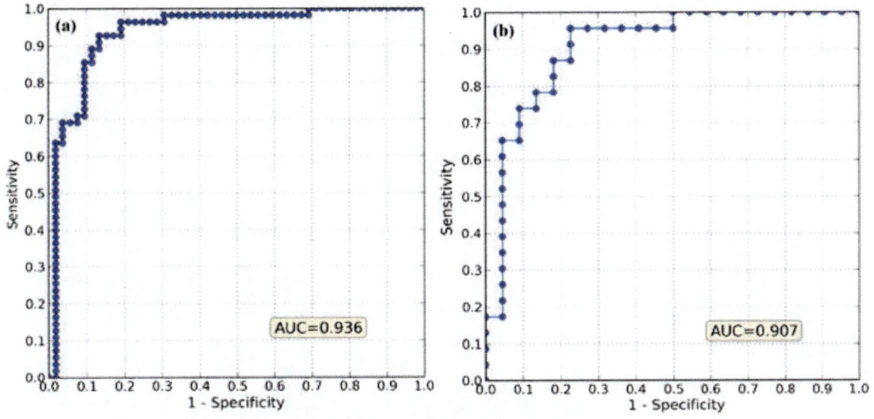


**Fig. 15.5** ROC curves: **a** training step and **b** validation step

step, results indicated that the RF model showed an excellent predictive performance with an AUC of 0.907. Other evaluation criteria confirmed this result ($E = 0.777$, TSS = 0.559, TPR = 0.809, FPR = 0.25). Therefore, RF can successfully identify landslide-prone areas even in data-scarce regions (Fig. 15.5).

### *15.4.3 Variable Importance Ranking*

The contribution of the landslide conditioning factors was investigated, and lithology was found to be the most important factor, followed by distance from road (Fig. 15.6). Since faults often contribute to landslide occurring, distance from faults stood on the third rank. Soil and profile curvature were the least important among predictive factors. However, it should be mentioned that all factors contributed to the modeling and improved the accuracy of the predictions.

## 15.5 Concluding Remarks

This chapter not only investigated the capability of the RF to predict the landslide susceptibility, but also assessed the importance of predictive factors. Both goodness-of-fit and predictive performance of models were quantitatively evaluated using different performance metrics. According to the achievements, RF obtained an outstanding performance for spatial modeling of landslide susceptibility. It can be considered as a practical and promising tool for landslide prediction and control. Our study also demonstrates that construction of roads in the study area had a significant contribution to the landslide occurrence.

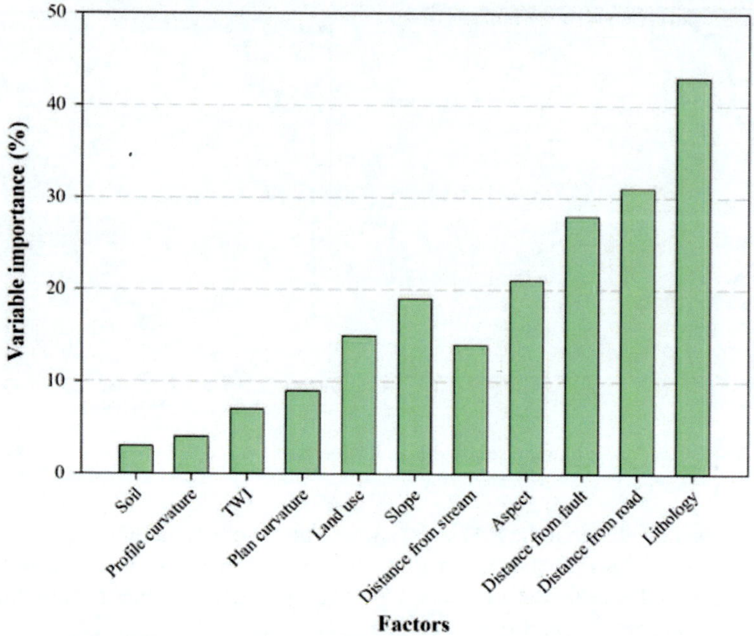


**Fig. 15.6** Importance of the landslide conditioning factors using the RF model

## References


Bachmair S, Weiler M (2012) Hillslope characteristics as controls of subsurface flow variability. Hydrol Earth Syst Sci 16(10):3699–3715

Breiman L (2001) Random forests. Mach Learn 45(1):5–32

Calle ML, Urrea V (2010) Letter to the editor: stability of random forest importance measures. Briefings Bioinf 12(1):86–89

Catani F, Lagomarsino D, Segoni S, Tofani V (2013) Landslide susceptibility estimation by random forests technique: sensitivity and scaling issues. Nat Hazards Earth Syst Sci 13(11):2815–2831

Frattini P, Crosta G, Carrara A (2010) Techniques for evaluating the performance of landslide susceptibility models. Eng Geol 111(1–4):62–72

Hungr O, Leroueil S, Picarelli L (2014) The Varnes classification of landslide types, an update. Landslides 11(2):167–194

Pourghasemi HR, Kerle N (2016) Random forests and evidential belief function-based landslide susceptibility assessment in western Mazandaran Province. Iran Environ Earth Sci 75(3):185

Prasad AM, Iverson LR, Liaw A (2006) Newer classification and regression tree techniques: bagging and random forests for ecological prediction. Ecosystems 9(2):181–199

Rahmati O, Kornejady A, Samadi M, Deo RC, Conoscenti C, Lombardo L, Bui DT (2019) PMT: New analytical framework for automated evaluation of geo-environmental modelling approaches. Sci Total Environ 664:296–311

Van Westen CJ, Van Asch TW, Soeters R (2006) Landslide hazard and risk zonation—why is it still so difficult? Bull Eng Geol Environ 65(2):167–184

# Chapter 16
# Artificial Neural Networks for Prediction of Steadman Heat Index

**Bhuwan Chand, Thong Nguyen-Huy, and Ravinesh C. Deo**

## Acronyms and Abbreviations

| | |
|---|---|
| ACT | Australian Capital Territory |
| ANN | Artificial Neural Network |
| ARIMA | Autoregressive Integrated Moving Average |
| AR | Autoregressive |
| ARW | Advanced Research Core of the Weather Research and Forecasting |
| BGFS | Broyden–Fletcher–Gold Farb-Shanno |
| CFS v2 | Climate Forecast System Version 2 |
| CPU | Central Processing Unit |
| (d) | Willmott's Index of Agreement |
| E | Legates and McCabe Index |
| $E_{NS}$ | Nash–Sutcliffe Coefficient |
| HI | Heat Index |
| Kpa | Kilo Pascal |
| LM | Levenberg- Marquardt |
| MA | Moving Average |
| MAE | Mean Absolute Error |
| MATLAB | Matrix Laboratory |
| MK | Man-Kendall |
| ML | Machine Learning |

B. Chand (✉) · R. C. Deo
School of Sciences, University of Southern Queensland, Springfield Central, QLD 4300, Australia
e-mail: bhuwan.schand@gmail.com

T. Nguyen-Huy
Centre for Applied Climate Sciences, University of Southern Queensland, Toowoomba, QLD 4350, Australia

© Springer Nature Singapore Pte Ltd. 2021

R. C. Deo et al. (eds.), *Intelligent Data Analytics for Decision-Support Systems in Hazard Mitigation*, Springer Transactions in Civil and Environmental Engineering,
https://doi.org/10.1007/978-981-15-5772-9_16

| | |
|---|---|
| MLP | Multi-Linear Perceptron |
| MLR | Multiple Linear Regression |
| NSW | New South Wales |
| NT | Northern Territory |
| PE | Prediction Error |
| PI | Performance Indicators |
| QLD | Queensland |
| RMSE | Rot Mean Squared Error |
| SA | South Australia |
| SVR | Support Vector Regression |
| TAS | Tasmania |
| VIC | Victoria |
| WA | Western Australia |

## 16.1 Introduction

Exacerbated and disastrous nature of heatwaves plays a key role in the loss of human lives, crop damage, wildfires, and interruptions to industrial and social activities, and so, it is a crucial research subject. Hyperthermia is a consequence of the heatwave resulting from the metabolic heat decrement at a temperature above 35 °C, affecting the wellbeing of especially the old people and children (Dodla et al. 2017; Sherwood and Huber 2010). Heatwave cannot be exactly defined due to regionally confined and spatially variant, daily surface temperature (Meehl and Tebaldi 2004).

Although it can be summarized as a chronic extreme scenario of very low rainfall and high temperatures persisting consecutively for many days, having a negative impact through physiological stress on plants, animals, and on the medically demanding people, infants, and elderly people. The characteristics of heat waves may differ, for example, the European heatwave of 2003 persisted for 3 months of June, July, and August (Fink et al. 2004) whereas the US heatwaves of 1995 and 1999 have been restricted only for a few days in July (Dodla et al. 2017; Palecki et al. 2001). As per historical records, extremely high atmospheric temperature in eastern Australia between 1st and 22nd February triggered the "most significant medical emergency on record" creating a massive load on the health sector (Deo et al. 2007). McMichael et al. (2003) reported that, on average, about 100 heat-related deaths per year occur in Australian cities alone. Air temperature and relative humidity are found to be a major factor in generating the consequences of heatwave.

A useful Heat Index (HI) for such disaster assessments is the Steadman Heat Index (SHI), that incorporates an ambient air temperature and relative humidity in its formulation to establish a relationship among all important variables (human body, clothing, heat transfer, and ventilation from skin) responsible for the heatwaves. The HI is developed to raise public awareness of a weather hazard that affects some groups more than others, using real physical parameters that have known and detrimental impacts on the human body. Accounting for all the factors above, the assumptions are summarized as the final HI equation (Ta or HI) that is most widely used (i.e., by the National Weather Service and most private-sector entities) where the computational methods, multiple regression analysis, is used with a statistical error of ±17.0556 °C (°C) or 1.3 °F (°F). In addition, for the error minimization and higher accuracy of results, a machine learning (ML) model is an efficient tool in the calculation of HIs. An essence of ML models as a data intelligent technique is a necessity as it is better than many statistical formulas and procedures. Furthermore, the performance error and computational time are reasonably low for the ML techniques than the empirical methods (Khald et al. 2015). In Australia, still the performance of data analytic techniques being an ML method is a necessity to predict and forecast the heatwaves for the sake of minority groups who used to suffer from the severity of heatwaves in various regions of the country.

This chapter aims to design intelligent data analytic models based on Artificial Neural Networks (ANN) for heatwave prediction with specific case studies in Australia. Scientists and engineers, medical researchers, flora and fauna scientists, and other climate researches must have knowledge of HI to give appropriate and most efficient predictive models, patterns, and trends of heatwaves across the various regions. Therefore, predictive models based on HI are useful in trend analysis, disaster risk mitigation, and decisions made for safety precautions and hazard management.

## 16.2 Literature Review

### *16.2.1 Australian Bureau of Meteorology Definition of Heatwaves*

The World Meteorological Organization defines a heatwave as five or more consecutive days of prolonged heat in which the daily maximum temperature is higher than the average maximum temperature by 9° F (−12.7778 °C) or more (Hutter et al. 2007). The Australian Bureau of Meteorology defines a heatwave as "three days or more of maximum and minimum temperatures that are unusual for the location" (Tong et al. 2010). In the United States, definitions also vary by region; however, a heatwave is usually defined as a period of at least two or more days of excessively hot weather (Aubrecht and Özceylan 2013).

### *16.2.2 Steadman Heat Index for Heatwave Monitoring*

The study of (Steadman 1979) proposed a general phenomenon of characterizing a heatwave by considering any prolonged period of excessively hot conditions followed by high atmospheric humidity maintained for at least three consecutive days. Steadman applied this non-linear principle to convert temperature and humidity into Heat Index by an empirical formula. The empirical formulae of SHI in °C is designed by (Steadman 1979). This HI value contains assumptions about the human body mass, clothing, amount of physical activity, individual heat tolerance, sunlight and ultraviolet radiation exposure, and the wind speed. The significant deviations from these will result in HI values which do not accurately reflect the perceived temperature (Rothfusz 1990). According to a study, global warming boosts the probability of extreme weather events like heatwaves, far more than it boosts more moderate events (Hansen 1989).

### *16.2.3 Statistical Heatwave Modeling*

For any atmospheric temperature and water vapor pressure, Steadman Heat Index (SHI) is defined as the temperature that would produce equivalent stress on the body if the water vapor pressure were changed to a predetermined reference pressure (specifically, 1.6 kPa, which is the saturation vapor pressure of water at 14 °C). For example, an air temperature of 30 °C with a relative humidity of 60 percent translates to an apparent temperature (Heat Index) of 33 °C (Delworth et al. 1999). The resulting equation could be considered as a HI equation although it is obtained in an approximate way.

Even though temperature and relative humidity are the only two variables in the HI equation as shown in Fig. 16.1 above, all the variables on the lists above are implied (Rothfusz and Headquarters 1990). We have access to readily available datasets of temperature and humidity datasets, but we do not have access to HI datasets for the predictive purpose (Bortolotti 2018) in any available data repositories. The target role of HI is to predict the extreme climatic conditions like heatwaves. Hence, to get a HI, we must perform some mathematical calculations as above-described to derive the SHI for the prediction algorithms that have never been described in any kind of research paper works.

Global heatwaves have been studied, for the European regions (García-Herrera et al. 2010), the USA (Peterson et al. 2013), and Australia (Jeng-Rung et al. 2013). Global Circulation Model (GCM) studies have been used both for validating simulated present-day climates and analyzing possible future climates including the dry spells and air temperature (Huth et al. 2000). For example, (Gregory et al. 1997) dealt with dry spells and their return periods in the transient experiment with the Hadley Center GCM.

The high temporal resolution of climate data provided (i.e., daily fields of $T_{max}$ and $T_{min}$) which is necessary for assessing the consequences of extreme events is the major cause behind the selection of this model (Huth et al. 2000). Huth et al. (2000) outlined that global 1.125° gridded datasets of daily maximum and minimum temperatures from the GCM at the National Institute for Environmental Studies have been already used for simulations. In contrast (Teixeira et al. 2013) suggested that

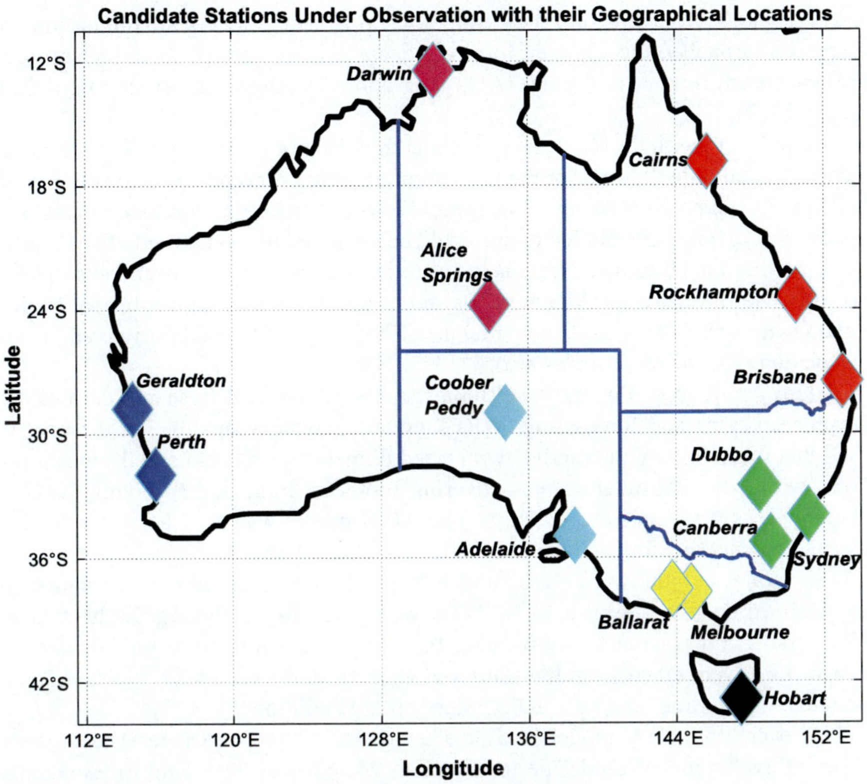


Fig. 16.1 Spatial map of study regions showing selected stations

available GCM outputs were mainly for monthly climate data which are less suitable for studying impacts of extreme temperature episodes, as extreme daily temperatures are fluctuating.

In other words, GCMs represent the most satisfactory approach to predicting future changes in climate, but their present low spatial resolution (a few hundred km) makes their output problematic to use in impact studies (Karl et al. 1990; Winkler et al. 1997). Despite the various simulation use of GCM model in climate forecast, it is still not implemented to predict the massive daily heatwave occurrence in various parts of Australia. Hence, a new model such as a higher version of the GCM model is an essence of state-of-art.

Alternatively, a physical model called CFSv2 did give clear indications of a warmer summer, several months in advance (Luo and Zhang 2012). CFSv2 hindcast data had a T126 spatial resolution (roughly 100 km) and included several near-surface variables at a 6-hourly temporal resolution. Since the launch of CFSv2 in operational forecasting in late March 2011, most of the forecast runs predicted that the number of days with a temperature higher than the 90th, 95th, and 99th percentile thresholds and it would be doubled from the climatological expectation. As the initial time of forecast approached the summer, forecasts became more certain about the summer heatwave with respect to the intensity, geographic locations, and timing (Luo and Zhang 2012).

The problem with the above model is that they cannot be localized. It means that the model can only be used for the prediction of extreme events at the resolution of 100 km. *So, what about the inner local regions to be predicted*? ML models can be a good solution that those hidden points could be localized for better predictive results. Downscaling the HI to local scale helps us to predict more precisely and accurately. For this, we must use the ML methods in the observed Scientific Information for Land Owners (SILO) datasets to model the CFSv2 so that daily 6 hourly predictions can be done for the local scales also.

Additionally, in one of the case studies, ARW (Advanced Research core of the weather research and forecasting (WRF) modeling system version 3.6.1) model is used that has two-way interactive two nested domains with 9-km and 3-km resolutions, with the 3-km inner domain covering southeast India and adjoining the Bay of Bengal, as a region for prediction. The ARW model was used to generate 72-h weather predictions for each day during May 17–31, 2015.

The model outputs were stored at 3-h interval, and the outputs as a maximum temperature corresponding to 0900 UTC were used for analyzing the heatwave. The initial conditions and time-varying boundary conditions necessary for model integrations were taken from the statistical analysis and forecasting files which are available at 3-h time interval and 0.25 horizontal resolution.

However, this ARW model-predicted temperatures at 2-m (meters) level were used for evaluation of model performance at 24, 48, and 72-h lead times (Dodla et al. 2017). Still, there is a necessity to predict heatwaves in a low resolution and the inner nominal regions more accurately with higher accuracy. Since the maximum and minimum temperature and the air humidity dataset are used in the calculation of SHI for the optimal model accuracy, the ARW approach could not meet the requirement of SHI calculation.

Finally, as per the available state-of-art, with the application of data intelligent algorithms, it can be predicted that the HI at the inner grid points with the low spatial resolution with the help of available temperature and humidity datasets (Trigo and Palutikof 1999). The support vector regression (SVR) is appealing algorithms for a large variety of regression problems since they do not only consider the error approximation to the data but also the generalization of the model, i.e., its capability to improve the prediction of the model when a new dataset is evaluated by it. Although there are several versions of the SVR, the classical model, SVR, described in detail (Smola and Schölkopf 2004). Common model derivation can help to predict the heatwave in the lowest possible spatial resolution in various locations.

The ANNs are based on a feed-forward configuration of the multilayer perceptron that has been excessively used by a growing number of authors (Trigo and Palutikof 1999). A multilayer perceptron (MLP) consists of several computing single elements (neurons) whose connections are adjusted by constructing an input-output mapping in a learning process so that it can predict samples that were not done before. In this model, an MLP is trained and validated by using, respectively, a training set and usually a validation set to stop the training process without overfitting. An MLP can learn in the sense it will be able to predict samples different from those used in the design process. This is the so-called ability to generalize and is the capability of why MLPs are known to be universal approximators of a large class of functions. MLPs have been successfully applied to a huge amount of non-linear prediction and classification problems (Bishop and Bishop 1995; Haykin 1998; Salcedo-Sanz et al. 2016). Thus, MLP could be a useful technique for the prediction of non-linear trend of heatwave occurring at various geographical locations.

For example, the performance of ML algorithms for monthly mean air temperature prediction was compared and evaluated from the previously measured values in observational stations of Australia and New Zealand, and climate indices of importance in the region (Salcedo-Sanz et al. 2016). The high-quality mean temperature dataset in each observational station using two different training and test sets were used to train the data-driven methods. First, the training set corresponding with months from January 1900 up to December 1930 (the test set is the remaining data from 1931 up to 2010). Second, a longer training set from 1900 up to 1970 was considered, with the test set from 1971 up to 2010. The performance of intelligent data analytic approaches was evaluated by mean absolute error (MAE) in the test set.

None of the research work has ever used ML in prediction of daily heatwave indices in Australia up to available literature reviews. With the help of Scientific Information for Land Owners (SILO) data, HI can be calculated using some statistical tools. SILO is a database of Australian climate data from 1889. It provides daily datasets for a range of climate variables in ready to use formats suitable for research and climate applications (Wallace et al. 2006). In this research, observed values are extracted from the SILO datasets for the HI calculation purposes.

With CFSV2 model, formulation on SILO data, SHI can be calculated through the downscaling process. Hence, ML algorithm can be useful in generalizing the model to predict the low-resolution gridded dataset to predict the heatwave more accurately and precisely.

### *16.2.4 Heatwave Trend Analysis Using Statistical Mean Method*

Based on mean climatic conditions, researchers built their analysis based on daily temperature and spatially variant relative humidity. The daily temperature time series has been employed for the statistical analysis to detect trends in heat-related extreme events. In Australian continent, from Western Australia to Queensland for the period (1967–1973), (Tucker 1975) observed that the change in annual mean temperature varied by some 3 °C, from 2 °C to −0.5 °C, respectively.

Similarly, (Torok and Nicholls 1996) noticed trends in annual maximum temperatures, ranging from 0.15 °C per decade (Western Australia) to −0.15 °C per decade in New South Wales. A 30 year (1946–1975) trend calculated by Coughlan (1979) reflected an increase in mean maximum temperature throughout Southern Australia but a decrease by a similar amount over tropical regions in Northern Territory.

The spatial distribution of change in HIs over more than last six decades for Australian continent need to be observed for better trend analysis and results. These outcomes will help in determining the areas of significant increase and decrease of heatwave with the severe effect on human beings, domestic and wild animals along with plants and vegetation.

## 16.3 Materials and Method

### *16.3.1 Study Area and Model Input Data*

For the prediction of daily SHI, two input (or predictor) variables that describe climatic attributes of 15 candidate stations (Fig. 16.1) are considered in the entire continent of Australia. The candidate stations are chosen such that entire Australia would be analyzed with respect to the heatwave analysis as per the most populous area of that state or territory. In addition, from the state-of-art, it is found that the meteorological data in these sites is more accurate and all of them are point dataset despite gridded dataset as extracted from the SILO.

The site locations like Sydney, Dubbo, Cairns, Adelaide, Melbourne, and Brisbane are already explored and analyzed for a small period of daily surface mean temperature and relative humidity to predict the heatwave trends (Deo et al. 2007). Therefore, it will be much beneficial to make a research study in updated years with a range of 68 years in these locations for verification and validation purpose. Additionally, sites like Darwin and Hobart are chosen such that there is a huge spatial difference in the geographical location. As, sites like Hobart and Ballarat are near to the south pole and Darwin, near to the equatorial region. This will also simplify the analysis of heatwaves with respect to global warming perspectives. For that propose, the eight different states location are chosen as Queensland (Cairns, Rockhampton, Brisbane), New South Wales (Dubbo, Sydney), Australian Capital Territory (Canberra), Victoria (Ballarat, Melbourne), South Australia (Adelaide, Coober Peddy), Northern Territory (Alice Springs, Darwin), Western Australia (Perth, Geraldton), and Tasmania (Hobart).

The site-specific inputs would be the station ID, year, month, latitude, longitude, and station elevation, with meteorological inputs as the maximum temperature, minimum temperature. The spatial locations can be seen in Table 16.1.

To design an ANN model, the maximum temperature and maximum relative humidity data for 1950–2017 are acquired. These climate data are obtained from the SILO data set. These datasets are originally collected from hourly observations at daily time-scales from the period 1950 A.D to 2017 A.D. The pictorial representation of the spatial map of study locations for stations considered in this study is

**Table 16.1** Spatial locations of candidate sites with geographic characteristics used in prediction of the Steadman Heat Index

| Geographic characteristics | | | | | |
|---|---|---|---|---|---|
| Station number | Station id | Station name | Latitude (°E) | Longitude (°S) | Elevation (m) |
| 1 | 8050 | Geraldton Town | −28.777 | 114.605 | 3 |
| 2 | 9159 | Perth West | −31.95 | 115.85 | 54 |
| 3 | 14,015 | Darwin Airport | −12.424 | 130.8925 | 30.4 |
| 4 | 15,590 | Alice Springs Airport | −23.795 | 133.889 | 546 |
| 5 | 16,007 | Coober Pedy | −29.004 | 134.7564 | 215 |
| 6 | 31,011 | Cairns Aero | −16.874 | 145.7458 | 2.2 |
| 7 | 23,034 | Adelaide Airport | −34.952 | 138.5204 | 2 |
| 8 | 39,264 | Rockhampton | −23.381 | 150.5172 | −1.4 |
| 9 | 40,913 | Brisbane | −27.481 | 153.0389 | 8.1 |
| 10 | 65,030 | Dubbo (Mentone) | −32.519 | 148.5187 | 330 |
| 11 | 66,037 | Sydney Airport AMO | −33.947 | 151.1731 | 6 |
| 12 | 70,282 | Canberra City | −35.267 | 149.1167 | 564 |
| 13 | 86,282 | Melbourne Airport | −37.666 | 144.8321 | 113.4 |
| 14 | 89,002 | Ballarat Aerodrome | −37.513 | 143.7911 | 435.2 |
| 15 | 94,008 | Hobart Airport | −42.834 | 147.5033 | 4 |

shown in Fig. 16.1. Data splitting is done as the training-1950–1990 (60%), then next validation-1991–2004 (20%) and remaining as a testing 2005–2017 (20%).

### 16.3.2 Calculation of Steadman's Heat Index

The approximate version of (Steadman 1979) formula to calculate SHI using relative humidity and air temperature is:

$$\begin{aligned} T_a(°\mathrm{C}) = (5/9)\Big[&(-42.4 + 2.0 \times T_{\mathrm{air}} + 10.1 \times \mathrm{RH} - 0.2 \times (T_{\mathrm{air}}\mathrm{RH}) - 6.8 \times 10^{-3} \times T_{\mathrm{air}}^2 \\ &- 5.5 \times 10^{-2} \times \mathrm{RH}^2 + 1.2 \times 10^{-3} \times (T_{\mathrm{air}}\mathrm{RH})^2 + 8.5 \times 10^{-4} \times (T_{\mathrm{air}}\mathrm{RH})^2 \\ &- 2.0 \times 10^{-6} \times (T_{\mathrm{air}}\mathrm{RH})^2) - 32\Big] \end{aligned} \tag{16.1}$$

where

$T_{\mathrm{air}}$ Ambient dry bulb temperature (°C)
RH Relative Humidity (integer percentage)

The range of Ta (apparent body temperature) values for corresponding relative humidity (%) and air temperature (°C) are shown in the appendix. It is noted that at an apparent temperature of Ta of:

- 32–40: Heat cramps or heat exhaustion possible,
- 41–54: Heat cramps or heat exhaustion likely, heatstroke possible,
- 54-higher: Heatstroke highly likely.

It is also stated that exposure to full sunshine can increase the HI value by up to 8 °C (Deo et al. 2007).

### *16.3.3 Trend Analysis Theory*

**Mann–Kendal Test**

According to (Hirsch and Slack 1984), the time series of heatwave indices are subjected to a univariate, Mann–Kendall (MK) test in order to detect monotonic trends. The MK test accounts for missing values, serial correlation, and numbers below a detection limit that commonly confound trend detection procedures in time series analysis (Hirsch and Slack 1984).

The nonparametric MK test is commonly employed to detect monotonic trends in a series of environmental data, climate data, or hydrological data. The null hypothesis, H0, is that the data come from a population with independent realizations, and are identically distributed. Moreover, there is no monotonic trend in the series. An alternative hypothesis, HA, is that the data follow a monotonic trend. In addition, a monotonic trend does exist with a positive, negative, or non-null value. The MK test statistic is calculated according to (AL-Ataby 2019):

$$S = \sum_{k=1}^{n-1} \sum_{j=k+1}^{n} \operatorname{sgn}(X_j - X_k) \tag{16.2}$$

with

$$\operatorname{sgn}(x) = \left\{ \begin{array}{l} 1 \text{ if } x > 0 \\ 0 \text{ if } x = 0 \\ -1 \text{ if } x < 0 \end{array} \right\} \tag{16.3}$$

The mean of $S$ is $E[S] = 0$ and the variance is

$$\sigma^2 = \left\{ n(n-1)(2n+5) - \sum_{j=1}^{p} t_j(t_j - 1)(2t_j + 5) \right\} \Big/ 18 \tag{16.4}$$

where $p$ is the number of the tied groups in the data set and $t_j$ is the number of data points in the $j$th tied group. The statistic $S$ is approximately normal distributed if the following $Z$-transform is employed:

$$Z = \left\{ \begin{array}{c} \frac{S-1}{\sigma} \text{ if S >0} \\ 0 \text{ if } S = 0 \\ \frac{S-1}{\sigma} \text{ if } S > 0 \end{array} \right\} \tag{16.5}$$

The statistic $S$ is closely related to Kendall's $\tau$ as given by:

$$\tau = \frac{S}{D} \tag{16.6}$$

where

$$D = \left[ \frac{1}{2}n(n-1) - \frac{1}{2}\sum_{j=1}^{p} t_j(t_j - 1) \right]^{\frac{1}{2}} \left[ \frac{1}{2}n(n-1) \right]^{\frac{1}{2}} \tag{16.7}$$

The MK test can be used to find the trends for as few as four samples. However, with only a few data points, the test has a high probability of not finding a trend when one would be present if more points were provided. The more data points, the more likely the test is going to find a true trend.

### *16.3.4 The Sen's Slope*

The Sen's slope is a nonparametric alternative method used to estimate the magnitude of slope for univariate time series, the HIs. This approach involves computing slopes for all the pairs of ordinal time points and then using the median of these slopes as an estimate of the overall slope. Since Sen's slope is insensitive to outliers, it provides us a realistic measure of the trends in the data series (Deo et al. 2007). The derived trends and slopes are further analyzed and verified with ANN algorithms and some statistical measures like MLR and ARIMA.

This test computes both the slope (i.e., linear rate of change) and intercepts according to Sen's method. First, a set of linear slopes is calculated as follows (AL-Ataby 2019):

$$d_k = \frac{Xj - Xi}{j - i} \tag{16.8}$$

for ($1 \leq i < j \leq n$), where $d$ is the slope, $X$ denotes the variable, $n$ is the number of data points, and $i, j$ are indices. Sen's slope is then calculated as the median from all slopes: $b$ = Median $d_k$. The intercepts computed for each time step t are given by:

$$a_t = X_t - b * t \tag{16.9}$$

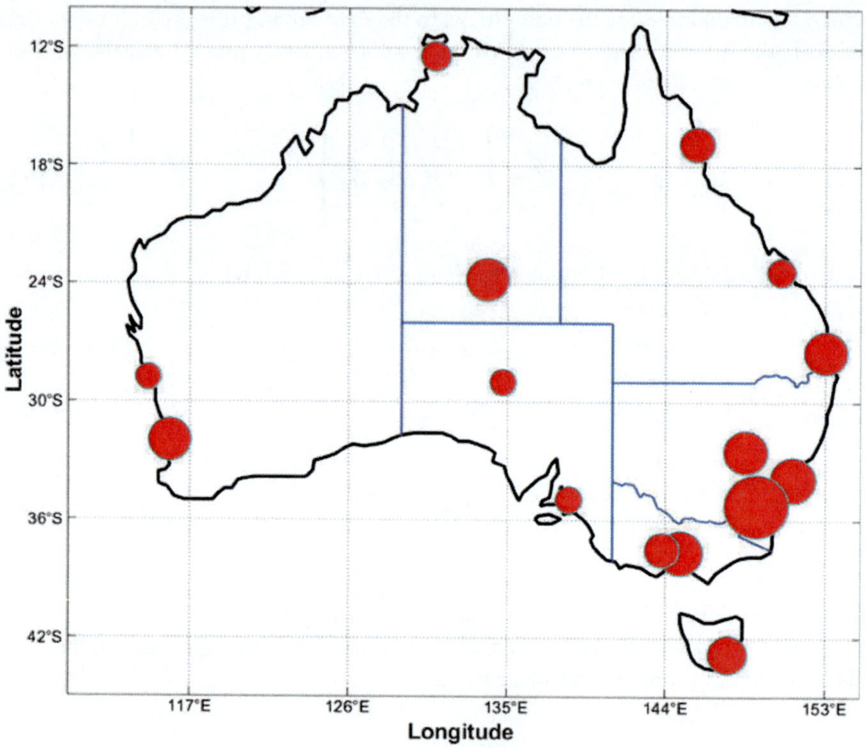


**Fig. 16.2** Geographical representation of the magnitude of Sen's slope for each of the 15 candidate stations. The greater the diameter, the larger is the magnitude. Refer to Fig. 16.1 for the individual site location where the magnitude is shown

and the corresponding intercept is the median of all intercepts.

Since the Sen's slope is insensitive to outliers and any missing data, it is more rigorous than the usual regression slopes and thus provides a realistic measure of the trends in the data series. The magnitude of yearly average HIs for all the 15 candidate sites can be shown as in Fig. 16.2.

## 16.4 Predictive Modeling Theory

### *16.4.1 Artificial Neural Network*

In this chapter, a standalone ANN model with maximum air temperature as a primary predictor and maximum relative humidity as a supplementary predictor are adopted for predictive modeling of the SHI.

The descriptive statistics of the data set are given in Table 16.2 for the training and testing phases. The dataset is free of any missing values, outliers, or invalid values. Thus, it can be observed from the descriptive statistics as shown in Table 16.3, that the test dataset is nearly identical to the training period dataset. It means the higher chances of model accuracy can be obtained from the predictive development procedures.

The aim of ANN model is to extract patterns (predictive features) contained in $x$ time series to forecast the objective variables, Ta.

Figure 16.3 outlines the schematic view of the model. An ANN model is a non-linear modeling technique with a network architecture that mimics the biological structure of our nervous system (McCulloch and Pitts 1943). It has interconnected that are related to the Ta and is able to transmit information through weighted connections (i.e., functional neurons) to map nonlinearly the predictor data features to a high dimensional hyper-plane (Deo and Şahin 2017b).

Mathematically, the ANN algorithm can be written as (Deo and Şahin 2015, 2016; Kim and Valdés 2003):

$$y(x) = F\left(\sum_{i=1}^{L} w_i(t) * x_i(t) + b\right) \tag{16.10}$$

where $x_{i(t)}$ = predictor (input) variables in discrete time-space $t$, $y(x)$ = forecasted Ta in cross-validation (test) data set, $L$ = hidden neurons determined iteratively, $w_i$ $(t)$ = weight that connects the $i$th neuron in the input layer, $b$ = neuronal bias and $F(\cdot)$ is the hidden transfer function.

ANN model does not identify the training algorithm in an explicit manner without an iterative model identification process. The superior model is selected after a series of a trial of several algorithms in this research work. As there is no mathematical formula to determine the neuronal structure in the hidden layer of the ANN model, the number of neurons in the hidden layer needs to be decided by trial and error method (Sahin 2012). A maximum appropriate number of neurons (30) are tested for the development of the network architecture. To determine the optimum architecture, combinations of the input, hidden layer, and output neurons need to be tried one by one. ANN model does not identify the training algorithm in an explicit manner without an iterative model identification process.

The superior model is selected after a series of a trial of several algorithms in this research work (Deo and Şahin 2015). MATLAB based algorithms used in this research are classified in three categories: the quasi-Newton (Huang 1970) (that utilizes *trainlm* and *trainbgf* functions), the gradient descent (Harte et al. 2014) (*traingdx*) and the conjugate gradient (Ali and Smith 2006) (*trainscg, traincgf, and traincgp*). The quasi-Newton method is based on the Levenberg–Marquardt (LM) and the Broyden–Fletcher–Goldfarb–Shanno (BFGS) algorithm (Dennis Jr and Schnabel 1996; Marquardt 1963) that minimize the mean square error whereas the LM algorithm locates the minimum of an input data that is expressed as the sum of squares of non-linear real-valued functions.

Both exhibit a memory overhead issue due to the gradient and Hessian matrix that needs to be calculated and this is especially a disadvantage for very large networks

**Table 16.2** Descriptive statistics of testing and training dataset from 1950 to 2017 data

| Descriptive statistics for Ta over the period (1950–2017) | | | | | |
|---|---|---|---|---|---|
| Data | Minimum | Maximum | Mean | Skewness | Kurtosis |
| *Cairns Aero* | | | | | |
| Training | 18.7 | 45.912,066 | 30.9068 | 0.16441 | −0.7048247 |
| Testing | 20.3 | 46.716005 | 31.6787 | 0.13768 | −0.6768489 |
| *Rockhampton* | | | | | |
| Training | 12.7 | 50.939732 | 29.0651 | 0.30623 | −0.3190134 |
| Testing | 12.3 | 49.898054 | 29.5401 | 0.30741 | −0.1725037 |
| *Brisbane* | | | | | |
| Training | 10.7 | 45.721302 | 26.0551 | 0.30301 | −0.2070829 |
| Testing | 12.6 | 45.409757 | 27.0078 | 0.29044 | 0.0147351 |
| *Dubbo (Mentone)* | | | | | |
| Training | 5.3 | 45.697556 | 23.1187 | 0.13545 | −0.7944909 |
| Testing | 7.7 | 47.140239 | 24.2337 | 0.16319 | −0.7015937 |
| *Sydney Airport* | | | | | |
| Training | 9.1 | 49.511859 | 22.1448 | 0.73215 | 0.8975933 |
| Testing | 11.6 | 51.22089 | 23.3131 | 0.69296 | 0.7963786 |
| *Canberra City* | | | | | |
| Training | 4.1 | 43.068929 | 19.3646 | 0.27885 | −0.7737069 |
| Testing | 4.3 | 43.281147 | 20.8065 | 0.30963 | −0.6853975 |
| *Ballarat Aerodrome* | | | | | |
| Training | 3.2 | 40.688543 | 17.2026 | 0.62953 | −0.3544427 |
| Testing | 4.6 | 42.541842 | 18.08 | 0.56271 | −0.5215875 |
| *Melbourne Airport* | | | | | |
| Training | 5.7 | 43.823793 | 19.2015 | 0.75311 | 0.0980141 |
| Testing | 8.1 | 45.052734 | 20.2575 | 0.71955 | 0.0508454 |
| *Adelaide Airport* | | | | | |
| Training | 9.8 | 48.177819 | 21.0909 | 0.68535 | −0.1031823 |
| Testing | 10.2 | 44.340013 | 21.8409 | 0.65246 | −0.0744881 |
| *Coober Peddy* | | | | | |
| Training | 9.3 | 50.608528 | 26.511 | 0.18462 | −0.7230072 |
| Testing | 10 | 47.748608 | 26.9136 | 0.17124 | −0.7349973 |
| *Alice Spring Airport* | | | | | |
| Training | 8.1 | 46.0672 | 27.4638 | −0.12433 | −0.6794036 |
| Testing | 8.5 | 45.603982 | 28.2933 | −0.11535 | −0.6869787 |
| *Darwin Airport* | | | | | |
| Training | 21.1 | 48.193924 | 35.8138 | −0.36583 | −0.4858765 |

(continued)

**Table 16.2** (continued)

| Descriptive statistics for Ta over the period (1950–2017) | | | | | |
|---|---|---|---|---|---|
| Data | Minimum | Maximum | Mean | Skewness | Kurtosis |
| Testing | 21.9 | 47.120473 | 36.4406 | −0.44729 | −0.5731137 |
| *Perth West* | | | | | |
| Training | 9.8 | 44.62738 | 23.2505 | 0.7006 | −0.0629591 |
| Testing | 11.6 | 43.973461 | 24.2342 | 0.62303 | −0.1681512 |
| *Geraldton Town* | | | | | |
| Training | 14.3 | 51.230233 | 25.9744 | 0.73062 | −0.017314 |
| Testing | 14.8 | 46.333953 | 26.5959 | 0.65194 | 0.0200816 |
| *Hobart Airport* | | | | | |
| Training | 4.5 | 39.566359 | 17.18 | 0.62018 | 0.359803 |
| Testing | 7.4 | 38.968461 | 17.9978 | 0.59367 | 0.22758 |

**Table 16.3** Steadman Heat Index for each month of 2017 data for Brisbane Airport, Canberra City, Sydney Airport Amo, Coober Peddy, and Darwin Airport

| Month | Brisbane Airport | Canberra City | Sydney Airport Amo | Dubbo Airport | Coober Peddy | Darwin Airport |
|---|---|---|---|---|---|---|
| January | 33.8312 | 32.4351 | 31.1407 | 36.5618 | 37.7224 | 38.4351 |
| February | 34.3194 | 29.9168 | 30.2144 | 35.8388 | 35.1169 | 37.5411 |
| March | 32.4515 | 25.7936 | 26.7899 | 30.2945 | 33.0566 | 39.1427 |
| April | 26.2252 | 19.9133 | 23.4726 | 23.4081 | 25.4301 | 36.4675 |
| May | 24.8239 | 16.3548 | 21.0903 | 20.2742 | 22.8721 | 34.6203 |
| June | 22.8267 | 13.5633 | 18.0933 | 17.6533 | 18.8933 | 30.8237 |
| July | 22.9226 | 12.7097 | 19.3608 | 16.8387 | 20.8962 | 33.926 |
| August | 24.5688 | 13.729 | 19.3871 | 17.6194 | 21.8555 | 34.3396 |
| September | 27.0101 | 17.8705 | 23.5937 | 22.3963 | 25.2769 | 36.8257 |
| October | 27.3034 | 22.9878 | 24.7278 | 25.9492 | 28.5829 | 39.4469 |
| November | 27.3685 | 24.1475 | 24.9399 | 27.9388 | 31.0809 | 39.8911 |
| December | 31.6168 | 27.5494 | 30.583 | 32.8648 | 32.4271 | 40.4316 |

These sites show a severe heatwave trend with respect to other sites

(Pham and Sagiroglu 2001). The gradient descent along with momentum and adaptive learning rate (*traingdx*) combines adaptive learning with momentum training where the momentum coefficient is included as a training parameter. The Bayesian regularization (*trainbr*) uses the Jacobian matrix to upgrade weight/biases to attain the best generalization of the input/target dataset (Battiti 1992; MacKay 1995; Vogl et al. 1988).

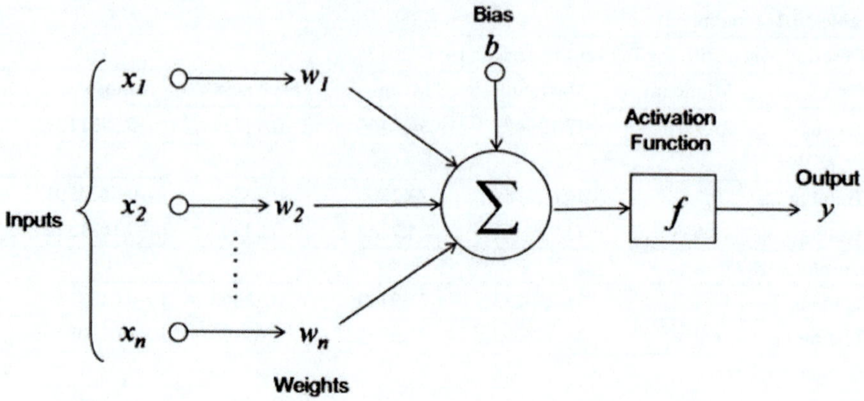


**Fig. 16.3** ANN architecture adopted for predictive modeling of SHI (Ta)

The main task of any modeler is to determine the appropriate transfer function that is not known a priori. A series of MATLAB equations, F(·) available in MATLAB toolbox can be tested (Vogl et al. 1988):

Tangent Sigmoid:

$$f(x) = \frac{2}{1 + \exp(-2x)} - 1 \tag{16.11}$$

Log Sigmoid:

$$f(x) = \frac{1}{1 + \exp(-x)} \tag{16.12}$$

SoftMax:

$$f(x) = \frac{\exp(x)}{\text{sum}\,(\exp(x))} \tag{16.13}$$

Hardlimit:

$$\mathrm{F(x) = 1,\ \ if\ x > 0} \tag{16.14}$$

Positive Linear:

$$\mathrm{F(x) = x,\ \ if\ x \geq 0,\ \ or\ 0,\ \ otherwise.} \tag{16.15}$$

Triangular Basis:

$$\mathrm{F(x) = 1 - abs(x),\ if\ -1 \le x \le 1,\ or\ 0\ otherwise} \tag{16.16}$$

Radial Basis:

$$\mathrm{F(x) = exp(-x^2)} \tag{16.17}$$

Symmetric Hardlimit:

$$\mathrm{F(x) = 1,\ if\ x \ge 0} \tag{16.18}$$

Saturating Linear:

$$\begin{aligned} &: \mathrm{F(x) = 0,\ if\ x \le 0,} \\ &: \mathrm{F(x) = x,\ if\ 0 < x \le 1} \\ &: \mathrm{F(x) = 1,\ if\ x \ge 1} \end{aligned} \tag{16.19}$$

Symmetric Saturating Linear:

$$\begin{aligned} &: \mathrm{F(x) = -1,\ if\ x \le -1,} \\ &: \mathrm{F(x) = x,\ if\ -1 < x \le 1} \\ &: \mathrm{F(x) = 1,\ if\ x > 1} \end{aligned} \tag{16.20}$$

where $x$ is the predictor dataset analyzed in accordance with the function $F(x)$ that can map the predictive features to create a hidden layer weight for the suitable model.

### *16.4.2 Multiple Linear Regression*

MLR model stands for multiple linear regression models. The decency of ANN model is evaluated with MLR model, which is a statistical technique that examines the cause and effect relationship between objective ($y$ = G) and predictor variables ($x$). The main purpose of MLR model is to explain the variations in the predictor dataset to determine their corresponding regression coefficients. The regression equation for $N$ observations for $k$ predictor variables is of the form (Montgomery et al. 2012):

$$Y = C + \beta_1 X_1 + \beta_2 X_2 + \cdots \beta_k X_k \tag{16.21}$$

where $Y$ ($N \times 1$) is a matrix of objective variable ($G$), $X$ ($n \times k$) is a vector of predictor variables, $C$ is the y-intercept and $\beta$ is the multiple regression coefficient for each regressor variable.

For prediction purposes, the multiple linear equations is fitted to a model with a set of $Y$ and $X$ matrix in the data-training period. Then, in the testing period, the fitted MLR model by its coefficient and the $y$-intercept, are used to generate the forecasts of $Y$ values with an additional set of $X$ values. For more details on MLR modeling process, readers can refer to the work of Draper and Smith (1998) and Montogomery and Peck(2012).

### *16.4.3 Autoregressive Integrated Moving Average*

ARIMA stands for Auto Regression Integrated Moving Average model. It is a widely used and well-known statistical method for time series forecasting. An ARIMA model is a class of statistical models that explicitly caters to a suite of standard structures in time series data for analyzing and forecasting time series data. This study has adopted the ARIMA model to validate the ANN model.

This acronym is descriptive, capturing the key aspects of the model itself. Briefly, they are as follows:

- AR: *Autoregression.* A model that uses the dependent relationship between an observation and some number of lagged observations.
- I: *Integrated.* The use of differencing of raw observations (e.g., subtracting an observation from observation at the previous time step) to make the time series stationary.
- MA: *Moving Average.* A model that uses the dependency between an observation and a residual error from a moving average model applied to lagged observations.

Each of these components is explicitly specified in the model as a parameter. The standard notation is used for it as ARIMA ($p$, $d$, $q$), where the parameters are substituted with integer values to quickly indicate the specific ARIMA model being used. The parameter p, indicates the number of lag observations included in the model, also called the lag order, d indicates the number of times that the raw observations are differenced, also called the degree of differencing and q indicates the size of the moving average window, also called the order of moving average.

ARIMA modeling procedure includes the prediction () function, which performs a one-step prediction using the model. The training dataset can be split into train and test sets, where the model is fitted using a training dataset, and generate a prediction for each element on the test set.

The next values of the variable are assumed as a combination of a linear function of past values and random error. ARIMA models can be generalized by the following equation:

$$y_t = \theta_o + \phi_1 y_{t-1} + \phi_2 y_{t-2} + \cdots + \phi_p y_{t-p} + \varepsilon_t - \theta_1 \varepsilon_{t-1} - \theta_2 \varepsilon_{t-2} - \cdots \theta_q \varepsilon_{t-q} \tag{16.22}$$

where $Y_t$ is the actual value, $\varepsilon_t$ is a random error at the time period $t$. $\emptyset_j(j = 0, 1, \ldots q)$ and $\emptyset_i(i = 0, 1, \ldots p)$ are model parameters (Zhang 2003).

Configuring the ARIMA model includes model identification, parameter estimation (use a fitting procedure, maximum log-likelihood, to find the coefficients of the regression model) and model checking (use plots and statistical tests of the residual errors to determine the amount and type of temporal structure not captured by the model). The model identification finalized the differencing parameter (d) by autocorrelation and partial autocorrelation to check whether a differencing is necessary in the case of the non-stationary dataset. Akaike's Information Criterion (AIC), is used to establish ARIMA model, considering the magnitude of maximum log-likelihood and the variance and correlation coefficients collectively assessed in the training data as (Pappas et al. 2008):

$$\text{AIC} = -2\log(L) + 2(p + q + k + 1) \tag{16.23}$$

where $L$ is a log-likelihood of data, $k = 1$ if $c \# 0$ and $k = 0$ if $c = 0$. The process is repeated until either a desirable level of fit is achieved on the in-sample or out-of-sample observations (e.g., training or test datasets).

## 16.5 Performance Evaluation

Only the last 20% of entire time series (i.e., test set) is normally used for model performance assessment, 60% of those for training, and 20% of those for validation. After developing ANN model, the accuracy of the model must be assessed by a comparison of the simulated (*Data.sim*) and observed values (Data.obs) within the test dataset.

Let us suppose,

- $\text{Data.sim}_i$ and $\text{Data.obs}_i$ are simulated and observed values of respective ith time series,
- $\langle\text{Data.sim}\rangle$ and $\langle\text{Data.obs}\rangle$ are the overall mean values within the entire time series
- $\text{Data.sim}^{\text{peak}}$ and $\text{Data.obs}^{\text{peak}}$ are the peak values within the entire time series
- $i$ is the respective datum point in the test set
- $N$ is the length of the test data (e.g., if we have 100 data points in the test set, then $N = 100$ and $i$ will vary from $i = 1, 2, 3 \ldots N(=100)$.

A statistical evaluation is performed using the metrics as follows along with their importance on model performance:

Coefficient of Determination ($r^2$), expressed as

$$\mathbf{r^2} = \left(\frac{\sum_{i=1}^{N}(\text{Data.obs}_i - \langle\text{Data.obs}\rangle)(\text{Data.sim}_i - \langle\text{Data.sim}\rangle)}{\sqrt{\sum_{i=1}^{N}(\text{Data.obs}_i - \langle\text{Data.obs}\rangle)^2}\sqrt{\sum_{i=1}^{N}(\text{Data.sim}_i - \langle\text{Data}_{\text{sim}}\rangle)^2}}\right)^2 \tag{16.24}$$

The coefficient of determination is the square of the correlation between the predicted scores in a data set versus the actual set of scores. The coefficient of determination is used to explain how much variability of one factor can be caused by its relationship to another factor (Murphy 1995). The coefficient of determination is simply the ratio of the model variation to total variation, where total variation is expressed as a function of the deviations (specifical sum of the squared deviations) from the grand mean. In other words, $r^2$ describes how much the model explains the total variation (Taylor 1990).

Regardless of representation, an R-squared equal to 0 means that the dependent variable cannot be predicted using the independent variable. Conversely, if it equals 1, it means that the dependent of a variable is always predicted by the independent variable. A higher value of *R*-Square indicates a model can predict the response variables with a lower error. A coefficient of determination that falls within this range measures the extent that the dependent variable is predicted by the independent variable. An R-squared of 0.20, for example, means that 20% of the dependent variable is predicted by the independent variable.

However, R-squared is unable to determine whether the data points or predictions are biased. It also doesn't tell the analyst or user whether the coefficient of determination value is good or not. A low R-squared is not bad, for example, and it's up to the person to decide based on the R-squared number (Ozer 1985). The coefficient of determination should not be interpreted naively. For example, if a model's R-squared is reported at 75%, the variance of its errors is 75% less than the variance of the dependent variable, and the standard deviation of its errors is 50% less than the standard deviation of the dependent variable. The standard deviation of the model's errors is about one-third the size of the standard deviation of the errors that you would get with a constant-only model (Taylor1990).

Finally, even if an R-squared value is large, there may be no statistical significance of the explanatory variables in a model, or the effective size of these variables may be very small in practical terms.

Nash–Sutcliffe Coefficient ($\mathrm{E_{NS}}$), expressed as

$$\mathbf{E_{NS}} = 1 - \left[\frac{\sum_{i=1}^{N}(\text{Data.obs}_i - \text{Data.sim}_i)^2}{\sqrt{\sum_{i=1}^{N}(\text{Data.obs}_i - \langle\text{Data.obs}\rangle)^2}}\right], \infty - \mathrm{E_{NS}} \leq 1 \quad (16.25)$$

Nash–Sutcliffe model efficiency coefficient ($\mathrm{E_{NS}}$) is used to quantify how well a model simulation can predict the outcome variable. The model may be calibrated, but the predicted values of the outcome variable $\text{Data.sim}_i$ are not inferred from the observed values. Unlike with a statistical model, the sum of squares of the model error, $\sum_{i=1}^{N}(\text{Data.obs}_i - \text{Data.sim}_i)^2$, maybe greater than the total sum of squares, $\sum_{i=1}^{N}(\text{Data.obs}_i - \langle\text{Data.obs}\rangle)^2$, and the coefficient can, therefore, be negative.

Usually, this coefficient is used to assess the predictive power of hydrological models. The normalization of the variance of the observation series results in relatively higher values of $\mathrm{E_{NS}}$ in catchments with higher dynamics and lower values of

$E_{NS}$ in catchments with lower dynamics. To obtain comparable values of $E_{NS}$ in a catchment with lower dynamics the prediction must be better than in a basin with high dynamics. The range of $E_{NS}$ lies between 1.0 (perfect fit) and $-\infty$. An efficiency of lower than zero indicates that the mean value of the observed time series would have been a better predictor than the model. The largest disadvantage of the Nash–Sutcliffe efficiency is the fact that the differences between the observed and predicted values are calculated as squared values. As a result, larger values in a time series are strongly overestimated whereas lower values are neglected (Legates and McCabe 1999). For the quantification of runoff predictions, this leads to an overestimation of the model performance during peak flows and an underestimation during low flow conditions. Similar to R-Square, the Nash–Sutcliffe is not very sensitive to systematic model over or under prediction especially during low flow periods. A test significance for $E_{NS}$ to assess its robustness has been proposed whereby the model can be objectively accepted or rejected based on the probability value of obtaining $E_{NS}$ > threshold (0.65 or other selected by the user) (Moriasi et al. 2007).

Nash–Sutcliffe coefficient has been reported in the scientific literature for model simulations of discharge, water quality constituents such as sediment, nitrogen, and phosphorus loading. Other applications are the use of Nash–Sutcliffe coefficients to optimize parameter values of geophysical models, such as models to simulate the coupling between isotope behavior and soil evolution (Campforts et al. 2016).

Willmott's Index of Agreement (*d*), expressed as

$$\mathbf{d} = 1 - \left[ \frac{\sum_{i=1}^{N} (\text{Data.obs}_i - \text{Data.sim}_i)^2}{\sum_{i=1}^{N} (|\text{Data.sim}_i - \langle \text{Data.obs} \rangle| + |\text{Data.obs}_i - \langle \text{Data.obs} \rangle|)^2} \right], \quad 0 \le \text{d} \le 1 \tag{16.26}$$

The Index of Agreement (*d*) developed by Willmott (1981) as a standardized measure of the degree of model prediction error and varies between 0 and 1. A value of 1 indicates a perfect match, and 0 indicates no agreement at all (Willmott 1981).

(Willmott 1981) demonstrated that the correlation coefficient can be a misleading measure of accuracy—"*r*" between very dissimilar model-predicted variable and observed one can easily approach 1. He discussed other drawbacks of "*r*" and "R-Square" and proposed an "Index of Agreement (*d*)". He noted that the index is intended to be a descriptive measure, and it is both a relative and bounded measure which can be widely applied for cross-comparison between models. Application of the index of the agreement shows that the relatively high values of "*d*" may be obtained even for a poor model fit.

The index of agreement can detect additive and proportional differences in the observed and simulated means and variances; however, it is overly sensitive to extreme values due to the squared differences (Legates and McCabe 1999).

Percentage Peak Deviation ($P_{dv}$; MJ $m^{-2)}$, expressed as:

$$\mathbf{P_{dv}} = 100 \sum_{i=1}^{N} \frac{\text{Data.sim}^{\text{peak}} - \text{Data.obs}^{\text{peak}}}{\text{Data.obs}^{\text{peak}}} \quad (16.27)$$

Percentage peak deviation can also refer to how much the mean of a set of data differs from a known or theoretical value. This can be useful, for instance, when comparing data gathered from a lab experiment to a known weight or density of a substance. To find this type of percent deviation, subtract the observed peak value from the simulated peak value, divide the result by the known observed peak value and multiply by 100.

The negative sign in our answer signifies that our mean is lower than the expected mean. If the percent deviation is positive, it signifies our mean is higher than expected. This percentage peak deviation is not as robust as other assessment metrics in the model performance evaluation process (Gohel et al. 2000).

Root Mean Square Error (RMSE; MJ $m^{-2)}$, expressed as

$$\text{RMSE} = \sqrt{\frac{1}{N} \sum\nolimits_{i=1}^{N} (\text{Data.sim}_i - \text{Data.obs}_i)^2} \quad (16.28)$$

The Root Mean Square Error (RMSE) (also called the root mean square deviation, RMSD) is a frequently used measure of the difference between values predicted by a model and the values observed from the environment that is being modeled. The RMSE serves to aggregate the magnitudes of the errors in predictions for various times into a single measure of predictive power. RMSE is a measure of accuracy to compare forecasting errors of different models for a dataset and not between datasets, as it is scale-dependent.

RMSE is always non-negative, and a value of 0 (almost never achieved in practice) would indicate a perfect fit to the data. In general, a lower RMSD is better than a higher one. However, comparisons across different types of data would be invalid because the measure is dependent on the scale of the numbers used. It is also sensitive to outliers (Hyndman 2006). The squared nature of this metric helps to deliver more robust results that prevent canceling the positive and negative error values. In other words, this metric displays the plausible magnitude of the error term. Reconstruction of error for an abundant sample is more reliable with RMSE. The value of RMSE lies between 0 to $\infty$

RMSE is used to see how effectively a mathematical model predicts the behavior of the atmosphere in meteorology. In GIS, it is one of the measures used to assess the accuracy of spatial analysis and remote sensing. Similarly, in computational neuroscience, the RMSE is used to assess how well a system learns a given model. In the simulation of energy consumption of buildings, the RMSE is used to calibrate models to measured building performance (Armstrong and Collopy 1992).

However, it is important to note that, RMSE is computed on the squared difference that may lead to performance biased in favor of the peaks and high magnitude events, exhibiting greater error and insensitive to lower magnitude sequence (Hyndman 2006).

Mean Absolute Error (MAE; MJ $m^{-2)}$, expressed as

$$\mathrm{MAE} = \frac{1}{N}\sum_{i=1}^{N} |((\mathrm{Data.sim}_i - \mathrm{Data.obs}_i)| \tag{16.29}$$

In ML, Mean Absolute Error (MAE), is commonly used for supervised learning predictive modeling problems. While performing any analytical work, if there is a difference between the continuous expected value and the actual values, the discrepancy could be captured in the form of an absolute error, called MAE. Mean absolute error has a clear interpretation as the average absolute difference between simulated data and observed data. Many researchers want to know this average difference because its interpretation is clear, but researchers frequently compute and misinterpret the RMSE, which is not the average absolute error. MAE does not require the use of squares or square roots.

The use of squared distances hinders the interpretation of RMSE. MAE is simply the average absolute vertical or horizontal distance between each point in a scatter plot and the $Y = X$ line. In other words, MAE is the average absolute difference between $X$ and $Y$. MAE is fundamentally easier to understand than the square root of the average of the sum of squared deviations. Furthermore, each error contributes to MAE in proportion to the absolute value of the error, which is not true for RMSE. It is a linear scoring rule that describes only the average errors, ignoring their direction of variation from the measured values, and it is used for non-Gaussian cases. MAE takes values from 0 (perfect fit) to $\infty$ (worse fit) (Willmott and Matsuura 2005).

Legates and McCabe Index ($E$)

Legates and McCabe Index has a value between 0 and 1. It is a modified version of Willmott Index, that provides integral information on overall model performance. For large dataset predictions like a large range of decades of year, Legates and McCabe Index is more informative than Willmott's index and NSE. The main reason behind its prime use is that poorly fitted models have errors that are not augmented by the square of the predicted errors as in Willmott's index and NSE. E is expressed as

$$\mathbf{E} = 1 - \frac{\sum_{i=1}^{n} (\mathrm{data.obs}_i - \mathrm{data.sim}_i)}{(\mathrm{data.obs}_i - \mathrm{data.obs}_i)} \tag{16.30}$$

However, a weakness of this assessment metric is that these are expressed in their absolute units, and thus should not be solely used to compare a model performance at geographically diverse sites (Chai and Draxler 2014).

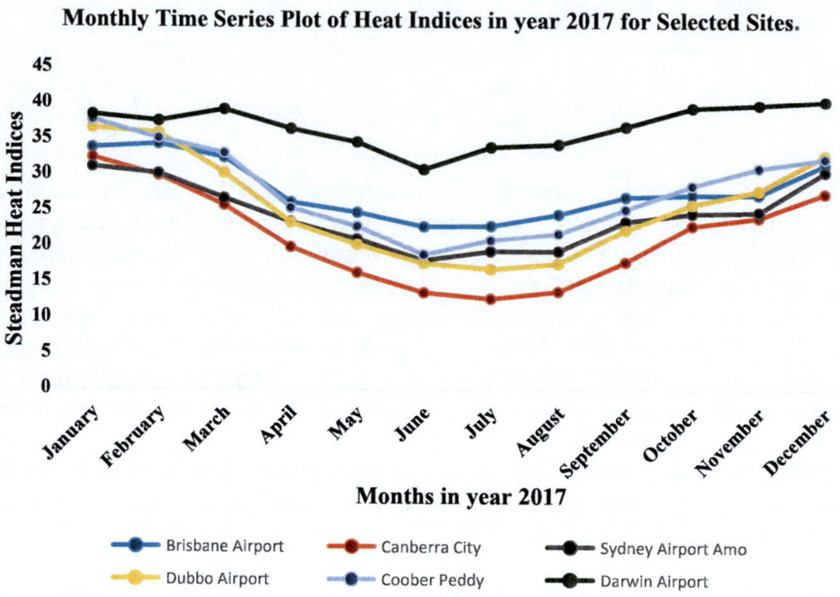


**Fig. 16.4** Time series of Heat Indices of Brisbane Airport for June from 1950 to 2017

## 16.6 Results and Discussion

### *16.6.1 Interpretation of Heat Index*

Time series of the average annual heatwave indices (Ta) for 15 stations across Australia is shown as in Fig. 16.4.

The mean Ta and Sen's slope estimates are also shown in Fig. 16.5.

Since January is considered as the month of Heatwave in Australia, the major susceptible sites with heatwave severity are selected for and visually represented monthly with respect to the average monthly heat indices values as shown in Table 16.2.

It is clearly depicted from the figure that in recent year (2017) Coober Peddy and Darwin Airport shows an increasing severe heatwave trend with respect to other selected sites. Darwin Airport in an average outperforms in the severity of heat indices with the highest values in each month of the year 2017.

The time series plot of all the heat indices in each month of the year 2017 is illustrated in Fig. 16.4.

In the year 2017, the heat indices trend shows somewhat different from the average annual heatwave indices trend. In the year 2017, Darwin Airport and Brisbane Airport trends line show an outperforming performance with respect to other sites where the value of heat indices is comparatively low. Similarly, Canberra City and Dubbo (Mentone) heatwave trends are decreasing if compared with the average annual time

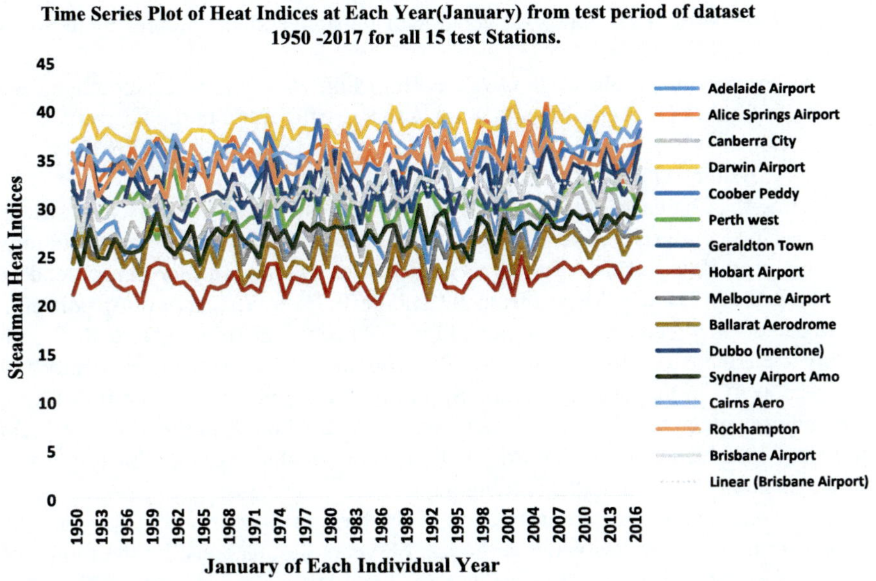


**Fig. 16.5** Time series plot of monthly averaged Heat Index of every month of January for each year from 1950 to 2050 dataset for all the 15 sites all over Australia. Every candidate site graph is depicted with separate colors. The month of January is usually called a month of heatwaves in Australia. The trend line is also plotted resonant with the trend of Canberra City in the month of January

series analysis of SHI. The month of January is seemingly severe for sites like Darwin Airport, Adelaide Airport, Rockhampton, and Alice Springs Airport with relatively higher values of heatwave indices over the period of 1950–2017. Hobart Airport and Ballarat Aerodrome show comparatively cooler Januaries when observed over a period of 6.5 decades HIs dataset.

There is a considerable increasing trend in each of the selected sites. Some locations show gradual growth over the past 68 years while few sites remain comparatively less emerging with respect to the heat indices. Form Fig. 16.5, it can be clearly observed from the time series of average annual heat indices from the year 2050 to the year 2017 that all the sites in Queensland NSW, ACT and Tasmania experienced an increasing trend of heatwaves with the value of heat indices from increasing with a value of 2.5–3.0 °C (±0.1).

In this increasing trend of heatwaves, ACT (Canberra) along with NT (Alice Springs Airport and Darwin) suffered most severely with the rise of 3 °C within the period of 68 years (1950–2017). However, the sites from South Australia went through relatively small growth in the trend with still an increasing trend. Adelaide Airport possesses an increment of 1 °C from the year 1950 to 2006, having decrement in the next 10 years with 0.5 °C. In addition, Coober Peddy faced an increasing trend from 1950 to 1990 with an increment of 2 °C and a fall of 0.5 °C in the next 27 years.

The trend of heatwaves is comparatively much constant and less increasing in South Australia.

If we look at the graphs for HIs values from Fig. 16.6, all the 12 months in the period of 68 years (1950–2017) are facing an increasing trend in the heatwave. The highest values of heat indices are observed in the month of January with 32 °C and the least experienced in the month of July with 13 °C.

The month of January is regarded as the month of heatwaves. Figure 16.4 illustrates the time series analysis of heat indices in the month of January for all the 15 stations from the year 1950–2017. All the years, each site is facing an increasing trend of heatwave except Darwin Airport from NT and Perth from WA. Darwin Airport time series depicts that there is an incremental trend of heatwaves from 1950 to 2012 with the highest HI value of 40 °C in the year 2012 but it started decreasing gradually and becomes 38 °C in the next five years. Similarly, for the Perth West, until 2012 the heat indices values are shown increasing up to 33 °C whereas in the next five years up to 2017 the values decreased to 29 °C. It signifies that there are bit less warm Januaries in WA and NT in the coming days.

Regarding the graphical interpretation of Brisbane in Queensland, it also faces an increasing trend in the heatwave in the last 68 years. Let us consider the month of June; it is a comparatively less warm and windy month in Brisbane. From Fig. 16.6, it is depicted that there is an increasing trend in the value of heat indices from the year 1950 to 1990. From 1990 to 2000 there seems to be a decrement in the heat indices value from 23.5 °C to 21 °C. This HI value again rises to 23.5 °C until the year 2004. Then again, this value decreases up to 20.5 until 2008, and from that year to the end of 2017, there seems a gradual increase in the value of HIs. It is observed from the graph that the trend is still increasing in further years too.

### *16.6.2 Trend Analysis and Significance Testing*

The positive or negative trend is determined by the positive and negative values ($z$ values) of MK-statistics as shown in Table 16.4.

The significance of the trend is determined by the confidence interval of 0.05. If the $p$-value obtained is below this level of significance, then data is insignificant for the trend analysis. All the values for the significance test, MK-statistics, $p$-values, and Sen's slope magnitude for all the candidate sites are tabulated in Table 16.4.

For comparison, the thresholds of SHI, showing the ranges of Ta, which could cause a potential health risk, are summarized in Appendix 6.

The minimum threshold of apparent body temperature that would initiate physiological stress, causing heat cramps or heat exhaustion is about Ta = 32 °C. In this research work, it is shown that over the period of 1950–2017, the mean value of Ta is mostly higher than this minimum threshold. Rather, most stations have mean Ta between the next higher threshold of 41–54 °C, which is the range at which heat stroke is possible if a person is exposed to the outdoor environment for a prolonged period.

The results, therefore, indicate that a possible increase in the incidences of heatwaves is posing a serious threat to human health across much of Australia. This is evident from the recent events where a severe heatwave, reportedly killed several

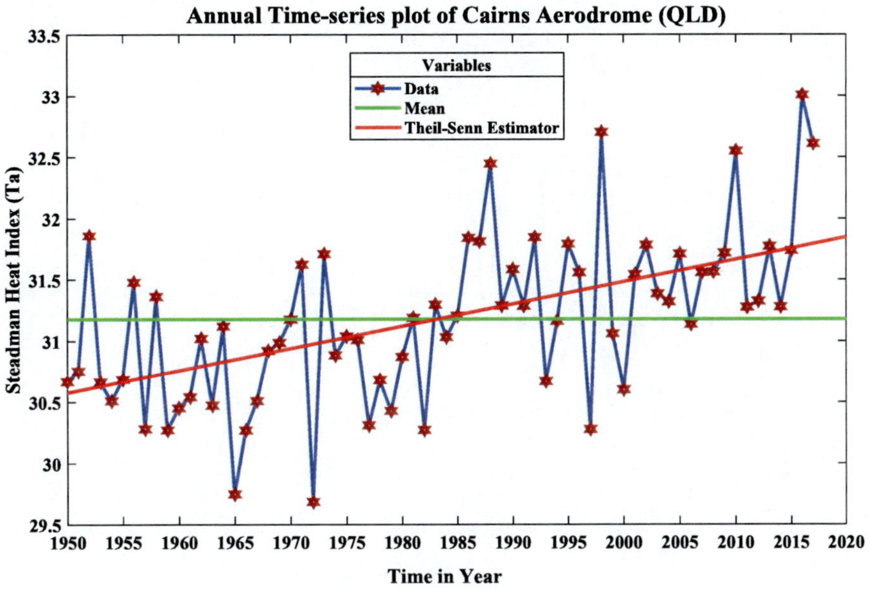


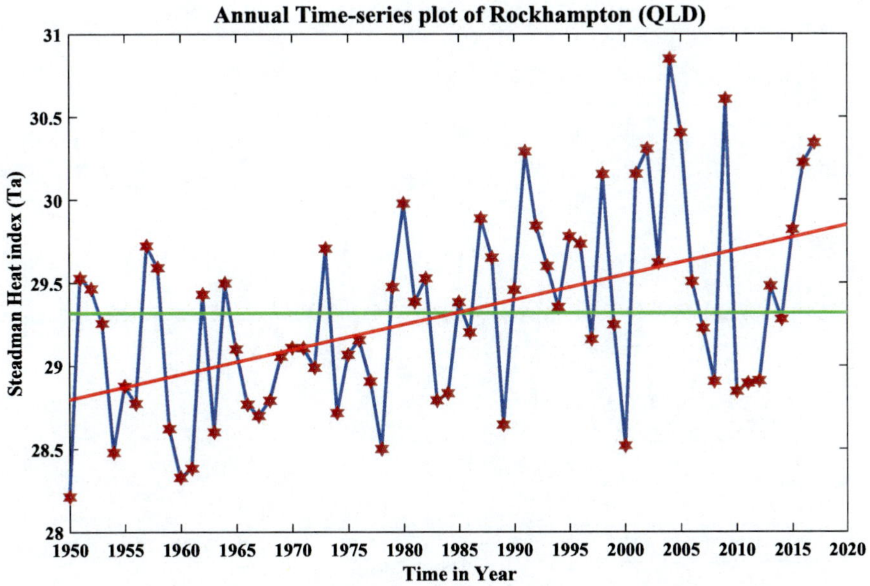


**Fig. 16.6** Time series plot analysis with Theil-Sen's estimate and Steadman Heat Indices plotted against time from 1950 to 2017. The Abbreviation in brackets on the heading of each graph represents the States of Australia as listed in the Abbreviation section of this report. As per legend, the blue line represents the time-series plot of heat indices against each ear (1950–2017), the red line represents the Theil-Sen's estimate and the green line represents the mean of the heat indices data from 1950 to 2017 in each of the candidate site

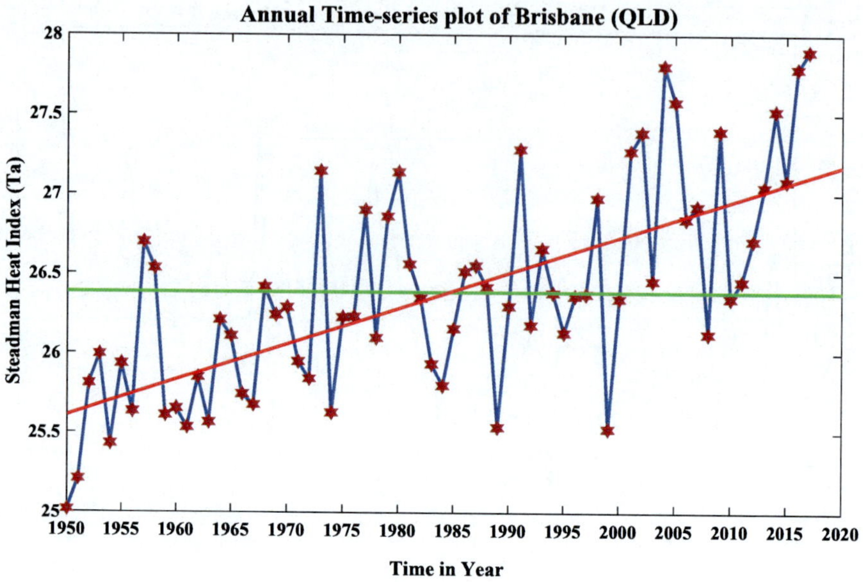


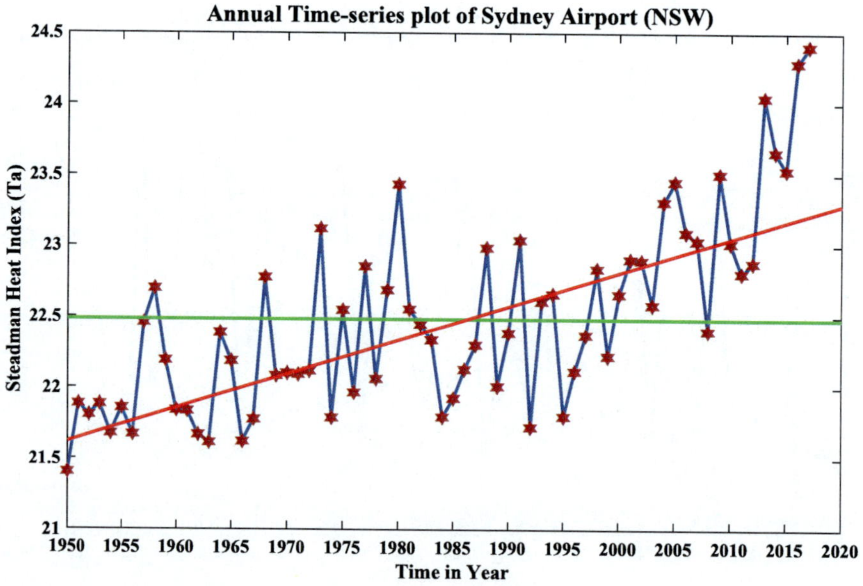


**Fig. 16.6** (continued)

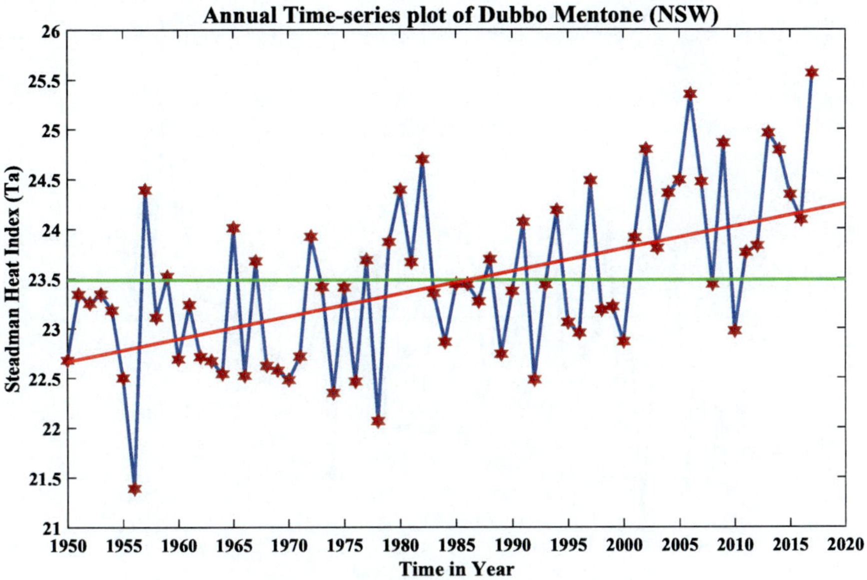


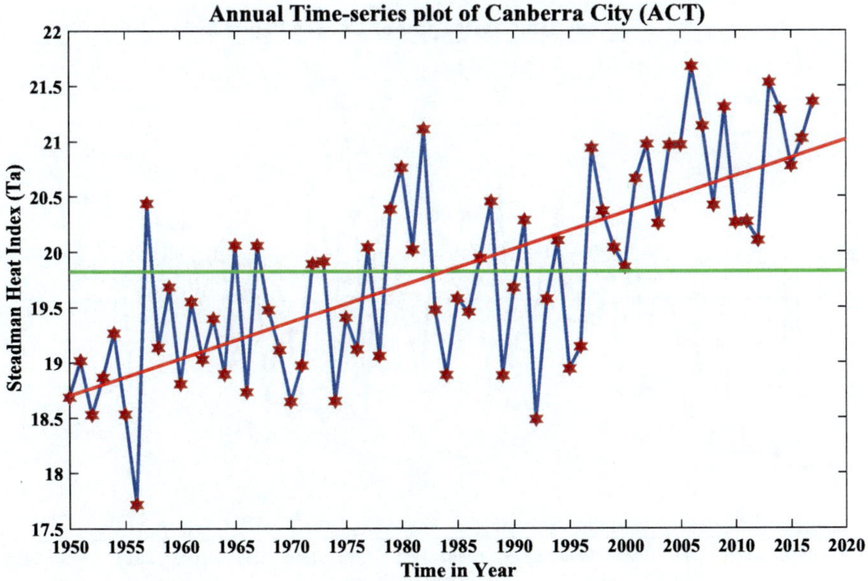


**Fig. 16.6** (continued)

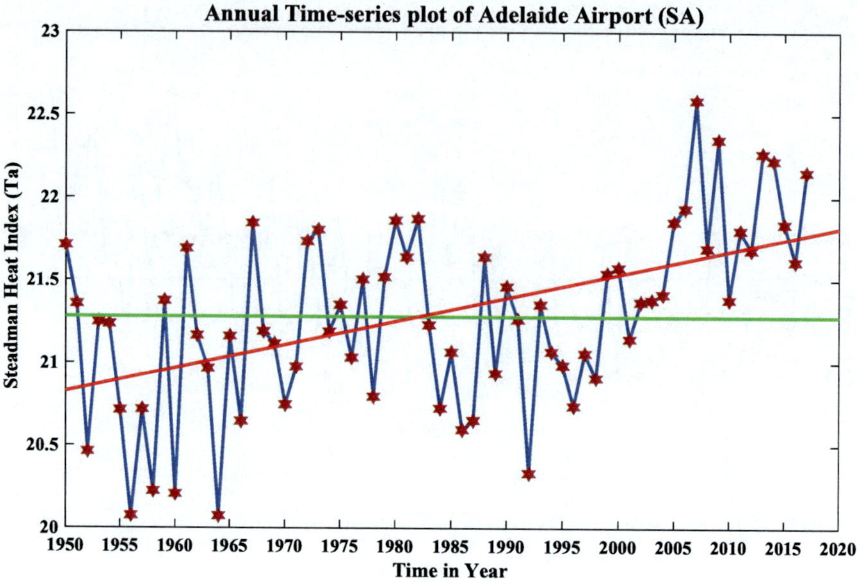


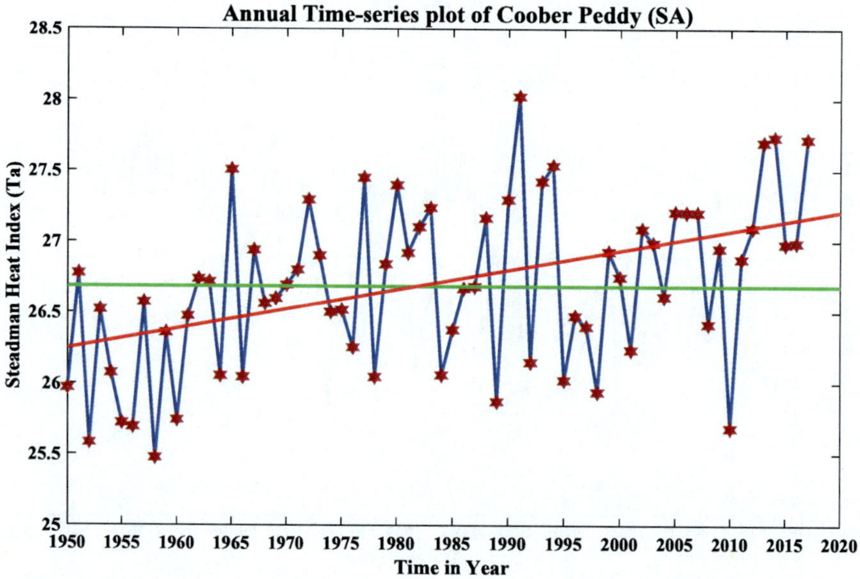


**Fig. 16.6** (continued)

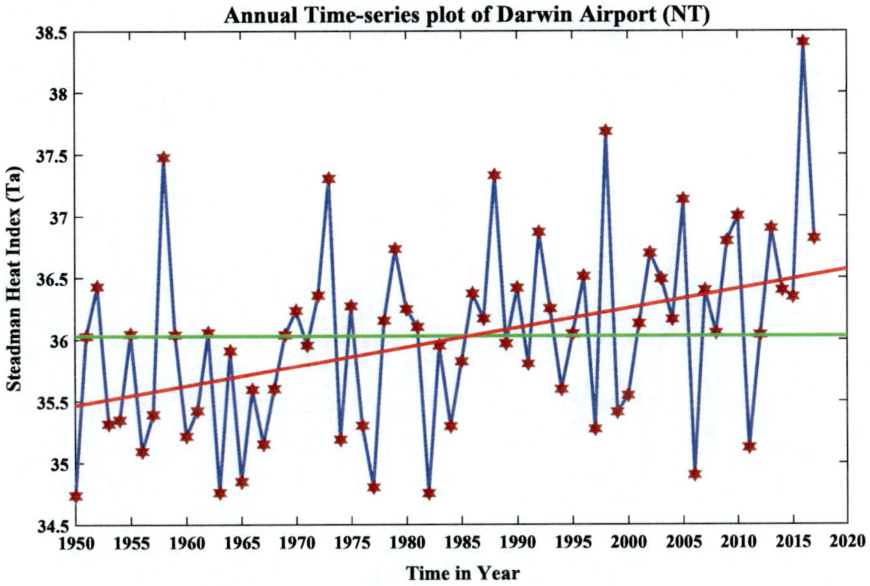


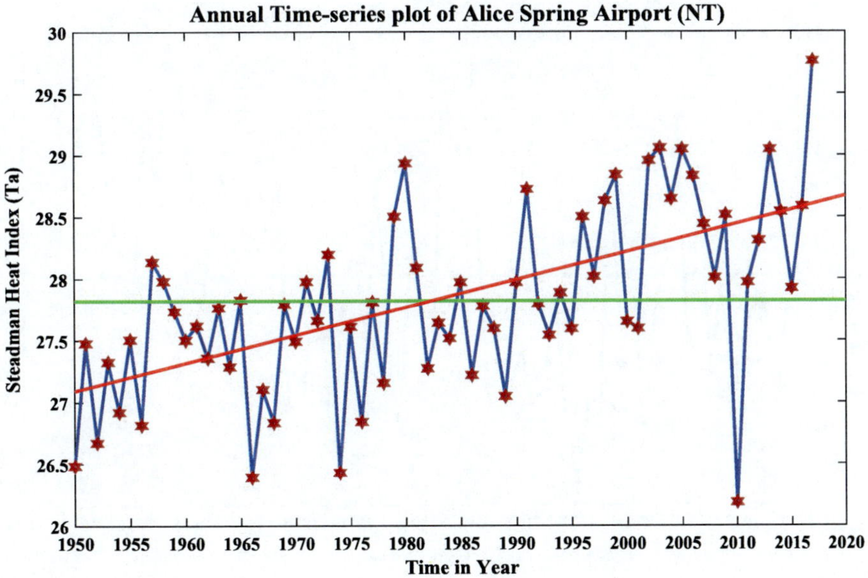


**Fig. 16.6** (continued)

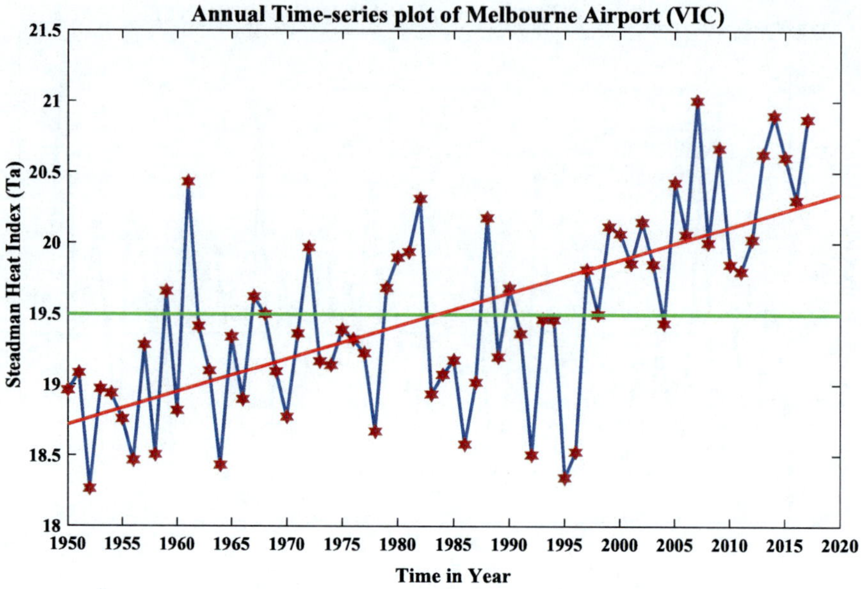


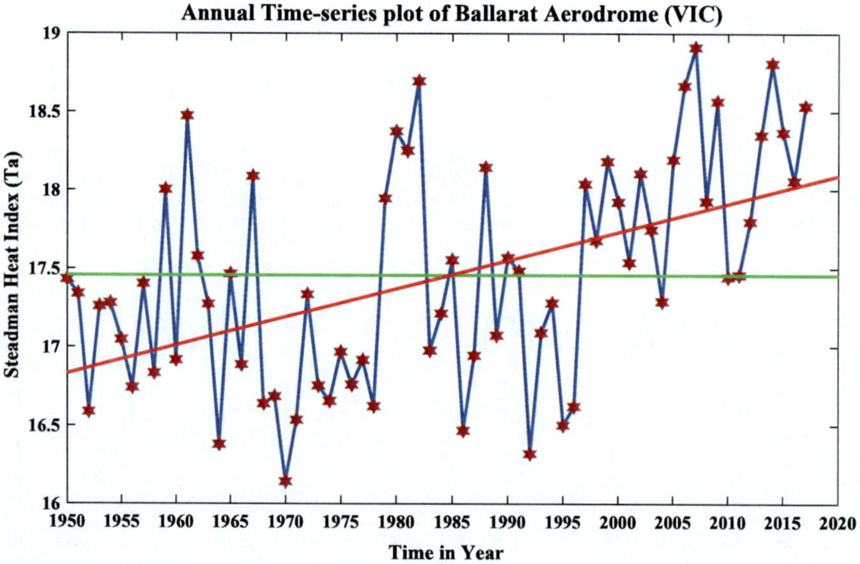


**Fig. 16.6** (continued)

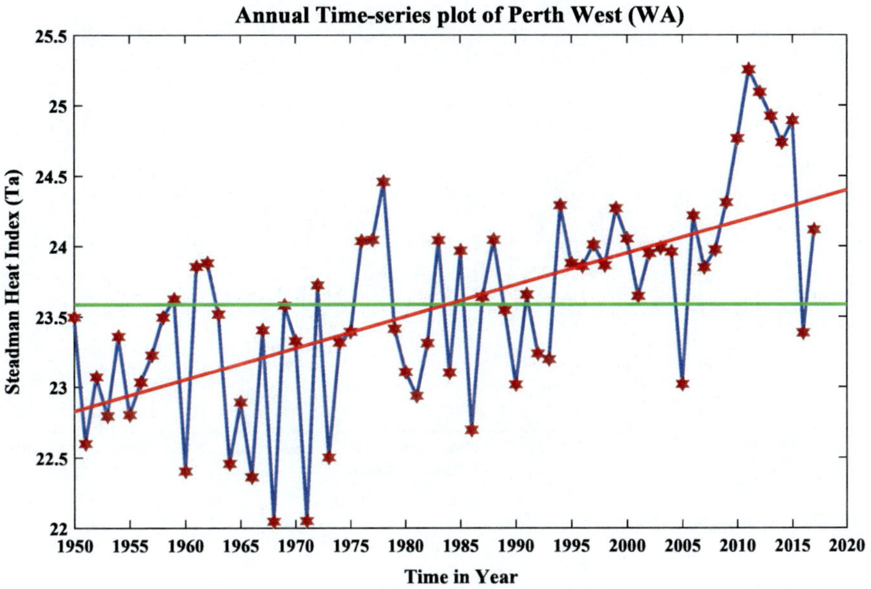


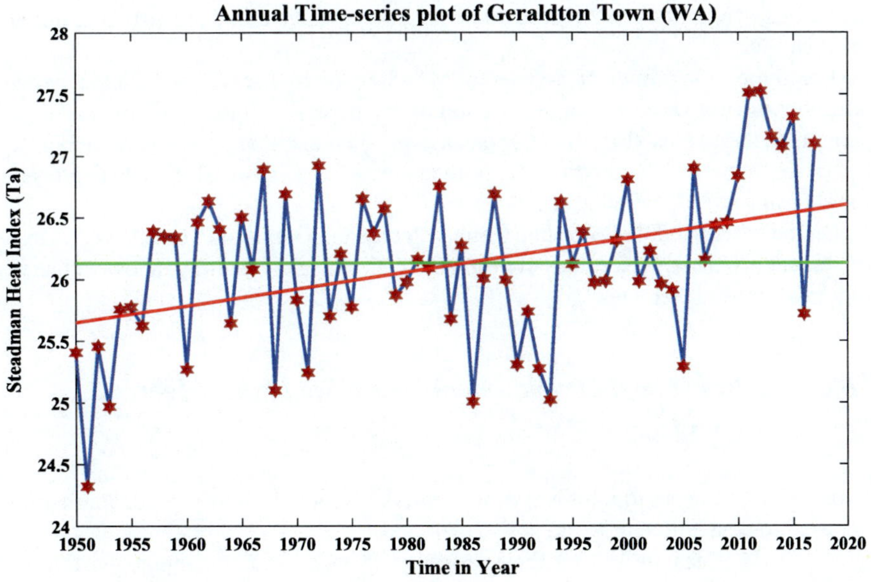


**Fig. 16.6** (continued)

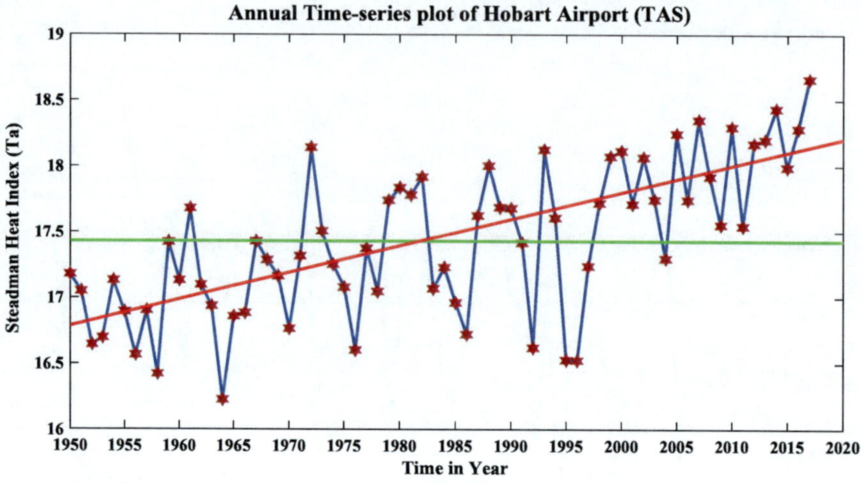


**Fig. 16.6** (continued)

people who were living alone behind closed doors. According to a study (Coates et al. 2014), from 1844 to 2010, extreme heat events have killed at least 5332 people in Australia.

The highest magnitude of Z-value for MK-Statistic is observed as 6.28 for Sydney Airport Amo and there is a large variation in the increasing trend of HI by 0.4 °C per year in Canberra City (Fig. 16.7). It means, on one hand, there is a severe increasing trend of heatwaves in Canberra City with an annual growth of HI by 0.4 °C per year (or we can say 4 °C per decade).

On the other hand, the sites like Coober Peddy and Geraldton Town have the least increment in heat indices values of nearly 0.17 °C per year, resulting in lower chances of heatwaves in those areas.

### *16.6.3 ANN Model Development Using Steadman Heat Indices*

Figure 16.8 displays a topological structure of ANN model where three-layer neurons are used to design the network.

Figure 16.9 despite testing the sensitivity of each predictor input, the training algorithm, hidden transfer function, and output transfer function are varied systematically to design the best ANN model (e.g., Sahin 2012; Shahin 2013) while the two predictors (maximum air temperature and maximum humidity) are considered. Thus, the number of inputs in the input layer are pre-selected as two-layered, namely, temperature, humidity, respectively.

**Table 16.4** MK-Statistics and Sen slope value

| Station number | Station name | Lat (°S) | Long (°E) | Elevation (m) | (Z-value) | Trend (±) | *p*-value | Significant (Y/N) | Slope (°C per yr.) |
|---|---|---|---|---|---|---|---|---|---|
| 1 | Cairns Aero | −16.8736 | 145.7458 | 2.2 | 4.7693547 | (+) | 1.85E−06 | Y | 2.604 |
| 2 | Rockhamptn | −23.3811 | 150.5172 | −1.4 | 3.9435841 | (+) | 8.03E−05 | Y | 2.16 |
| 3 | Brisbane | −27.4808 | 153.0389 | 8.1 | 5.9339031 | (+) | 2.96E−09 | Y | 3.228 |
| 4 | Dubbo | −32.5192 | 148.5187 | 330 | 4.5470319 | (+) | 5.44E−06 | Y | 3.264 |
| 5 | Sydney Airport Amo | −33.9465 | 151.1731 | 6 | 6.2832676 | (+) | 3.32E−10 | Y | 3.432 |
| 6 | Canberra City | −35.2667 | 149.1167 | 564 | 6.2197467 | (+) | 4.98E−10 | Y | 4.752 |
| 7 | Ballarat Aerodrome | −37.5127 | 143.7911 | 435.2 | 3.9541709 | (+) | 7.68E−05 | Y | 2.592 |
| 8 | Melbourne Airport | −37.6655 | 144.8321 | 113.4 | 5.5316045 | (+) | 3.17E−08 | Y | 3.36 |
| 9 | Adelaide Airport | −34.9524 | 138.5204 | 152.4 | 4.0917993 | (+) | 4.28E−05 | Y | 2.04 |
| 10 | Coober Peddy | −29.004 | 134.7564 | 215 | 3.7000876 | (+) | 2.16E−04 | Y | 1.98 |
| 11 | AliceSprings Airport | −23.7951 | 133.889 | 546 | 5.0975456 | (+) | 3.44E−07 | Y | 3.24 |
| 12 | Darwin Airport | −12.4239 | 130.8925 | 30.4 | 3.5730460 | (+) | 0.000353 | Y | 2.268 |
| 13 | Perth West | −31.95 | 115.85 | 54 | 5.5104309 | (+) | 3.58E−08 | Y | 3.228 |
| 14 | Geraldton Town | −28.7769 | 114.605 | 3 | 3.2236815 | (+) | 0.001266 | Y | 1.968 |
| 15 | Hobart Airport | −42.8339 | 147.5033 | 4 | 5.7751010 | (+) | 7.69E−09 | Y | 2.916 |

(+) denotes positive trends and (−) is a negative trend. The positive or negative *z*-value determines the nature of the trend (supplied it must be significant)

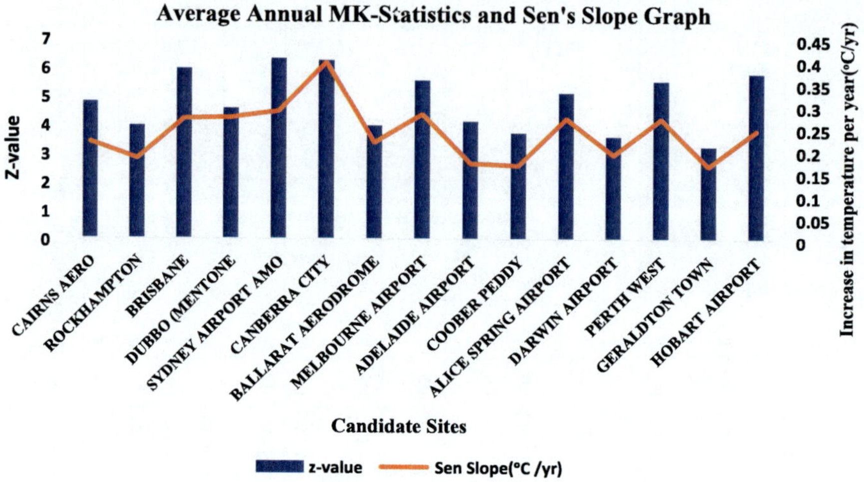


**Fig. 16.7** Average Annual MK-statistics (*Z*-value) and Sen Slope', value in °C per year. All the candidate stations *Z* values are located at the left-hand side of the graph and the magnitude of the trend in the right-hand side. The magnitude gives the annual increment in heat indices value

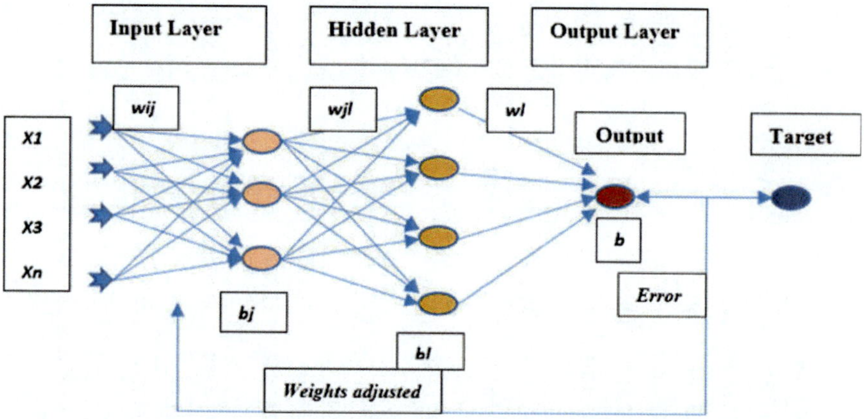


**Fig. 16.8** Topological structure of ANN model where three-layer neurons are used to design the network. *Wij is weights between each input variable and each node in the first layer, bl is biased (with unit inputs) for each node in the second layer and Error describes backpropagation of error signal*

As there is no mathematical formula to determine the neuronal structure in the hidden layer of the ANN model, the number of neurons in the hidden layer is decided by trial and error (Şahin 2012) method.

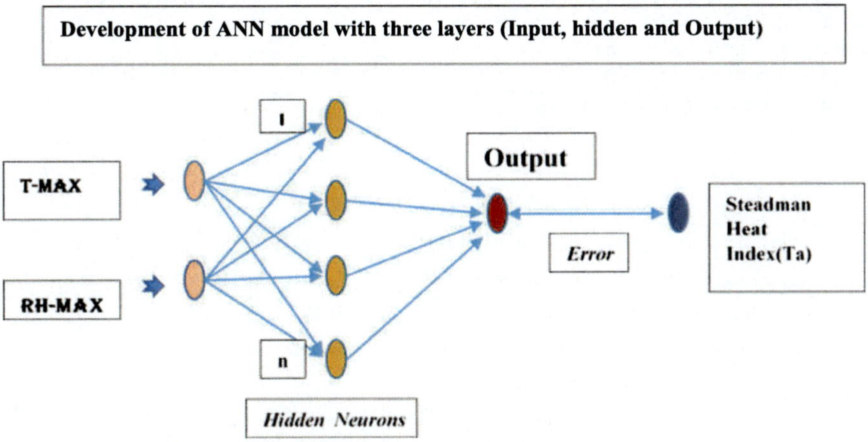


**Fig. 16.9** ANN architecture adopted for predictive modeling of Steadman Heat Indices (Ta) with two input variables (maximum air temperature and maximum humidity) observed on a daily basis with 250 test neurons. The best neuron among the 250 neurons gives the best output value. T-Max is maximum daily air temperature and RH-Max is maximum daily relative humidity

**Table 16.5** ANN model performance with training data with individual units with combinations of training function (*trainlm, trainbgf, traindx, trainscg, traincgf, traincgp*), output layer and activation functions (*purelin and logsig*)

| Station number | ANN parameter combination | Training values of performance metrics | | | | |
|---|---|---|---|---|---|---|
| | | *r* | *d* | ENS | RMSE (MJ $m^{-2}$) | MAE (MJ $m^{-2}$) |
| 1 | ({'logsig','purelin'},'trainlm') | 0.9990 | 0.999 | 0.998 | 0.213 | 0.119 |
| 2 | ({'purelin','logsig'},'trainlm') | 0.9992 | 0.991 | 0.985 | 0.776 | 0.587 |
| 3 | ({'logsig','purelin'},'trainbfg') | 0.9130 | 0.89 | 0.831 | 2.559 | 2.079 |
| 4 | ({'purelin','logsig'},'trainbgf') | 0.2290 | 0.274 | −1.83 | 7.23 | 6.24 |
| 5 | ({'logsig','purelin'},'traingdx') | −0.410 | 0.307 | −1.947 | 8.541 | 6.85 |
| 6 | ({'purelin','logsig'},'traingdx') | 0.4520 | 0.084 | −36.25 | 31.26 | 29.41 |
| 7 | ({'logsig','purelin'},'trainscg') | −0.370 | 0.375 | −2.25 | 8.265 | 7.55 |
| 8 | ({'purelin','logsig'},'trainscg') | 0.5610 | −0.018 | −29.46 | 33.25 | 21.56 |
| 9 | ({'logsig','purelin'},'traincgf') | 0.9180 | 0.891 | 0.856 | 2.491 | 2.038 |
| 10 | ({'purelin','logsig'},'traincgf') | 0.3830 | 0.119 | −4.25 | 11.25 | 9.21 |
| 11 | ({'logsig','purelin'},'traincgp') | 0.9190 | 0.907 | 0.844 | 2.14 | 1.221 |
| 12 | ({'purelin','logsig'},'traincgp') | 0.3570 | 0.196 | −3.21 | 13.25 | 9.871 |

A maximum number of 250 neurons are trail-performed and tested for the development of the network architecture. Tables 16.5 and 16.6 show the parameters of the ANN model architecture used for the prediction of the SHI (Ta).

**Table 16.6** ANN model performance with test data with individual units with combinations of training function (*trainlm, trainbgf, traindx, trainscg, traincgf, traincgp*), output layer and activation functions (*purelin and logsig*)

| Station number | ANN parameter combination | Testing values of performance metrics | | | | |
|---|---|---|---|---|---|---|
| | | *r* | *d* | ENS | RMSE ($MJ/m^2$) | MAE ($MJ/m^2$) |
| 1 | ({'logsig','purelin'},'trainlm') | 0.999 | 0.999 | 0.998 | 0.227 | 0.117 |
| 2 | ({'purelin','logsig'},'trainlm') | 0.1 | 0.194 | −23.18 | 25.837 | 18.99 |
| 3 | ({'logsig','purelin'},'trainbfg') | 0.908 | 0.891 | 0.997 | 0.226 | 0.117 |
| 4 | ({'purelin','logsig'},'trainbgf') | −0.08 | 0.103 | −56.24 | 38.241 | 31.59 |
| 5 | ({logsig','purelin'},'traingdx') | −3.21 | 0.101 | −1.824 | 6.256 | 5.22 |
| 6 | ({'purelin','logsig'},'traingdx') | 0.452 | 0.084 | −56.23 | 39.21 | 31.53 |
| 7 | ({logsig','purelin'},'trainscg') | 0.215 | 0.09 | −1.25 | 8.52 | 6.981 |
| 8 | ({'purelin','logsig'},'trainscg') | 0.215 | 0.09 | −56.28 | 36.56 | 29.86 |
| 9 | ({'logsig','purelin'},'traincgf') | 0.941 | 0.89 | 0.702 | 2.31 | 1.802 |
| 10 | ({'purelin','logsig'},'traincgf') | 9.00E−04 | 0.1 | −56.84 | 46.25 | 33.59 |
| 11 | ({'logsig','purelin'},'traincgp') | 0.3903 | 0.884 | 0.761 | 2.523 | 1.985 |
| 12 | ({'purelin','logsig'},'traincgp') | −0.01 | 0.103 | −75.32 | 46.251 | 38.98 |

Among the 15 sites studied, the dataset from the Canberra City is chosen to select the best combination of the hidden transfer function, output function, and training algorithm for the model development process. The training performance metrics of ANN for the training period of Canberra City are shown in Table 16.5. It is observed for training dataset the combination of (logsig, purelin, trainlm) as hidden transfer function, output function, and training algorithm outperforms the others model development parameters with a low RMSE and MAE values of 0.213 $MJ/m^2$ and 0.1198 $MJ/m^2$, respectively.

The correlation coefficient, d, and ENS have also significantly greater values, i.e., 0.999, 0.999, and 0.998, respectively. Since the higher value of d, r, and NSE and low values of RMSE and MAE represents that the model performance is good (Deo and Şahin 2017a), a model can be built on these datasets since the model is performing well on the training data. The model is tested on with the test datasets, the performance metrics are shown in Table 16.6. Again, the combination of (purelin, logsig, trainlm) as hidden transfer function, output function, and training algorithm outperforms the combination of other model development parameters. The RMSE and MAE for the optimal neural network are 0.227 MJ/mDeo and Şahin, and 0.117 MJ/m^Deo and Şahin, 2, respectively. In addition, the r, d, and ENS were observed 0.999, 0.999, and 0.998, respectively.

One of the major disadvantages of this way of finding optimal parameters is that it is an iterative process and the model must go through different combinations of parameters to find the best model and it is time-consuming too. The major disadvantage of ANN is its working algorithm is not known and is considered as a "Black

Box" and also causes overfitting issues in some cases and has a higher computational burden (Al-Fatlawi et al. 2015).

The Levenberg–Marquardt and Broyden–Fletcher–Goldfarb–Shanno (BFGS) quasi-Newton backpropagation algorithms are utilized to train the network, tangent, and logarithmic sigmoid equations are to be used as the activation functions and the linear, logarithmic and tangent sigmoid equations as the output function. A final weight matrix is obtained after training the networks that are further applied to the independent inputs in the test data set. Finally, the outcomes are compared with the calculated values of the daily SHI. Furthermore, MLR and ARIMA can also be used as a data intelligent technique to verify the consequences of the outputs as a HI. A 250-fold ANN was developed using the best parameters (logsig, purelin, trainlim) as hidden transfer function, output function, and training algorithm to find the optimal number of neurons.

A 250-fold ANN is developed using the best parameters (logsig, purelin, trainlm) as hidden transfer function, output function, and training algorithm to find the optimal number of neurons. As shown in Table 16.5, while applying the ANN model in the training dataset, it is observed that the model is best suited for the test dataset for the prediction purpose. Then, for the dataset in the testing period (last 20% of the dataset), all the 15 sites undergo the process of ANN model performance to provide the best-fitted neuron with the best model and optimum performance metrics parameters. The ANN is run for 250 times and the performance is evaluated for each neuron, the model with the highest Legates and McCabe Index (E) as the best model from the 250-fold ANN algorithm is considered as the optimal neuron with the best possible outcome of the model performance metrics. The performance metrics along with the best neuron for each of the 15 spatial locations all over Australia are listed as in Table 16.7.

The scatterplot of observed and simulated data in the test period of the best performing site indicates the model accuracy and purity of the present data set (Fig. 16.10).

We can conclude that the ANN model performs the best prediction with respect to the calculation and prediction of HIs in the given period of 1950–2017.

## 16.7 Model Comparison

### *16.7.1 ANN Versus MLR*

MLR is not an ML method so there is no parameter identification process involved.

It is the relationship between the target variable and explanatory variables by fitting a linear equation through the set of data (Kristopher et al. 2006). MLR performs well when the variables are linear in nature, for non-linear data it does not perform well, and it cannot be used in the dataset having missing data. The model development parameters for the linear equation are shown in Table 16.8.

**Table 16.7** Performance metrics output from ANN model development over the test data period

| | ANN performance metrics | | | | | | | |
|---|---|---|---|---|---|---|---|---|
| Site name | *r* | RMSE MJ/m$^2$ | MAE MJ/m$^2$ | MAPE (%) | RRMSE (%) | *d* | ENS | ***E*** |
| Cairns Aero | 0.9989 | 0.19644 | 0.1044190 | 0.3620 | 0.62012 | 0.99895 | 0.99783 | 0.97057 |
| Rockhamptn | 0.99901 | 0.22741 | 0.1171165 | 0.4314 | 0.76988 | 0.99907 | 0.99799 | 0.97186 |
| Brisbane | 0.99909 | 0.19006 | 0.1082628 | 0.4247 | 0.70378 | 0.99914 | 0.99817 | 0.97012 |
| Dubbo | 0.99961 | 0.20272 | 0.1372962 | 0.6271 | 0.83656 | 0.99962 | 0.99921 | 0.97745 |
| Sydney Airport | 0.99658 | 0.43657 | 0.1256222 | 0.5166 | 1.87238 | 0.99703 | 0.99314 | 0.97008 |
| Canberra City | 0.99961 | 0.19175 | 0.1201852 | 0.5969 | 0.92163 | 0.99964 | 0.99921 | 0.97923 |
| Ballarat Aero | 0.99939 | 0.23783 | 0.1255471 | 0.7273 | 1.31550 | 0.99948 | 0.99878 | 0.97813 |
| Melbourne | 0.99877 | 0.31284 | 0.1220417 | 0.6139 | 1.54430 | 0.99898 | 0.99755 | 0.97632 |
| Adelaide Airport | 0.99959 | 0.17122 | 0.1157355 | 0.5284 | 0.78402 | 0.99965 | 0.99918 | 0.97656 |
| Coober Peddy | 0.99963 | 0.19282 | 0.1332528 | 0.5341 | 0.71649 | 0.99963 | 0.99924 | 0.97700 |
| Alice Springs | 0.99961 | 0.18826 | 0.1320227 | 0.4948 | 0.66539 | 0.99959 | 0.99922 | 0.97635 |
| Darwin Airport | 0.99956 | 0.13346 | 0.0719501 | 0.2015 | 0.36622 | 0.99939 | 0.99900 | 0.97945 |
| Perth West | 0.99944 | 0.18941 | 0.1082543 | 0.4384 | 0.78159 | 0.99953 | 0.99888 | 0.97674 |
| Geraldton Town | 0.99950 | 0.19241 | 0.1194960 | 0.4358 | 0.72341 | 0.99951 | 0.998817 | 0.97325 |
| Hobart Airport | 0.99927 | 0.18992 | 0.100972276 | 0.55509 | 1.055202 | 0.999358 | 0.9985 | 0.974432 |

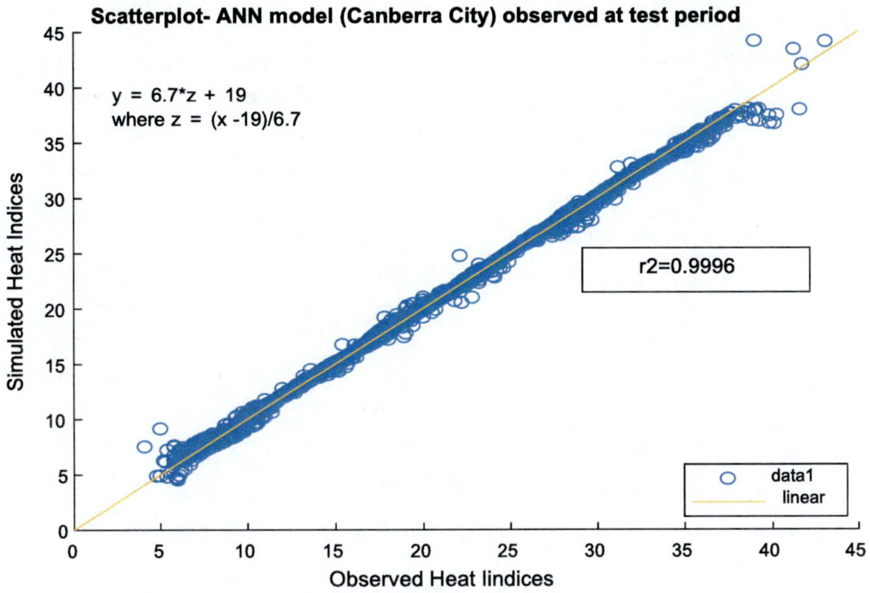


**Fig. 16.10** Scatterplot of observed versus simulated data for the best site observed at test period over the period (1950–2017)

**Table 16.8** MLR parameters over test dataset with the coefficient and magnitude of maximum air temperature, maximum relative humidity, and *y*-intercept representing the constant term in MLR model development

| Station number | Site name | Predictor variable (input), *x* | | Daily prediction |
|---|---|---|---|---|
| | | Name | Coefficient | Magnitude |
| 1 | Cairns Aero | T_max | β1 | −0.2369 |
| | | RH_max | β2 | 1.0683 |
| | | Y-Intercept | C | 0.2192 |
| 2 | Rockhampton | T_max | β1 | −0.138142901 |
| | | RH_max | β2 | 1.054541927 |
| | | Y-Intercept | C | 0.115238707 |
| 3 | Brisbane Airport | T_max | β1 | −0.072924454 |
| | | RH_max | β2 | 0.958246062 |
| | | Y-Intercept | C | 0.059210578 |
| 4 | Canberra City | T_max | β1 | −0.001878151 |
| | | RH_max | β2 | 0.935520078 |
| | | Y-Intercept | C | 0.015692686 |

(continued)

**Table 16.8** (continued)

| Station number | Site name | Predictor variable (input), $x$ | | Daily prediction |
|---|---|---|---|---|
| | | Name | Coefficient | Magnitude |
| 5 | Sydney Airport Amo | T_max | β1 | −0.017486219 |
| | | RH_max | β2 | 0.914342271 |
| | | Y-Intercept | C | 0.01916732 |
| 6 | Dubbo (Mentone) | T_max | β1 | −0.019491704 |
| | | RH_max | β2 | 0.907536547 |
| | | Y-Intercept | C | 0.041836328 |
| 7 | Melbourne Airport | T_max | β1 | −0.002450302 |
| | | RH_max | β2 | 1.016015466 |
| | | Y-Intercept | C | 0.01788927 |
| 8 | Ballarat Aerodrome | T_max | β1 | 0.002423353 |
| | | RH_max | β2 | 1.011353305 |
| | | Y-Intercept | C | 0.008781054 |
| 9 | Adelaide Airport | T_max | β1 | −0.006797018 |
| | | RH_max | β2 | 0.857001887 |
| | | Y-Intercept | C | 0.029532691 |
| 10 | Coober Peddy | T_max | β1 | −0.023616996 |
| | | RH_max | β2 | 0.876792656 |
| | | Y-Intercept | C | 0.091386549 |
| 11 | Alice Springs Airport | T_max | β1 | −0.03237783 |
| | | RH_max | β2 | 0.873788101 |
| | | Y-Intercept | C | 0.130830396 |
| 12 | Darwin Airport | T_max | β1 | −0.464188067 |
| | | RH_max | β2 | 1.127914417 |
| | | Y-Intercept | C | 0.581947419 |
| 13 | Perth West | T_max | β1 | −0.018513314 |
| | | RH_max | β2 | 0.929091275 |
| | | Y-Intercept | C | 0.044263072 |
| 14 | Geraldton Town | T_max | β1 | −0.065289484 |
| | | RH_max | β2 | 0.919849593 |
| | | Y-Intercept | C | 0.112692695 |
| 15 | Hobart Airport | T_max | β1 | 0.002377737 |
| | | RH_max | β2 | 0.978160014 |
| | | Y-Intercept | C | 0.004181857 |

Table 16.8 also shows the performance metric of MLR model for its training and testing period. The testing performance shows a high correlation and low RMSE and MAE values, thus the model can be used to test the dataset. The RMSE and MAE of MLR model are 2.81 and 2.13, respectively, which are relatively higher than the ANN and RF models. The predictor variables along with their daily prediction with maximum temperature, maximum relative humidity, and y-intercept magnitude values are listed in Table 16.8.

The comparative performance indices for ANN and MLR models are tabulated in Table 16.9.

The major contributing factors in the model performance are measured with RMSE, MAE, $E$, and $r$.

To testify the validity of the ANN model, firstly it is compared with MLR model to check the level of agreement between the simulated and observed values.

Performance metrics are the best tools to testify and assess predictive accuracy with the error percentage values compared for each of the performance metrics (r, RMSE, MAE, MAPE, RRMSE, $d$, ENS, $E$). Among the eight-performance metrics, the best values of Legates and McCabe Index (E) are observed to describe the best model. Here, we have 15 sites and each site have their own E values along with other performance metrics. The site, which has the highest E value and lowest RMSE, MAE values, is regarded as the best-fitted modeled site. In addition, all the values of performance indices of ANN model are compared with MLR for the sake of validation of ANN model and to check the accuracy.

From Table 16.10, for the ANN model, the site, Cabrera City has the highest value of $E$, i.e., 0.979 with respect to other sites.it means the Canberra City has the best data set for the prediction of daily HIs with the highest accuracy and minimal error. The $r$, MAE, and RMSE values for the Canberra City in ANN model are 0.999, 0.120, and 0.192, respectively. It represents the best-fitted dataset for the ANN model development process. If we look at the MLD model development for the Canberra site, it has the legate value of 0.959 with MAE, RMSE and r values of 0.234, 0.392, and 0.998, respectively. From all the perspective values of MLR, ANN model has become the best model for Canberra City. Likewise, if we look at the value of E for Cairns Aero from Table 16.8, above, it is observed as 0.971 for ANN and 0.78 for MLR model.

Also, the performance metrics like r, RMSE, MAE, MAPE, RRMSE, d, ENS for ANN are 0.999, 0.196, 0.104, 0.362, 0.62, 0.999, 0.998, respectively, and for MLR (0.971, 1.025,0.781, 2.476, 3.237, 0.968, 0.941, respectively). Again, for the Cairns Aero, which has the least Legates value among the 15 sites, outperforms the MLR model with respect to the performance metrics values. Again, if we look at the performance metrics value of Hobart Airport as shown in Table 16.10, the performance parameter values are similar for the ANN and MLR model. The values of r, RMSE, MAE, MAPE, RRMSE, $d$, ENS, and $E$ for ANN and MLR model are 0.999, 0.19, 0.101, 0.555, 0.999, 0.999, 0.974, and 0.999, 0.225, 0.104, 1.252, 0.999, 0.998, and 0.974, respectively.

The difference between the ANN and MLR model development parameter with performance error is depicted in Fig. 16.11. A clear demarcation of performance error is observed in all sites except Hobart Airport that has similar performance metric

**Table 16.9** Performance metrics values (ANN versus MLR) for 15 sites over test period (1950–2017) with *r*, RMSE, MAE, MAPE, *d*, ENS, and *E* values

Performance metrics over the test period (1950–2017)

| Model | r | RMSE ($MJ/m^2$) | MAE ($MJ/m^2$) | MAPE (%) | RRMSE (%) | *d* | ENS | *E* |
|---|---|---|---|---|---|---|---|---|
| *Cairns Aero* | | | | | | | | |
| ANN | 0.999 | 0.19644336 | 0.104419 | 0.362 | 0.62 | 0.99895 | 0.998 | 0.971 |
| MLR | 0.9706 | 1.02529387 | 0.7812286 | 2.476 | 3.24 | 0.96793 | 0.941 | 0.78 |
| *Rockhampton* | | | | | | | | |
| ANN | 0.999 | 0.2274145 | 0.1171166 | 0.431 | 0.77 | 0.99908 | 0.998 | 0.972 |
| MLR | 0.9752 | 1.12899798 | 0.8670574 | 2.924 | 3.82 | 0.97501 | 0.951 | 0.792 |
| *Brisbane* | | | | | | | | |
| ANN | 0.9991 | 0.19006962 | 0.1082629 | 0.425 | 0.7 | 0.99915 | 0.998 | 0.97 |
| MLR | 0.9829 | 0.82310395 | 0.5792461 | 2.052 | 3.05 | 0.98259 | 0.966 | 0.84 |
| *Dubbo (Mentone)* | | | | | | | | |
| ANN | 0.9996 | 0.20272412 | 0.1372962 | 0.627 | 0.84 | 0.99962 | 0.999 | 0.977 |
| MLR | 0.996 | 0.64934384 | 0.4352467 | 1.673 | 2.68 | 0.99606 | 0.992 | 0.929 |
| *Sydney Airport AMO* | | | | | | | | |
| ANN | 0.9966 | 0.4365761 | 0.1256223 | 0.517 | 1.87 | 0.99703 | 0.993 | 0.97 |
| MLR | 0.9926 | 0.64161274 | 0.3359892 | 1.241 | 2.75 | 0.99345 | 0.985 | 0.92 |
| *Canberra City* | | | | | | | | |
| ANN | 0.9996 | 0.19175277 | 0.1201853 | 0.597 | 0.92 | 0.99964 | 0.999 | 0.979 |
| MLR | 0.9984 | 0.39230466 | 0.2342104 | 0.998 | 1.89 | 0.9985 | 0.997 | 0.96 |
| *Ballarat Aerodrome* | | | | | | | | |
| ANN | 0.9994 | 0.23783344 | 0.1255472 | 0.727 | 1.32 | 0.99949 | 0.999 | 0.978 |
| MLR | 0.9982 | 0.41427941 | 0.2462248 | 1.211 | 2.29 | 0.99847 | 0.996 | 0.957 |
| *Melbourne Airport* | | | | | | | | |
| ANN | 0.9988 | 0.31284454 | 0.1220417 | 0.614 | 1.54 | 0.99899 | 0.998 | 0.976 |
| MLR | 0.9975 | 0.44735071 | 0.2678622 | 1.152 | 2.21 | 0.99798 | 0.995 | 0.948 |
| *Adelaide Airport* | | | | | | | | |
| ANN | 0.9996 | 0.17122301 | 0.1157355 | 0.528 | 0.78 | 0.99966 | 0.999 | 0.977 |
| MLR | 0.9955 | 0.57302042 | 0.3655092 | 1.52 | 2.62 | 0.99624 | 0.991 | 0.926 |
| *Coober peddy* | | | | | | | | |
| ANN | 0.9996 | 0.1928207 | 0.1332529 | 0.534 | 0.72 | 0.99964 | 0.999 | 0.977 |
| MLR | 0.9897 | 1.00578869 | 0.6943445 | 2.549 | 3.74 | 0.99018 | 0.979 | 0.88 |
| *Alice Spring Airport* | | | | | | | | |
| ANN | 0.9996 | 0.18826013 | 0.1320227 | 0.495 | 0.67 | 0.9996 | 0.999 | 0.976 |
| MLR | 0.9835 | 1.22477379 | 0.8882323 | 3.166 | 4.33 | 0.98227 | 0.967 | 0.841 |

(continued)

**Table 16.9** (continued)

| Model | r | RMSE ($MJ/m^2$) | MAE ($MJ/m^2$) | MAPE (%) | RRMSE (%) | *d* | ENS | *E* |
|---|---|---|---|---|---|---|---|---|
| Performance metrics over the test period (1950–2017) | | | | | | | | |
| *Darwin Airport* | | | | | | | | |
| ANN | 0.9996 | 0.13346108 | 0.0719501 | 0.202 | 0.37 | 0.9994 | 0.999 | 0.979 |
| MLR | 0.9602 | 1.18505523 | 0.9365391 | 2.661 | 3.25 | 0.94338 | 0.921 | 0.733 |
| *Perth West* | | | | | | | | |
| ANN | 0.9994 | 0.18941532 | 0.1082543 | 0.438 | 0.78 | 0.99953 | 0.999 | 0.977 |
| MLR | 0.9948 | 0.57792937 | 0.3770145 | 1.426 | 2.38 | 0.99565 | 0.99 | 0.919 |
| *Geraldton Town* | | | | | | | | |
| ANN | 0.9995 | 0.19241669 | 0.1194961 | 0.436 | 0.72 | 0.99951 | 0.999 | 0.973 |
| MLR | 0.9873 | 0.89432276 | 0.6201121 | 2.208 | 3.36 | 0.98919 | 0.974 | 0.861 |
| *Hobart Airport* | | | | | | | | |
| ANN | 0.9993 | 0.18992224 | 0.1009723 | 0.555 | 1.06 | 0.99936 | 0.999 | 0.974 |
| MLR | 0.999 | 0.22533635 | 0.1038511 | 0.49 | 1.25 | 0.99911 | 0.998 | 0.974 |

values for ANN and MLR models. Hence, it can be concluded that an ML model (ANN) outperforms the statistical approach (MLR) in every value of performance metrics to best predict the result.

### 16.7.2 ANN Versus ARIMA

Table 16.10 displays the ARIMA model architecture, derived parameters, correlation coefficient, sigma 2 values, and the respective goodness-of-fit tests performed to construct the predictive model.

Table 16.11 describes the comparative indices for the ANN and ARIMA model.

To testify the validity of the ANN model, finally, it is compared with ARIMA model to check the level of agreement between the simulated and observed values. From Table 16.11, for the ANN model, the site, Cabrera City has the highest value of *E*, i.e., 0.979 with respect to other sites.it means the Canberra City has the best data set for the prediction of daily heat indices with the highest accuracy and minimal error. The r, MAE, and RMSE values for the Canberra City in ANN model are 0.999, 0.120, and 0.192, respectively. It represents the best-fitted dataset for the ANN model development process.

If observed at the ARIMA model development for the Canberra site, it has the legate value of −0.00548 with MAE, RMSE, and r values of 5.8204, 6.9915, and 0.0475, respectively. From all the perspective values of ARIMA, ANN model has become par better the best model for Canberra City. Likewise, if considered at the value of E for Cairns Aero from Table 16.11, above, it is observed as 0.971 for

**Table 16.10** ARIMA with structure ($p$, $d$, $q$) with $d$ = differencing, $p$ and $q$ = order of autoregressive (AR) and moving average (MA) term

| Site name | ARIMA structure | | | | | | | ARIMA parameters | | | | | |
|---|---|---|---|---|---|---|---|---|---|---|---|---|---|
| | $P$ | $d$ | $q$ | AIC | Log-Likelihood | Sigma2 | r | AR1 | AR2 | AR3 | intercept | MA1 | MA2 |
| Cairns Aero | 14 | 0 | 0 | 61678.9 | −30823.5 | 3.671 | −0.0912 | 0.58 | 0.086 | 0.03 | 30.906 | NA | NA |
| Rockhampton | 20 | 0 | 0 | 6828.31 | −34121.5 | 5.717 | −0.0307 | 0.74 | −0.07 | 0.03 | 29.065 | NA | NA |
| Brisbane | 17 | 0 | 0 | 66207.2 | −33084.6 | 4.973 | −0.0342 | 0.60 | 0.007 | 0.026 | 26.055 | NA | NA |
| Dubbo | 13 | 0 | 0 | 73087.7 | −36522.9 | 7.894 | 0.0522 | 0.77 | −0.09 | 0.01 | 23.118 | NA | NA |
| Sydney | 20 | 0 | 0 | 77303.2 | −38629.6 | 10.47 | 0.0381 | 0.44 | 0.018 | 0.038 | 22.145 | NA | NA |
| Canberra | 13 | 0 | 0 | 72795.1 | −36382.6 | 7.741 | 0.0474 | 0.70 | −0.08 | 0.05 | 19.364 | NA | NA |
| Ballarat | 16 | 0 | 0 | 78326.1 | −39145.1 | 11.22 | 0.0648 | 0.67 | −0.13 | 0.056 | 17.202 | NA | NA |
| Melbourne | 17 | 0 | 0 | 79731.2 | −39846.6 | 12.33 | 0.0623 | 0.60 | −0.11 | 0.05 | 19.201 | NA | NA |
| Adelaide | 1 | 0 | 0 | 77530.6 | −38760.3 | 10.64 | 0.0552 | 0.99 | NA | NA | 21.0907 | −0.4 | −0.4 |
| Coober Peddy | 16 | 0 | 0 | 75198.6 | −37581.3 | 9.094 | −0.0935 | 0.77 | −0.13 | 0.036 | 26.511 | NA | NA |
| Alice Springs | 1 | 0 | 0 | 74922.1 | −37456.1 | 8.934 | 0.0471 | 0.97 | NA | NA | 27.464 | −0.2 | −0.3 |
| Darwin | 3 | 0 | 0 | 66041.3 | −33013.7 | 4.922 | 0.0688 | 1.23 | −0.09 | −0.1 | 35.814 | −0.7 | −0.2 |
| Perth West | 17 | 0 | 0 | 72055.3 | −36008.7 | 7.364 | −0.0397 | 0.71 | −0.16 | 0.02 | 23.25 | NA | NA |
| Geraldton | 19 | 0 | 0 | 73886.3 | −36922.2 | 8.326 | −0.0412 | 0.78 | −0.14 | 0.00 | 25.973 | NA | NA |
| Hobart | 20 | 0 | 0 | 76054.09 | −38005 | 9.629 | −0.0485 | 0.493 | −0.0 | 0.03 | 17.18 | NA | NA |

The parameters are obtained as per the model development and the values of the dataset over the test period. Autoregressive (AR) and Moving Average (MA) values are not obtained in every candidate site, so those values are indicated with NA. Log-Likelihood, sigma 2 and correlation coefficient determines the perfect model fit

*Note* Akaike's Information Criterion (AIC) used to identify the model in conjunction with log-likelihood, variance and correlation coefficient

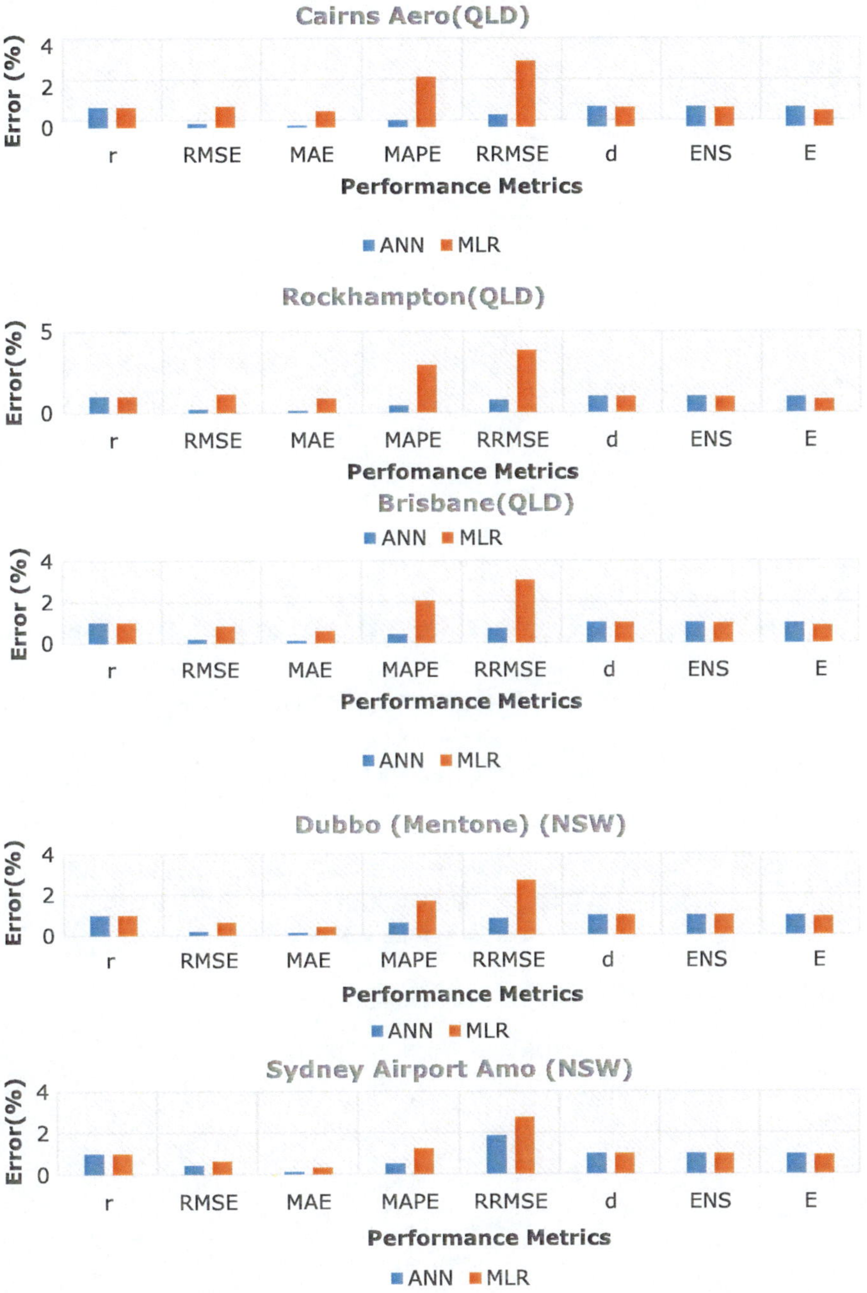


**Fig. 16.11** An error distribution comparison of the simulation error of ANN and MLR model for each of the 15 sites across Australia over the test period (1950–2017) based on *r*, RMSE, MAE, MAPE, RRMSE, *d*, ENS, and E

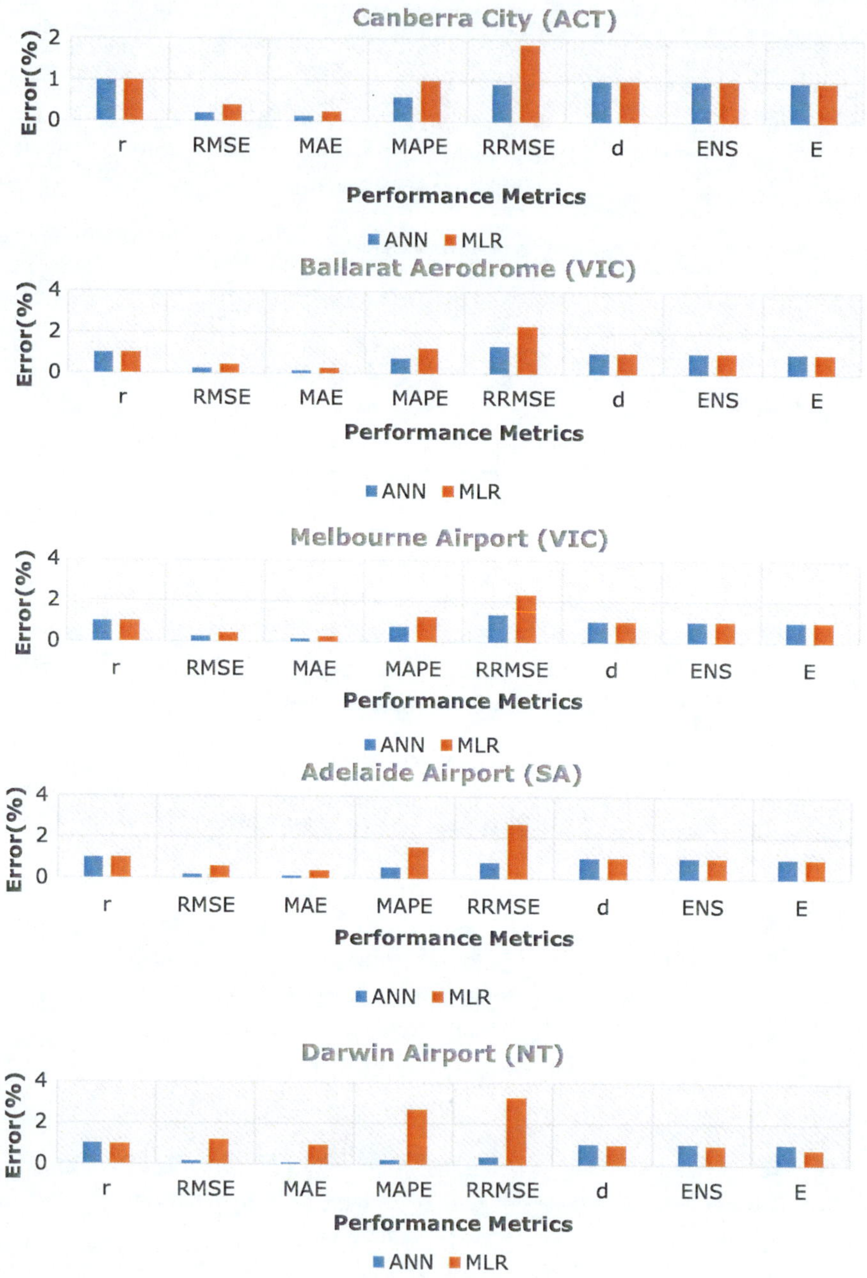


**Fig. 16.11** (continued)

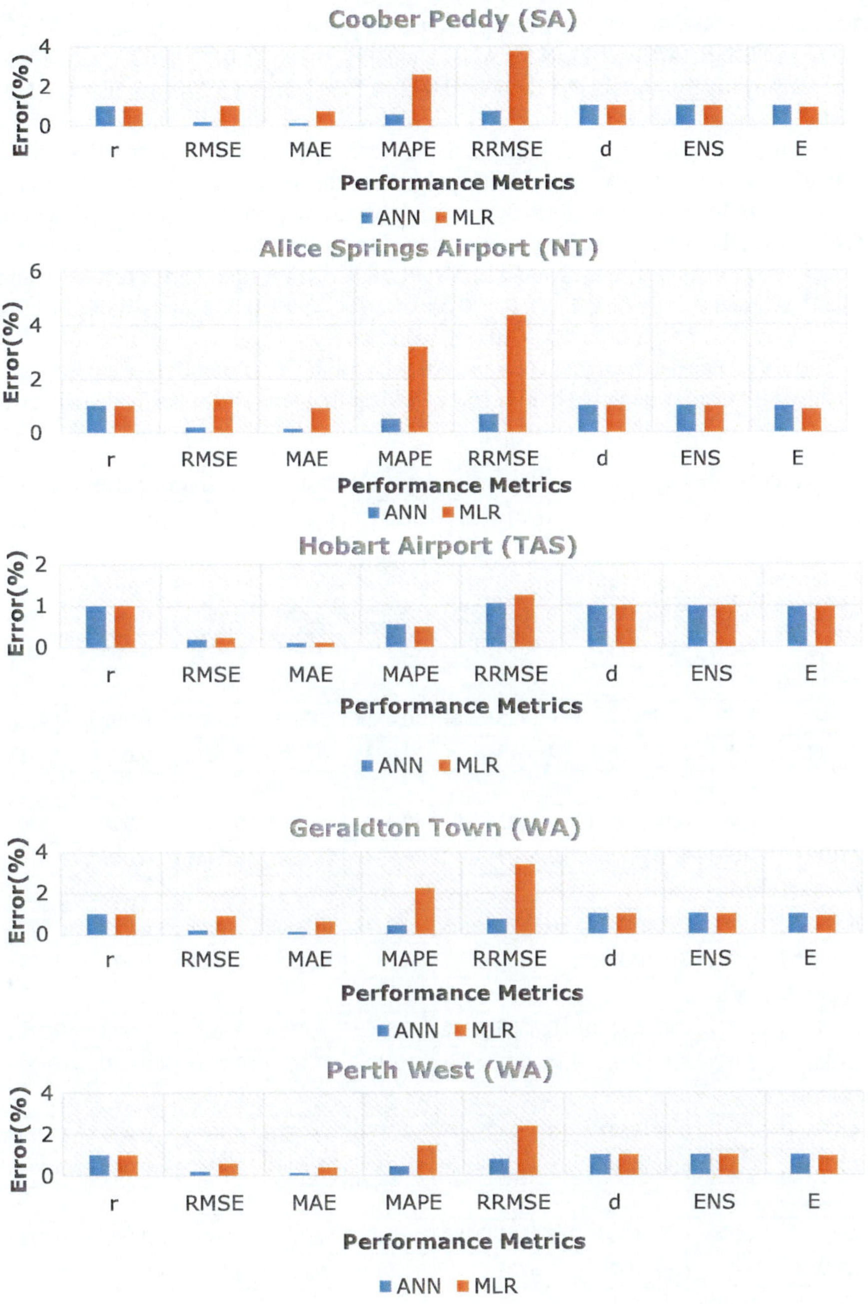


**Fig. 16.11** (continued)

ANN and −0.0042 for ARIMA model. Also, the performance metrics like r, RMSE, MAE, MAPE, RRMSE, *d*, ENS for ANN are 0.999, 0.196, 0.104, 0.362, 0.62, 0.999, 0.998, respectively, and for ARIMA (−0.092, 4.298, 3.563, 11.240, 13.567, 0.2238, −0.0355, respectively).

Again, for the Cairns Aero, which has the least Legates value among the 15 sites, outperforms the ARIMA model with respect to the performance metrics values. Again, if we look at the performance metrics value of Hobart Airport as shown in Table 16.11, the performance parameter values are similar for the ANN and ARIMA model. The values of r, RMSE, MAE, MAPE, RRMSE, d,ENS and E for ANN and ARIMA model are 0.999, 0.19, 0.101, 0.555, 0.999, 0.999, 0.974 and −0.048, 4.895, 3.892, 22.710, 27.502, 0.1682, −0.0185, respectively.

The difference between the ANN and ARIMA model development parameter with performance error is depicted in Fig. 16.12. A clear demarcation of performance error

**Table 16.11** Performance metrics: ANN versus MLR for 15 sites over test period (1950–2017)

| Model | *r* | RMSE ($MJ/m^2$) | MAE ($MJ/m^2$) | MAPE (%) | RRMSE (%) | *d* | ENS | *E* |
|---|---|---|---|---|---|---|---|---|
| *Cairns Aero* | | | | | | | | |
| ANN | 0.999 | 0.196443357 | 0.104419024 | 0.362 | 0.6201 | 0.999 | 0.998 | 0.9706 |
| ARIMA | −0.091 | 4.297973 | 3.563419 | 11.239 | 13.568 | 0.2239 | −0.035 | −0.0042 |
| *Rockhampton* | | | | | | | | |
| ANN | 0.999 | 0.227414495 | 0.117116564 | 0.4314 | 0.7699 | 0.9991 | 0.998 | 0.9719 |
| ARIMA | −0.031 | 5.108826 | 4.167386 | 14.423 | 17.295 | 0.1272 | −0.012 | −0.0013 |
| *Brisbane* | | | | | | | | |
| ANN | 0.999 | 0.190069617 | 0.108262887 | 0.4247 | 0.7038 | 0.9991 | 0.998 | 0.9701 |
| ARIMA | −0.034 | 4.551875 | 3.660959 | 13.594 | 16.854 | 0.2441 | −0.047 | −0.0102 |
| *Dubbo (Mentone)* | | | | | | | | |
| ANN | 1 | 0.202724122 | 0.137296206 | 0.6271 | 0.8366 | 0.9996 | 0.999 | 0.9775 |
| ARIMA | 0.052 | 7.320853 | 6.146907 | 28.133 | 30.21 | 0.2039 | −0.023 | −0.0093 |
| *Sydney Airport AMO* | | | | | | | | |
| ANN | 0.997 | 0.436576101 | 0.125622272 | 0.5167 | 1.8724 | 0.997 | 0.993 | 0.9701 |
| ARIMA | 0.038 | 5.400996 | 4.209104 | 17.962 | 23.164 | 0.2535 | −0.049 | −0.0024 |
| *Canberra City* | | | | | | | | |
| ANN | 1 | 0.19175277 | 0.120185297 | 0.5969 | 0.9216 | 0.9996 | 0.999 | 0.9792 |
| ARIMA | 0.047 | 6.991494 | 5.820478 | 30.347 | 33.604 | 0.2567 | −0.044 | −0.0055 |
| *Ballarat Aerodrome* | | | | | | | | |
| ANN | 0.999 | 0.237833444 | 0.125547198 | 0.7274 | 1.3155 | 0.9995 | 0.999 | 0.9781 |
| ARIMA | 0.065 | 6.883002 | 5.672531 | 34.609 | 38.071 | 0.1832 | −0.015 | 0.012 |
| *Melbourne Airport* | | | | | | | | |
| ANN | 0.999 | 0.312844536 | 0.122041741 | 0.6139 | 1.5443 | 0.999 | 0.998 | 0.9763 |

(continued)

**Table 16.11** (continued)

| Model | *r* | RMSE ($MJ/m^2$) | MAE ($MJ/m^2$) | MAPE (%) | RRMSE (%) | *d* | ENS | *E* |
|---|---|---|---|---|---|---|---|---|
| ARIMA | 0.062 | 6.406119 | 5.084406 | 25.761 | 31.623 | 0.2159 | −0.027 | 0.0137 |
| *Adelaide Airport* | | | | | | | | |
| ANN | 1 | 0.171223009 | 0.11573551 | 0.5285 | 0.784 | 0.9997 | 0.999 | 0.9766 |
| ARIMA | 0.055 | 6.039928 | 4.904222 | 23.038 | 27.657 | 0.1703 | −0.015 | 0.0071 |
| *Coober Peddy* | | | | | | | | |
| ANN | 0.901 | 3.013871 | 2.251183 | 8.9661 | 11.368 | 0.9458 | 0.812 | 0.6092 |
| ARIMA | −0.093 | 6.980649 | 5.744329 | 23.265 | 25.908 | 0.0932 | −0.007 | −0.0017 |
| *Alice Spring Airport* | | | | | | | | |
| ANN | 1 | 0.18826013 | 0.132022744 | 0.4949 | 0.6654 | 0.9996 | 0.999 | 0.9764 |
| ARIMA | 0.047 | 6.791157 | 5.584853 | 21.746 | 23.95 | 0.1761 | −0.017 | −0.0097 |
| *Darwin Airport* | | | | | | | | |
| ANN | 1 | 0.133461075 | 0.071950122 | 0.2016 | 0.3662 | 0.9994 | 0.999 | 0.9795 |
| ARIMA | 0.069 | 4.105173 | 3.449204 | 9.7718 | 11.293 | 0.1784 | −0.013 | −0.0227 |
| *Perth West* | | | | | | | | |
| ANN | 0.999 | 0.189415316 | 0.108254331 | 0.4385 | 0.7816 | 0.9995 | 0.999 | 0.9767 |
| ARIMA | −0.04 | 5.726119 | 4.61537 | 19.217 | 23.765 | 0.1973 | −0.03 | 0.0059 |
| *Geraldton Town* | | | | | | | | |
| ANN | 1 | 0.192416691 | 0.119496079 | 0.4359 | 0.7234 | 0.9995 | 0.999 | 0.9733 |
| ARIMA | −0.041 | 5.601124 | 4.446581 | 17.09 | 21.241 | 0.1057 | −0.011 | 0.0028 |
| *Hobart Airport* | | | | | | | | |
| ANN | 0.999 | 0.189922242 | 0.100972276 | 0.5551 | 1.0552 | 0.9994 | 0.999 | 0.9744 |
| ARIMA | −0.049 | 4.895113 | 3.861772 | 22.71 | 27.502 | 0.1681 | −0.019 | 0.0051 |

for all the sites is observed except Hobart Airport that has similar performance metrics values for ANN and ARIMA models. Hence, it can be concluded that an ML model (ANN) outperforms the statistical approach (ARIMA) in every value of performance metrics to best predict the result.

### *16.7.3 ANN Versus (MLR and ARIMA) Time Series Analysis*

To check how each accurate are each of datum points in simulated data, a time series of observed data (Data.Obs) and simulated data (Data.Sim) are plotted.

Figure 16.13 shows a time series of observed and predicted SHI with the model error encountered for each datum point, for ANN, MLR, and ARIMA model.

While the overall agreement between the observed and predicted SHI is good, the model error (shown on the right-hand side of each graph in Fig. 16.14) is different for

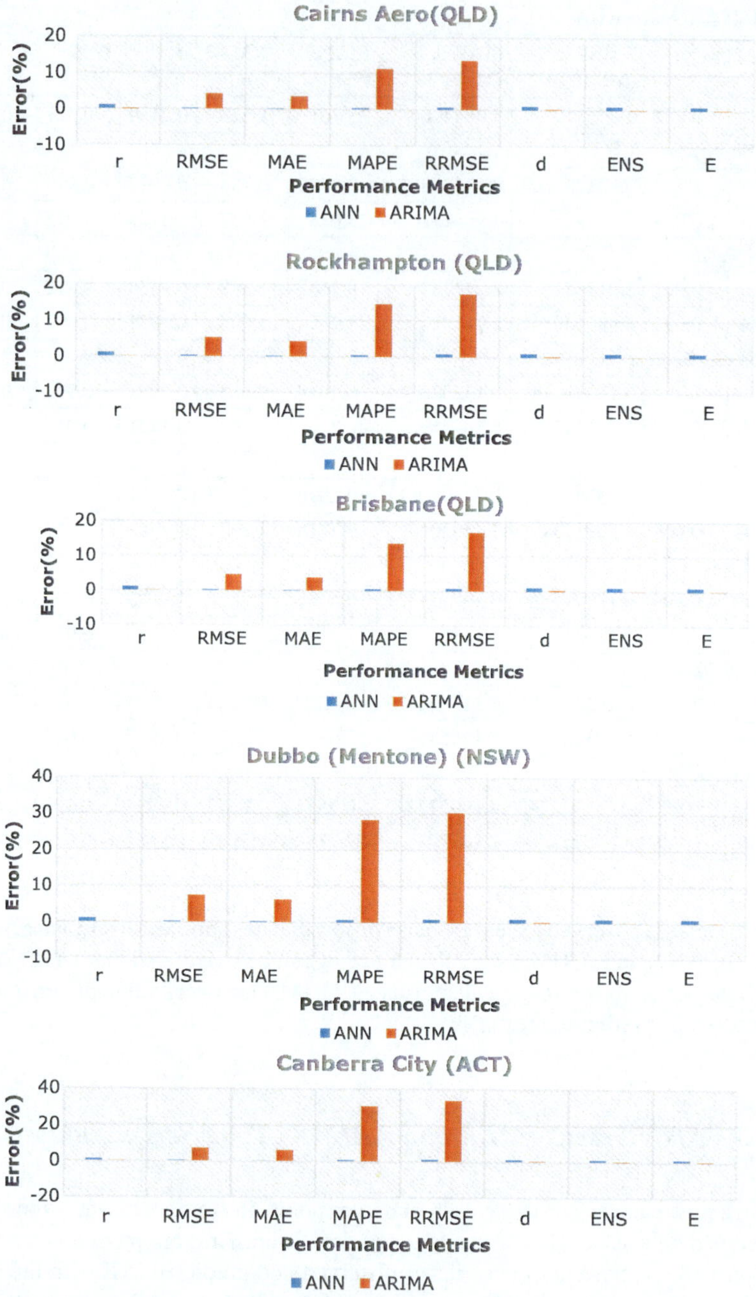


**Fig. 16.12** Error distribution comparison of ANN and ARIMA model

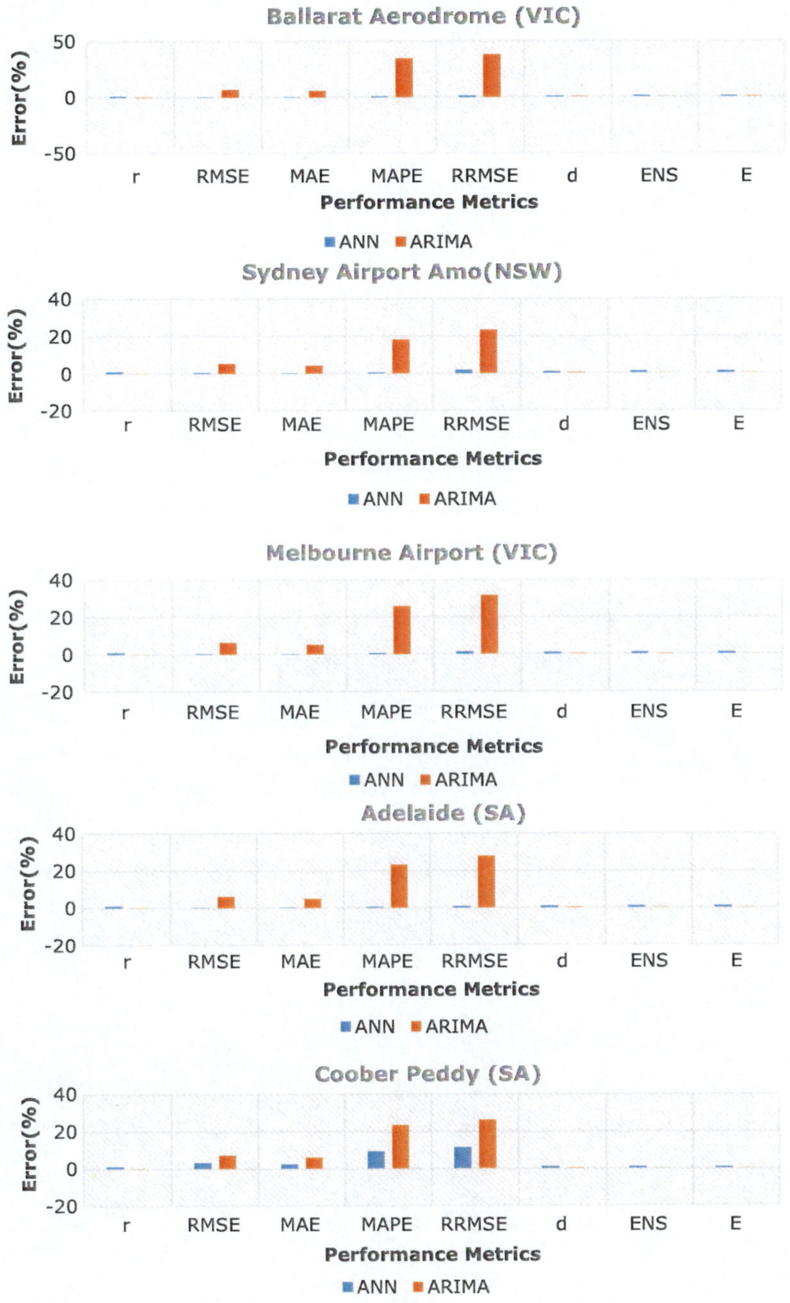


**Fig. 16.12** (continued)

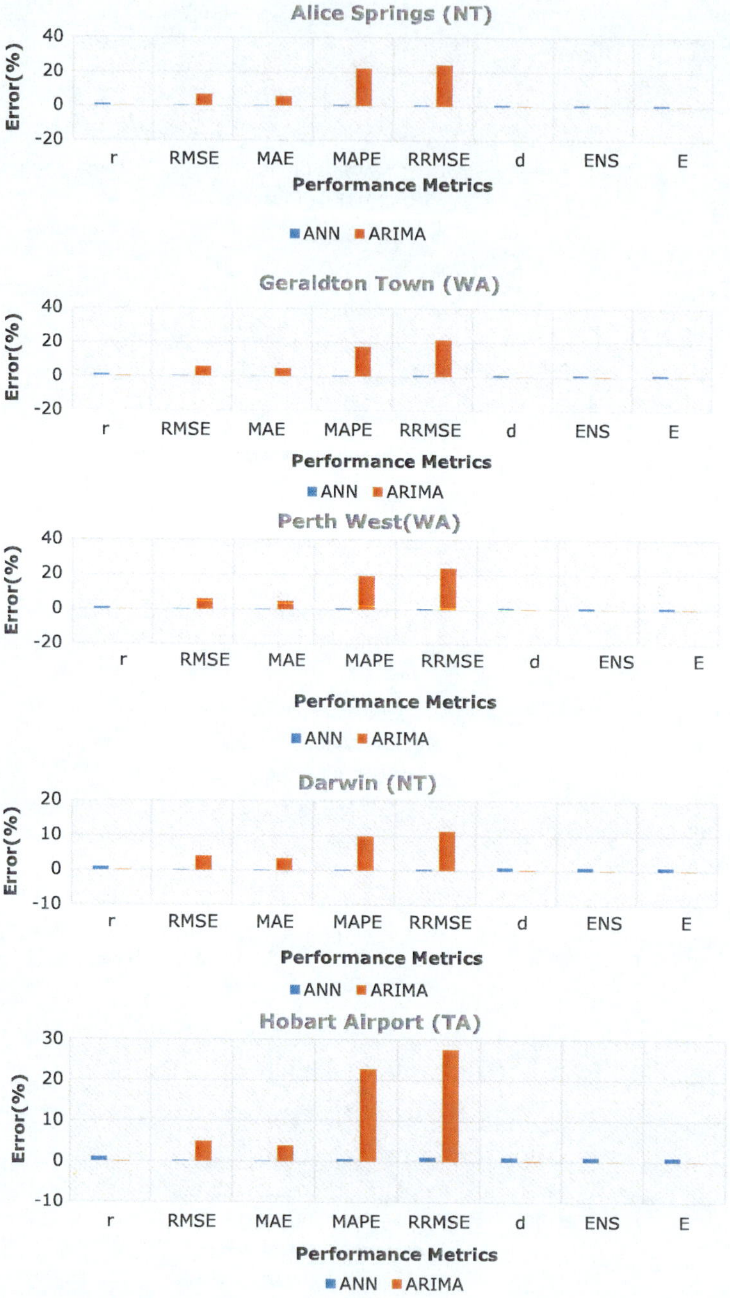


**Fig. 16.12** (continued)

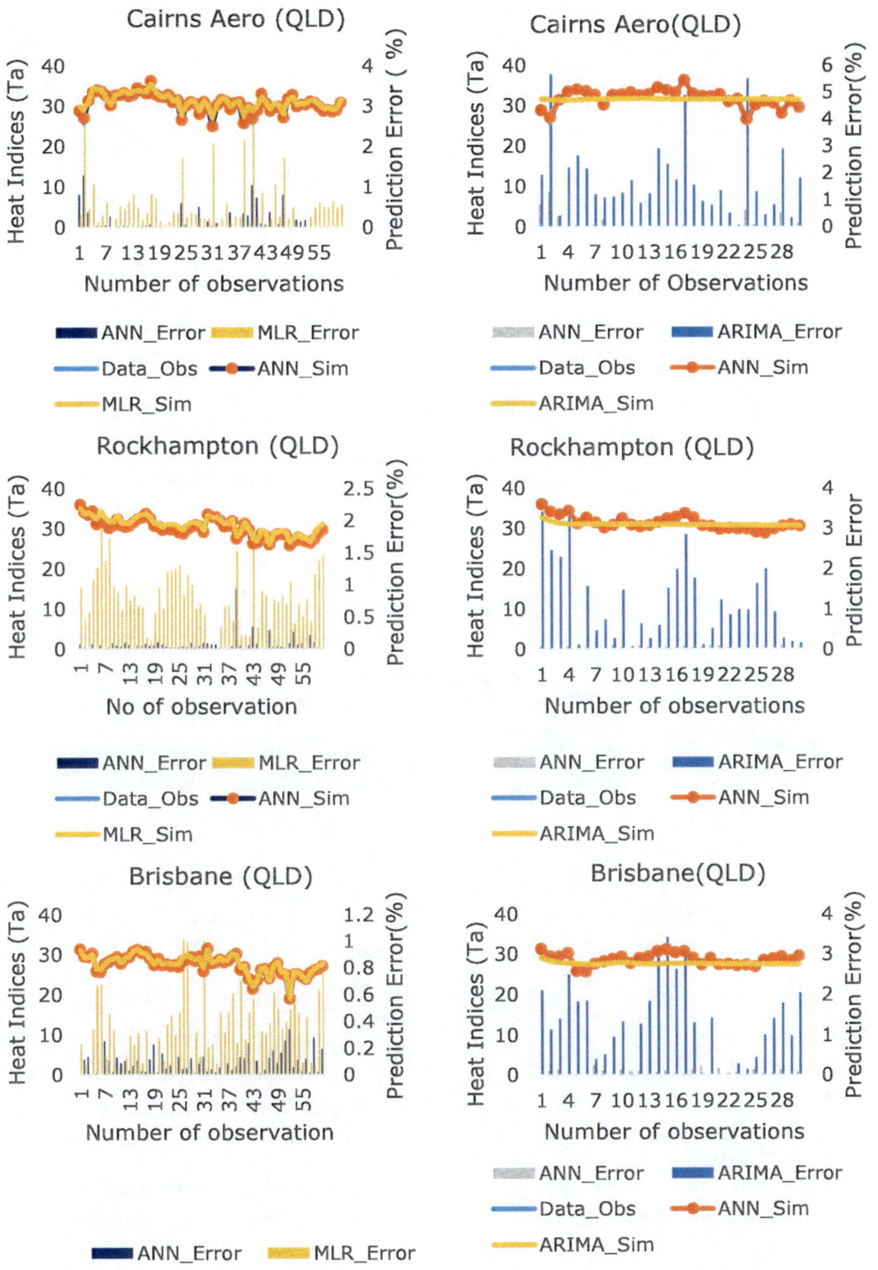


**Fig. 16.13** Time series comparison of ANN versus MLR and ARIMA. The observed and predicted Steadman HI values plotted with corresponding prediction error (PE) over the test period (1950–2017). The left-hand side depicts the comparison between ANN and MLR and the right-hand column represents the comparison of ANN and ARIMA model

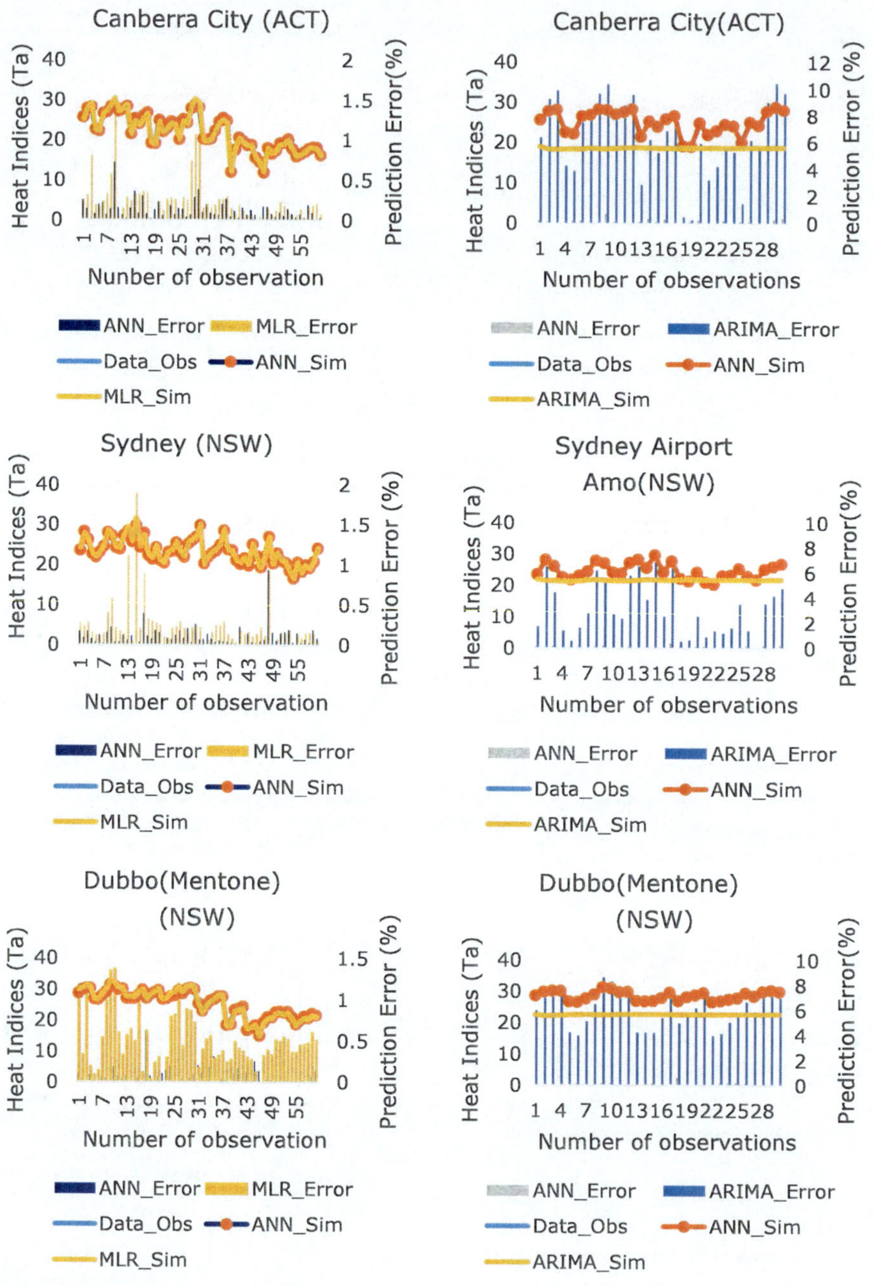


**Fig. 16.13** (continued)

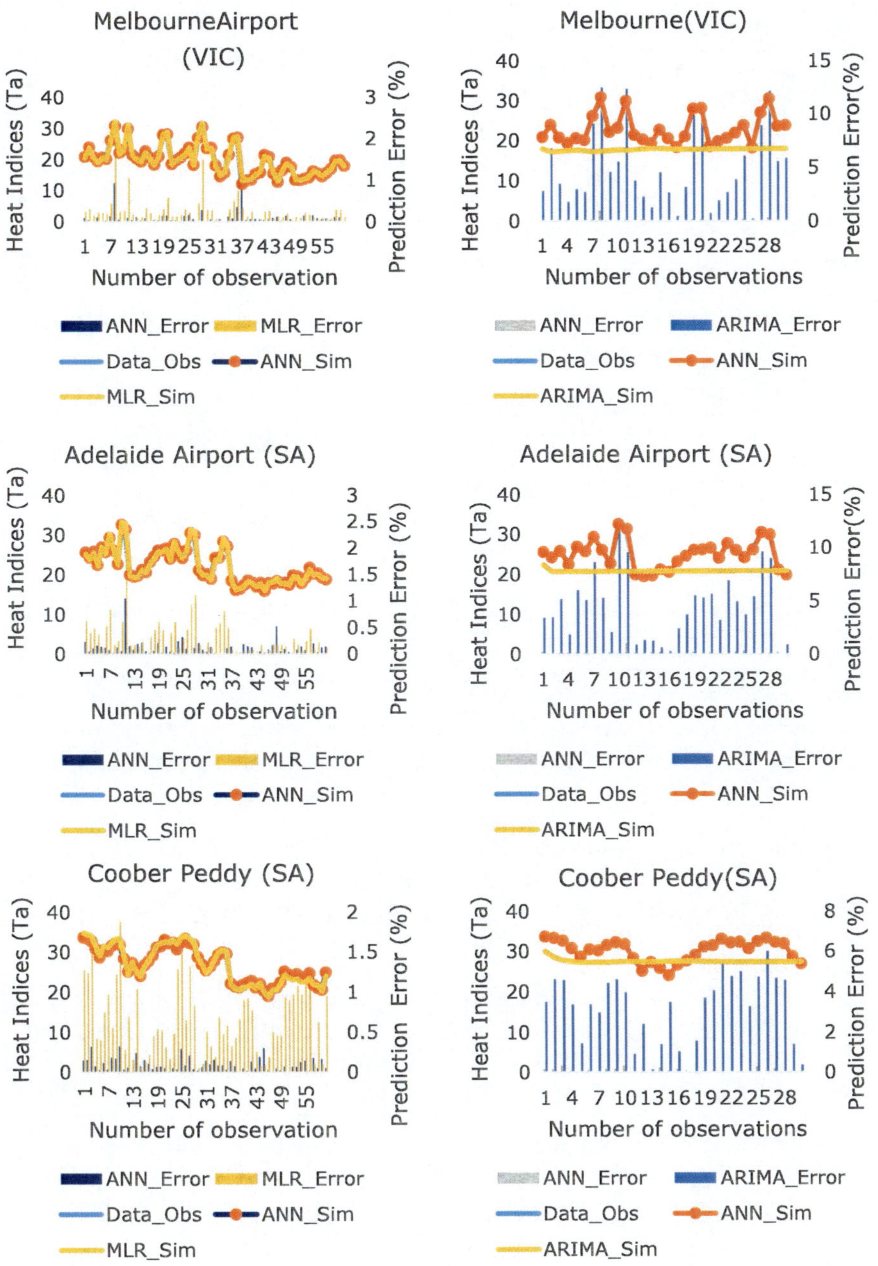


**Fig. 16.13** (continued)

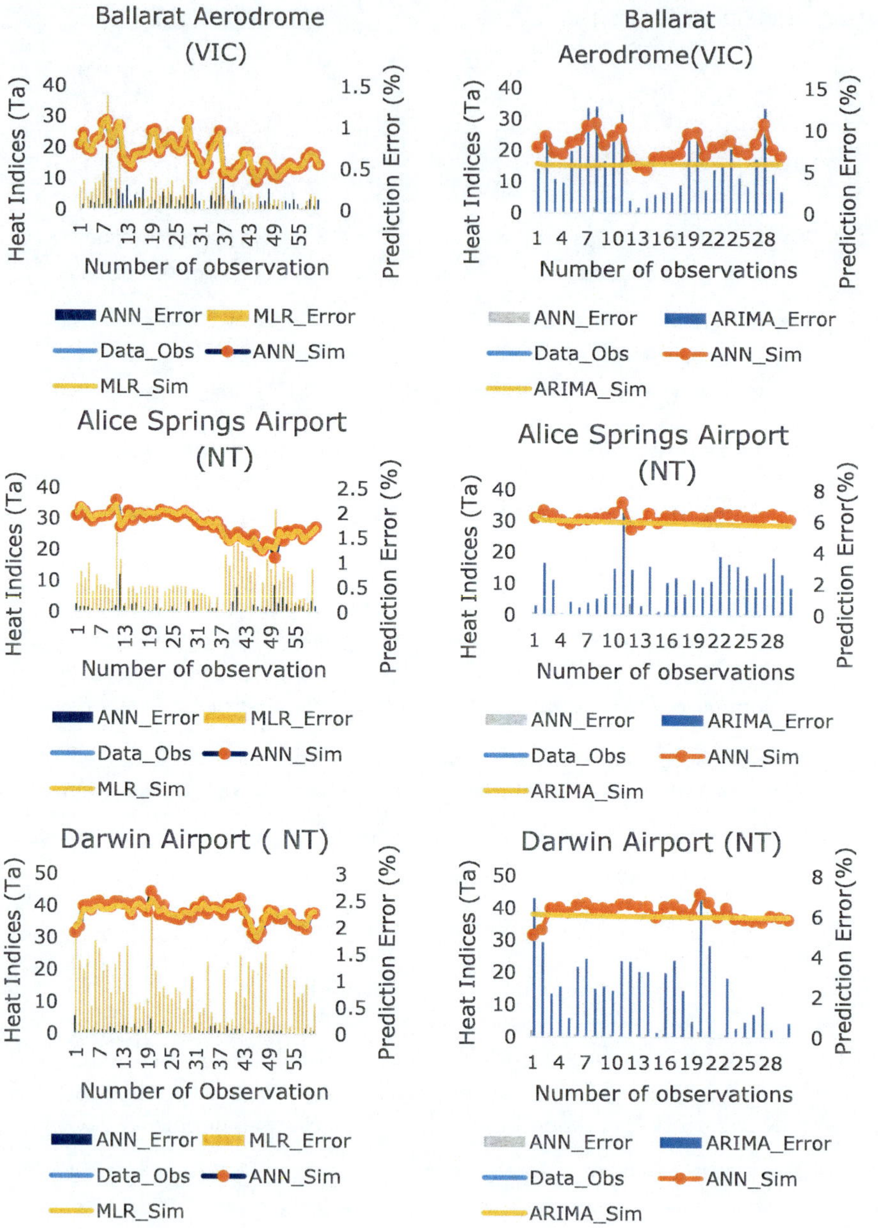


**Fig. 16.13** (continued)

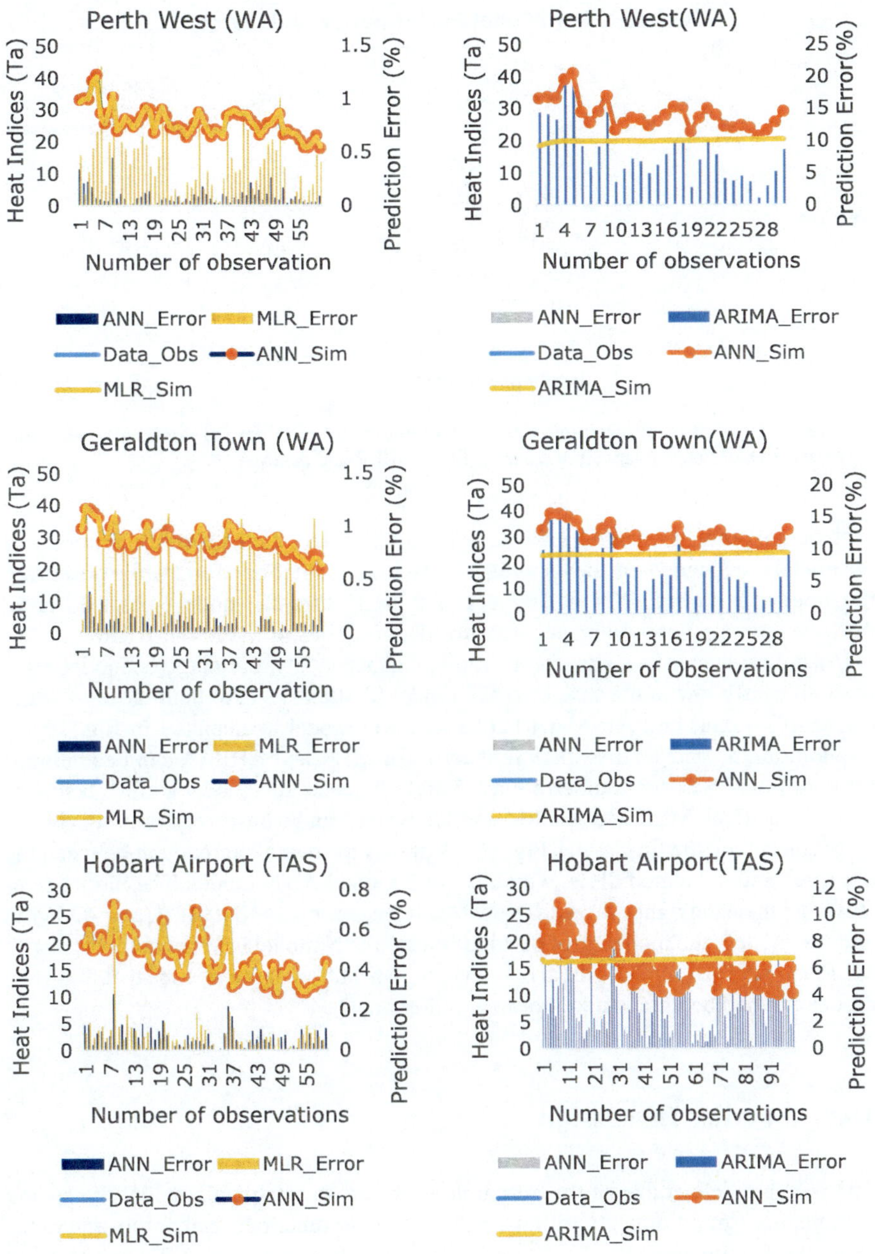


**Fig. 16.13** (continued)

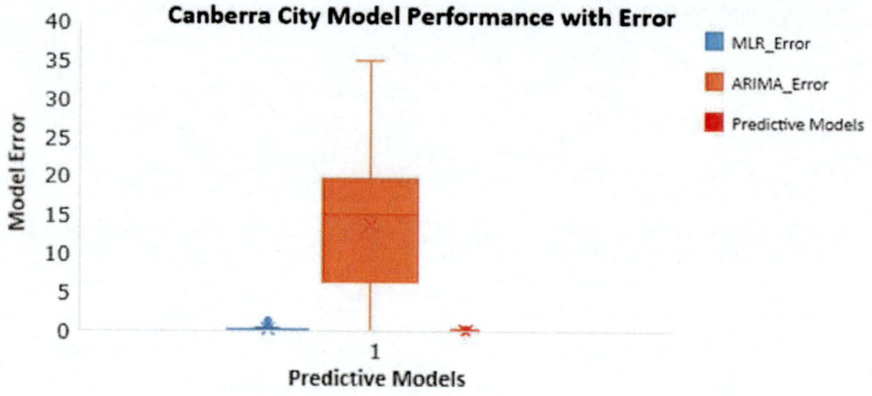


**Fig. 16.14** Boxplot of distribution of Canberra City model error for (**a**) ANN, a good model (**b**) extreme values have relatively large magnitudes (MLR and ARIMA)

each datum point in a model (ANN or MLR or ARIMA). This shows the accuracy of the model is dependent on the location of the datum point. Therefore, while observing the time series plot of ANN model, in each of the 15 sites, the model error is different for each datum point with respect to error datum points of MLR and ARIMA.

Also, if analyzed at a site, for example, Canberra City, the range of model error is significantly low in the case of ANN model (0.1964–1.9298) than those of MLR (1.58E-05–4.054) and ARIMA (0.2651–43.281) model as depicted in Fig. 16.13. In addition, the uneven distribution of error for MLR and ARIMA can be analyzed through Fig. 16.11 for Canberra City. Similarly, each site shows similar behavior toward the ANN. MLR and ARIMA model, which can be observed from Fig. 16.13.

Furthermore, the time series Fig. 16.13, shows the overall agreement between the observed and simulated SHIs, which is quite a good ANN model. The model error (PE) shown on the right side is different for all the models (ANN, MLR, and ARIMA) and it is evident the model error distribution for ANN model is lesser compared with MLR and ARIMA model, which has a wider distribution. It is evidence that the ANN model attained better accuracy for heat indices data.

## 16.8 Further Discussion

The possible deliverable of this research work was to establish a model for testing combinations of training algorithms, hidden transfer functions, and output equations resulting in comparatively small prediction errors. In this regard, ANN outperforms the MLR and ARIMA model in the least prediction error with the highest values of Legates. Thus, we can conclude that the ML models are better than the statistical approach models in terms of encapsulating the non-linear relationship between the input and output variables.

Exploratory data analysis of all the three models with respect to the corresponding prediction error is depicted by the box-plots as shown in Fig. 16.14. Here, ANN model outperforms all the comparative models with the least model error with the mean, median, and mode values of error. The exploratory data analysis of model error can be seen from the boxplot as shown in Fig. 16.14, for Canberra City. In this, boxplot, it is clearly depicted that, ANN model has the least model error distribution in comparison with MLR and ARIMA model. MLR model is comparatively competitive with the ANN model but the ARIMA model has a very wide range of error with significantly higher magnitude.

## 16.9 Conclusions

This chapter investigated the spatial patterns of Steadman heatwave trends in average annual heatwave indices at 15 sites across Australia between 1950 and 2017.

The magnitude of trends is estimated from the Sen's slopes and statistical significance is tested using the Mann–Kendall's test. The resulted trends are further verified with the predictive model development process. For the predictive model development, data intelligent ML models (ANN) and statistical predictive techniques (MLR and ARIMA) are performed and analyzed for the trend verification and model performance evaluation.

Of the 15 sites over the observation period of annual heat indices, all the sites show an increasing trend in heatwave incidences since 1950. However, Adelaide Airport and Coober Peddy show a small decrease in the heatwave magnitude (0.5 °C per year) with a decreasing trend from the year 2006 and 1990, respectively, until 2017. The major sites in New South Wales (Sydney Airport Amo) and the Australian Capital Territory (Canberra) along with Norther Territory sites (Alice Springs Airport and Darwin Airport) shows a significant increase in the magnitude of a trend. Whereas, the sites from Queensland (Cairns Aero, Rockhampton, Brisbane), Tasmania (Hobart Airport), Western Australia (Perth West and Geraldton Town) are confined to a smaller positive trend with a magnitude of trend increases varying across the continent. Large increases of about 3.5 °C per year have occurred at Alice Springs and Darwin (NT), about 3 °C per year at Brisbane Airport (QLD), Canberra City (ACT), and Sydney Airport Amo (NSW). All the tests are statistically significant at a 95% confidence interval.

To enhance the preciseness of the ANN model, several training algorithms and hidden transfer functions are performed for hit and trial, such that the Levenberg–Marquardt training algorithm and logarithmic sigmoid function is finally adopted for the prediction of the SHIs (Ta). To meet the objectives of this research work, ANN model is benchmarked with multiple linear regression (MLR) and autoregressive moving average (ARIMA) models. The performance of the ANN model in terms of its Legates and McCabe's Index is dramatically higher than the MLR and ARIMA model for the sites with severe trends of heatwaves such as Canberra City, Darwin Airport, Alice Springs Airport, Sydney Airport, and Brisbane, respectively. By an analysis of model error between different 15 sites, optimum performances are observed with ANN model while comparing with MLR and ARIMA model. In the case of Canberra City, the range of model error is significantly low in the case of ANN model than those of MLR model. A similar range of error is seen for all of the 15 sites with the quite outperformance of ANN model with respect to ARIMA model and a bit of similarity with MLR model. Despite the small similarity, ANN is able to project the best results in terms of model performance metrics compared with statistical models like ARIMA and MLR models.

Mitigation of heatwaves may be possible by reducing urban heat island effects and reducing greenhouse gas emissions to slow climate change. In terms of adaptation to heatwaves and heat stress, adaptive strategies have been implemented in all sectors, although most attention must be given to the health sector. Adaptation of infrastructure, lifestyle choices, and emergency response systems are essential to maintain a healthy life and a healthy society.

During the twentieth century, heatwaves were the major cause of death than other natural calamities in Australia. A heatwave alone resulted in 437 deaths and several casualties in Australia along with a severe loss to livestock, crops, roads, railways, and bridges (Management 2018). In consequence, the prediction of heatwaves is of great significance these days to the entire living organisms on this planet. In this study, a little effort is given to enhance further efficient results with the help of ANN as an applied ML algorithm for predicting the Steadman's HI using the various available meteorological variables.

Finally, the significant better prediction performance is expected when comparing to various other relative measures of prediction for the HI as per available state-of-art, in various spatial locations in Australia. This study set a foundation for the potential of using more extensive predictor data products such as solar radiation, wind speed, diverse spatial locations with a huge dataset with a range of many decades, where possible for heatwave prediction and modeling. Partial Autocorrelation Function (PACF) can be used to find the time lags to perform the forecast of heat indices. The preliminary steps in the PACF are shown in Appendix 1, where the value of the correlation coefficient shows the accuracy and data variation and shows how well they are correlated with the trend.

**Acknowledgements** The conceptualization and compilation of this chapter is a result of concerted efforts of various individuals without whose help this work would not be possible. We are not able to list all the contributors here due to space constraints, but we value the significance of each contribution especially from the members of the *Advanced Data Analytics*: Environment Modeling and Simulation Research Group and colleagues, their seminars, discussions, ideas, and feedback. In terms of the current book chapter, we would like to express our gratitude for the contribution of all our family members and close friends for their intense support and motivation toward the completion of this work.

## References

AL-Ataby IK (2019) Trend analysis for some climate variables of selected stations in Iraq. J Edu Pure Sci-Univ of Thi-Qar 9(2):253–258

Al-Fatlawi A, Abdul Rahim N, Rahman S, Ward T (2015) Improving solar energy prediction in complex topography using artificial neural networks: case study Peninsular Malaysia. 34

Ali S, Smith KA (2006) On learning algorithm selection for classification. Appl Soft Comput 6(2):119–138

Armstrong JS, Collopy F (1992) Error measures for generalizing about forecasting methods: empirical comparisons. Int J Forecast 8(1):69–80

Aubrecht C, Özceylan D (2013) Identification of heat risk patterns in the US national capital region by integrating heat stress and related vulnerability. Environ Int 56:65–77

Battiti R (1992) First-and second-order methods for learning: between steepest descent and Newton's method. Neural Comput 4(2):141–166

Bishop C, Bishop CM (1995) Neural networks for pattern recognition. Oxford University Press

Bortolotti R (2018) Tutorial K—data prep 2-2: dummy coding category variables A2—Nisbet, Robert. In: Miner G, Yale K (eds) Handbook of statistical analysis and data mining applications, 2nd edn. Academic Press, Boston, pp 497–514

Campforts B et al (2016) Simulating the mobility of meteoric 10Be in the landscape through a coupled soil-hillslope model (Be2D). Earth Planet Sci Lett 439:143–157

Chai T, Draxler RR (2014) Root mean square error (RMSE) or mean absolute error (MAE)?– Arguments against avoiding RMSE in the literature. Geosci Model Development 7(3):1247–1250

Coates L, Haynes K, O'Brien J, McAneney J, De Oliveira FD (2014) Exploring 167 years of vulnerability: an examination of extreme heat events in Australia 1844–2010. Environ Sci Policy 42:33–44

Coughlan M (1979) Recent variations in annual-mean maximum temperatures over Australia. Q J Royal Meteorological Soc 105(445):707–719

Delworth TL, Mahlman J, Knutson TR (1999) Changes in heat index associated with $CO_2$-induced global warming. Clim Change 43(2):369–386

Dennis Jr JE, Schnabel RB (1996) Numerical methods for unconstrained optimization and nonlinear equations, 16. Siam

Deo R, McAlpine C, Syktus J, McGowan H, Phinn S (2007) On Australian heat waves: time series analysis of extreme temperature events in Australia, 1950–2005. In: Proceedings of the international congress on modelling and simulation (MODSIM07). modelling and simulation Society of Australia and New Zealand Inc., pp 626–635

Deo R, Şahin M (2017a) Forecasting long-term global solar radiation with an ANN algorithm coupled with satellite-derived (MODIS) land surface temperature (LST) for regional locations in Queensland. 72, 828–848 pp

Deo RC, Şahin M (2015) Application of the artificial neural network model for prediction of monthly standardized precipitation and evapotranspiration index using hydrometeorological parameters and climate indices in eastern Australia. Atmos Res 161:65–81

Deo RC, Şahin M (2016) An extreme learning machine model for the simulation of monthly mean streamflow water level in eastern Queensland. Environ Monit Assess 188(2):90

Deo RC, Şahin M (2017b) Forecasting long-term global solar radiation with an ANN algorithm coupled with satellite-derived (MODIS) land surface temperature (LST) for regional locations in Queensland. Renew Sustain Energy Rev 72:828–848

Dodla VB, Satyanarayana GC, Desamsetti S (2017) Analysis and prediction of a catastrophic Indian coastal heat wave of 2015. Nat Hazards 87(1):395–414

Draper NR, Smith H (1998) Applied regression analysis, 326. Wiley

Fink AH et al (2004) The 2003 European summer heatwaves and drought–synoptic diagnosis and impacts. Weather 59(8):209–216

García-Herrera R, Díaz J, Trigo RM, Luterbacher J, Fischer EM (2010) A review of the European summer heat wave of 2003. Crit Rev Environ Sci Technol 40(4):267–306

Gohel MC, Panchal MK, Jogani VV (2000) Novel mathematical method for quantitative expression of deviation from the Higuchi model. AAPS Pharm Sci Tech 1(4):43–48
Gregory JM, Mitchell J, Brady A (1997) Summer drought in northern midlatitudes in a time-dependent $CO_2$ climate experiment. J Clim 10(4):662–686
Hansen JE (1989) The greenhouse effect: Impacts on current global temperature and regional heat waves. In: The challenge of global warming. Island Press, Washington, DC, pp 35–43, 3 fig, 7 ref
Harte T, Bruce GD, Keeling J, Cassettari D (2014) Conjugate gradient minimisation approach to generating holographic traps for ultracold atoms. Opt Express 22(22):26548–26558
Haykin S (1998) Neural Networks: a comprehensive foundation. Prentice Hall PTR, Upper Saddle River, NJ, USA
Hirsch RM, Slack JR (1984) A nonparametric trend test for seasonal data with serial dependence. Water Resour Res 20(6):727–732
Huang H-Y (1970) Unified approach to quadratically convergent algorithms for function minimization. J Optim Theory Appl 5(6):405–423
Huth R, Kyselý J, Pokorná L (2000) A GCM simulation of heat waves, dry spells, and their relationships to circulation. Clim Change 46(1–2):29–60
Hutter H-P, Moshammer H, Wallner P, Leitner B, Kundi M (2007) Heatwaves in Vienna: effects on mortality. Wien Klin Wochenschr 119(7–8):223–227
Hyndman RJ (2006) Another look at forecast-accuracy metrics for intermittent demand. Foresight Int J Appl Forecast 4(4):43–46
Jeng-Rung H et al (2013) Application of multivariate empirical mode decomposition and sample entropy in EEG signals via artificial neural networks for interpreting depth of anesthesia. Entropy 15(9):3325–3339
Karl TR, Wang W-C, Schlesinger ME, Knight RW, Portman D (1990) A method of relating general circulation model simulated climate to the observed local climate. Part I: Seasonal statistics. J Clim 3(10):1053–1079
Khald AI, Aboalayon W, Almuhammadi S, Faezipour M (2015) A comparison of different machine learning algorithms using single channel EEG signal for classifying human sleep stages. IEEE
Kim T-W, Valdés JB (2003) Nonlinear model for drought forecasting based on a conjunction of wavelet transforms and neural networks. J Hydrol Eng 8(6):319–328
Kristopher JP, Patrick JC, Daniel JB (2006) Computational tools for probing interactions in multiple linear regression, multilevel modeling, and latent curve analysis. J Edu Behavioral Statistics 31(4):437–448
Legates DR, McCabe GJ (1999) Evaluating the use of "goodness-of-fit" measures in hydrologic and hydroclimatic model validation. Water Resour Res 35(1):233–241
Luo L, Zhang Y (2012) Did we see the 2011 summer heat wave coming? Geophys Res Lett 39(9)
MacKay DJ (1995) Probable networks and plausible predictions—a review of practical Bayesian methods for supervised neural networks. Netw Comput Neural Syst 6(3):469–505
Management AE (2018) Heatwaves are perhaps our most under-rated and least studied natural hazard. In: N.T.E. service (Editor)
Marquardt DW (1963) An algorithm for least-squares estimation of nonlinear parameters. J Soc Ind Appl Math 11(2):431–441
McCulloch WS, Pitts W (1943) A logical calculus of the ideas immanent in nervous activity. The Bull Mathe Biophys 5(4):115–133
McMichael AJ et al (2003) Human health and climate change in oceania: a risk assessment. canberra: commonwealth department of health and ageing
Meehl GA, Tebaldi C (2004) More intense, more frequent, and longer lasting heat waves in the 21st century. Science 305(5686):994–997
Montgomery DC, Peck EA, Vining GG (2012) Introduction to linear regression analysis 821. Wiley
Moriasi DN et al (2007) Model evaluation guidelines for systematic quantification of accuracy in watershed simulations. Trans ASABE 50(3):885–900

Murphy AH (1995) The coefficients of correlation and determination as measures of performance in forecast verification. Weather forecasting 10(4):681–688

Ozer DJ (1985) Correlation and the coefficient of determination. Psychol Bull 97(2):307

Palecki MA, Changnon SA, Kunkel KE (2001) The nature and impacts of the July 1999 heat wave in the midwestern United States: learning from the lessons of 1995. Bull Am Meteor Soc 82(7):1353–1368

Pappas SS et al (2008) Electricity demand loads modeling using AutoRegressive Moving Average (ARMA) models. Energy 33(9):1353–1360

Peterson TC et al (2013) Monitoring and understanding changes in heat waves, cold waves, floods, and droughts in the United States: state of knowledge. Bull Am Meteor Soc 94(6):821–834

Pham DT, Sagiroglu S (2001) Training multilayered perceptrons for pattern recognition: a comparative study of four training algorithms. Int J Mach Tools Manuf 41(3):419–430

Rothfusz LP (1990) The heat index equation. National Weather Service Technical Attachment (SR 90–23)

Rothfusz LP, Headquarters NSR (1990) The heat index equation (or, more than you ever wanted to know about heat index). Fort Worth, Texas: National Oceanic and Atmospheric Administration, National Weather Service, Office of Meteorology, 9023

Şahin M (2012) Modelling of air temperature using remote sensing and artificial neural network in Turkey. Adv Space Res 50(7):973–985

Salcedo-Sanz S, Deo R, Carro-Calvo L, Saavedra-Moreno B (2016) Monthly prediction of air temperature in Australia and New Zealand with machine learning algorithms. Theoret Appl Climatol 125(1–2):13–25

Shahin MA (2013) Artificial intelligence in geotechnical engineering: applications, modeling aspects, and future directions. Metaheuristics Water Geotech Transp Eng 169204

Sherwood SC, Huber M (2010) An adaptability limit to climate change due to heat stress. Proc Natl Acad Sci 107(21):9552–9555

Smola AJ, Schölkopf B (2004) A tutorial on support vector regression. Statistics Comput 14(3):199–222

Steadman RG (1979) The assessment of sultriness. Part I: a temperature-humidity index based on human physiology and clothing science. J Appl Meteorol 18(7):861–873

Taylor R (1990) Interpretation of the correlation coefficient: a basic review. J Diagnostic Med Sonography 6(1):35–39

Teixeira EI, Fischer G, van Velthuizen H, Walter C, Ewert F (2013) Global hot-spots of heat stress on agricultural crops due to climate change. Agric For Meteorol 170:206–215

Tong S, Wang XY, Barnett AG (2010) Assessment of heat-related health impacts in Brisbane, Australia: comparison of different heatwave definitions. PLoS ONE 5(8):e12155

Torok S, Nicholls N (1996) A historical annual temperature dataset. Australian Meteorol Mag 45(4)

Trigo RM, Palutikof JP (1999) Simulation of daily temperatures for climate change scenarios over Portugal: a neural network model approach. Climate Res 13(1):45–59

Tucker G (1975) Climate: is Australia's changing?

Vogl TP, Mangis J, Rigler A, Zink W, Alkon D (1988) Accelerating the convergence of the back-propagation method. Biol Cybern 59(4–5):257–263

Wallace J, Williamson I, Rajabifard A, Bennett R (2006) Spatial information opportunities for Government. J Spatial Sci 51(1):79–99

Willmott CJ (1981) On the validation of models. Phys Geogr 2(2):184–194

Willmott CJ, Matsuura K (2005) Advantages of the mean absolute error (MAE) over the root mean square error (RMSE) in assessing average model performance. Climate Res 30(1):79–82

Winkler JA, Palutikof JP, Andresen JA, Goodess CM (1997) The simulation of daily temperature time series from GCM output. Part II: sensitivity analysis of an empirical transfer function methodology. J Clim 10(10):2514–2532

Zhang GP (2003) Time series forecasting using a hybrid ARIMA and neural network model. Neurocomputing 50:159–175

# Chapter 17
# Daily Flood Forecasts with Intelligent Data Analytic Models: Multivariate Empirical Mode Decomposition-Based Modeling Methods

**Ramendra Prasad, Dhrishna Charan, Lionel Joseph, Thong Nguyen-Huy, Ravinesh C. Deo, and Sanjay Singh**

## 17.1 Background Review

Floods are destructive natural disasters that cause massive damage to human life, infrastructure, and agriculture. Floods accounted for 30% of the total number of natural disasters globally in the periods of 1900–2006, causing 19% of all deaths resulting from these disasters, and accounting for 26% of all economic damages from water-related disasters (Adikari and Yoshitani 2009). The changing climate is expected to have a profound impact on the frequency of floods in many parts of the globe. The radiative effect of the changing atmospheric composition is projected to cause an intensification of the global water cycle which consequently increases flood risks (Milly et al. 2002). Increased precipitation and reduced evapotranspiration will lead to increases in river discharge on a global scale (Hirabayashi et al. 2008), which poses a serious risk to human life and can extensively damage infrastructure.

Floods in Australia are predominately caused by heavy rainfall. A total of 253 major flood events occurred during the period 1860–2013 in the coastal catchments in Eastern Australia from Brisbane in Queensland to Eden in New South Wales (NSW) (Callaghan and Power 2014). A study of data from 491 stations across Australia revealed that 30% of the stations show trends in annual maxima flood series data with upward trends shown in the Northern part of Australia (Ishak et al. 2010). A

R. Prasad (✉) · D. Charan · L. Joseph · S. Singh
Department of Science, School of Science and Technology, The University of Fiji, Saweni, Lautoka, Fiji
e-mail: ramendrap23@gmail.com; ramendrap@unifiji.ac.fj

R. C. Deo
School of Sciences, University of Southern Queensland, Springfield Central, QLD 4300, Australia

T. Nguyen-Huy
Centre for Applied Climate Sciences, University of Southern Queensland, Toowoomba, Australia

© Springer Nature Singapore Pte Ltd. 2021

R. C. Deo et al. (eds.), *Intelligent Data Analytics for Decision-Support Systems in Hazard Mitigation*, Springer Transactions in Civil and Environmental Engineering,
https://doi.org/10.1007/978-981-15-5772-9_17

total of 73 fatalities directly related to floods occurred in Australia in the period 1997–2008, with an average of six fatalities per year (FitzGerald et al. 2010). The largest number of these fatalities was recorded in NSW and Queensland (FitzGerald et al. 2010). Brisbane, the largest city in Queensland is also tattered by flood events. The Brisbane River and its major tributaries have a long history of flooding with large flood events recorded in 1893 and 1974 (Barton et al. 2015). Recent devastating events in Queensland include the 2011 major floods that occurred through most of the Brisbane River catchment (van den Honert and McAneney 2011) and the 2019 Townsville flood in the city of Townsville and the surrounding areas (Adekunle et al. 2019), with damage bills in the billions. The 2011 Queensland flooding caused over 2 billion (AUD) of infrastructure damage alone along with multiple other indirect costs to the economy (Johnson et al. 2016).

The floods associated with a strong La Niña event were especially severe in the Lockyer region, where a combination of intense rainfall, saturated ground, and steep topography produced a flash flood event (Okada et al. 2014) in which more than 20 people lost their lives and over 120 homes were destroyed, resulting in a flood damage bill of over $176 million (Lockyer Valley Local Disaster Management Group 2014). The Lockyer Valley was also severely affected by the January 2013 flood event in Queensland (Lokuge and Setunge 2013). Events of increasingly extreme rainfall are projected to continue throughout the twenty-first century in Australasia (Reisinger et al. 2014) making studies of hydrological extremes especially at short time scales become indispensable (Oki and Kanae 2006).

Flood forecasting is an essential component of flood warning systems which continues to be one of the most important tasks in hydrology. The early flood warning systems would allow people and civil protection authorities the much-needed preparation time to reduce the flooding impacts (Penning-Rowsell et al. 2000). However, flood forecasting is often identified as one of the most challenging tasks in hydrology. A source of complexity associated with forecasting floods lies in the mathematical modeling of physical processes (Mosavi et al. 2018), and this has driven a move from using physical models to using data-driven techniques for flood modeling. Studies have suggested this shift could be a result of the gap in the short-term prediction capacity of physical models (Costabile and Macchione 2015). Evidence of the inadequacy of physical models for flood forecasting is exemplified in the case of the 2010–2011 Queensland floods. Van den Honert and McAneny (2011) in their study of the causes and implications of the recent Queensland floods found that the uncertainty in Bureau of Meteorology (BoM) rainfall forecasts which was based on statistical models and lack of rain gauges, clouded the reliability of the precipitation forecast leading to bad decision making. They further noted that the rainfall forecasts which were particularly bad during the floods can be improved by using tools such as geostatistical techniques of kriging and by using decision trees among others.

The BoM, which is the national agency for flood warning services and streamflow forecasting in Australia, made a transition toward an updated system of forecasting based on the Predictive Ocean Atmosphere Model for Australia (POAMA 2) in 2013 (Abbot and Marohasy 2014). BoM currently makes predictions of future river levels using hydrological forecasting models and other tools which include

POAMA and the Bayesian joint probability model (BJP) for seasonal streamflow forecasts and probabilistic seasonal rainfall outlooks (White et al. 2015). The use of such a numerical weather prediction model to make short-term precipitation forecasts requires a complex and meticulous stimulation of physical equations (Xingjian et al. 2015). Difficulties associated with modeling such extensive physical processes can be resolved with data-driven machine learning (ML) models. ML models have considerable ability for early flood warning systems because of its great flexibility and scalability in extracting the more significant features from complex data sets to make predictions (Chang et al. 2019).

Data-driven ML models such as artificial neural networks (ANNs), neuro-fuzzy, adaptive neuro-fuzzy inference systems (ANFIS), wavelet neural networks (WNN), multilayer perceptron (MLP), and decision trees (DT) are becoming increasingly popular in flood forecasting due to its complete dependence on historical data and its ability to function without knowledge about the underlying physical processes (Mosavi et al. 2018). Data-driven techniques detect and learn patterns in the data without attempting to stimulate the physical processes as in the case of physical models (Sadler et al. 2018). Globally, ANN has been a popular learning algorithm for hydrological modeling since the 1990s (Adnan et al. 2012). Designed to mimic the information processing and knowledge acquisition method of the human brain (Wu and Chau 2006), ANNs have come to be established as a reliable tool for constructing black box models of complex and nonlinear relationships associated with rainfall and flood modeling (Sulaiman and Wahab 2018). But despite its popularity, ANN has some drawbacks in flood modeling particularly related to data handling and interpretation of the modeled systems (Mosavi et al. 2018) including poor predictions beyond the range of trained data (Wu and Chau 2006) and unsuitability for modeling temporal sequences (Coulibaly et al. 2001) resulting in suboptimal solutions, as in the case of (Schoof and Pryor 2001), where daily precipitation prediction by ANN was vastly inferior to monthly prediction.

Decision tree (DT)-based modeling approaches, in particular, can help resolve the problems faced by ANNs. DT models have been reported to be efficient and robust in their capability to predict floods (Mosavi et al. 2018). Using a tree of decisions from branches to the target values of leaves, DT considers all the characteristics of the variables by following a precise set of rules (Tehrany et al. 2013) and is especially well established in the case of sequential decision problems providing an intuitively appealing means of developing flood management strategies with the ability to identify actions that need to be urgently taken and those that can be delayed (Sayers et al. 2012). The commonly used DT modeling approach is the M5 model tree, which is a machine learning technique. The M5 model tree combines the features of classification and regressions and splits parameter space into subareas subsequently building linear regression models in respective subspaces (Adarsh et al. 2018). Solomatine and Xue (2004) built ANN and M5 model tree models for predicting the flood discharge of the Xixian station located in the upper reach of the Huai River in China using data from 17 rainfall stations and three evaporation stations. They found that the M5 model tree was considerably easier to use and had greater transparency than the ANN model, with both models demonstrating comparable levels of accuracy.

Singh et al. (2010) evaluated the potential of using backpropagation neural network (BNN) and M5 model tree to estimate mean annual flood for 93 Indian catchments and found correlation coefficients of 0.975 and 0.994 for the BNN and MT models, respectively. M5 model tree was demonstrated as having certain advantages over ANN which included better insights into the generated models, greater acceptability by decision-makers, and superior efficiency in training.

The use of M5 model tree models has also been popular for streamflow forecasting and has proved to be easier to use and more accurate compared to conventional models (Štravs and Brilly 2007; Londhe and Charhate 2010; Sattari et al. 2013). Other hydrological applications of M5 model tree include rainfall–runoff modeling (Solomatine and Dulal 2003) and predicting flow discharge in compound channels (Zahiri and Azamathulla 2014). Results from some studies involving stage–discharge modeling found that the performance of M5 model tree to be superior to ANN modeling (Bhattacharya and Solomatine 2003; Ajmera and Goyal 2012; Onyari and Ilunga 2013).

DT-based modeling approach gaining prominence is random forest (RF). RF algorithm comprises an ensemble of simple tree predictions for classification and regression (Breiman 2001). It is based on a combination of tree predictors with each tree dependent on the values of a random vector sampled independently (Breiman 2001). It has been gaining traction in the area of flood modeling for which a multitude of decision trees is constructed to explain the relationship between flood occurrences and related factors (Lee et al. 2017). Short-term flash flood forecasting was carried out by Muñoz et al. (2018) using a parsimonious model based on RF for an Andean Mountain Catchment, with validation efficiencies of 0.761 for the 4-h predictions to 0.384 for the 24-h predictions. Zhao et al. (2018) used a RF model to characterize the relationship for flooding occurrence with twelve explanatory factors in mountainous regions in China and found it to be more effective at identifying flood-prone areas compared to ANN and support vector machine (SVM) methods. Wang et al. (2015) used an assessment model based on RF for flood hazard risk assessment for the Dongjiang River Basin in China using 11 risk indices and found the RF method to be a highly successful approach for flood hazard risk assessment. In another study, Li et al. (2019) compared four different machine learning algorithms for flood risk assessment of global watersheds based on 13 conditioning factors and found RF to be best suited out of the three others which included logistic regression, Naïve Bayes, and AdaBoost. RF has been applied for flood analysis in urban areas including modeling coastal flood severity for Norfolk (Sadler et al. 2018), evaluating the importance of contributing discharges in Nechako River for the city of Prince George (Albers et al. 2016) and creating flood susceptibility maps (Feng et al. 2015; Lee et al. 2017). RF for other hydrological applications includes rainfall forecast (Herman and Schumacher 2018; Monira et al. 2010; Yu et al. 2017), forecasting streamflow (Liang et al. 2018; Papacharalampous and Tyralis 2018; Petty and Dhingra 2018; Shortridge et al. 2016; Lima et al. 2015), and for forecasting water levels (Yang et al. 2017; Nguyen et al. 2015). The advantages of RF include ease of use, ability to handle large data set (Zhao et al. 2018), robust and ability to deal with complex data structures effectively (Muñoz et al. 2018). Despite having the abovementioned advantages, the stand-alone models including ANN, M5 Tree, RF, etc., may not be able to capture all the relevant features in the time series during the training process. Hence, the performance of machine learning models can be improved by hybridization (Mosavi et al. 2018). Hybridization of decision trees can help overcome some of its shortcomings which include sudden and inappropriate changes in class due to small changes in attribute values and missing or imprecise information preventing the classification of a class (Quinlan 1987). Hybridization allows for higher accuracy and robustness to overcome these drawbacks.

To develop a hybrid model, a data preprocessing phase by means of an apt multi-resolution analysis (MRA) tool able to extract and isolate the embedded information within the non-stationary flood index time series is necessary. Recently, the empirical mode decomposition (EMD) (Huang et al. 1998)-based MRA tools gained popularity among scholars and practitioners. The key advantage of EMD over Fourier and wavelet transforms is that EMD is able to demarcate a given higher-frequency input series into resolved narrowband oscillatory modes called the intrinsic mode functions (IMFs) that essentially emerge naturally from the embedded oscillatory frequencies within the signal (Huang et al. 1998; Ur Rehman et al. 2013) making it self-adaptable. In addition, there is no information loss and the decomposed features represent the physical structure of the data (Wu et al. 2011). However, the success of EMD highlighted two main issues of mode mixing and aliasing that were gradually addressed in the successors of EMD variants including ensemble EMD (EEMD) (Wu and Huang 2009), complete ensemble empirical mode decomposition with adaptive noise (CEEMDAN) (Torres et al. 2011), and improved complete ensemble empirical mode decomposition with adaptive noise (ICEEMDAN) (Colominas et al. 2014)). Nonetheless, the EMD and its variants can only be applied to a univariate data series. When applied to multivariate data, complications such as non-uniformity and scale alignment arise that invalidate its applications (Ur Rehman et al. 2013). Hence, in order to simultaneously demarcate multivariate data, the utilization of multivariate EMD (MEMD) is necessary. The MEMD utilizes a significantly different underlying conceptualization that accurately performs MRA of multichannel and nonlinear dynamical processes (Rehman and Mandic 2009). The MEMD essentially surmounts the mode alignment issue in the joint analysis of multiple oscillatory components within a higher-dimensional signal (Looney and Mandic 2009) and has been successfully applied in forecasting of evapotranspiration (Adarsh et al. 2017), soil water (Hu and Si 2013), crude oil price (He et al. 2016), iceberg drift forecast (Andersson et al. 2017), solar radiation (Prasad et al. 2019), and standardized precipitation index (Ali et al. 2019).

This chapter demonstrates the potential utility of the MEMD-based M5 tree models. The method incorporates four multichannel predictor inputs (i.e., effective drought index (EDI), daily precipitation (PCN), available water resources index (AWRI), and precipitation return to normal (PRN) to emulate the flood index (FI)) utilizing the respective IMFs and residual components by MEMD process. After feature selection process, daily FI is forecasted in the flood-prone Lockyer Valley region of Queensland. The performances of hybrid models are appraised with single

M5 tree and RF models to demonstrate the importance of the method for disaster risk mitigation.

## 17.2 Theoretical Overview of Predictive Models

### *17.2.1 Multivariate Empirical Mode Decomposition (MEMD)*

Despite the advantages of EMD, an early version of MEMD, including its intelligent data analytic nature, a compact decomposition, and its inherent ability to process non-stationary data, EMD only caters for signals with a sufficient number of local extrema (Ur Rehman et al. 2013). The MEMD, on the other hand, is designed to deal with multi-dimensional signals in addition to overcoming the problem of mode mixing via utilization of Gaussian white noise (Rehman and Mandic 2009; Ur Rehman and Mandic 2011).

In univariate EMD, the oscillation concept is the underpinning technique in demarcating the signals, while in MEMD, the rotation concept is utilized to accommodate the correlations and dependence among the signals from different channels that allow for simultaneous resolving into respective IMFs and the residual component (Ur Rehman et al. 2013). Essentially, multivariate data are viewed as fast rotations superimposed on a slow rotation, where intrinsic models are jointly determined and scale aligned (He et al. 2016).

In mathematical terms whereby $e^{\theta_k}(t)$ is referred to the multivariate envelop curves and $s$ is the length of the vectors, the mean denoted by $m(t)$ can be computed as follows:

$$M(t) = \frac{1}{s}\sum_{k=1}^{s} e^{\theta_k}(t) \tag{17.1}$$

While the remaining sub-frequency can be found as follows:

$$R(t) = X(t) - M(t) \tag{17.2}$$

This remaining sub-frequency is a set of multivariate IMFs that satisfy the stopping criterion; or else, the process would continue until the remainder $R(t)$ is acquired. MEMD-based signal processing and forecasting have successfully been applied in analyzing signals (Huang et al. 2013; Mandic et al. 2013), soil water prediction (Hu and Si 2013; She et al. 2015), monthly solar radiation (Prasad et al. 2019), and standardized precipitation index (Ali et al. 2019). The concept of rotation adopted in the multivariate EMD during the characterization of the intrinsic mode allows for synchronous and coherent treatment of multivariate data in addition to giving MEMD the capability to explore the underlying causality of respective events.

### 17.2.2 M5 Tree Model

M5 tree model, developed by (Quinlan 1992), is inspired by the conventional decision tree framework. The learning algorithm is a class of hierarchical models with linear regression functions at the leaves, constructed primarily to deal with continuous-class learning problems. The construction of the M5 model tree requires two phases: the growing phase and pruning phase. During the growing phase, the algorithm commences with one node and recursively tries to split the input/output data into subsets/subspaces and then in each subspace, it builds a local specialized linear regression model.

Splitting is done by applying 'divide-and-conquer' lemma at the nodes, of the training data set whereby data points are either associated with a leaf or the standard deviation of the class values in training set is treated as a measure of the error by the regulator, which calculates the expected reduction in error as a result of testing each attribute at that node.

The attribute that maximizes the expected error reduction/standard deviation reduction is selected for splitting at the node. The splitting process ceases when the class values of the instances that reach a node vary just slightly or that only a few instances remained. In this study, the process ceased when the SD was less than 5% of the SD of the original instance set.

After the first phase, a linear model is in place for each interior node and a large overfitting model tree is generated. To reduce overfitting, a second phase called pruning is carried out whereby the tree is pruned back from leaves as long as the expected estimated error decreases. Finally, the smoothing process is employed to eliminate sharp discontinuities between adjacent linear models at the leaves of the pruned trees (Wang and Witten 1997; Witten et al. 2011).

### 17.2.3 Random Forest

The random forest (RF), introduced by Breiman (2001) as an extension of bagging (Breiman 1996), is a regression tree-based ensemble modeling technique of the form classification and regression trees (CART). The algorithm employs the bootstrap aggregation (bagging) approach during training to overcome overfitting.

During the training process, '*n*' number of bootstrap replicas are taken from the training data set, using random sampling with replacement. Then, a single tree is constructed on every separate '*N-replicas*' of training data set, and subsequently, the out-of-bag (OOB) samples errors of respective trees using the data that were not used during training are computed. Then, all single regression trees are put together, forming an ensemble, and the forecasted output is averaged over this ensemble of trees. RF requires manual tuning of three parameters including (1) *m*, the number of randomly assigned predictor variables at each node, (2) $J$, the number of trees grown in the forest, and (3) tree size, measured by the maximum number of terminal nodes/leaf. For this study, $J = 200$ and tree size $= 5$ together with $m$ as one-third of the total number of variables were used.

## 17.3 Materials and Method

### 17.3.1 Study Region and Data Description

The study region, Lockyer Valley in Queensland, Australia, is extremely important as this region is very much drought-prone and experiences severe flooding events. The Lockyer Creek is a major tributary of the Brisbane River, which is the longest river (309 km) in southeast Queensland, with a catchment area of around 13,570 $km^2$ (van den Honert and McAneney 2011). Around 26% of the creek's length has experienced some forms of geomorphic adjustment since the first half of the nineteenth century due to flood events (Fryirs et al. 2015). The Lockyer Valley has a population of 41,011 and agriculture serves as one of its main employing industries (Australian Bureau of Statistics 2018). The valley is one of the most fertile farming areas in the world and is one of Queensland's most important regions of diversified agriculture (Apan et al. 2002). Past flood events have caused severe damage to the agriculture and infrastructure in Lockyer region, and hence, flood forecasting can be controlled and minimized the negative impacts of future floods.

Several approaches with different levels of uncertainty and forecasting lead times are used for river flood forecasting (Bartholmes and Todini 2005). The severity of flooding can be expressed by FI, which is computed from specific characteristics of past flood hydrographs (Bhattacharya et al. 2016). Defined as the expected value of maximum annual flood peaks, FI represents the merging of statistical and physical hydrology and is crucial for flood forecasting (Bocchiola et al. 2003). Deo et al. (2015) investigated the practicality of using FI based on PR N, EDI, and AWRI (mm) for flood detection in Brisbane and Lockyer Valley and found it to be a robust utilitarian for flood risk monitoring. Hence, in this study, daily flood forecasting for the Lockyer Valley region is undertaken with PRN, EDI, AWRI (mm), and significant lags of FI as inputs.

The data from January 01, 1950, till December 31, 2012, were acquired from the Australian BoM. In the initial stage, data quality was checked and all missing data points were imputed. Then, the statistical evaluation of the primary and predictor variables was performed and presented in Table 17.1.

The preliminary analysis of the primary variable, FI, showed that the skewness and the kurtosis values were 0.977 and 4.766, respectively. The positive skewness indicates that the data are moderately skewed to the right (since 0.977 falls between 0.5 and 1), i.e., data above the average are larger than data below the average, and

**Table 17.1** Climate statistics: flood Index (FI) and its predictor variables denoted as the effective drought index (EDI), daily precipitation (PCN), available water resources index (AWRI), and precipitation return to normal (PRN) for the present study site

| Primary variables | Acronym | Annual climatic statistics (1950–2012) | | | | |
|---|---|---|---|---|---|---|
| | | Min. | Mean | Max. | Skewness | Kurtosis |
| Objective variable: *Flood index* | FI | – 3.100 | – 1.652 | 3.620 | 0.977 | 4.766 |
| Effective drought index | EDI | – 2.370 | 0.182 | 7.160 | 0.823 | 4.494 |
| Daily precipitation (mm) | PCN | 0.000 | 2.183 | 199.400 | 7.989 | 108.421 |
| Available water resources index (mm) | AWRI | 26.100 | 123.070 | 476.530 | 0.977 | 4.766 |
| Precipitation return to normal | PRN | – 335.410 | – 4.970 | 132.050 | – 0.579 | 4.557 |

the data set has a heavier tail relative to a normal distribution since 4.766 > 3 (the kurtosis of normal distribution is 3).

Also, a maximum FI of 3.620 was recorded while the least FI was −3.100. The mean FI was −1.652 (i.e., a dimensionless value). The distributions of EDI and AWRI are similar to FI. The skewness of PRN is −0.579 indicating the left-handed tail will typically be larger than the right-handed tail. The exceptional values of skewness and kurtosis of PCN reflect the nature of the data containing many zero values, and hence, the outliers strongly affect the distribution.

### *17.3.2 Model Development*

The inputs and corresponding target data consisting of 22,966 data points are separated into training (70%), validation (15%), and testing (15%). An independent approach was adopted to prevent inclusion of future data (Deo et al. 2016a; Kim and Valdes 2003). All data were normalized prior to the modeling stages that can be summarized as follows:

1. The input time series were resolved into simpler and lower-frequency components via MEMD. A total of 68 input sub-series resulted with 16 IMFs + 1 residual component for each input (i.e., total sub-series = 17 × 4 = 68).
2. *Input selection*: Significant input lags of respective IMFs and the residual components were determined using cross-correlation function. Finally, the significant lags together with undecomposed FI at $t - 1$ lag were collated as the input matrix for modeling.
3. *Forecasting*: Using this input matrix, the future FI values were forecasted using the M5 tree and RF models. The optimal models were averred based on $r$, RMSE, and MAE during validation phases with the least mean square error (MSE) giving further confirmation.

Figure 17.1 displays the schematic of the model development for clarity.

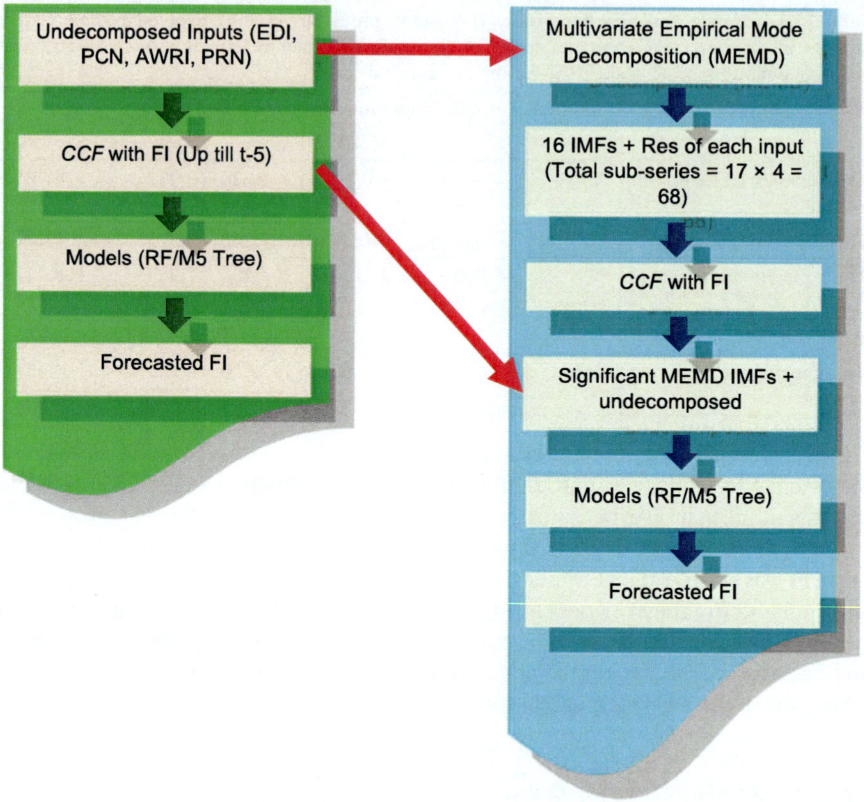


**Fig. 17.1** Schematic of different models

### *17.3.3 Accuracy Evaluation*

An independent test data set was used to assess the forecasting accuracy of the MEMD-M5 tree, MEMD-RF, and stand-alone M5 tree and RF models. The performances of these models were measured via statistical metrics. The mathematical representation of these metrics where observed magnitudes are represented as $\mathrm{FI}_i^{\mathrm{OBS}}$ and $\mathrm{FI}_i^{\mathrm{FOR}}$ represents forecasted *i*th magnitudes of flood indices, while the observed and forecasted mean of FI values are represented by $\mathrm{FI}_i^{\mathrm{OBS}}$ and $\mathrm{FI}_i^{\mathrm{FOR}}$, respectively, are given below (ASCE 1993, 2000; Yen 1995) and (Dawson et al. 2007; Deo et al. 2016b; Legates and McCabe 1999; Willmott 1981, 1982, 1984):

1. Coefficient of correlation ($r$):

$$r = \left( \frac{\sum_{i=1}^{N} \left( \mathrm{FI}_i^{\mathrm{OBS}} - \overline{\mathrm{FI}_i^{\mathrm{OBS}}} \right) \left( \mathrm{FI}_i^{\mathrm{FOR}} - \overline{\mathrm{FI}_i^{\mathrm{FOR}}} \right)}{\sqrt{\sum_{i=1}^{N} \left( \mathrm{FI}_i^{\mathrm{OBS}} - \overline{\mathrm{FI}_i^{\mathrm{OBS}}} \right)^2} \sqrt{\sum_{i=1}^{N} \left( \mathrm{FI}_i^{\mathrm{FOR}} - \overline{\mathrm{FI}_i^{\mathrm{FOR}}} \right)^2}} \right) \tag{17.3}$$

2. Willmott's index (WI) defined as follows:

$$\mathrm{WI} = 1 - \left[ \frac{\sum_{i=1}^{N} \left(\mathrm{FI}_i^{\mathrm{FOR}} - \mathrm{FI}_i^{\mathrm{OBS}}\right)^2}{\sum_{i=1}^{N} \left( \left|\mathrm{FI}_i^{\mathrm{FOR}} - \overline{\mathrm{FI}_i^{\mathrm{OBS}}}\right| + \left|\mathrm{FI}_i^{\mathrm{OBS}} - \overline{\mathrm{FI}_i^{\mathrm{OBS}}}\right| \right)^2} \right], \; 0 \le \mathrm{WI} \le 1 \tag{17.4}$$

3. Nash–Sutcliffe efficiency ($E_{\mathrm{NS}}$) value:

$$E_{\mathrm{NS}} = 1 - \left[ \frac{\sum_{i=1}^{N} \left(\mathrm{FI}_i^{\mathrm{OBS}} - \mathrm{FI}_i^{\mathrm{FOR}}\right)^2}{\sum_{i=1}^{N} \left(\overline{\mathrm{FI}_i^{\mathrm{OBS}}} - \overline{\mathrm{FI}_i^{\mathrm{FOR}}}\right)^2} \right], 0 \le E_{\mathrm{NS}} \le 1 \tag{17.5}$$

4. Root mean square error (RMSE) is mathematically derived as follows:

$$\mathrm{RMSE} = \sqrt{\frac{1}{N} \sum_{i=1}^{N} \left(\mathrm{FI}_i^{\mathrm{FOR}} - \mathrm{FI}_i^{\mathrm{OBS}}\right)^2} \tag{17.6}$$

5. Mean absolute error (MAE):

$$\mathrm{MAE} = \frac{1}{N} \sum_{i=1}^{N} \left| \left(\mathrm{FI}_i^{\mathrm{FOR}} - \mathrm{FI}_i^{\mathrm{OBS}}\right) \right| \tag{17.7}$$

6. Legates and McCabe's ($L$):

$$L = 1 - \left[ \frac{\sum_{i=1}^{N} \left|\mathrm{FI}_i^{\mathrm{FOR}} - \mathrm{FI}_i^{\mathrm{OBS}}\right|}{\sum_{i=1}^{N} \left|\mathrm{FI}_i^{\mathrm{OBS}} - \overline{\mathrm{FI}_i^{\mathrm{OBS}}}\right|} \right], 0 \le L \le 1 \tag{17.8}$$

7. Relative root mean squared error (RRMSE; %) is stated as

$$\mathrm{RRMSE} = \frac{1}{N} \sum_{i=1}^{N} \left| \frac{\left(\mathrm{FI}_i^{\mathrm{FOR}} - \mathrm{FI}_i^{\mathrm{OBS}}\right)}{\mathrm{FI}_i^{\mathrm{OBS}}} \right| \times 100 \tag{17.9}$$

The diagnostic plots including the box plot analysis and forecasting error histogram further reveal that the MEMD-M5 tree model has a greater resemblance to that of the observed data supporting the outcomes of statistical evaluations.

## 17.4 Results and Discussion

This section presents the results attained in evaluating the proposed MEMD-M5 tree against the comparative MEMD-RF, stand-alone M5 tree, and stand-alone RF models, using statistical metrics, diagnostic plots, and error distributions between the simulated and observed daily FI values in Lockyer Valley, southeast Queensland, Australia.

Table 17.2 presents the outcomes of the model training phase on the basis of $r$, RMSE, and MAE. During model creation, 316,076 data points were used for training and initial fitting of the parameters in the models is used for this study.

The proposed MEMD-M5 tree (and the comparative models) was validated and then used to predict an unseen data during the testing phase, with a total of 3445 data points. The outcomes of these tests are furnished in Table 17.3 in which case seven different selection criteria were used in appraising model performances.

During the testing phase, the proposed MEMD-M5 tree outperformed the comparative models by attaining maximum values of $r = 0.990$, WI = 0.992, $E_{NS} = 0.979$, and $L = 0.920$. In terms of the error metrics, RMSE and MAE, the proposed model again performed well, yet standalone RF performed equally well with RMSE = 0.11 and MAE = 0.05. Interestingly, stand-alone M5 tree also registered the same MAE value. The selection criteria of each of the abovementioned metrics evidently

**Table 17.2** Standalone RF and M5 tree versus decomposed intelligent data analytic model (MEMD-RF and MEMD-M5 tree) in training phase. Correlation coefficient ($r$), root mean square error (RMSE), and mean absolute error (MAE)

| Predictive models | Training performance | | |
|---|---|---|---|
| | $r$ | RMSE | MAE |
| Stand-alone RF | 0.992 | 0.09 | 0.04 |
| Stand-palone M5 tree | 0.989 | 0.11 | 0.04 |
| MEMD—RF | 0.992 | 0.09 | 0.04 |
| MEMD—M5 tree | 0.988 | 0.11 | 0.05 |

**Table 17.3** Proposed MEMD-M5 tree with respect to MEMD-RF and standalone (RF and M5 tree) models during the testing phase. Correlation coefficient ($r$), Willmott's index (WI), Nash–Sutcliffe efficiency ($E_{NS}$), Legates–McCabe's index ($L$), root mean square error (RMSE), mean absolute error (MAE), and relative root mean square (RRMSE) error in %

| Predictive models | Testing performance | | | | | | |
|---|---|---|---|---|---|---|---|
| | $r$ | WI | $E_{NS}$ | $L$ | RMSE | MAE | RRMSE |
| MEMD—M5 tree | **0.990** | **0.992** | **0.979** | **0.920** | **0.11** | **0.05** | **−6.38** |
| MEMD—RF | 0.987 | 0.989 | 0.972 | 0.884 | 0.13 | 0.07 | −7.39 |
| Stand-alone M5 tree | 0.989 | 0.991 | 0.977 | 0.917 | 0.12 | **0.05** | −6.71 |
| Stand-alone RF | 0.989 | **0.992** | 0.978 | 0.916 | **0.11** | **0.05** | −6.52 |

indicate that MEMD-M5 tree model is better at forecasting daily FI over the comparative models, including MEMD-RF, stand-alone M5 tree, and stand-alone RF at the selected study site. In addition, the relative error measurement tool, RRMSE, further affirms the better performance of the proposed MEMD-M5 tree model in forecasting daily FI at Lockyer Valley (Table 17.3).

To further assess the suitability of MEMD-M5 tree in daily FI forecasting, diagnostic plots were used to overcome the shortcomings of objective metrics. Figure 17.2 shows the scatter plots of the observed and forecasted FI during the test phase for all four models at the selected site. The linear fit equation in the scatter plots and the coefficient of determination ($r^2$) are in the range = (0, +1); with the ideal value of +1. Figure 17.2 shows that MEMD-M5 tree model achieved the greatest $r^2$ value of 0.9794 (i.e., closest to +1). On the contrary, MEMD-RF attained the lowest $r^2$ value of 0.9747 compared to the four models tested; thus, revealing that the proposed MEMD-M% tree model could emulate 97.94% of daily FI values outperforming the other models under consideration.

In addition to previous tests for the preciseness of models, the model evaluations were carried out via the box plots as well, which are illustrated in Fig. 17.3. Being nonparametric, box plots do not make any assumptions about the underlying statistical distributions (Prasad et al. 2017). This helps in better understanding the degree of spread of the observed and forecasted values with respect to quartiles, while the whiskers indicate the variability of data points outside of the lower (i.e.,

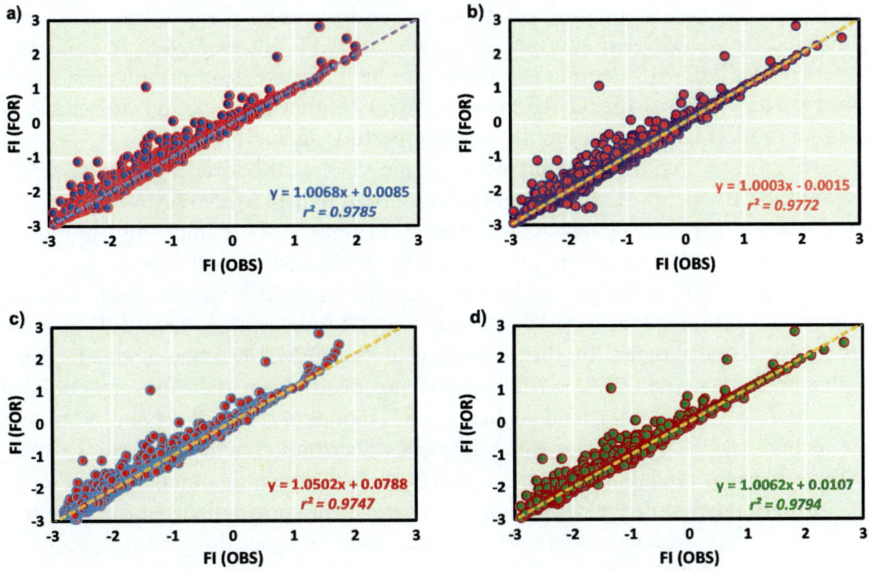


**Fig. 17.2** Scatter plots of observed FI (OBS) and forecasted FI (FOR) flood index (FI) for **a** stand-alone RF, **b** M5 tree model, and the decomposed **c** MEMD-RF and **d** MEMD-M5 tree models (*Note* the dashed lines in all four scatter plots are the least-squares fit lines)

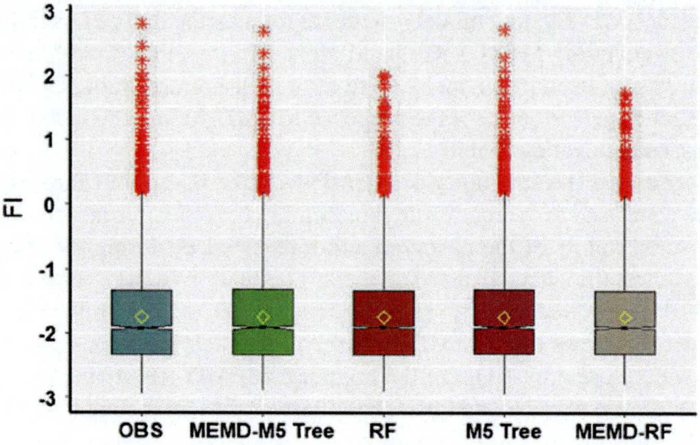


**Fig. 17.3** Box plots of observed, flood index (FI) compared with forecasted standalone RF and M5 tree models, and the decomposed MEMD-RF and MEMD-M5 tree models (in order from best-performing to worst-performing models with respect to the observed data)

25th percentile) and upper quartiles (i.e., 75th percentile), and the centric points show the median values. For the selected site, Lockyer Valley, the spreads of all four models closely resemble the spread of the observed FI values.

The difference shown is not significant as all the models performed very closely, but some differences could be observed in the forecasted distribution of outliers registered by MEMD-RF and that of the observed FI values. However, if closely observed, MEMD-M5 tree model is found to have the greatest resemblance to the observed data, while MEMD-RF shows a slightly higher degree of difference out of the four models tested revealing its superior performance.

Furthermore, Fig. 17.4 elaborates an ample view of the forecasting accuracy in terms of the frequency distribution of the proposed hybrid MEMD-M5 tree model's forecasting error |FE| against other models generated for comparison purposes. Prasad et al. (2018) stated that an ideal value of forecast error |FE| must be equivalent to zero. Thus, a better model has frequencies of |FE| values closer to zero. The results presented in Fig. 17.4 reveal that all four models have |FE| in close proximity to one another. But the |FE| values of the MEMD-M5 tree model display a higher percentage (i.e., 89.01%) of the frequency distribution in the first error bracket (i.e., the first longest bar). The standalone RF model performed second-best, with the second-highest percentage of the frequency distribution in the first error bracket, while it was observed that MEMD-RF was the worst-performing model shown under the histogram diagnostic plot as well. Therefore, error spread distribution also illustrates better performance of MEMD-M5 tree over the other three models tested in daily FI forecasting.

Further statistical evaluation of observed and forecasted daily FI values during the testing phase is displayed in Table 17.4. It can be visualized that most of the predicted values of minimum, lower quartile, median, mean, upper quartile, and

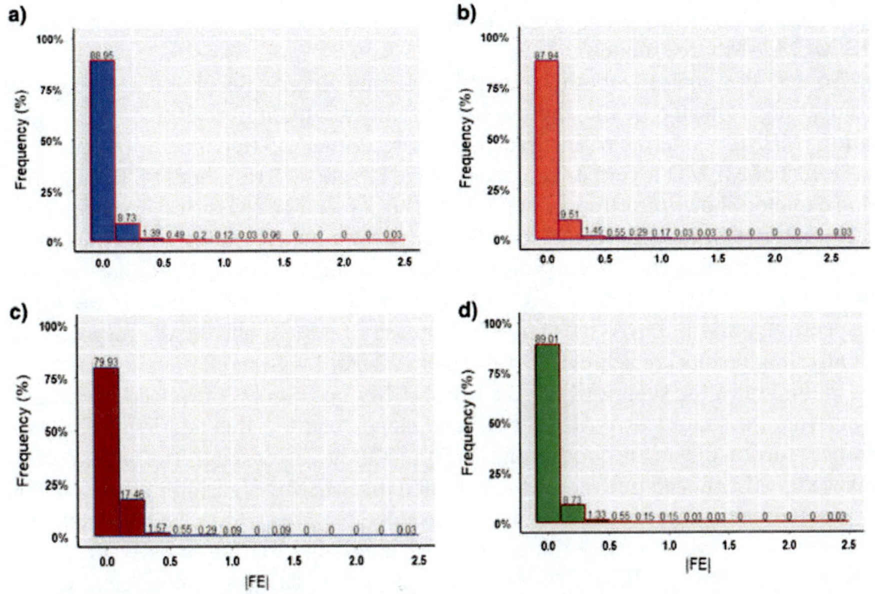


**Fig. 17.4** Histogram illustrating the frequency (in percentage) of absolute forecasting errors |FE| of the **a** stand-alone RF, **b** M5 tree models, and decomposed **c** MEMD-RF and **d** MEMD-M5 tree models

**Table 17.4** Statistical comparison of the observed (FI) against the forecasted (FI), using the stand-alone RF, M5 tree models, and the decomposed MEMD-RF and MEMD-M5 tree models (in order from best-performing to worst-performing models with respect to the observed data)

| Statistical property | OBS | MEMD—M5 Tree | Stand-alone RF | Stand-alone M5 Tree | MEMD—RF |
|---|---|---|---|---|---|
| Minimum | −2.950 | −2.906 | −2.922 | **−2.954** | −2.805 |
| Lower quartile | −2.340 | **−2.331** | **−2.331** | −2.323 | −2.306 |
| Median | −1.920 | **−1.912** | −1.905 | −1.901 | −1.903 |
| Mean | −1.749 | **−1.749** | −1.746 | −1.747 | −1.741 |
| Upper quartile | −1.330 | −1.326 | −1.330 | −1.322 | **−1.330** |
| Maximum | 2.800 | 2.684 | 2.008 | **2.710** | 1.746 |

maximum daily FI are closest to the observed data, achieved via the MEMD-M5 tree model (which were lower quartile, median, and mean). For other models, including stand-alone RF and stand-alone M5 tree, only (lower quartile and upper quartile) and (minimum and maximum) of FI values were similar to the observed values, respectively. The result tabulated in Table 17.4 further reinforces the outcomes of the evaluation metrics presented in Table 17.3 in supporting better performance of the MEMD-M5 tree model.

So far, the analysis (both tabulated data and diagnostic plots) has provided compelling evidence of the superiority of multivariate empirical mode decomposition (MEMD) in terms of accuracy of the prescribed M5 tree model. In Fig. 17.5, the effect of MEMD is further analyzed for the RF and M5 tree models. The illustration depicted shows the effect of MEMD on the percentage change of error measures including $r$, WI, $E_{NS}$, $L$, RMSE, and MAE for both the RF and M5 tree models. It clearly depicts that the hybridization of M5 tree model with MEMD has resulted in 0.101%, 0.101%, 0.205%, and 0.327% increase in $r$, WI, $E_{NS}$, and $L$ measures, respectively, and 8.333% and 3.999% decrease in RMSE and MAE, respectively. Contrasting results are achieved for the hybrid MEMD-RF model.

In this study, the superiority of the MEMD-M5 tree model is revealed in comparison with the comparative MEMD-RF and standalone M5 tree and RF models for daily FI forecasting in the flood-prone Lockyer Valley, Queensland. The better performance of MEMD-M5 tree model has revealed that MEMD algorithm was beneficial in resolving and unveiling the relevant embedded features within the predictor time series to assist the M5 model three in better emulating the future daily FI values. The incorporation of MEMD algorithm for concurrent data preprocessing of climatological predictors was the major accomplishment of the current research. The key benefit of MEMD is its ability to identify concurrently the signal's main frequency

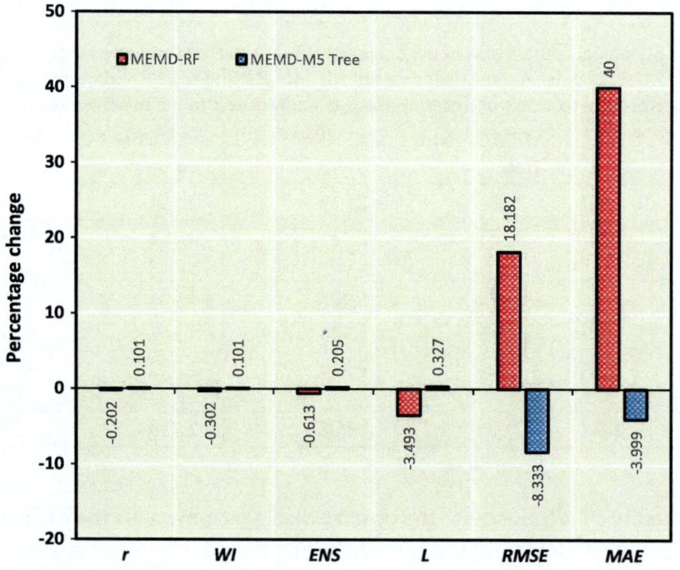


**Fig. 17.5** Percentage change showing the effect of multivariate empirical mode decomposition (MEMD) on RF and M5 tree models (*Note* all error measures used; $r$, WI, $E_{NS}$, $L$, RMSE, and MAE are unitless)

to capture the respective features unlike the empirical mode decomposition (EMD), ensemble empirical mode decomposition (EEMD), complete ensemble empirical mode decomposition (CEEMDAN), or improved complete ensemble empirical mode decomposition (ICEEMDAN) that can only demarcate one signal at a time. On top of that MEMD is data-dependent or self-adaptive in nature, which does not require any a priori knowledge and selection of parameters like in wavelet decomposition where mother wavelets need to be carefully selected. As such, the predicting ability of the stand-alone M5 tree model improved with the incorporation of the MEMD algorithm.

Despite the MEMD-M5 tree being an effective and versatile daily FI forecasting tool, there are few limitations to this modeling approach, which could be explored in later studies. In particular, several factors such as land use, hydrology, meteorological events, and topology of the land affect the risks of floods. In the newly proposed modeling framework, the forecasting was only based on PRN, EDI, AWRI (mm), and antecedent lags of FI only, and a more robust model could include the land use, hydrology, meteorological events, and topology.

## 17.5 Concluding Remarks

Intelligent data analytics used to design machine learning forecasts is an exhaustive and technologically advanced approach resulting in accurate forecasts. Due to advancements in ML, hydrologists adopt this for precise and effective forecasting results. This study presented the prediction of daily FI, which is of quite significant interest for the welfare of and property protection of people. The study attempted to develop an effective data-driven machine learning model for predicting daily FI using meteorological data sets from Australian Bureau of Meteorology. The proposed MEMD-M5 tree was evaluated against MEMD-RF, stand-alone M5 tree, and RF models. Based on our results, hybrid MEMD-RF is seen to be the most effective and accurate model for forecasting daily FI in comparison with MEMD-RF, stand-alone M5 tree, and RF models. An extensive statistical evaluation together with diagnostic plots and error distributions between the simulated and observed daily FI values were utilized. The hybrid MEMD-RF tree model outperformed the other tested models in relation to prediction metrics inclusive of $r$, WI, $E_{NS}$, $L$, RMSE, MAE, and RRMSE, forecasting accuracy in terms of the frequency distribution and forecasting error. Interestingly, the hybrid MEMD-RF tree model registered the highest coefficient of determination ($r^2$) value of 0.9794 while the lowest $r^2$ value was obtained for MEMD-RF model. Advanced and robust flood prediction models are indeed imperative for civil protection in emergencies and would be effective early warning and risk reduction systems. These models would also allow decision-makers in formulating effective policies for disaster risk reduction and reduced casualties and fatalities in such extreme weather events.

**Acknowledgements** The authors are grateful to the Australian Bureau of Meteorology for providing the relevant meteorological data for the study region.

## References

Abbot J, Marohasy J (2014) Input selection and optimisation for monthly rainfall forecasting in Queensland, Australia, using artificial neural networks. Atmos Res 138:166–178. https://doi.org/10.1016/j.atmosres.2013.11.002

Adarsh S, John AP, Anagha RN, Abraham A, Afiya MP, Arathi KK, Azeem A (2018) Developing stage-discharge relationships using multivariate empirical mode decomposition-based hybrid modeling. Appl Water Sci 8:230. https://doi.org/10.1007/s13201-018-0874-8

Adarsh S, Sanah S, Murshida KK, Nooramol P (2017) Scale dependent prediction of reference evapotranspiration based on Multi-Variate Empirical mode decomposition. Ain Shams Eng J. https://doi.org/10.1016/j.asej.2016.10.014

Adekunle AI, Adegboye OA, Rahman KM (2019) Flooding in townsville, North Queensland, Australia, in February 2019 and its effects on mosquito-borne diseases. Int J Environ Res Public Health 16:1393. https://doi.org/10.3390/ijerph16081393

Adikari Y, Yoshitani J (2009) Global trends in water-related disasters: an insight for policymakers. United Nations Educational, Scientific and Cultural Organization, Paris

Adnan R, Ruslan FA, Samad AM, Zain ZM (2012) Flood water level modelling and prediction using artificial neural network: case study of Sungai Batu Pahat in Johor. Paper presented at the IEEE Control and System Graduate Research Colloquium, Shah Alam, Malaysia, 17 July 2012

Ajmera TK, Goyal MK (2012) Development of stage–discharge rating curve using model tree and neural networks: an application to Peachtree Creek in Atlanta. Expert Syst Appl 39:5702–5710. https://doi.org/10.1016/j.eswa.2011.11.101

Albers SJ, Déry SJ, Petticrew EL (2016) Flooding in the Nechako river basin of Canada: a random forest modeling approach to flood analysis in a regulated reservoir system. Can Water Resour J 41:250–260

Ali M, Deo RC, Maraseni T, Downs NJ (2019) Improving SPI-derived drought forecasts incorporating synoptic-scale climate indices in multi-phase multivariate empirical mode decomposition model hybridized with simulated annealing and kernel ridge regression algorithms. J Hydrol 576:164–184. https://doi.org/10.1016/j.jhydrol.2019.06.032

Andersson LE, Aftab MF, Scibilia F, Imsland L (2017) Forecasting using multivariate empirical mode decomposition-applied to iceberg drift forecast. Paper presented at the IEEE conference on control technology and applications (CCTA), Kohala Coast, Hawai'i, USA

Apan AA, Raine, SR, Paterson MS (2002) Mapping and analysis of changes in the riparian landscape structure of the Lockyer Valley catchment, Queensland, Australia. Landscape Urban Plann 59:43–57. https://doi.org/10.1016/S0169-2046(01)00246-8

ASCE (1993) Criteria for evaluation of watershed models. J Irrig Drainage Eng 119:429–442

ASCE (2000) Artificial neural networks in hydrology. II: Hydrologic applications. J Hydrol Eng 5:124–137

Australian Bureau of Statistics (2018) Census QuickStats [Online]. https://itt.abs.gov.au/itt/r.jsp?databyregion. Accessed 7 November 2019

Barton C, Wallace S, Syme B, Wong WT, Onta P (2015) Brisbane River catchment flood study: comprehensive hydraulic assessment overview. Paper presented at the Floodplain Management Association National Conference, Brisbane Convention & Exhibition Centre, Brisbane, 19–22 May 2015

Bartholmes and Todini (2005) Coupling meteorological and hydrological models for flood forecasting. Hydrol Earth Syst Sci 9(4):333–346

Bhattacharya B, Islam T, Masud S, Suman A, Solomatine DP, Lang M, Klijn F, Samuels P (2016) The use of a flood index to characterise flooding in the north-eastern region of Bangladesh. E3S Web Conferences, vol 7, p 10003

Bhattacharya B, Solomatine DP (2003) Neural Networks and M5 model trees in modeling water level-discharge relationship for an Indian river. In: Proceedings of the European Symposium on Artificial Neural Networks Bruges (ESANN), Belgium, 23–35 April 2003

Bocchiola D, De Michele C, Rosso R (2003) Review of recent advances in index flood estimation. Hydrol Earth Syst Sci 7(3):283–296

Breiman L (1996) Bagging predictors. Mach Learn 24:123–140. https://doi.org/10.1023/a:1018054314350

Breiman L (2001) Random forests. Mach Learn 45:5–32

Callaghan J, Power S (2014) Major coastal flooding in southeastern Australia, associated deaths and weather systems. Aust Meteorol Oceanogr J 64(3):183–213

Chang LC, Chang F-J, Yang S-N, Kao I-F, Ku Y-Y, Kuo C-L, Ir. Amin (2019) Building an intelligent hydroinformatics integration platform for regional flood inundation warning systems. Water 11(1):9

Colominas MA, Schlotthauer G, Torres ME (2014) Improved complete ensemble EMD: a suitable tool for biomedical signal processing. Biomed Signal Process Control 14:19–29. https://doi.org/10.1016/j.bspc.2014.06.009

Costabile P, Macchione F (2015) Enhancing river model set-up for 2-D dynamic flood modelling. Environ Modell Softw 67:89–107. https://doi.org/10.1016/j.envsoft.2015.01.009

Coulibaly P, Anctil F, Aravena R, Bobée B (2001) ANN modeling of water table depth fluctuations. Water Resour Res 37:885–896. https://doi.org/10.1029/2000WR900368

Dawson CW, Abrahart RJ, See LM (2007) HydroTest: a web-based toolbox of evaluation metrics for the standardised assessment of hydrological forecasts. Environ Modell Softw 22:1034–1052

Deo RC, Byun HR, Adamowski JF, Kim DW (2015) A real-time flood monitoring index based on daily effective precipitation and its application to Brisbane and Lockyer Valley flood events. Water Resour Manage 29(11):4075–4093. https://doi.org/10.1007/s11269-015-1046-3

Deo RC, Tiwari MK, Adamowski JF, Quilty JM (2016a) Forecasting effective drought index using a wavelet extreme learning machine (W-ELM) model. Stoch Environ Res Risk Assess. https://doi.org/10.1007/s00477-016-1265-z

Deo RC, Wen X, Qi F (2016b) A wavelet-coupled support vector machine model for forecasting global incident solar radiation using limited meteorological dataset. Appl Energy 168:568–593

Feng Q, Liu J, Gong J (2015) Urban flood mapping based on unmanned aerial vehicle remote sensing and random forest classifier – a case of Yuyao, China. Water 7(4):1437–1455. https://doi.org/10.3390/w7041437

FitzGerald G, Du W, Jamal A, Clark M, Hou XY (2010) Flood fatalities in contemporary Australia (1997–2008). Emerg Med Australas 22(2):180–186. https://doi.org/10.1111/j.1742-6723.2010.01284.x

Fryirs K, Lisenby P, Croke J (2015) Morphological and historical resilience to catastrophic flooding: the case of Lockyer Creek, SE Queensland, Australia. Geomorphology 241:55–71. https://doi.org/10.1016/j.geomorph.2015.04.008

Herman GR, Schumacher RS (2018) Money doesn't grow on trees, but forecasts do: Forecasting extreme precipitation with random forests. Mon Weather Rev 146(5):1571–1600. https://doi.org/10.1175/MWR-D-17-0250.1

He K, Zha R, Wu J, Lai K (2016) Multivariate EMD-based modeling and forecasting of crude oil price sustainability 8:387. https://doi.org/10.3390/su8040387

Hirabayashi Y, Kanae S, Emori S, Oki T, Kimoto M (2008) Global projections of changing risks of floods and droughts in a changing climate. Hydrol Sci J 53(4):754–772. https://doi.org/10.1623/hysj.53.4.754

Hu W, Si BC (2013) Soil water prediction based on its scale-specific control using multivariate empirical mode decomposition. Geoderma 193–194:180–188. https://doi.org/10.1016/j.geoderma.2012.10.021

Huang J-R, Fan S-Z, Abbod M, Jen K-K, Wu J-F, Shieh J-S (2013) Application of multivariate empirical mode decomposition and sample entropy in EEG signals via artificial neural networks for interpreting depth of anesthesia. Entropy 15:3325–3339

Huang NE et al (1998) The empirical mode decomposition and the Hilbert spectrum for nonlinear and non-stationary time series analysis. Proc Royal Soc A 454:903–995

Hudson D, Alves O, Hendon HH, Lim EP, Liu G, Luo JJ, MacLachlan C, Marshall AG, Shi L, Wang G, Wedd R (2017) ACCESS-S1: the new Bureau of Meteorology multi-week to seasonal prediction system. J South Hemisphere Earth Syst Sci 67:132–159

Ishak EH, Rahman A., Westra S, Sharma A, Kuczera G (2010) Preliminary analysis of trends in Australian flood data. In: Proceeding of the World Environmental and Water Resources Congress 2010: Challenges of Change, Rhode Island, 16–20 May 2010, pp 115–124. https://doi.org/10.1061/41114(371)14

Johnson F, White CJ, van Dijk A, Ekstrom M, Evans JP, Jakob D, Kiem AS, Leonard M, Rouillard A, Westra S (2016) Natural hazards in Australia: floods. Climatic Change 139(1):21–35

Kim T-W, Valdes JB (2003) Nonlinear model for drought forecasting based on a conjunction of wavelet transforms and neural networks. J Hydrol Eng 8:319–328. https://doi.org/10.1061//ASCE/1084-0699/2003/8:6/319

Lee S, Kim JC, Jung HS, Lee MJ, Lee S (2017) Spatial prediction of flood susceptibility using random-forest and boosted-tree models in Seoul metropolitan city, Korea. Geomatics Nat Hazards Risk 8(2):1185–1203. https://doi.org/10.1080/19475705.2017.130897

Legates DR, McCabe GJ (1999) Evaluating the use of "goodness of fit" measures in hydrologic and hydroclimatic model validation. Water Resour Res 35:233–241

Li X, Yan D, Wang K, Weng B, Qin T, Liu S (2019) Flood risk assessment of global watersheds based on multiple machine learning models. Water 11(8):1654. https://doi.org/10.3390/w11081654

Liang Z, Tang T, Li B, Liu T, Wang J, Hu Y (2018) Long-term streamflow forecasting using SWAT through the integration of the random forests precipitation generator: case study of Danjiangkou Reservoir. Hydrol Res 49(5):1513–1527. https://doi.org/10.2166/nh.2017.085

Lima AR, Cannon AJ, Hsieh WW (2015) Nonlinear regression in environmental sciences using extreme learning machines: a comparative evaluation. Environ Modell Softw 73:75–188. https://doi.org/10.1016/j.envsoft.2015.08.002

Lockyer Valley Local Disaster Managment Group (2014) Lockyer Valley Disaster Management Plan Version 4.2. Gatton. https://www.lockyervalley.qld.gov.au/our-services/disastermanagement/Documents/Disaster%20Management%20Plan/lockyer%20valley%20local%20disaster%20management%20plan%20-%20version%204.2%20-%20web%20version.pdf. Accessed 7 November 2019

Looney D, Mandic DP (2009) Multiscale image fusion using complex extensions of EMD. IEEE Trans Signal Process 57:1626–1630. https://doi.org/10.1109/TSP.2008.2011836

Lokuge W, Setunge S (2013) Evaluating disaster resilience of bridge infrastructure when exposed to extreme natural events. In: 3rd International Conference on Building Resilience: Individual, Institutional and Societal Coping Strategies to Address the Challenges Associated with Disaster Risk, 17–19 September 2013, Heritance Ahungalla, Sri Lanka

Londhe S, Charhate S (2010) Comparison of data-driven modelling techniques for river flow forecasting. Hydrol Sci J 55(7):1163–1174. https://doi.org/10.1080/02626667.2010.512867

Mandic DP, ur Rehman N, Wu Z, Huang NE (2013) Empirical mode decomposition-based time-frequency analysis of multivariate signals: the power of adaptive data analysis. IEEE Signal Process Mag 30:74–86

Milly PCD, Wetherald RT, Dunne KA, Delworth TL (2002) Increasing risk of great floods in a changing climate. Nature 415(6871):514. https://doi.org/10.1038/415514a

Monira SS, Faisal ZM, Hirose H (2010) Comparison of artificially intelligent methods in short term rainfall forecast. In 2010 13th International Conference on Computer and Information Technology (ICCIT). IEEE. Dhaka, Bangladesh, 23–25 December 2010. https://doi.org/10.1109/ICCITECHN.2010.5723826

Mosavi A, Ozturk P, Chau KW (2018) Flood prediction using machine learning models: literature review. Water 10(11):1536. https://doi.org/10.3390/w10111536

Muñoz P, Orellana-Alvear J, Willems P, Célleri R (2018) Flash-flood forecasting in an Andean mountain catchment—Development of a step-wise methodology based on the random forest algorithm. Water 10(11):1519. https://doi.org/10.3390/w10111519

Nguyen TT, Huu QN, Li MJ (2015). Forecasting time series water levels on Mekong river using machine learning models. In 2015 Seventh International Conference on Knowledge and Systems Engineering (KSE) (pp. 292-297). IEEE. Ho Chi Minh City, Vietnam .8–10 October 2015. https://doi.org/10.1109/KSE.2015.53

Okada T, Haynes K, Bird D, van den Honert R, King D (2014) Recovery and resettlement following the 2011 flash flooding in the Lockyer Valley. Int J Disaster Risk Reduction 8:20–31. https://doi.org/10.1016/j.ijdrr.2014.01.001

Oki T, Kanae S (2006) Global hydrologic cycles and world water resources. Science 313:1068–1072. https://doi.org/10.1126/science.1128845

Onyari EK, Ilunga FM (2013) Application of MLP neural network and M5P model tree in predicting streamflow: a case study of Luvuvhu catchment, South Africa. Int J Innov Manage Technol 4(1):11

Papacharalampous GA, Tyralis H (2018) Evaluation of random forests and Prophet for daily streamflow forecasting. Adv Geosci 45:201–208. https://doi.org/10.5194/adgeo-45-201-2018

Penning-Rowsell E, Tunstall S, Tapsell S, Parker D (2000) The benefits of flood warnings: real but elusive, and politically significant. J Chartered Inst Water Environ Manage 14:7–14. https://doi.org/10.1111/j.1747-6593.2000.tb00219.x

Petty TR, Dhingra P (2018) Streamflow Hydrology Estimate Using Machine Learning (SHEM). J Am Water Resour Assoc 54:55– 68. https://doi.org/10.1111/1752-1688.12555

Prasad R, Ali M, Kwan P, Khan H (2019) Designing a multi-stage multivariate empirical mode decomposition coupled with ant colony optimization and random forest model to forecast monthly solar radiation. Appl Energy 236:778–792. https://doi.org/10.1016/j.apenergy.2018.12.034

Prasad R, Deo RC, Li Y, Maraseni T (2017) Input selection and performance optimization of ANN-based streamflow forecasts in the drought-prone Murray Darling Basin region using IIS and MODWT algorithm. Atmos Res 197:42–63. https://doi.org/10.1016/j.atmosres.2017.06.014

Prasad R, Deo RC, Li Y, Maraseni T (2018) Soil moisture forecasting by a hybrid machine learning technique: ELM integrated with ensemble empirical mode decomposition. Geoderma 330:136–161. https://doi.org/10.1016/j.geoderma.2018.05.035

Quinlan JR (1987) Decision trees as probabilistic classifiers. In: Proceedings of the Fourth International Workshop on Machine Learning (pp 31-37). Morgan Kaufmann

Quinlan JR (1992) Learning with continuous classes. In: Sterling A (ed) 5th Australian joint conference on artificial intelligence. Singapore, pp 343–348

Rehman N, Mandic DP (2009) Multivariate empirical mode decomposition. Proc Royal Soc A Math Phys Eng Sci 466:1291–1302. https://doi.org/10.1098/rspa.2009.0502

Reisinger A, Kitching RL, Chiew F, Hughes L, Newton PCD, Schuster SS, Tait A, Whetton P (2014) Australasia. In: climate change 2014: impacts, adaptation, and vulnerability. Part B: Regional aspects. Contribution of Working Group II to the Fifth Assessment Report of the Intergovernmental Panel on Climate Change. In Barro VR, Field CB, Dokken DJ, Mastrandrea MD, Mach KJ, Bilir TE, Chatterjee M, Ebi KL, Estrada YO, Genova RC, Girma B, Kissel ES, Levy AN, MacCracken S, Mastrandrea PR, White LL (eds). Cambridge University Press, Cambridge, United Kingdom and New York, NY, USA, pp 1371–1438

Sadler JM, Goodall JL, Morsy MM, Spencer K (2018) Modeling urban coastal flood severity from crowd-sourced flood reports using poisson regression and random forest. J Hydrol 559:43–55. https://doi.org/10.1016/j.jhydrol.2018.01.044

Sattari MT, Pal M, Apaydin H, Ozturk F (2013) M5 model tree application in daily river flow forecasting in Sohu Stream, Turkey. Water Resour 40(3):233–242. https://doi.org/10.1134/S0097807813030123

Sayers PB, Galloway GE, Hall JW (2012) Robust decision-making under uncertainty–towards adaptive and resilient flood risk management infrastructure. In: Flood Risk Planning, Design and

Management of Flood Defence Infrastructure, ICE Publishing, pp 281–302. https://doi.org/10.1680/fr.41561.281. Chapter 11

Schoof JT, Pryor SC (2001) Downscaling temperature and precipitation: a comparison of regression-based methods and artificial neural networks. Int J Climatol: A J Roy Meteorol Soc 21(7):773–790. https://doi.org/10.1002/joc.655

She D, Zheng J, Ma Shao, Timm LC, Xia Y (2015) Multivariate empirical mode decomposition derived multi-scale spatial relationships between saturated hydraulic conductivity and basic soil properties. CLEAN–Soil, Air, Water 43:910–918

Shortridge JE, Guikema SD, Zaitchik BF (2016) Machine learning methods for empirical streamflow simulation: a comparison of model accuracy, interpretability, and uncertainty in seasonal watersheds. Hydrol Earth Syst Sci 20:2611–2628. https://doi.org/10.5194/hess-20-2611-2016

Singh KK, Pal M, Singh VP (2010) Estimation of mean annual flood in Indian catchments using backpropagation neural network and M5 model tree. Water Resour Manage 24(10):2007–2019. https://doi.org/10.1007/s11269-009-9535-x

Solomatine DP, Dulal KN (2003) Model trees as an alternative to neural networks in rainfall—runoff modelling. Hydrol Sci J 48(3):399–411

Solomatine DP, Xue Y (2004) M5 model trees and neural networks: application to flood forecasting in the upper reach of the Huai River in China. J Hydrol Eng 9(6):491–501. https://doi.org/10.1061/(ASCE)1084-0699(2004)9:6(491)

Štravs L, Brilly M (2007) Development of a low-flow forecasting model using the M5 machine learning method. Hydrol Sci J 52(3):466–477

Sulaiman J, Wahab SH (2018) Heavy Rainfall Forecasting Model Using Artificial Neural Network for Flood Prone Area. In: Kim K, Kim H, Baek N (eds) IT Convergence and Security 2017. Lecture Notes in Electrical Engineering, vol 449. Springer, Singapore

Tehrany MS, Pradhan B, Jebur MN (2013) Spatial prediction of flood susceptible areas using rule-based decision tree (DT) and a novel ensemble bivariate and multivariate statistical models in GIS. J Hydrol 504:69–79. https://doi.org/10.1016/j.jhydrol.2013.09.034

Torres ME, Colominas MA, Schlotthauer G, Flandrin P (2011) A complete ensemble empirical mode decomposition with adaptive noise. In: 2011 IEEE international conference on acoustics, speech and signal processing (ICASSP), 22–27 May 2011, pp 4144–4147. https://doi.org/10.1109/icassp.2011.5947265

Ur Rehman N, Mandic DP (2011) Filter bank property of multivariate empirical mode decomposition. IEEE Trans Signal Process 59:2421–2426

Ur Rehman N, Park C, Huang NE, Mandic DP (2013) EMD Via MEMD: multivariate noise-aided computation of standard EMD. Adv Adapt Data Anal 05. https://doi.org/10.1142/s1793536913500076

van den Honert RC, McAneney J (2011) The 2011 Brisbane floods: causes, impacts, and implications. Water 3(4):1149–1173. https://doi.org/10.3390/w3041149

Wang Y, Witten IH (1997) Inducing model trees for continuous classes. In: European conference on machine learning, Prague, pp 128–137

Wang Z, Lai C, Chen X, Yang B, Zhao S, Bai X (2015) Flood hazard risk assessment model based on random forest. J Hydrol 527: 1130-1141. https://doi.org/10.1016/j.jhydrol.2015.06.008

White CJ, Franks SW, McEvoy D (2015) Using subseasonal-to-seasonal (S2S) extreme rainfall forecasts for extended-range flood prediction in Australia. In: Proceedings of the International Association of Hydrological Sciences 370:229–234

Willmott CJ (1981) On the validation of models. Phys Geogr 2:184–194

Willmott CJ (1982) Some comments on the evaluation of model performance. Bull Am Meteor Soc 63:1309–1313

Willmott CJ (1984) On the evaluation of model performance in physical geography. In: Spatial statistics and models. Springer, pp 443–460

Witten IH, Frank E, Hall MA (2011) Data mining—practical machine learning tools and techniques, 3rd edn. Morgan Kaufmann Publishers, United States

Wu CL, Chau KW (2006) A flood forecasting neural network model with genetic algorithm. Int J Env Pollut 28(3/4):261–273. https://doi.org/10.1504/IJEP.2006.011211

Wu Z, Huang NE (2009) Ensemble empirical mode decomposition: a noise-assisted data analysis method. Adv Adapt Data Anal 1:1–41. https://doi.org/10.1142/S1793536909000047

Wu Z, Huang NE, Wallace JM, Smoliak BV, Chen X (2011) On the time-varying trend in global-mean surface temperature. Clim Dyn 37:759–773. https://doi.org/10.1007/s00382-011-1128-8

Xingjian SHI, Chen Z, Wang H, Yeung DY, Wong WK, Woo WC (2015) Convolutional LSTM network: A machine learning approach for precipitation nowcasting. In Cortes C, Lawrence ND, Lee DD, Sugiyama M, Garnett R (eds) Proceeding of Advances in neural information processing systems, pp 802–810

Yang JH, Cheng CH, Chan CP (2017) A time-series water level forecasting model based on imputation and variable selection method. Comput Intell Neurosci 2017. https://doi.org/10.1155/2017/8734214

Yen BC (1995) Discussion and closure: criteria for evaluation of watershed models. J Irrig Drainage Eng 121:130–132

Yu PS, Yang TC, Chen SY, Kuo CM, Tseng HW (2017) Comparison of random forests and support vector machine for real-time radar-derived rainfall forecasting. J Hydrol 552:92–104. https://doi.org/10.1016/j.jhydrol.2017.06.020

Zahiri A, Azamathulla HM (2014) Comparison between linear genetic programming and M5 tree models to predict flow discharge in compound channels. Neural Comput Appl 24(2):413–420. https://doi.org/10.1007/s00521-012-1247-0

Zhao G, Pang B, Xu Z, Yue J, Tu T (2018) Mapping flood susceptibility in mountainous areas on a national scale in China. Sci Total Environ 615:1133–1142. https://doi.org/10.1016/j.scitotenv.2017.10.037

# Chapter 18
# Machine Learning Method in Prediction Streamflow Considering Periodicity Component

**Rana Muhammad Adnan, Mohammad Zounemat-Kermani, Alban Kuriqi, and Ozgur Kisi**

## 18.1 Introductory Note

Streamflow data monitoring and prediction is a fundamental task for sustainable water resources management. In the eve of increasing influence from the climate change and uncontrolled anthropogenic activities related to water abstraction particularly from freshwater systems has profoundly altered the natural flow regime and as a result threatens the biotic community and human life as well (Ali et al. 2019; Kuriqi et al. 2019; Adnan et al. 2019b). Also, the intensity and timing of the extreme events have been changing significantly (Tongal and Booij 2018). Therefore, it is indispensable to have a robust streamflow monitoring and database in order to conduct more accurate prediction regarding the different impacts induced either from natural phenomenon or anthropogenic factors. Nevertheless, in many regions, because of the different reasons, streamflow data monitoring is very scarce or not available at all (Tongal and Booij 2018; Mosavi et al. 2018). In these circumstances, during decades, alternative estimation approaches are applied to reconstruct and predict streamflow

R. M. Adnan
State Key Laboratory of Hydrology-Water Resources and Hydraulic Engineering, Hohai University, 210098 Nanjing, China
e-mail: rana@hhu.edu.cn

M. Zounemat-Kermani
Department of Water Engineering, Shahid Bahonar University of Kerman, Kerman, Iran
e-mail: zounemat@uk.ac.ir

A. Kuriqi
CERIS, Instituto Superior Técnico, Universidade de Lisboa, Lisbon, Portugal
e-mail: alban.kuriqi@tecnico.ulisboa.pt

O. Kisi (✉)
Department of Civil Engineering, Ilia State University, 0162 Tbilisi, Georgia
e-mail: ozgur.kisi@iliauni.edu.ge

© Springer Nature Singapore Pte Ltd. 2021

R. C. Deo et al. (eds.), *Intelligent Data Analytics for Decision-Support Systems in Hazard Mitigation*, Springer Transactions in Civil and Environmental Engineering,
https://doi.org/10.1007/978-981-15-5772-9_18

data in the lack of direct monitoring. In this regard, there are developed many empirical and physical-based hydrological models. However, both of these categories of models have several limitations concerning the diversity and quantity of the data requirements, transferability, and scale-issues applicability (Beven 2002). Among others, data-driven models represent an innovative category of the models which are gaining increasingly great attention and applicability in different fields, including water resources management as well. Several studies show that data-driven models are practical tools, easy to apply, and robust in terms of the accuracy (Tongal and Booij 2018; Mosavi et al. 2018; Rodriguez-Galiano et al. 2014; Chen et al. 2012; Muhammad Adnan et al. 2017) by overcoming, therefore, some of the limitations of the empirical and physical-based hydrological models. Data-driven or so-called artificial intelligence (AI) models although considered as "black box" approaches are efficient and robust in modeling of different natural phenomenon (Yaseen et al. 2016; Yuan et al. 2018; Adnan et al. 2019c).

In comparison with empirical and physical-based hydrological models, AI models are less demanding in terms of data accusation and quantity. Within the field of waters resources management, data-driven models are being intensively applied for predicting extreme events such as drought modeling (Deo and Şahin 2015; Lee et al. 2017), pan evaporation estimation (Adnan et al. 2019d), floods (Tongal and Booij 2018; Singh et al. 2017), water quality (Adnan et al. 2019a), sediment prediction (Kratzert et al. 2018), rainfall prediction (Taormina and Chau 2015; Liang et al. 2017), and streamflow prediction as well (Yaseen et al. 2016; Makkeasorn et al. 2008; Shortridge et al. 2016; Nourani et al. 2014; Kisi et al. 2018). Among several data-driven models, artificial neural network (ANN), random forest (RF), long short-term memory (LSTM), and extreme learning machines (ELM) are the most popular models (Xu and Niu 2018). The last three models, in combination with ANN and other evolutionary algorithms, are widely applied in the field of earth and planetary sciences and also in other sub-fields such as water science. (Deo and Şahin 2015) applied RF model for droughts prediction, and they concluded that RF provides better results comparing to other models. Similar performance regarding the drought prediction was reported by Lee et al. (2017) in case of ELM model application. Wang et al. (2015) applied LSTM to predict landslide occurrence in Baijiabao, China, and they found that the LSTM model could provide more accurate results than static models and also some other dynamic models. According to Sadler et al. (2018), the RF model was able to run efficiently a sophisticated database which was composed of different parameters used to predict flood hazard in Dongjiang River, China. Further, they found that the error rate of RF could be easily reduced by increasing the samples size and discretization of the classification trees.

Regarding the flood severity prediction, (Kong et al. 2018) reported that RF performed better than Poisson regression and had the lowest error rate. As reported by Liang et al. (2018), RF shows better performance compared to other AI models in prediction of the water infiltration rate in the soil, and groundwater quality prediction as well (Le et al. 2019). While related to the applicability of the LSTM in water resources/sciences-related issues, there are several studies concerning remote sensing (Barzegar et al. 2017), water level prediction (Papacharalampous and Tyralis

2018), and floods (Sahoo et al. 2019) where LSTM shows quite accurate results. Also, successful applicability of ELM has been reported at several studies. (Adnan et al. 2019e) applied ELM to model rainfall–runoff by comparing the performance of ELM to other AI models, and they found that ELM shows highly accurate results while being noticeably faster than the other models. Similar to previous models, ELM shows robust performance in water quality prediction as well (Yadav et al. 2016), whereas although scarce, there are also few studies which applied RF (Mosavi et al. 2018; Nourani et al. 2014; Yaseen et al. 2019), LSTM (Shortridge et al. 2016; Zounemat-Kermani 2016), and ELM (Yaseen et al. 2016; Kisi et al. 2017; Hochreiter and Schmidhuber 1997; Chung et al. 2014) by comparing with other AI models, for streamflow prediction at different resolution (i.e., daily, monthly) and climate conditions. Nevertheless, to the best of our knowledge, there is no any study which compares the performance of these three models among each other. Therefore, considering this gap in knowledge, in this study, we attempt to bring some new findings related to the applicability of these three models for streamflow prediction. Furthermore, since all of these three models seem to be very promising in predicting different natural phenomenon where among them streamflow prediction, this chapter investigates the potential of these three models in long-term streamflow data prediction and also compares their performance against each other. The rest of this chapter is organized as follows: Section 18.2 presents the study case and provides a detailed description of the methodology, including governing equations, models configuration, and statistical indicators used to assess the performance of the three applied models. The result interpretations and discussion are presented in Sect. 18.3. Finally, the main findings resulted from this study and the concluding remarks are summarized in Sect. 18.4.

## 18.2 Materials and Method

### *18.2.1 Case Study*

Two streamflow gauging stations, namely Garhihabibullah and Kohala Stations from the Jhelum River Basin (JRB) of Pakistan, are selected as case study areas in this study. The drainage area of the JRB is 33,342 km$^2$. JRB catchment spread area is from 33 to 35 longitude and 73 to 75.62 latitude (Fig. 18.1). The average of altitude, yearly precipitation, and temperature of JRB is 2094 m a.s.l., 1202 mm, and 13.7 °C, respectively. Streamflow prediction of JRB is very vital due to having an installed hydropower plant of 1000 MW, namely Mangla hydropower plant on downstream of this basin. Due to this hydropower plant importance, this basin also called as Mangla reservoir catchment. This reservoir is the 2nd biggest reservoir of the country, and it helps to irrigate 6 million hectares of agriculture land and in addition provides 6 percent electricity production of whole country production.

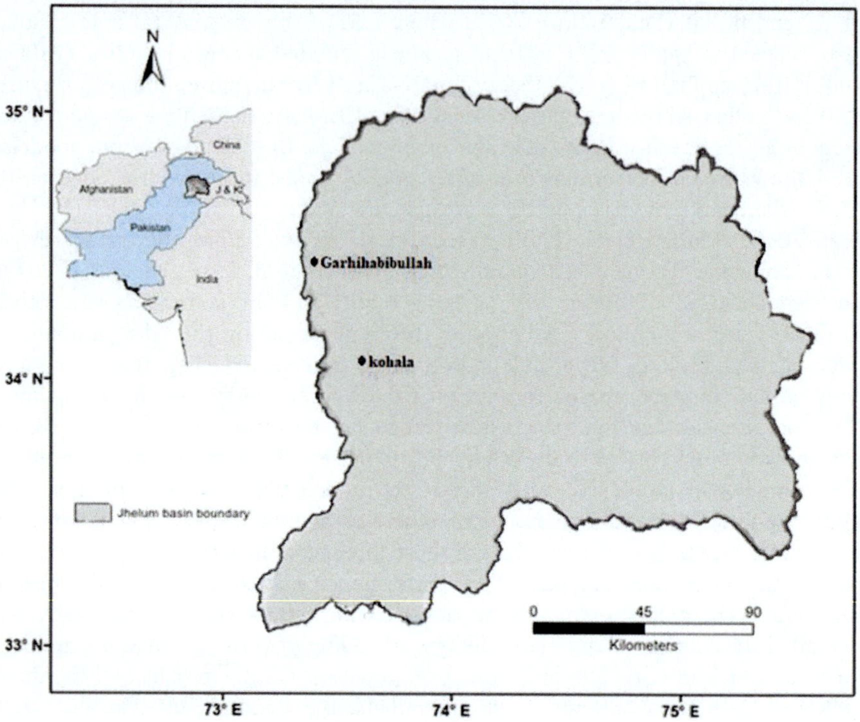


**Fig. 18.1** Location map of the Garhihabibullah and Kohala Stations in Jhelum Basin

The electricity production of this reservoir aids to minimize the electricity shortage problem of this developing country.

### *18.2.2 Long Short-Term Memory Network (LSTM)*

During the past decades, artificial neural networks (ANNs in the sequel) have proven their high capabilities in simulating complex dynamic systems, such as hydrological time series modeling and prediction (Zounemat-Kermani 2016; Kisi et al. 2017). Contrary to the standard feedforward neural networks (FFNNs), such as multi-layer perceptrons (MLPs), recurrent neural networks (RNNs) get the advantages of their ability in using contextual information, in terms of applying an internal state of the network, in modeling time series problems and simulating sequences of datasets suing stacked information layers. Having a sequential input vector $x = [x[1], x[2], \ldots, x[T]]$ with $T$ consecutive time steps, the internal state of an RNN, namely as $h[t]$, can be defined as below:

$$h[t] = f(Wx[t] + Uh[t-1] + b) \quad (18.1)$$

where $W, U$ are adjustable weight matrices, and $b$ is the bias vector. $f(.)$ is the activation function, for instance, hyperbolic tangent. As an illustration, Fig. 18.1 shows the internal operation of a one-layer RNN structure. The final prediction of the network ($y$) can be attained from the output layer, so-called the dense layer, based on the calculated state at the last time step $h[T]$(Xu and Niu 2018).

$$y = g(W_d h[T] + b_d) \quad (18.2)$$

where $g(.)$ denotes the dense layer's activation function, and $W_d$ and $b_d$ represent the weight matrix and bias vector of the dense layer, respectively.

Standard RNNs mostly suffer from a serious shortcoming in their backpropagating phase of their training process, which is the vanishing behavior of the learning gradient. That is to say, in RNNs with several recurrent layers, the decaying feature of the backpropagation procedure makes the first hidden layers of the RNNs less effective or even ineffective during the training process. Hence, regular RNNs might encounter problems for dealing with long-term dependencies, such as modeling streamflow and flood in watersheds, between the independent input variables and the dependent output values. To cope with this downside feature of traditional RNNs (i.e., the vanishing gradient problem), several modified RNNs have been suggested and developed. Examples include the long short-term memory networks (LSTM; (Hochreiter and Schmidhuber 1997)) and gated recurrent unit networks (GRU; (Chung et al. 2014)) (Fig. 18.2).

LSTM neural networks are categorized as RRNs that contain specific types of memory blocks formed in a stacked architecture for storing information of long time periods. Each block contains one or more memory cells as well as three types of dedicated units (gates), including the input gate, the output gate, and the forget gate. These gates regulate and control the information flow through the structure of the LSTM network (Hochreiter and Schmidhuber 1997). The three mentioned

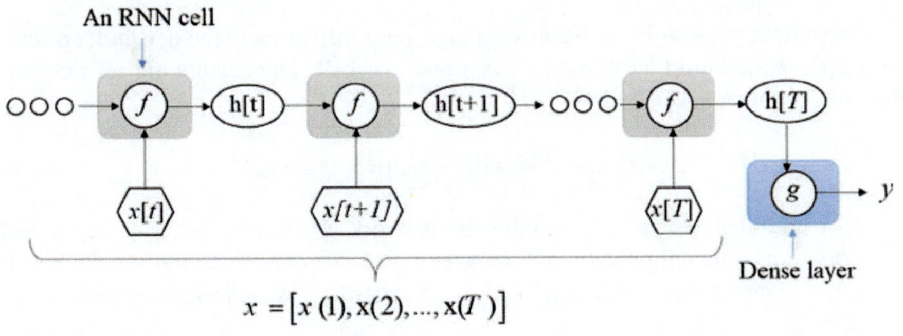


**Fig. 18.2** Unfold structure of a standard RNN model; *Note* the same structure equipped with memory cells, as shown in Fig. 18.3, forms an LSTM network

gates can be presented as nonlinear summation units, which use the sigmoid and hyperbolic tangent activation functions. Figure 18.3 illustrates the conceptual sketch of a memory block with a single cell in an LSTM network. As can be seen in Fig. 18.3, the gates (units) construct a multiplicative configuration that able the memory blocks to store long-term information and consequently have access to them even at the earliest hidden layers of an RNN. It should be noted that the general structure of an LSTM network is similar to the standard RNNs, except that in the LSTM, the memory blocks replace the summation units in the standard RNNs (see Fig. 18.3).

The LSTM network can be formed based on the calculated values of the cell memory (cell state) as below:

$$c[t] = f[t] \odot c[t-1] + i[t] \odot g[t] \tag{18.3}$$

where $f[t]$, $i[t]$, and $g[t]$ are the forget gate (the first gate), the input gate (the second gate), and the cell update, respectively. $c[t-1]$ is the cell state at the antecedent time step.$\odot$ is the Hadamard product (element-wise) multiplication. Values for the input gate, forget gate, and cell update are calculated based on the following formulae:

$$i[t] = \text{sig}(W_i x[t] + U_i h[t-1] + b_i); \quad 1 \le t \le T \tag{18.4}$$

$$f[t] = \text{sig}\left(W_f x[t] + U_f h[t-1] + b_f\right) \tag{18.5}$$

$$g[t] = \tanh\left(W_g x[t] + U_g h[t-1] + b_g\right) \tag{18.6}$$

In the above formulae, sig(cot) and $\tanh(\cdot)$ are the sigmoid and hyperbolic tangent activation functions, respectively. $W$, $U$, and $b$ are adjustable weight matrices, and bias vector related to each gate should be trained and tuned during the training process of the network. The training procedure of the original LSTM is based on a hybrid learning algorithm for approximating the error gradient. The hybrid learning combined the real-time recurrent learning (RTRL) technique and backpropagation (Hochreiter and Schmidhuber 1997).

The output gate, which controls the outgoing information to the new hidden state cell $h[t]$, is the third unit in the LSTM's memory cell. The calculating formula for the output gate is given below:

$$o[t] = \text{sig}(W_o x[t] + U_o h[t-1] + b_o) \tag{18.7}$$

As mentioned earlier, $x[t]$ denotes the network input vector at time step t, and $o[t]$ represents the output gate. $h[t-1]$ represents the recurrent hidden state input. Finally, the new hidden state ($h[t]$(is defined according to the following formula:

$$h[t] = o[t] \odot \tanh(c[t]) \tag{18.8}$$

In the LSTM, the cell states ($c[t]$), which get modified by the forget gate, input gate, and cell update, characterize the memory of the network (Kratzert et al. 2019). The final output of the LSTM network (e.g., predicted values for the streamflow) can be attained from the last layer (dense) in the LSTM similar to Eq. (18.2).

### 18.2.3 Extreme Learning Machine (ELM)

During the past decades, feedforward neural networks (FFNN), such as multi-layer perceptron, have been successfully employed in plenty of scientific fields. The FFNNs are universal approximators that can map the nonlinear nature of sample vectors via a black-box strategy. Despite their remarkable achievements in modeling various areas in engineering, traditional FFNNs mostly suffer from a slow training process when it comes to simulating high-dimensional and complex problems. Two main features of the learning procedure of traditional neural networks might cause this shortcoming, including the employed slow gradient-based learning algorithms and the iterative nature of network's parameter tuning (Huang et al. 2006). In a traditional FFNN, all the network's learning parameters, such as synaptic weights and biases, need to be adjusted via a backpropagation procedure. In the last decades, the majority of FFNNs have applied gradient descent-based algorithms (e.g., conjugate gradient) for their backpropagating phase. Although there exist some faster mathematical algorithms to be used in the FFNN learning process (e.g., the Levenberg–Marquardt algorithm), most of the commonly used ones require many iterative steps for reaching convergence, which makes their learning process generally slow.

In the light of enhancing the speed of traditional FFNN learning algorithm, (Huang et al. 2006) proposed a new version of a single-layer feedforward neural network (SLFN), so-called the extreme learning machine (ELM). The main intuition behind choosing an SLFN for constructing the proposed ELM model is that the SLFNs have been proven to have the required capability in approximating any function just based on adjusting the number of their hidden nodes. In other words, an SLFN can provide as accurate results as a multi-hidden-layer FFNN in simulating sophisticated phenomena (Huang and Babri 1998).

Considering an SLFN with $n$ explanatory samples and $N$ hidden neurons, the objective of the ELM model is to map the $n$ samples of the training input data ($x = [x_1, x_2, \ldots, x_n]$) into the output target based on the following function:

$$f(x) = \sum_{j=1}^{n}\sum_{i=1}^{N} \beta_i g(w_i x_j + b_i) \tag{18.9}$$

where $g(\cdot)$ is the transfer function, which presents the hidden neuron output. $w_i = [w_{i1}, w_{i2}, \ldots, w_{\text{in}}]$ and $\beta_i = [\beta_i, \beta_2, \ldots, \beta_N]$ are the input and output weight vectors, respectively. The input weights link the input layer to the hidden layer, while the output weights link the hidden nodes to output nodes. $b$ denotes the hidden layer bias vector. $N$ stands for the number of neurons in the hidden layer. In Fig. 18.3, the conceptual design for the ELM architecture is given.

In the ELM model, contrary to the training procedure in traditional FFNNs, there is no need for tuning all the network's parameters of the SLFN to build up a robust and effective neural network, including the input weight matrix and hidden layer biases. In this sense, the relation between the input variables ($x = [x_1, x_2, \ldots, x_n]$), the output variables ($y = [y_1, y_2, \ldots, y_n]$), and the target values ($t = [t_1, t_2, \ldots, t_n]$) can be defined based on the following equation:

$$y_j = \sum_{j=1}^{n} \sum_{i=1}^{N} \beta_i g(w_i x_j + b_i) = t_j + \varepsilon_j \tag{18.10}$$

where $\varepsilon$ is the residual (the network's error). The commonly used nonlinear differentiable mapping functions in the ELM method are the sigmoid and Gaussian functions. The sigmoid and Gaussian activation functions can be written as below:

$$g(x_i) = h(x_i) = \frac{e^{x_i}}{1 + e^{x_i}} \tag{18.11}$$

$$g(x_i) = h(a, c, x_i) = \exp(-a\|x_i - c\|^2) \tag{18.12}$$

where $a$ and $c$ are the variables for the Gaussian activation function. The ELM does not need an iterative process for updating the neurons' activation function parameters (the weight vector for the hidden layer, $w$, and the bias vector, $b$), instead, it initially assigns some random values to them. Consequently, the ELM tries to optimize the output weight vector ($\beta$) in a way that the error values ($\varepsilon$) in Eq. (18.11) to be zero

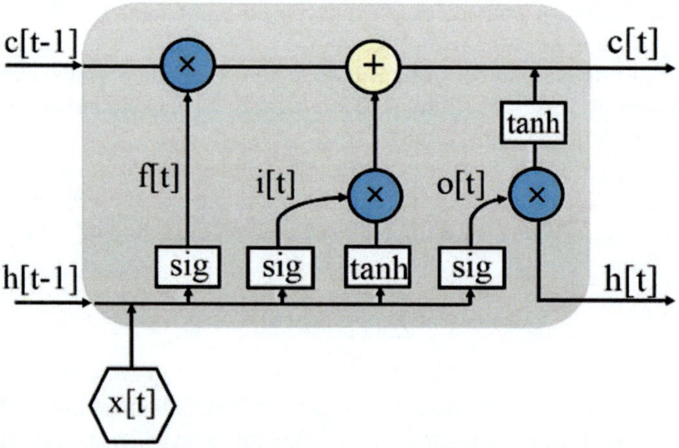


**Fig. 18.3** Internal components of a long short-term memory model cell

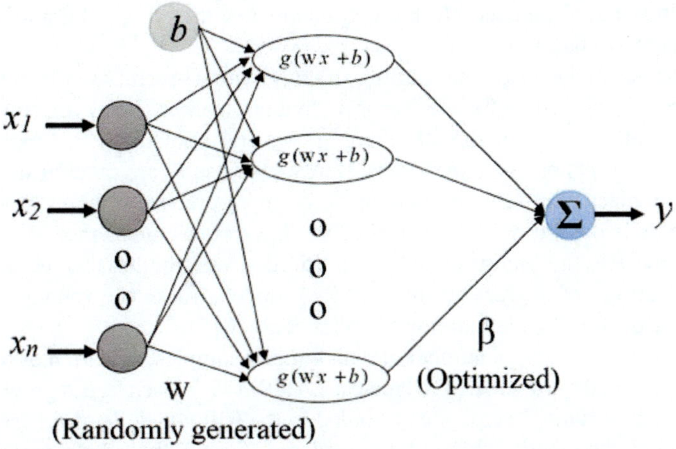


**Fig. 18.4** Illustrating the general structure of the extreme learning machine model

so that $\sum_{i=1}^{N} \|y_i - t_i\| = 0$. In that case, one can write the $N$ system of equations in Eq. (18.11) as follows:

$$\mathbf{H}\beta\mathbf{=T} \quad (18.13)$$

**H** is the randomized hidden layer output matrix, and **T** is the target matrix.

$$\mathbf{H} = \begin{bmatrix} g(x_{11}) & g(x_{12}) & \ldots & g(x_{1N}) \\ \vdots & \vdots & \cdots & \vdots \\ \vdots & \vdots & \cdots & \vdots \\ g(x_{n1}) & g(x_{n2}) & \ldots & g(x_{nN}) \end{bmatrix} \mathbf{T} = \begin{bmatrix} t_1^T \\ \vdots \\ \vdots \\ t_n^T \end{bmatrix} \quad (18.14)$$

Assuming $\beta = \mathrm{H}^{+}\mathbf{T}$ as a system of linear equation, the output weights can be calculated by solving the linear equation system for ($\beta$($\cdot\mathrm{H}^{+}$ is known as the Moore–Penrose generalized inverse of the hidden layer weight matrix (**H**). (Yadav et al. 2016) suggested the minimum norm least-squares (LS) solving process for calculating the output weight matrix.

## *18.2.4 Random Forests (RF)*

Soft computing models based on tree structure principles and tree-based learning are popular and practical machine learning methods that have been used in various fields of data mining projects. In general, according to the nature of the intended problem, tree-based models fall into two main categories of (i) classification trees and (ii) regression trees.

RF is based on the ensemble learning strategy, where several base-learners (weak models) are combined to form a strong individual learner (Breiman 2001). In the ensemble method used in the RF, there are several bootstrapped samples of the dataset training sets of tree-like models. Bootstrap samples are random samples that conduct a replacement selecting policy. To be more precise, in bootstrapping, any selected data is not left out of the original sample and has the chance to be selected again. In the RF, the binary recursive partitioning tree models, which are known as classification and regression tree (CART) models, make the foundation of the ensemble aspect of the methodology (Cutler et al. 2012).

In this respect, having *n* number of data training samples, *M* variables (e.g., 60% of the original data) are selected to create a CART model. Each tree is grown as large as possible; no pruning technique is needed in the RF model. Thus, in establishing an RF model, two stages should be considered. Firstly, the creation of *T* CART models based on the bootstrap (e.g., bagging) algorithm. Secondly, representing the final output of the built RF model according to the type of the problem, either as a classification or a regression type.

In each CART, the root node of the tree represents the classifier/predictor space. A binary partition (split) technique is used on one predictor variable to partition the input parameters for each node (leaf). Each node, except for the terminal nodes, splits into two descendants left and right nodes, based on the value of the predictor variable (see Fig. 18.5). In the regression context, the splitting procedure is done based on the minimized value for the mean squared error (MSE) at each terminal node (a node that is not split). The MSE values, at nodes, can be calculated as (Cutler et al. 2012):

$$\mathrm{MSE} = {}^{1}/_{n} \sum_{i=1}^{n} (y_i(t) - \bar{y}(t))^2 \tag{18.15}$$

where $y$ is the predicted value and $\bar{y}$ stands for the mean value of $y_i$. Detailed information about the theory and basics of CART models can be found in several available papers (e.g., (Steinberg 2009)). After training the RF model, the cross-validation strategy can be applied to evaluate the test set accuracy, which are not considered in the bootstrap of a CART yet. Once all of the tree-like models have been trained, the obtained results for the out-of-bag predictions, which are the unseen samples that do not occur in the bootstraps, are aggregated to achieve the output of the model. Considering a classification modeling problem, the output of the RF model is determined by a majority vote, while averaging is used for the regression problems. Taking into account a regressive RF model that is trained to predict the streamflow in a river, the final output of the RF model, $\widehat{y}$, is calculated by averaging the predictions from all the bootstrapped CART models as below:

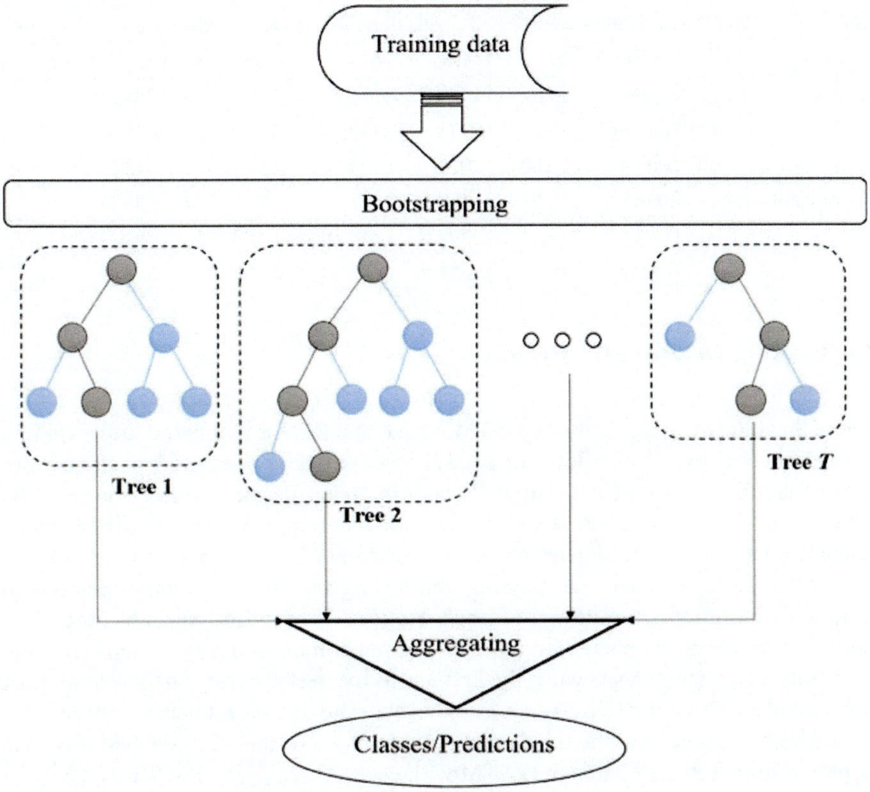


**Fig. 18.5** Conceptual sketch for the architecture of the random forests model

$$\widehat{y} = \frac{1}{T}\sum_{i=1}^{T} y_i \tag{18.16}$$

where $T$ is the number of bagging iterations (i.e., the number of the trees), and $y$ is the output of each CART base-learner (Ließ et al. 2012).

One of the major benefits of applying the RF model in simulating nonlinear and highly sophisticated problems can be related to its acceptable convergence speed in comparison to conventional bagging ensemble techniques, for instance. The reason for its convergence speed improvement lies in the fact that in the RF method, each CART uses a portion of the input parameters. In short, due to its potential ability to construct a great number of CART models in an ensemble structure, the RF algorithm can deal with high-dimensional data and appropriately handle complicated dynamic systems, such as modeling phenomena in surface hydrology (Table 18.1).

**Table 18.1** Monthly streamflow statistics of Garhihabibullah and Kohala Stations

| Station | Dataset | Mean | Max | Min | Skewness | S. deviation |
|---|---|---|---|---|---|---|
| Kohala | Whole | 772.4 | 2773 | 110.7 | 0.827 | 604.0 |
| | Training | 808.2 | 2773 | 112.3 | 0.759 | 625.8 |
| | Testing | 665.0 | 2014 | 110.7 | 0.962 | 518.8 |
| Garhihabibullah | Whole | 102.1 | 409.5 | 13.72 | 1.156 | 91.18 |
| | Training | 102.3 | 409.5 | 18.20 | 1.175 | 92.31 |
| | Testing | 101.3 | 358.0 | 13.72 | 1.096 | 87.71 |

## 18.3 Results and Discussion

Three machine learning methods, LSTM, ELM, and RF, are compared in streamflow prediction. Previous streamflows are used as inputs to the applied models. Periodicity input (month number of the output) is also included to see its effect on models' efficiency. Three evaluation statistics are root mean square error (RMSE), mean absolute error (MAE), and determination coefficient ($R^2$) for models' validation.

Table 18.2 summarizes the training and testing results of the three methods in prediction monthly streamflows of Kohala Station. As seen from the table, the ELM and RF have their best results from the third input combinations ($Q_{t-1}$, $Q_{t-2}$, $Q_{t-3}$) in both training and test stages while the LSTM has the best accuracy in the test stage of the second combination ($Q_{t-1}$, $Q_{t-2}$). α in the table shows the best periodicity component which is added into the third input combinations of the all three methods. It is apparent from Table 18.2 that the LSTM with $Q_{t-1}$, $Q_{t-2}$, $Q_{t-3}$, α (RMSE: 158.5 m$^3$/s, MAE: 96.8 m$^3$/s, $R^2$: 0.935 and RMSE: 204.7 m$^3$/s, MAE: 143.4 m$^3$/s, $R^2$: 0.871) performs superior to the ELM (RMSE: 163.1 m$^3$/s, MAE: 104.5 m$^3$/s, $R^2$: 0.932 and RMSE: 228.8 m$^3$/s, MAE: 157.7 m$^3$/s, $R^2$: 0.856) and RF (RMSE: 181.1 m$^3$/s, MAE: 120.1 m$^3$/s, $R^2$: 0.916 and RMSE: 300.4 m$^3$/s, MAE: 189.4 m$^3$/s, $R^2$: 0.828) having same inputs in both training and testing stages, respectively. The results clearly present that α component has positive influence on the models' efficiency. It improves the LSTM, ELM, and RF models' accuracy from 258.4 m$^3$/s, 275.3 m$^3$/s, and 311.8 m$^3$/s to 204.7 m$^3$/s, 218.8 m$^3$/s, and 300.4 m$^3$/s with respect to RMSE in the testing stage, respectively.

Time variation graphs of the observed and predicted streamflows by the best LSTM, ELM, and RF models in the test period are illustrated in Fig. 18.6 for the Kohala Station. It is observed that the LSTM follows the streamflow trend better than the other two models. RF cannot map the peak streamflows.

Figure 18.7 shows the scatter diagrams of the observed and predicted streamflows of three models. Less scattered predictions belong to the LSTM and it is followed by the ELM model. Residual histograms of the three machine learning models are shown in Fig. 18.8. As clearly observed from the histograms, LSTM residuals have generally

**Table 18.2** RMSE, MAE, and $R^2$ statistics of LSTM, ELM, and RF models using different river discharges input combinations—Kohala Station

| Models | Model inputs | Training period | | | Test period | | |
|---|---|---|---|---|---|---|---|
| | | RMSE | MAE | $R^2$ | RMSE | MAE | $R^2$ |
| *LSTM* | | | | | | | |
| | $Q_{t-1}$ | 326.5 | 241.4 | 0.727 | 311.2 | 236.7 | 0.672 |
| | $Q_{t-1}$, $Q_{t-2}$ | 185.9 | 113.5 | 0.911 | 251.3 | 184.5 | 0.818 |
| | $Q_{t-1}$, $Q_{t-2}$, $Q_{t-3}$ | 174.1 | 106.3 | 0.922 | 258.4 | 191.3 | 0.808 |
| | $Q_{t-1}$, $Q_{t-2}$, $Q_{t-3}$, α | 158.5 | 96.8 | 0.935 | **204.7** | **143.4** | **0.871** |
| | **Mean** | 211.3 | 139.5 | 0.874 | 256.4 | 189.0 | 0.792 |
| *ELM* | | | | | | | |
| | $Q_{t-1}$ | 355.6 | 278.9 | 0.677 | 316.0 | 240.3 | 0.656 |
| | $Q_{t-1}$, $Q_{t-2}$ | 220.0 | 153.3 | 0.876 | 291.8 | 190.0 | 0.786 |
| | $Q_{t-1}$, $Q_{t-2}$, $Q_{t-3}$ | 196.6 | 137.0 | 0.901 | 275.3 | 176.1 | 0.803 |
| | $Q_{t-1}$, $Q_{t-2}$, $Q_{t-3}$, α | 163.1 | 104.5 | 0.932 | **218.8** | **157.7** | **0.856** |
| | **Mean** | 233.8 | 168.4 | 0.847 | 284.0 | 192.5 | 0.771 |
| *RF* | | | | | | | |
| | $Q_{t-1}$ | 358.5 | 280.3 | 0.671 | 382.5 | 298.2 | 0.651 |
| | $Q_{t-1}$, $Q_{t-2}$ | 236.7 | 169.0 | 0.856 | 335.4 | 224.5 | 0.730 |
| | $Q_{t-1}$, $Q_{t-2}$, $Q_{t-3}$ | 230.0 | 169.8 | 0.864 | 311.8 | 216.4 | 0.762 |
| | $Q_{t-1}$, $Q_{t-2}$, $Q_{t-3}$, α | 181.1 | 120.1 | 0.916 | **300.4** | **189.4** | **0.828** |
| | **Mean** | 251.6 | 184.8 | 0.827 | 332.5 | 236.6 | 0.735 |

concentrated on the center (zero) while the RF has higher residuals compared to the other two models. For the range [−80,80], the total number residuals are, 63, 55, and 57 for the LSTM, ELM, and RF models, respectively.

Training and testing results of the implemented methods are reported in Table 18.3 or the Garhihabibullah Station. As observed from the table, the LSTM with $Q_{t-1}$, $Q_{t-2}$, $Q_{t-3}$, α (RMSE: 33.2 $m^3/s$, MAE: 20.5 $m^3/s$, $R^2$: 0.874) performs superior to the ELM (RMSE: 35.3 $m^3/s$, MAE: 22.5 $m^3/s$, $R^2$: 0.854) and RF (RMSE: 37.7 $m^3/s$, MAE: 25.9 $m^3/s$, $R^2$: 0.823) having same inputs in testing stage. There is a slight difference among the three models. In these stations, α component has also positive influence on the models' efficiency but not as much as the previous station. It improves the RMSE accuracies of the LSTM, ELM, and RF models from 37.7 $m^3/s$, 37.9 $m^3/s$, and 38.1 $m^3/s$ to 33.2 $m^3/s$, 35.3 $m^3/s$, and 37.7 $m^3/s$ in the testing stage, respectively.

Figure 18.9 demonstrates the time variation graphs of the observed and predicted streamflows by the best LSTM, ELM, and RF models in the test period for the Garhihabibullah Station. Similar trends are seen from the hydrograph, and graphs indicating the slight difference among the methods. Figure 18.10 illustrates the scat-

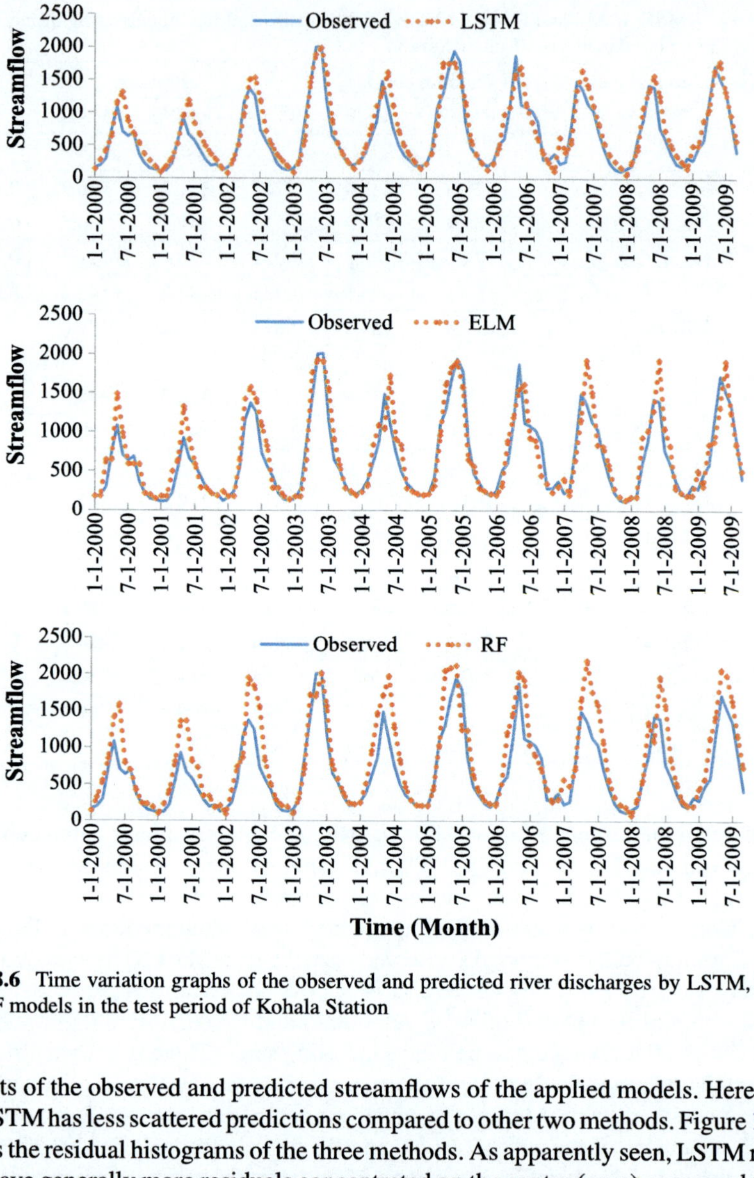


Fig. 18.6 Time variation graphs of the observed and predicted river discharges by LSTM, ELM, and RF models in the test period of Kohala Station

terplots of the observed and predicted streamflows of the applied models. Here also, the LSTM has less scattered predictions compared to other two methods. Figure 18.11 shows the residual histograms of the three methods. As apparently seen, LSTM residuals have generally more residuals concentrated on the center (zero) compared to the other two models. In the center, the number residuals in the center are 49, 40, and 40 for the LSTM, ELM, and RF models, respectively.

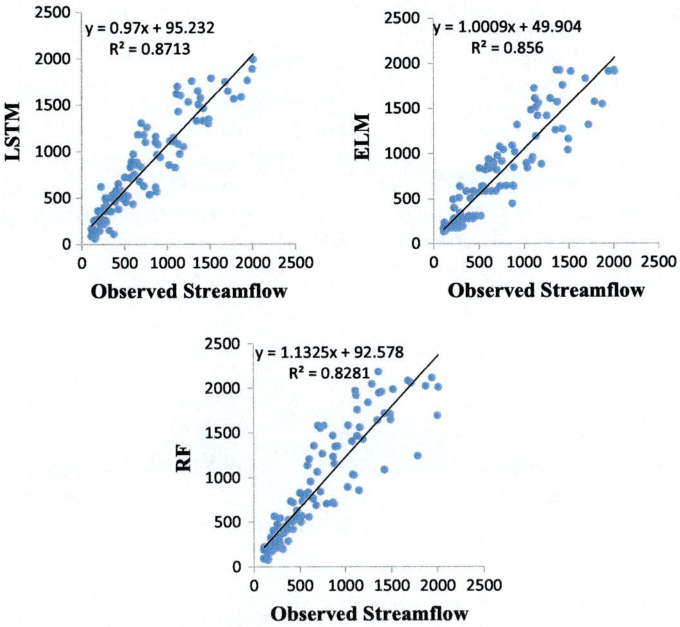


**Fig. 18.7** Time scatterplots of the observed and predicted river discharges by LSTM, ELM, and RF models in the test period of Kohala Station

## 18.4 Concluding Remarks

In the presented chapter, the ability of three machine learning methods, LSTM, ELM, and RF, was investigated in monthly streamflow prediction considering periodicity component. Data from two stations, Kohala and Garhihabibullah, Pakistan, were utilized. The following conclusions can be drawn from the outcomes of the applications.

- The LSTM with periodicity has generally provided better streamflow predictions in both stations.
- The RF model performed worse than the other two methods which must be because of the linear structure of this method.
- Importing periodicity component into the inputs considerably improves the models' efficiency for all the methods implemented.

In future studies, the implemented methods may be investigated using more data from different regions. The results can be improved by applying hybrid machine learning methods considering other effective parameters (e.g., rainfall, temperature, evaporation, etc., if available) periodicity input.

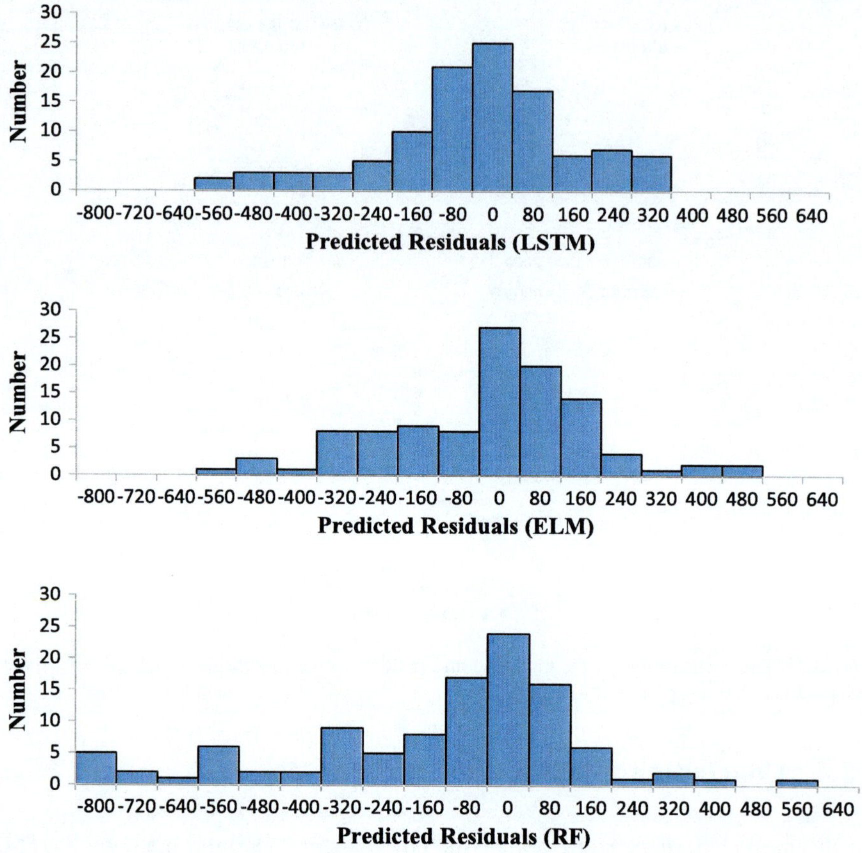


**Fig. 18.8** Residual histograms of the LSTM, ELM, and RF models in the test period of Kohala Station

**Table 18.3** RMSE, MAE, and $R^2$ statistics of LSTM, ELM, and RF models using different river discharges input combinations—Garhihabibullah Station

| Models | Model inputs | Training period | | | Test period | | |
|---|---|---|---|---|---|---|---|
| | | RMSE | MAE | $R^2$ | RMSE | MAE | $R^2$ |
| *LSTM* | | | | | | | |
| | $Q_{t-1}$ | 44.4 | 29.2 | 0.768 | 54.9 | 39.3 | 0.628 |
| | $Q_{t-1}$, $Q_{t-2}$ | 28.7 | 15.5 | 0.903 | 37.6 | 24.1 | 0.819 |
| | $Q_{t-1}$, $Q_{t-2}$, $Q_{t-3}$ | 27.4 | 14.1 | 0.911 | 37.7 | 25.3 | 0.828 |
| | $Q_{t-1}$, $Q_{t-2}$, $Q_{t-3}$, $\alpha$ | 27.4 | 15.5 | 0.912 | **33.2** | **20.5** | **0.874** |
| | **Mean** | 32.1 | 18.6 | 0.874 | 40.8 | 27.3 | 0.785 |
| *ELM* | | | | | | | |
| | $Q_{t-1}$ | 54.5 | 40.2 | 0.651 | 56.6 | 41.2 | 0.612 |

(continued)

**Table 18.3** (continued)

| Models | Model inputs | Training period | | | Test period | | |
|---|---|---|---|---|---|---|---|
| | | RMSE | MAE | $R^2$ | RMSE | MAE | $R^2$ |
| | $Q_{t-1}$, $Q_{t-2}$ | 29.9 | 18.1 | 0.894 | 39.6 | 27.1 | 0.808 |
| | $Q_{t-1}$, $Q_{t-2}$, $Q_{t-3}$ | 30.4 | 18.3 | 0.891 | 37.9 | 25.7 | 0.814 |
| | $Q_{t-1}$, $Q_{t-2}$, $Q_{t-3}$, $\alpha$ | 28.7 | 16.2 | 0.903 | **35.3** | **22.5** | **0.854** |
| | **Mean** | 35.9 | 23.2 | 0.835 | 42.4 | 29.1 | 0.774 |
| *RF* | | | | | | | |
| | $Q_{t-1}$ | 56.5 | 40.7 | 0.625 | 62.1 | 44.7 | 0.508 |
| | $Q_{t-1}$, $Q_{t-2}$ | 34.2 | 21.5 | 0.862 | 38.7 | 29.3 | 0.818 |
| | $Q_{t-1}$, $Q_{t-2}$, $Q_{t-3}$ | 32.1 | 20.1 | 0.879 | 38.1 | 27.4 | 0.808 |
| | $Q_{t-1}$, $Q_{t-2}$, $Q_{t-3}$, $\alpha$ | 23.9 | 12.6 | 0.933 | **37.7** | **25.9** | **0.823** |
| | **Mean** | 36.7 | 23.7 | 0.825 | 44.1 | 31.8 | 0.739 |

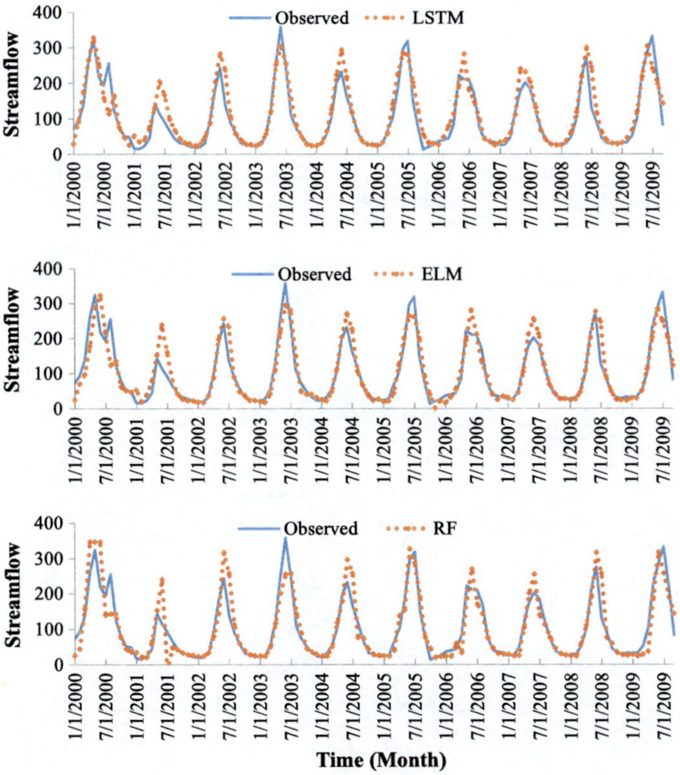


**Fig. 18.9** Time variation graphs of the observed and predicted river discharges by LSTM, ELM, and RF models in the test period of Garhihabibullah Station

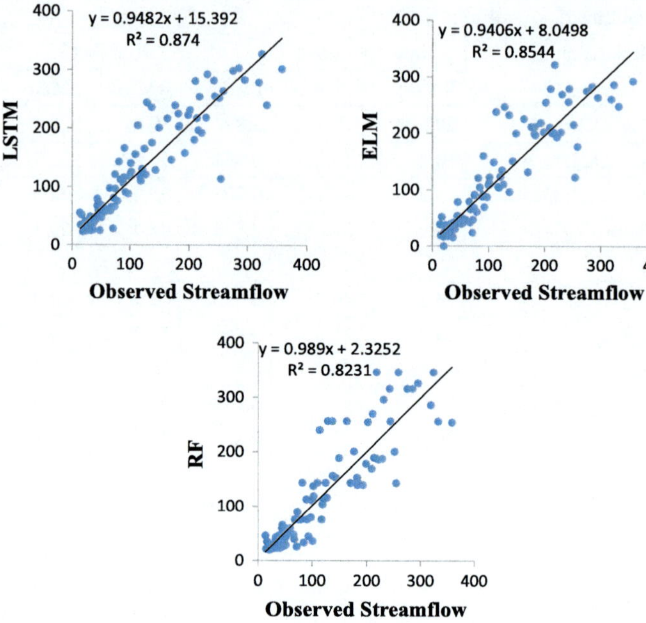


**Fig. 18.10** Time scatterplots of the observed and predicted river discharges by LSTM, ELM, and RF models in the test period of Garhihabibullah Station

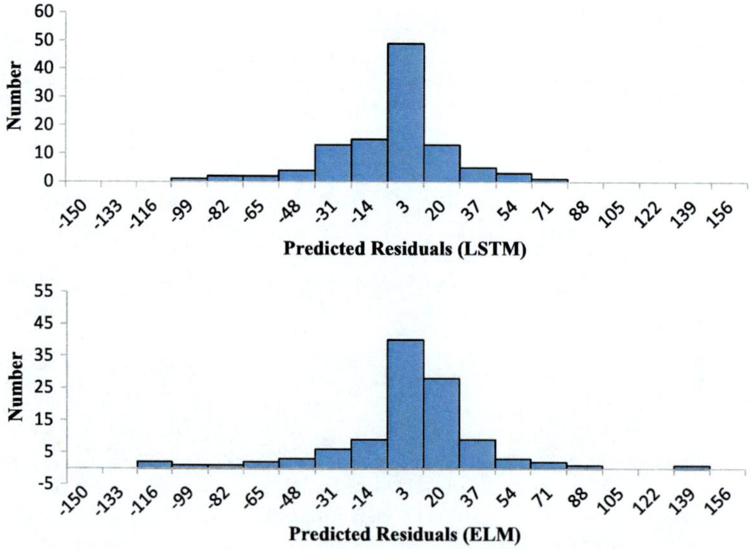


**Fig. 18.11** Residual histograms of the LSTM, ELM, and RF models in the test period of Garhihabibullah Station

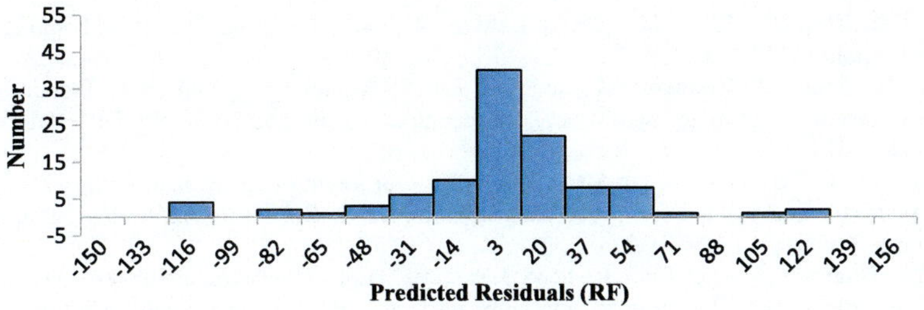


**Fig. 18.11** (continued)

# References

Adnan RM, Liang Z, El-Shafie A, Zounemat-Kermani M, Kisi O (2019a) Prediction of suspended sediment load using data-driven models. Water 11(10):2060

Adnan RM, Liang Z, Heddam S, Zounemat-Kermani M, Kisi O, Li B (2019b) Least square support vector machine and multivariate adaptive regression splines for streamflow prediction in mountainous basin using hydro-meteorological data as inputs. J Hydrol 124371

Adnan RM, Liang Z, Yuan X, Kisi O, Akhlaq M, Li B (2019c) Comparison of LSSVR, M5RT, NF-GP, and NF-SC models for predictions of hourly wind speed and wind power based on cross-validation. Energies 12(2):329

Adnan RM, Malik A, Kumar A, Parmar KS, Kisi O (2019d) Pan evaporation modeling by three different neuro-fuzzy intelligent systems using climatic inputs. Arab J Geosci 12(20):606

Adnan RM, Liang Z, Trajkovic S, Zounemat-Kermani M, Li B, Kisi O (2019e) Daily streamflow prediction using optimally pruned extreme learning machine. J Hydrol 577. https://doi.org/10.1016/j.jhydrol.2019.123981

Ali R, Kuriqi A, Abubaker S, Kisi O (2019) hydrologic alteration at the upper and middle part of the Yangtze River, China: towards sustainable water resource management under increasing water exploitation sustainability. 11, https://doi.org/10.3390/su11195176

Barzegar R, Asghari Moghaddam A, Adamowski J, Ozga-Zielinski B (2017) Multi-step water quality forecasting using a boosting ensemble multi-wavelet extreme learning machine model. Stoch Env Res Risk Assess 32:799–813. https://doi.org/10.1007/s00477-017-1394-z

Beven K (2002) Towards an alternative blueprint for a physically based digitally simulated hydrologic response modelling system. Hydrol Process 16:189–206. https://doi.org/10.1002/hyp.343

Breiman L (2001) Random forests. MachLearni 45(1):5–32

Chen J, Li M, Wang W (2012) Statistical uncertainty estimation using random forests and its application to drought forecast. Mathe Problems Eng 2012:1–12. https://doi.org/10.1155/2012/915053

Chung J, Gulcehre C, Cho K, Bengio Y (2014) Empirical evaluation of gated recurrent neural networks on sequence modeling. arXiv preprint arXiv:1412.3555

Cutler A, Cutler DR, Stevens JR (2012) Random forests. In Ensemble machine learning. Springer, Boston, MA, pp 157–175

Deo RC, Şahin M (2015) Application of the extreme learning machine algorithm for the prediction of monthly effective drought index in Eastern Australia. Atmos Res 153:512–525. https://doi.org/10.1016/j.atmosres.2014.10.016

Hochreiter S, Schmidhuber J (1997) Long short-term memory. Neural Comput 9(8):1735–1780

Huang GB, Babri HA (1998) Upper bounds on the number of hidden neurons in feedforward networks with arbitrary bounded nonlinear activation functions. IEEE Trans Neural Netw 9(1):224–229

Huang GB, Zhu QY, Siew CK (2006) Extreme learning machine: theory and applications. Neurocomputing 70(1–3):489–501

Kisi O, Alizamir M, Zounemat-Kermani M (2017) Modeling groundwater fluctuations by three different evolutionary neural network techniques using hydroclimatic data. Nat Hazards 87(1):367–381

Kisi O, Shiri J, Karimi S, Adnan RM (2018) Three different adaptive neuro fuzzy computing techniques for forecasting long-period daily streamflows. In: Big data in engineering applications. Springer, Singapore, pp 303–321

Kong Y-L, Huang Q, Wang C, Chen J, Chen J, He D (2018) Long short-term memory neural networks for online disturbance detection in satellite image time series remote sensing 10. https://doi.org/10.3390/rs10030452

Kratzert F, Klotz D, Brenner C, Schulz K, Herrnegger M (2018) Rainfall–runoff modelling using Long Short-Term Memory (LSTM) networks. Hydrol Earth Syst Sci 22:6005–6022. https://doi.org/10.5194/hess-22-6005-2018

Kratzert F, Klotz D, Shalev G, Klambauer G, Hochreiter S, Nearing G (2019) Towards learning universal, regional, and local hydrological behaviors via machine learning applied to large-sample datasets. Hydrol Earth Syst Sci 23(12):5089–5110

Kuriqi A, Pinheiro AN, Sordo-Ward A, Garrote L (2019) Influence of hydrologically based environmental flow methods on flow alteration and energy production in a run-of-river hydropower plant. J Clean Prod 232:1028–1042. https://doi.org/10.1016/j.jclepro.2019.05.358

Lee S, Kim J-C, Jung H-S, Lee MJ, Lee S (2017) Spatial prediction of flood susceptibility using random-forest and boosted-tree models in Seoul metropolitan city. Korea Geomatics, Nat Hazards Risk 8:1185–1203. https://doi.org/10.1080/19475705.2017.1308971

Le, Ho, Lee, Jung (2019) Application of long short-term memory (LSTM) neural network for flood forecasting water 11. https://doi.org/10.3390/w11071387

Liang Z, Tang T, Li B, Liu T, Wang J, Hu Y (2017) Long-term streamflow forecasting using SWAT through the integration of the random forests precipitation generator: case study of Danjiangkou Reservoir. Hydrol Res. https://doi.org/10.2166/nh.2017.085

Liang C, Li H, Lei M, Du aQ (2018) Dongting lake water level forecast and its relationship with the three gorges dam based on a long short-term memory network Water 10. https://doi.org/10.3390/w10101389

Ließ M, Glaser B, Huwe B (2012) Uncertainty in the spatial prediction of soil texture: comparison of regression tree and Random Forest models. Geoderma 170:70–79

Makkeasorn A, Chang NB, Zhou X (2008) Short-term streamflow forecasting with global climate change implications—A comparative study between genetic programming and neural network models J Hydrol 352:336–354. https://doi.org/10.1016/j.jhydrol.2008.01.023

Mosavi A, Ozturk P, Chau K-w (2018) Flood prediction using machine learning models: literature review water 10. https://doi.org/10.3390/w10111536

Muhammad Adnan R, Yuan X, Kisi O, Yuan Y, Tayyab M, Lei X (2017, October). Application of soft computing models in streamflow forecasting. In: Proceedings of the institution of civil engineers-water management, vol. 172, No. 3. Thomas Telford Ltd, pp 123–134

Nourani V, Hosseini Baghanam A, Adamowski J, Kisi O (2014) Applications of hybrid wavelet–Artificial Intelligence models in hydrology: a review. J Hydrol 514:358–377. https://doi.org/10.1016/j.jhydrol.2014.03.057

Papacharalampous GA, Tyralis H (2018) Evaluation of random forests and Prophet for daily streamflow forecasting. Adv Geosci 45:201–208. https://doi.org/10.5194/adgeo-45-201-2018

Rodriguez-Galiano V, Mendes MP, Garcia-Soldado MJ, Chica-Olmo M, Ribeiro L (2014) Predictive modeling of groundwater nitrate pollution using Random Forest and multisource variables related to intrinsic and specific vulnerability: a case study in an agricultural setting (Southern Spain) Sci Total Environ 476–477:189–206. https://doi.org/10.1016/j.scitotenv.2014.01.001

Sadler JM, Goodall JL, Morsy MM, Spencer K (2018) Modeling urban coastal flood severity from crowd-sourced flood reports using Poisson regression and Random Forest. J Hydrol 559:43–55. https://doi.org/10.1016/j.jhydrol.2018.01.044

Sahoo BB, Jha R, Singh A, Kumar D (2019) Long short-term memory (LSTM) recurrent neural network for low-flow hydrological time series forecasting Acta Geophysica 67:1471–1481. https://doi.org/10.1007/s11600-019-00330-1

Shortridge JE, Guikema SD, Zaitchik BF (2016) Machine learning methods for empirical streamflow simulation: a comparison of model accuracy, interpretability, and uncertainty in seasonal watersheds Hydrol Earth Syst Sci 20:2611–2628. https://doi.org/10.5194/hess-20-2611-2016

Singh B, Sihag P, Singh K (2017) Modelling of impact of water quality on infiltration rate of soil by random forest regression. Modeling Earth Syst Environ 3:999–1004. https://doi.org/10.1007/s40808-017-0347-3

Steinberg D (2009) CART: classification and regression trees. In The top ten algorithms in data mining. Chapman and Hall/CRC, pp 193–216

Taormina R, Chau K-W (2015) Data-driven input variable selection for rainfall–runoff modeling using binary-coded particle swarm optimization and Extreme Learning Machines. J Hydrol 529:1617–1632. https://doi.org/10.1016/j.jhydrol.2015.08.022

Tongal H, Booij MJ (2018) Simulation and forecasting of streamflows using machine learning models coupled with base flow separation. J Hydrol 564:266–282. https://doi.org/10.1016/j.jhydrol.2018.07.004

Wang Z, Lai C, Chen X, Yang B, Zhao S, Bai X (2015) Flood hazard risk assessment model based on random forest. J Hydrol 527:1130–1141. https://doi.org/10.1016/j.jhydrol.2015.06.008

Xu S, Niu R (2018) Displacement prediction of Baijiabao landslide based on empirical mode decomposition and long short-term memory neural network in Three Gorges area. China Comput Geosci 111:87–96. https://doi.org/10.1016/j.cageo.2017.10.013

Yadav B, Ch S, Mathur S, Adamowski J (2016) Discharge forecasting using an Online Sequential Extreme Learning Machine (OS-ELM) model: a case study in Neckar River. Germany Measure 92:433–445. https://doi.org/10.1016/j.measurement.2016.06.042

Yaseen ZM, Jaafar O, Deo RC, Kisi O, Adamowski J, Quilty J, El-Shafie A (2016) Stream-flow forecasting using extreme learning machines: a case study in a semi-arid region in Iraq. J Hydrol 542:603–614. https://doi.org/10.1016/j.jhydrol.2016.09.035

Yaseen ZM, Sulaiman SO, Deo RC, Chau K-W (2019) An enhanced extreme learning machine model for river flow forecasting: State-of-the-art, practical applications in water resource engineering area and future research direction. J Hydrol 569:387–408. https://doi.org/10.1016/j.jhydrol.2018.11.069

Yuan X, Chen C, Lei X, Yuan Y, Adnan RM (2018) Monthly runoff forecasting based on LSTM–ALO model. Stoch Env Res Risk Assess 32(8):2199–2212

Zounemat-Kermani M (2016) Assessment of several nonlinear methods in forecasting suspended sediment concentration in streams. Hydrol Res 48(5):1240–1252

# Chapter 19
# Empirical Model for the Assessment of Climate Change Impacts on Spatial Pattern of Water Availability in Nigeria

**Mohammed Sanusi Shiru, Eun-Sung Chung, and Shamsuddin Shahid**

## 19.1 Introduction

Climate change has serious potential impacts on the economic, environmental, social and agricultural sectors of any nation. Without a doubt, water resources and the agricultural sectors which are the most important to human existence are among the most affected sectors by the changing climate. There have been several reports of the impacts of climate change on water resources in different parts of the world (Wilhemi and Wilhite 2002; Piao et al. 2010; Ward 2014; Byakatonda et al. 2018). With continuous changes in the climatic variables and the effects they are having on our existence, understanding of the whole process from the causes to the changes that have occurred in the past to those that may happen in the future is crucial for preparation or mitigation of the impacts. The developing countries would be more affected by the impacts of climate change due to their lower adaptation capabilities (Abiodun et al. 2013, Collins et al. 2013). Most developing countries also have a higher density of population with less awareness of climate change (Lee et al. 2015). This implies that a significant population of the world is at the risk of one or more forms of the impacts of climate change.

M. S. Shiru (✉) · E.-S. Chung
Department of Civil Engineering,
Seoul National University of Science and Technology, Seoul 01811, Republic of Korea
e-mail: shiru.sanusi@gmail.com

M. S. Shiru
Department of Environmental Sciences, Faculty of Science, Federal University Dutse,
P.M.B 7156, Dutse, Nigeria

S. Shahid
Department of Water and Environmental Engineering, School of Civil Engineering,
Faculty of Engineering, Universiti Teknologi Malaysia (UTM), Johor Bahru 81310, Malaysia

© Springer Nature Singapore Pte Ltd. 2021

R. C. Deo et al. (eds.), *Intelligent Data Analytics for Decision-Support Systems in Hazard Mitigation*, Springer Transactions in Civil and Environmental Engineering,
https://doi.org/10.1007/978-981-15-5772-9_19

Water scarcity is the major issue that results from the impacts of climate change on natural systems (Salman et al. 2020). In line with this and due to the need for water for food security, countries are gradually directing focus to irrigated agriculture. Some recent studies have, however, noted that water resources would also face threats from climate change in the near future (Ranjan et al. 2006; Shahid et al. 2017; Salem et al. 2018; Kahsay et al. 2018). Precipitation pattern changes due to temperature rise will affect runoff (Cullen et al. 2002; Ionita et al. 2012) and consequently, the recharge of groundwater and TWS (Hanson et al. 2004; Holman et al. 2009; Tremblay et al. 2011; Perez-Valdivia et al. 2012). The decrease in soil moisture content could also reduce the recharge of groundwater and its availability as higher temperatures will increase evaporation and plant transpiration rates (Yu et al. 2015). The impact of climate change on water resources may be more severe in Africa.

The climate of Nigeria is changing in line with the global climatic change (Oloruntade et al. 2017; Shiru et al. 2018; Shiru et al. 2019a). This has had significant impacts on water resources in the country. Ayanlade et al. (2018) reported that rainfall is fluctuating, and there are reductions in rainfalls over the past 30 years as seen from the 5-year moving average. Negative impacts on crops and livestock were observed in Nigeria. As shown by studies, temperatures over Nigeria are generally increasing and some parts of Nigeria where rainfalls used to be high are witnessing decreasing trends in rainfall (Oguntunde et al. 2016; Oloruntade et al. 2017). For example, at the central and southeastern parts particularly in Benue state, which is known for its extensive agricultural practices, hence called "the food basket of the nation". Atedhor (2016) reported delayed rainfall onset in the Benue area resulting in the constriction of the growing seasons, and a negative deviation from normal precipitation (up to 35%) in some growing season months. This indicates a drying trend in the area and less availability of rainfall during growing seasons. The Benue area and the surrounding Kogi and Nassarawa states among some other states have been experiencing frequent violent clashes between farmers and herders due to resource competitions; mainly water and forage, for which climate change has been attributed to be one of the major causes (Merietu and Olarewaju 2009; Weezel 2017; Okoli and Atelhe 2014; Ubelejit 2016). If the dry trend continues as projected, in which Africa will be among the worst affected regions (Naumann et al. 2018), the conflicts may be aggravated as competition for these resources intensifies.

Understanding on-going changes and possible future changes in climate and their possible impacts are essential components of adaptive capacity and necessary in the development of effective climate change adaptation policies (Batisani and Yarnal 2010; Wang et al. 2016). This is particularly important as many countries of the globe now see groundwater as a buffer to water scarcity putting enormous pressure on the resource. Therefore, a reliable assessment of the changes in water resources due to climate change is very important for impact assessment and formulation of effective preparedness plans. However, the availability of reliable data is one major obstacle in the quantification of the impacts of climate change in many parts of the world, particularly on the African continent. Therefore, the choice of suitable gridded climate and hydrological data and the application of robust methods for the analysis of climate change impacts is critical for hydro-climatic studies in data-scarce regions such as Nigeria.

The objective of the present study was to project the impacts of climate change on water resources in Nigeria under different climate change scenarios. The study uses the Global Precipitation Climatology (GPCC) and Climate Research Unit (CRU) precipitation and temperature data, respectively, and Gravity Recovery and Climate Experiment (GRACE) TWS in assessing climate change impacts on water resources of Nigeria. In addition, downscaled future projections general circulation models (GCM) of precipitation and temperature of the models HadGEM2-ES, MRI-CGCM3, CESM1-CAM5 and CSIRO-Mk3-6-0 and their multi-model ensembles (MME) were used in assessing changes in water resources. Random forest (RF) and Support vector machine (SVM) were employed in the calibration and validation of empirical models after which performance indices were used in assessing the performances of the two methods. The best performing method was applied in projecting the changes in water resources of the study area due to climate change.

## 19.2 Description of Study Area

Covering an area of 923,000 km$^2$, Nigeria located in West Africa lies between latitude 4°15′–13°55′ N of the equator and longitude 2°40′ and 14°45′ E east of the Greenwich meridian. It is bounded to the north by the Niger Republic, northeast by Chad, to the west by the Benin Republic and to the east by Cameroon. The southern boundary is the Atlantic Ocean. Nigeria has 36 states including the federal capital territory. The states are divided into six geopolitical zones namely, the north-west, northeast, north-central, south-west, south-east and south-south. The population of Nigeria today stands at more than one hundred and eighty million people. The main contributors to the country's GDP are petroleum which provides about 80% and agriculture which contributes about 20% to the GDP (World Bank Group (WBG) 2019).

### *19.2.1 Climate of Nigeria*

Due to its large latitudinal extent (1,100 km), the climate of Nigeria is more varied than any other country in West Africa, thereby covering virtually all climatic belts of the region (Iloeje 1981). Precipitation over Nigeria is as a result of the progression of the West African Monsoon (WAM) over the Atlantic Ocean to the in-lands of the country which fundamentally arise as a result of thermodynamic contrasts between the land and the ocean surfaces (Balogun 1981; Olaniran and Sumner 1989; Thorncroft et al. 2011). The Atlantic cold tongue (cool water close to the equator between boreal spring and summer) and the Saharan heat low strongly influences the annual generation of the moisture fluxes, associated convergence and precipitation (Thorncroft et al. 2011).

Depending on the varying emphasis placed on cloud and precipitation development or temperature and humidity contrasts, the zone at the surface which separates these air streams has been variously called the intertropical convergence zone (ITCZ), the intertropical front (ITF) and the intertropical discontinuity (ITD) (Olaniran and Sumner 1989). The surface location of the ITD is determinant of the spatial and temporal precipitation variation pattern over Nigeria in which there is a general decrease both in the duration and amount from the coastline to the interior. The elevation across Nigeria ranges from zero meters in the south around the Atlantic Ocean to over two thousand meters in the north with the highest point being 2,419 m at Chappal Waddi in the northeastern part. This altitudinal variation alongside the aforementioned ITD location plays a significant role in the precipitation patterns and distribution over the country.

In terms of climate, Nigeria is divided into four zones from the south to the north: monsoon climate, tropical savanna climate, warm semi-arid climate and warm desert climate (Fig. 19.1). In addition, there are significant variations in the ecology of Nigeria with Sahel Savanna, Sudan Savanna, Guinea Savanna, Rainforest and Mangrove Swamp ecological types occurring from the north to the south in respective order.

As the climate varies from the north to the south, the onset of the seasons varies for the two regions. The rainy season in the semi-arid and arid north is between June and September while rainfall starts in April and occurs till October in the central and

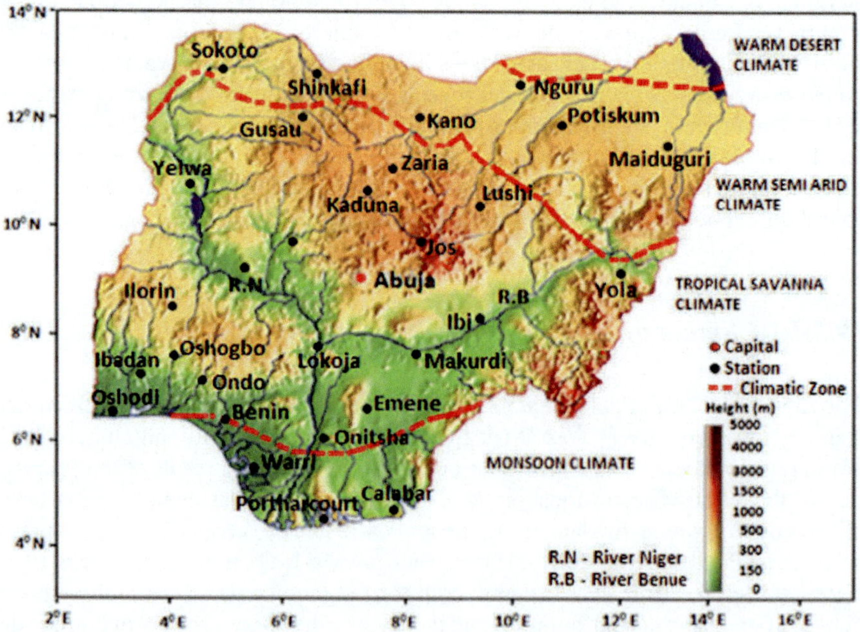


**Fig. 19.1** Study area map showing the climatic zones, gauge stations, and elevation

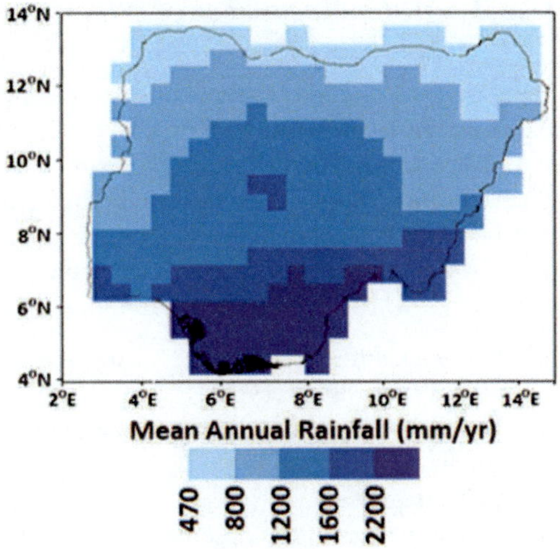


**Fig. 19.2** The spatial variation of mean annual rainfall (mm/yr) of Nigeria

southern parts of the country. The annual average rainfall in the country varies from more than 2,000 mm in the south to less than 500 mm in the north (Fig. 19.2). The daily maximum temperature in the south ranges from 30 to 37 °C. In the north, it can reach up to 45 °C during the dry season. The minimum temperatures during the dry cold season in the north are as low as 12 °C while it ranges from 17 to 24 °C in the south. The average temperature varies from less than 22 °C in high elevation areas to more than 28 °C across the country (Fig. 19.3).

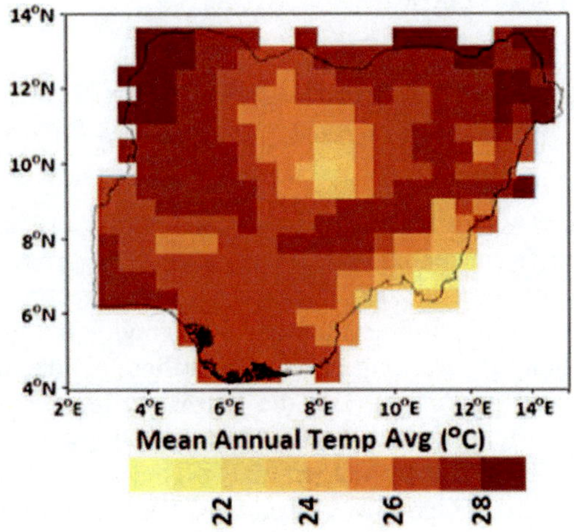


**Fig. 19.3** The spatial variation of mean daily average temperature (°C) over Nigeria

### 19.2.2 Water Resources of Nigeria

Many parts of Nigeria receive considerable amounts of rainfall; even the semi-arid and arid regions of the country receive better annual average rainfall than some other parts of the world where rainfalls are extremely scarce. Therefore, water resources in Nigeria compared to some other areas of the globe are not scarce in terms of quantity. It is, however, faced with several problems including pollution, mismanagement and over abstraction in many areas of the country.

Oteze (1981) divided the water resources of Nigeria under the following headings:

i. Surface sources: rivers and streams, springs, lakes and drainage areas funneling waters towards reservoirs
ii. Underground sources: eleven principal aquifers in the sedimentary basins of the country
iii. Coastal aquifers
iv. Groundwaters of the basement complex areas.

A close network of rivers and streams having their sources from the Precambrian basement complex and flow over the sediment in their lower reaches drain Nigeria (Oteze 1981). The drainage system of the country is dominated by the two main rivers: River Niger (RN) and River Benue (RB) (Fig. 19.1). The runoff quantities generated from the drainage basins varies from place to place and is affected by the rainfall intensity, climate and vegetation and geological and topographical features (Ojiako 1985).

## 19.3 Data and Sources

### 19.3.1 Gravity Recovery and Climate Experiment (GRACE) Terrestrial Water Storage (TWS) Data

GRACE was launched in the year 2002 and has a number of available products of different spatial scales that are processed routinely by different data centers (Chinnasamy et al. 2015). The gravity phenomenon on which GRACE is based has advantages of (1) direct link between gravity and mass storage which is independent of lithology and requires no calibration, (2) the distant effect of the satellite allowing deep penetration into the Earth and recording of mass storage in water systems (Castellazzi et al. 2016). The GRACE gridded data considered in this present study are available from April 2002 till the year 2017. GRACE monthly gridded terrestrial water storage (TWS) data have been used in many parts of the globe including in Africa for water resources studies (Bonsor et al. 2018; Hassan and Jin 2016).

This study uses the monthly changes in TWS data of the CSR RL05 GRACE Mascon solutions having a spatial resolution of $1.0° \times 1.0°$ in assessing the impacts of climate change on water resources of the study area during the period 2002–2016 for 80 grid points within Nigeria.

### *19.3.2 Global Precipitation Climatology Center (GPCC) Rainfall Data*

The monthly precipitation data of GPCC was developed from rainfall records obtained from various sources (Schneider et al. 2014). The data include available near real-time such as synoptic weather reports (SYNOP) and monthly climate reports (CLIMAT) distributed by national meteorological and hydrological services (NMHSs) through the World Meteorological Organization (WMO) and global telecommunication system (GTS), and non-real-time rainfall data obtained from over 85,000 rain-gauges located in about 190 countries (Schneider et al. 2011).

The GPCC, unlike some other gauge based gridded rainfall products that were unable to address the sparseness of observed data and quality control issues in data analysis, was able to address these problems (Schneider et al. 2014). It has a suite of global gridded precipitation products including the global precipitation climatology, the first guess product, the monitoring product and the full data reanalysis product in which the full data reanalysis product offers better accuracy compared to the other near real-time products (Schneider et al. 2011).

For gridding, the GPCC employs a smart interpolation technique that has the ability to consider the systematic relationship between elevation and station observations, and consequently, the ability to enhance estimation accuracies (Funk et al. 2007). Other advantages of using GPCC data include: (1) the data set quality is good enough for hydro-climatic analysis; (2) it is a climate model derived dataset which uses the highest number of observed rainfall records; and (3) timespan of data is long enough for conducting hydro-climatic studies (Spinoni et al. 2014). The GPCC has been used in a number of studies for hydro-climatic analysis in Africa (Dinku et al. 2008; Yang et al. 2014; McNally et al. 2017).

### *19.3.3 Climate Research Unit (CRU) Temperature Data*

The Climate Research Center (CRU) was established in the year 1972 in the school of Environmental Sciences at the University of East Anglia, Norwich. Owing to less investigation of past climatic changes and variability before the 1960s, the CRU came with the objective of establishing the past climatic record over as much of the world as possible and dating as far back as possible, and in enough details. Among their work, the production of the world's land-based, gridded temperature data set currently at 5° by 5° latitude/longitude boxes which started in 1978 has been of great significance to the international community with the probable largest impact. The most recent update of the CRU temperature data set is the HadCRUT4 which includes the addition of newly digitized measurement data at both sea and overland, new bias adjustment of sea surface temperature, and a more comprehensive error model for sea surface temperature measurements uncertainty description (Morice et al. 2012).

The CRU database was developed from gauge measurements obtained from about 4000 weather stations located around the world. All the collected data are passed through two-stage extensive manual and semi-automatic quality control measures to develop the gridded monthly temperature anomalies (Harris et al. 2014). An interpolation method known as angular distance-weighted was used to develop CRU gridded data from gauge measurements. A number of studies found that CRU was suitable in reconstructing the historical temperature of Africa (Simmons et al. 2004; Hao et al. 2013; Omondi et al. 2014).

### *19.3.4 Methodology*

#### 19.3.4.1 Modelling Climatic Influences on Water Resources

GRACE TWS anomaly, GPCC rainfall and CRU temperature were used to assess the climatic influences on water resources. In this study, TWS anomaly was modelled using only rainfall and temperature data. Data-driven models namely, RF and SVM were used for this purpose. The procedure used for this purpose is outlined below:

i. Calibration and validation of the developed models using historical data and the selection of the better performing model between the RF and the SVM.
ii. The projected MME rainfall and temperature data were used in the simulation of the TWS for each of the selected models under the four RCPs.
iii. Generation of an ensemble TWS projection model for all RCPs from the four selected models.
iv. Spatial assessment of the annual changes in TWS under all RCPs during the period 2010–2039, 2040–2069 and 2070–2099.
v. Spatial assessment of annual TWS anomaly during the whole period 2010–2099.

Details about the selection of GCMs, their downscaling, MME generation and projection of rainfall used in this study can be found in Shiru et al. (2019b). The same approach used in the study was applied for temperature.

## 19.4 Results and Discussion

### *19.4.1 Model Calibration and Validation*

The scatter plots of the calibration and validation of the developed model for simulation of TWS anomalies are given in Fig. 19.4. It could be seen that the RF has better calibration and validation results compared to the SVM and was therefore chosen for the modelling of the impacts of climate change on water resources for Nigeria.

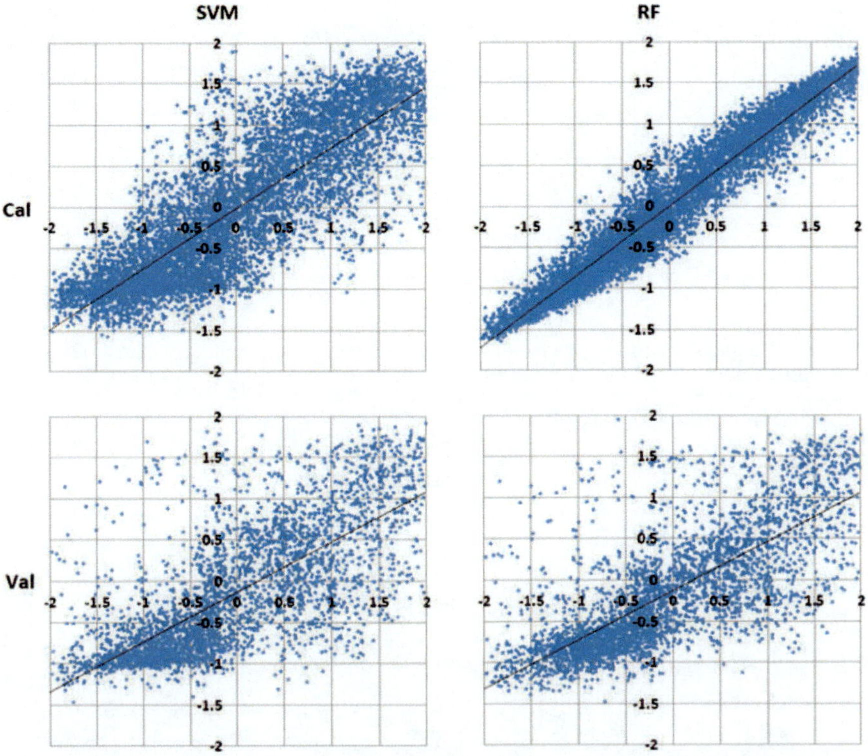


**Fig. 19.4** Scatter plots of the calibration and validation of the developed model for simulation of TWS changes

The performance of the calibration and validation models using RF was assessed using the performance metrics: Pbias, NRMSE, RSD, NSE and $R^2$ as presented in Figs. 19.5 and 19.6, respectively. The results showed good performance indicating the ability of the RF to model changes in water availability under a changing climate.

### *19.4.2 Changes in Seasonal Rainfall*

Seasonal changes of rainfall for the period 2070–2099 were assessed for the different zones within the study area by averaging the monthly rainfalls for all the grids points of Nigeria. The results, presented in Fig. 19.7, show that at zone 1, changes in rainfall will be higher during this period than the GPCC for all the RCPs. However, the RCPs 2.6, 4.5 and 8.5 shows lower changes in rainfall between the months of July and August at this zone. In zone 2, while rainfall increments are observed for most of the rainy months, changes in rainfall decreased between the months of July

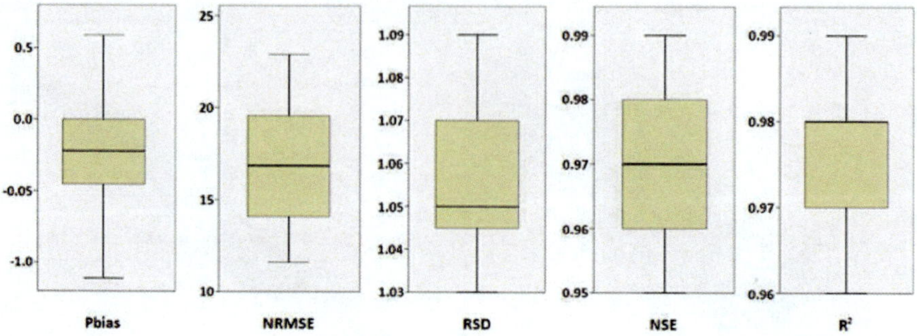


**Fig. 19.5** Boxplots of performance metrics for the calibration of the impacts of climate change on water resources

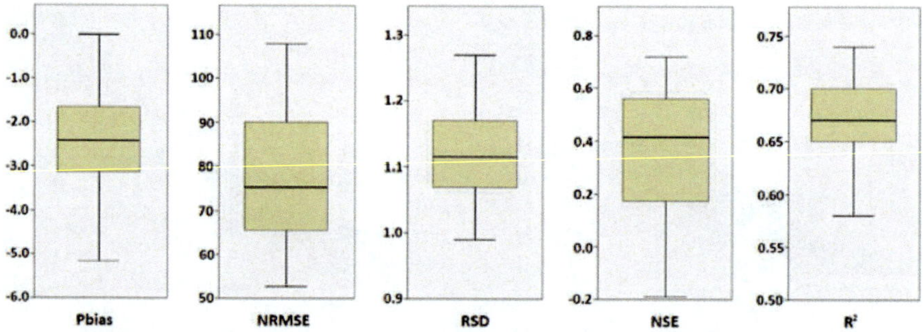


**Fig. 19.6** Boxplots of performance metrics of the models during the validation

and August for all the RCPs at this zone. Similarly, changes in rainfall decreased from middle June to the end of September in the zone 3 for all the RCPs. Other rainy periods for this zone show higher increases in projected rainfalls for all RCPs. In zones 4 and 5, the period between June and September also shows that rainfalls will decrease for the projections for all the RCPs. Other periods, however, show increases in rainfall projections. The rainfalls decrease to the greatest extent in the zones 3, 4 and 5 where the rainfalls are usually normally high. Decreases in rainfall for this period are highest for RCP 8.5 in all zones and especially in zone 5. Highest increases in rainfall are observed between April and June in zones 4 and 5 while increases are occurring in the semi-arid zone 1 and 2 during the month of June when the rainfalls are just starting. Other months after the peak of rainfalls, occurring in August and September in the north, and in July and August in the south shows slight increases in the projected rainfalls.

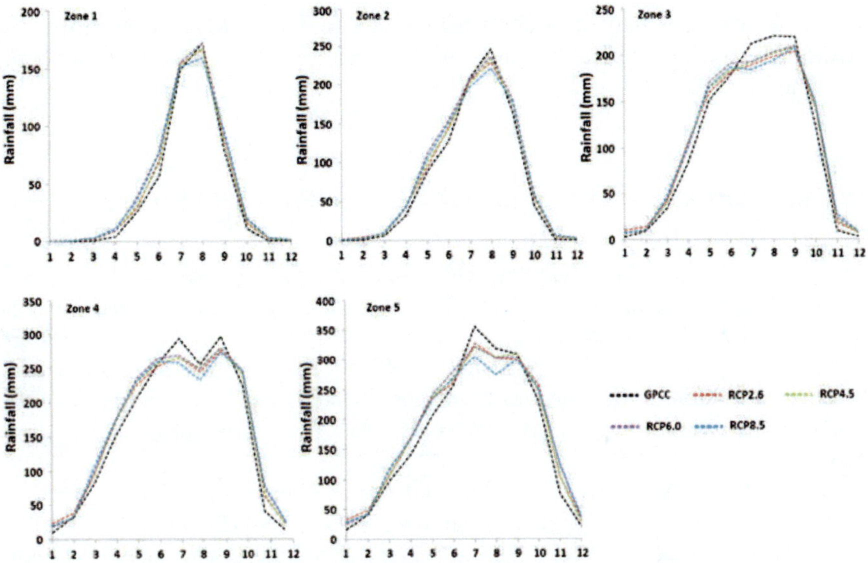


**Fig. 19.7** Projected changes (%) in monthly rainfall in different regions of Nigeria between 2070 and 2099 compared to GPCC rainfall for the base year (1971–1990)

## *19.4.3 Changes in Seasonal Maximum Temperature*

Seasonal changes of maximum temperature for the period 2070–2099 were assessed for the different zones within the study area by averaging the monthly temperatures for all the grid points. The changes in temperature for all the months show that they are highest for the RCP 8.5, followed by RCPs 6.0, 4.5 and 2.6 in respective orders for all zones (Fig. 19.8). While changes among the RCPs are lower between January and March in the zones 3, 4 and 5, they are higher in the zones 1 and 2. After these

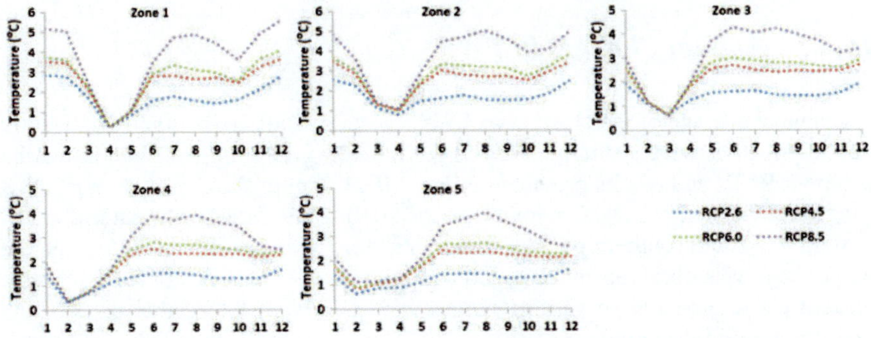


**Fig. 19.8** Projected changes (absolute) in monthly maximum temperature in different regions of Nigeria between 2070 and 2099 compared to CRU temperature for the based year (1961–1990)

months, changes are higher especially between the months of May and October. The smallest changes in temperature for the study area are expected for zone 1 for all the RCPs in the month of April.

### *19.4.4 Seasonal Changes in TWS Under Projected Climate*

The calibrated and validated models were used for the simulation of future change in TWS due to the changes in climate. The results of the projections of the seasonal changes in TWS for the future periods 2010–2039, 2040–2069 and 2070–2099 are given in Figs. 19.9, 19.10, 19.11 and 19.12 for RCPs 2.6, 4.5, 6.0 and 8.5, respectively. There are variations in the changes for all the selected models, the periods and the RCPs considered in this study. Variations also exist from zone to zone of the study area.

For RCP 2.6, there will be a significant decrease in ranging from 0 to 12 m in the TWS at zone 1 as projected by the models HadGEM2-ES and MRI-CGCM3 during the rainy season between June and September for the three considered periods. The models CESM1-CAM5 and CSIRO-Mk3-6-0 also showed decreases during the rainy season but lesser compared to the other models. Decreases during the peak rainy season at zones 2, 3, 4 and 5 are 8, 5, 4 and 5 m, respectively, for RCP 2.6 for the 2010–2039 periods. Other periods 2040–2069 and 2070–2099 also showed the highest declines in TWS were at the zones 1 and 2 while the minimum changes were at zone 3 of the study area.

For RCP 4.5, zone 1 also showed the greatest changes in TWS compared to observe during the month of September when the rainfall is at its peak. This was also observed to be highest for HadGEM2-ES and MRI-CGCM3 compared to CESM1-CAM5 and CSIRO-Mk3-6-0. Changes in TWS were lowest for this RCP at zone 3 for all the periods and especially for 2010–2039. Similar results were obtained for RCPs 6.0 and 8.5 with the highest changes of up to 13 m for the century.

### *19.4.5 Annual Changes in TWS*

The annual average of the changes in TWS for the periods 2010–2039, 2040–2069 and 2070–2099 were spatially assessed for the 323 grid points of Nigeria for the different RCPs and results presented in Fig. 19.13. Figure show decreases in TWS ranges between −3.0 and 0.0 m for the period 2010–2039 at the north-eastern, south-eastern and south southern parts of the country for RCPs 2.6, 4.5 and 8.5 while the same range was observed for RCP 6.0 and at the same areas except for the north-eastern parts. Increases of varying ranges were observed in the other parts of the country during this period.

During the period 2040–2069, decreases in TWS were highest for RCPs 4.5 and 8.5 reaching −4.0 m at the south-eastern parts of the country. Increases of up to 7.2 m

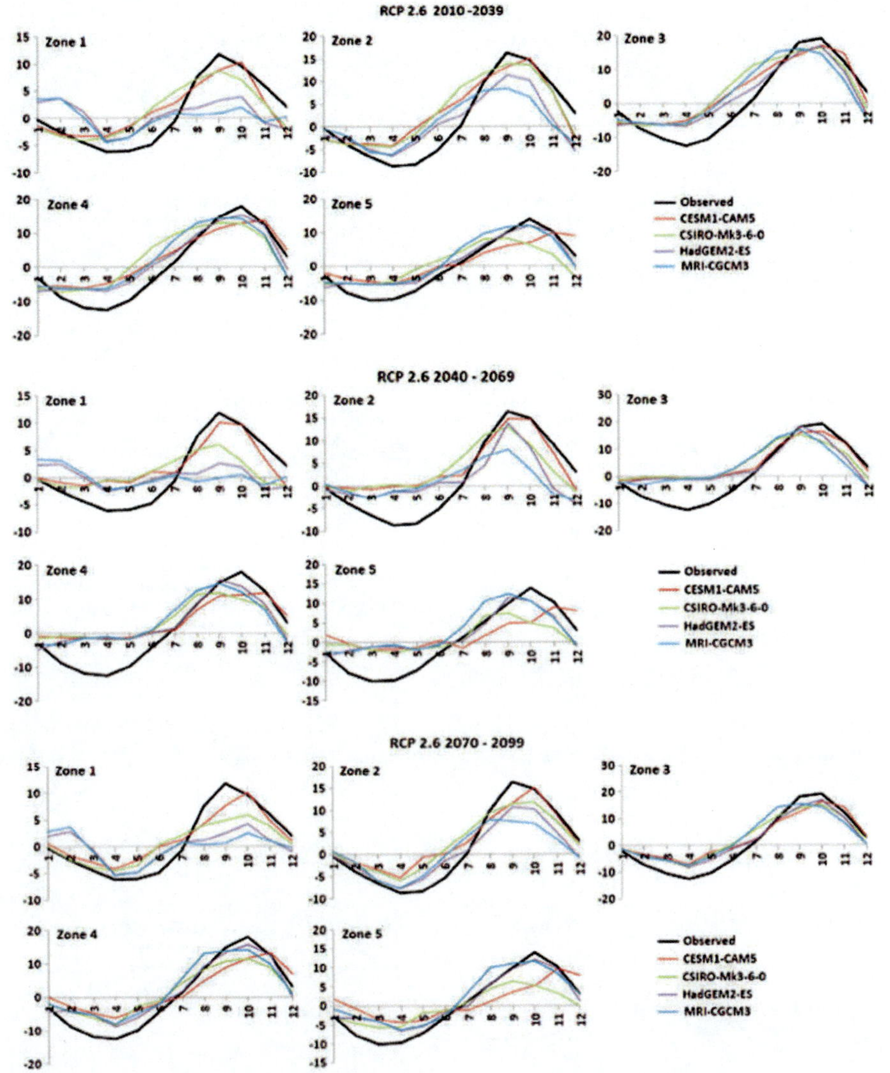


**Fig. 19.9** Projection of TWS in different months during 2010–2039, 2040–2069, and 2070–2099 under RCP 2.6 using selected GCMs

were observed at some locations within the study area. For the period 2070–2099, decreases in TWS were up to −1.5 m for the RCPs 2.6 and 8.5 while for the RCPs 4.5 and 6.0 decreases were up to −3.0 m. Decreases were most pronounced in the southeast and south-south of the country. Most northern parts of the country were found to experience increasing TWS. This is attributable to the increasing rainfalls expected to occur for the area.

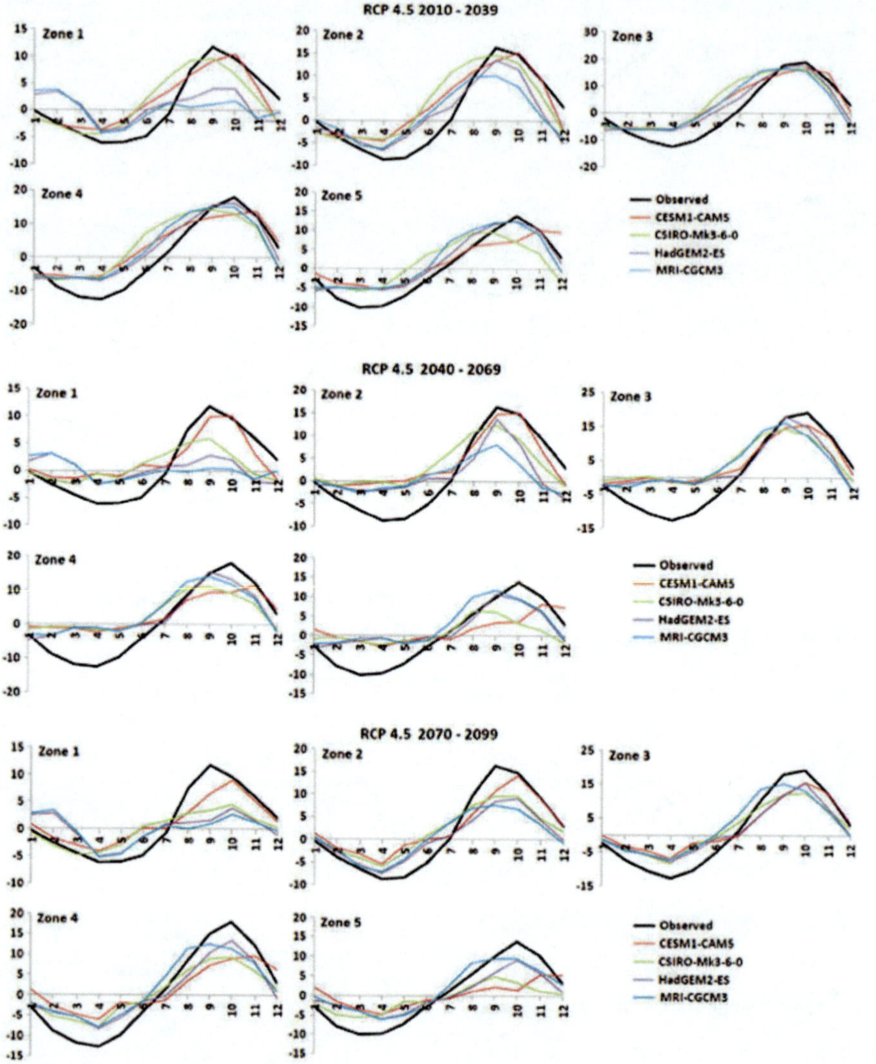


**Fig. 19.10** Projection of TWS in different months during 2010–2039, 2040–2069, and 2070–2099 under RCP 4.5 using selected GCMs

Yearly average TWS for all RCPs taken for the entire period 2010–2099 at 323 grid points of the country was also spatially mapped and results presented in Fig. 19.14. The figure shows that TWS would decrease at the north-eastern down to the south-eastern and south southern parts of the country for RCPs 2.6 and 8.5. For RCPs 4.5

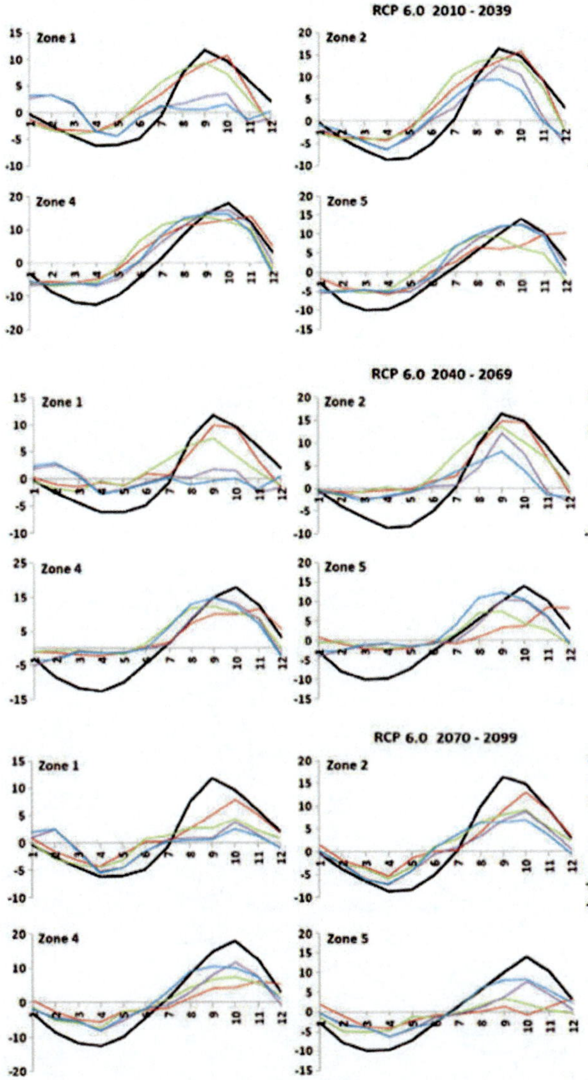


**Fig. 19.11** Projection of TWS in different months during 2010–2039, 2040–2069, and 2070–2099 under RCP 6.0 using selected GCMs

and 6.0, decreases in TWS were not observed in the northern parts of the country but in the south-eastern and south southern parts of the country. The highest decreases in TWS were observed for RCP 4.5.

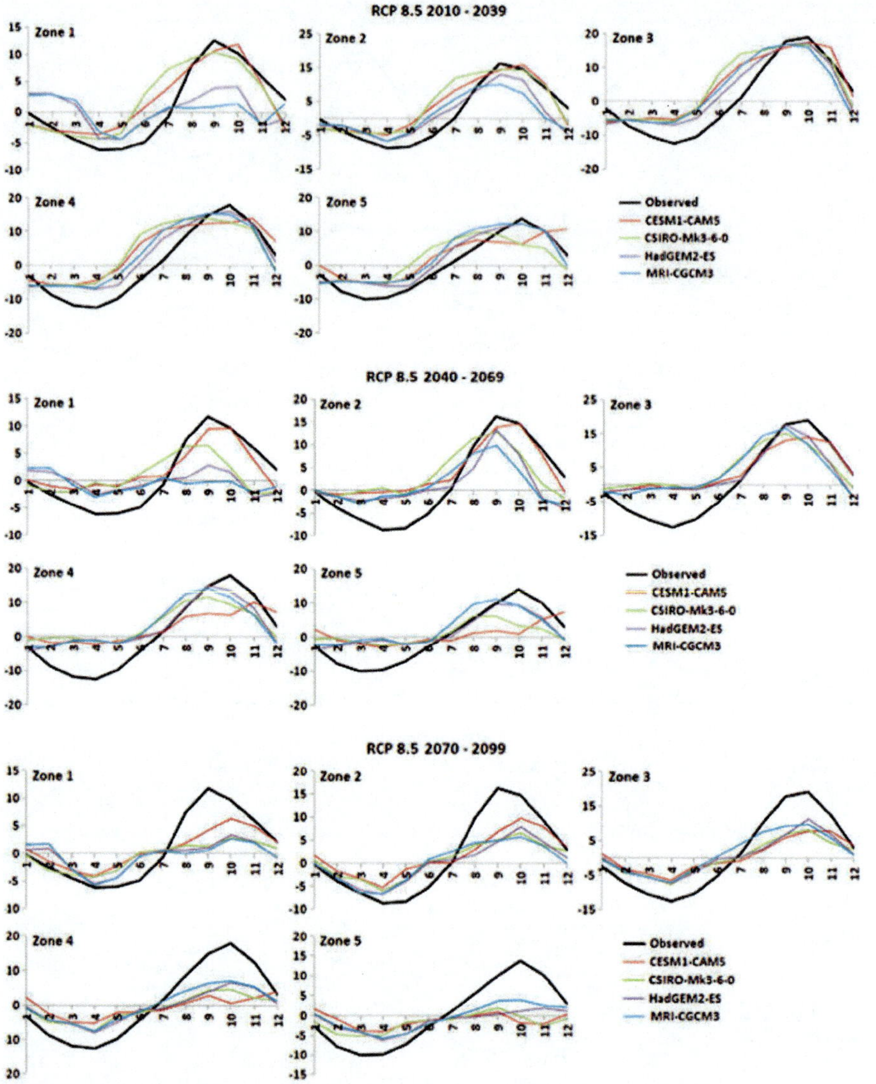


Fig. 19.12 Projection of TWS in different months during 2010–2039, 2040–2069, and 2070–2099 under RCP 8.5 using selected GCMs

## 19.5 Discussion

Freshwater resources which are mostly faced with deterioration of quality due to pollution in many parts of the globe have been further stressed by over-abstraction in many parts of the globe (Salem et al. 2017; Ahmed et al. 2019; Sediqi et al.

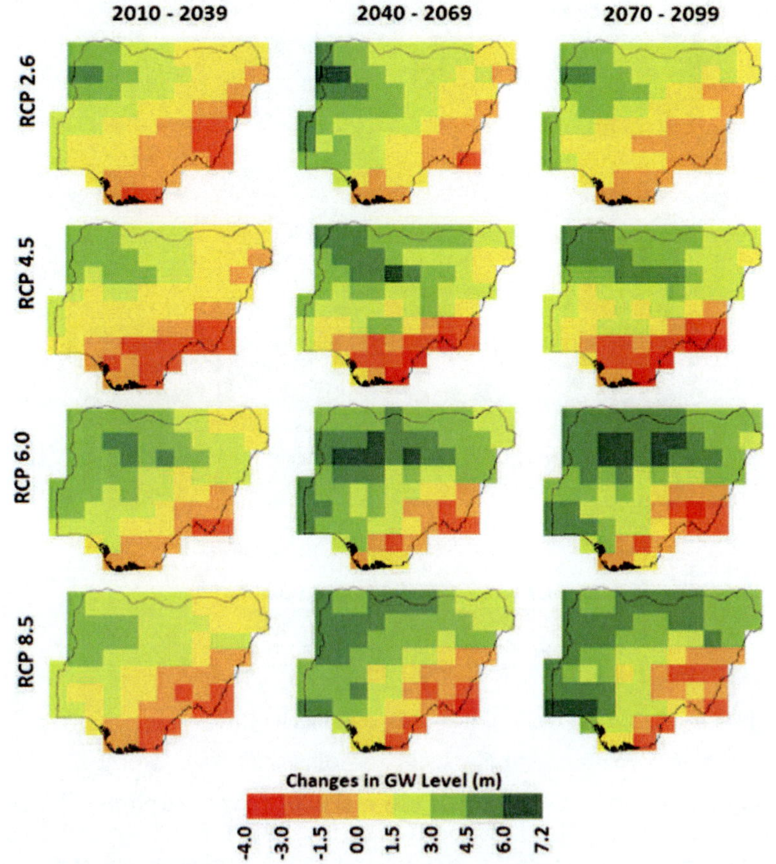


**Fig. 19.13** Changes in TWS for the periods 2010–2039, 2040–2069, and 2070–2099 for all RCPs

2019). This is due to the impacts of climate change which have caused an increase in temperature and decrease in rainfall leading to insufficient water for crops and agricultural practices in many areas especially areas practicing rain-fed agriculture (Qutbudin et al. 2019; Alamgir et al. 2019). As a result, groundwater is seen as a buffer for continuous water availability for agricultural practices and is overly drafted for irrigation purposes. This has been more severe in countries having a higher population or higher rates of population growth. For example, about 48 million hectares of lands in Bangladesh, Pakistan and India of the Indian sub-continent are cultivated using groundwater irrigation, constituting about 42% of the global groundwater-fed croplands (Watto 2015). The risks of water resource availability may be aggravated by future changes in climate change as shown from climate projections in many studies (Rashid et al. 2015; Sa'adi et al. 2019; Homsi et al. 2020; Shiru et al. 2019c).

In Africa, climate change impacts on water resources have been reported (Calow et al. 2010; Bokar et al. 2012) and projections have shown that water resources will be

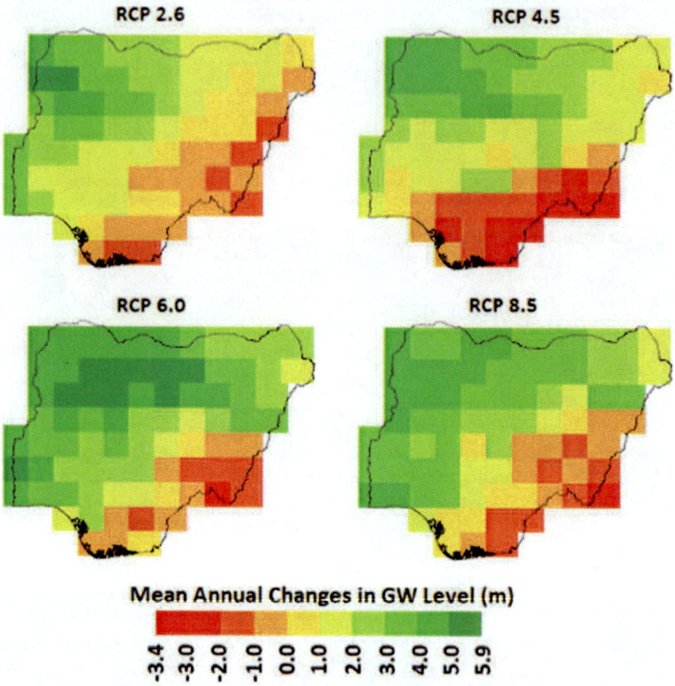


**Fig. 19.14** Changes in TWS for the period 2010–2099 for all RCPs

significantly impacted in the future. Kahsay et al. (2018) reported projected decreases in groundwater recharge by 3.4% for RCP 2.6 and 1.3% for RCP 4.5 and decreases in base flow by 1.5 and 0.55% for RCP 2.6 and RCP 4.5, respectively, at the Tekeze basin of Ethiopia in a study conducted to assess the impacts of climate change on the groundwater recharge and base flow of the basin. In Algeria, Meddi and Boucefiane (2013) conducted research on climate change impacts on the water resources of the Cheliff-Zahrez basin using the rain/infiltration approach. They concluded that the overexploitation of water resources of the basin has increased due to decreases in rainfall which affects water levels in dams. Their projections showed that the water resources of the region will decrease by 4.4–6.6% by 2020 and by 10–15% by 2050.

While studies on climate change relating to hazards such as droughts, floods and agriculture have been widely conducted in Nigeria, climate change impacts on water resources have not been widely assessed particularly those relating to projections of climate change impacts on water resources. Macdonald et al. (2005) reported that some areas of Nigeria particularly the semi-arid and arid northern regions are facing fast declination in groundwater. Similarly, an overall decline in water resources in Nigeria as a result of climate change has been suggested, with the most likely highest impacts in the North Niger and Chad basins (JICA 2014). This present study supports findings from previous studies on the already declining water resources and the future impacts of climate change on water resources at large.

## 19.6 Concluding Remarks

The impacts of climate change on water resources of Nigeria were assessed in this study using GPCC precipitation and CRU temperature data and GRACE TWS data at 323 grid points of Nigeria. Downscaled future projections GCM of precipitation and temperature of the models HadGEM2-ES, MRI-CGCM3, CESM1-CAM5 and CSIRO-Mk3-6-0 and their MME were used in assessing changes in water resources. Calibration and validation of empirical models were conducted using RF and SVM after which performance indices were used in assessing the performances of the two methods and the better performing one chosen.

The assessment of the impacts of climate change on water resources shows that the changes in rainfall patterns and the rise in temperature would affect TWS in Nigeria in the future. The effects may be more pronounced in the northeast, southeast and south-south of the country as observed from the spatial assessment of annual changes in water resources.

With continuous increase in population and dependency on rain-fed agriculture, development of reliable and mitigation measures against droughts are imperative for agricultural sustainability in Nigeria. Due to the rise in frequency and severity of droughts, surface waters may become less available in areas where they are mostly dependent on, and shifts from such usage to water resources may arise. This may result in over-abstraction of the resource in the future jeopardizing sustainable availability in water resources in many places. It is therefore holistically critical for Nigeria to implement better management strategies of water resources to sustain its water availability especially with its projected increase in population, a continually changing climate, and its efforts to transition into a more agricultural-based economy.

## References

Abiodun BJ, Lawal KA, Salami AT, Abatan AA (2013) Potential influences of global warming on future climate and extreme events in Nigeria. Reg Environ Change 13:477–491

Ahmed K, Shahid S, Demirel MC, Nawaz N, Khan N (2019) The changing characteristics of groundwater sustainability in Pakistan from 2002 to 2016. Hydrogeol J 12. https://doi.org/10.1007/s10040-019-02023-x

Alamgir M, Mohsenipour M, Homsi R, Wang X, Shahid S, Shiru MS, Alias NE, Yuzir A (2019) Parametric assessment of seasonal drought risk to crop production in Bangladesh. Sustainability 11:1442. https://doi.org/10.3390/su11051442

Atedhor GO (2016) Growing season rainfall trends, alterations and drought intensities in the Guinea Savanna belt of Nigeria: implications on agriculture. J Environ Earth Sci 6(3):13

Ayanlade A, Radeny M, Morton JF, Muchaba T (2018) Rainfall variability and drought characteristics in two agro-climatic zones: an assessment of climate change challenges in Africa. Sci Total Environ 630:728–737. https://doi.org/10.1016/j.scitotenv.2018.02.196

Balogun EE (1981) Seasonal and spatial variations in thunderstorm activity over Nigeria. Weather 36(1):192–218

Batisani N, Yarnal B (2010) Rainfall variability and trends in semi-arid Botswana: implications for climate change adaptation policy. Appl Geogr 30(4):483–489

Bokar H, Mariko A, Bamba F, Diallo D, Kamagaté B, Dao A (2012) Impact of climate variability on groundwater resources in Kolondieba Catchment Basin, Sudanese Climate Zone in Mali. Int J Eng Res Appl 2:1201–1210

Bonsor HC, Shamsudduha M, Marchant BP, MacDonald AM, Taylor RG (2018) Seasonal and decadal groundwater changes in African sedimentary aquifers estimated using GRACE products and LSMs. Remote Sens 10:904. https://doi.org/10.3390/rs10060904

Byakatonda J, Parida BP, Moalafhi DB, Kenabatho PK (2018) Analysis of long term drought severity characteristics and trends across semiarid Botswana using two drought indices. Atmos Res 213:492–508. https://doi.org/10.1016/j.atmosres.2018.07.002

Calow RC, MacDonald AM, Nicol AL, Robins NS (2010) ground water security and drought in Africa: linking availability, access, and demand. Ground Water 48(2):246–256

Castellazzi P, Martel R, Galloway DL, Longuevergne L, Rivera A (2016) Assessing groundwater depletion and dynamics using GRACE and InSAR: potential and limitations. Groundwater, 13. https://doi.org/10.1111/gwat.12453

Chinnasamy P, Maheshwari B, Prathapar S (2015) Understanding groundwater storage changes and recharge in Rajasthan, India through remote sensing. Water 7(10):5547–5565

Collins M, Knutti R, Arblaster J, Dufresne J-L, Fichefet T, Friedlingstein P, Gao X, Gutowski WJ, Johns T, Krinner G, Shongwe M, Tebaldi C, Weaver AJ, Wehner M (2013) Long-term climate change: projections, commitments and irreversibility. In: Stocker TF, Qin D, Plattner GK, Tignor M, Allen SK, Boschung J, Nauels A, Xia Y, Bex V, Midgley PM (eds) (2013) Climate change: the physical science basis. contribution of working group i to the fifth assessment report of the intergovernmental panel on climate change. Cambridge University Press, Cambridge, United Kingdom and New York

Cullen HM, Kaplan A, Arkin PA, Demenocal PB (2002) Impact of the North Atlantic oscillation on middle eastern climate and streamflow. Clim Change 55(3):315–338

Dinku T, Connor SJ, Ceccato P, Ropelewski CF (2008) Comparison of global gridded precipitation products over a mountainous region of Africa. Int J Climatol 28:1627–1638

Funk C, Husak G, Michaelsen J, Love T, Pedreros D(2007) Third generation rainfall climatologies: satellite rainfall and topography provide a basis for smart interpolation. In: Proceedings of the JRC—FAO Workshop, Nairobi, Kenya

Hanson RT, Newhouse MW, Dettinger MD (2004) A methodology to assess relations between climatic variability and variations in hydrologic time series in the southwestern United States. J Hydrol 287(1–4):252–269

Hao Z, AghaKouchak A, Phillips JT (2013) Changes in concurrent monthly precipitation and temperature extremes. Environ Res Lett 8, 034014:7. https://doi.org/10.1088/1748-9326/8/3/034014

Harris I, Jones P, Osborn T, Lister D (2014) Updated high-resolution grids of monthly climatic observations—the CRU TS3. 10 Dataset. Int J Climatol 34(3):623–642

Hassan A, Jin S (2016) Water storage changes and balances in Africa observed by GRACE and hydrologic models. Geodesy Geodyn 7(1):39–49. https://doi.org/10.1016/j.geog.2016.03.002

Holman IP, Rivas-Casado M, Howden NJK, Bloomfield JP, Williams AT (2009) Linking North Atlantic ocean–atmosphere teleconnection patterns and hydrogeological responses in temperate groundwater systems. Hydrol Proc 23:3123–3126

Homsi R, Shiru MS, Shahid S, Ismail T, Harun S, Al-Ansari N, Chau K-W, Yaseen ZM (2020) Precipitation projection using a CMIP5 GCM ensemble model: a regional investigation of Syria Engineering applications of computational fluid mechanics 14, pp 90–106. https://doi.org/10.1080/19942060.2019.1683076

Iloeje NP (1981) A new geography of Nigeria, new revised edition. Great Britain: Longman: In: Odekunle TO (2006) Determining rainy season onset and retreat over Nigeria from precipitation amount and number of rainy days. Theor Appl Climatol 83:193–201

Ionita M, Lohmann G, Rimbu N, Chelcea S (2012) Interannual variability of Rhine River streamflow and its relationship with large-scale anomaly patterns in spring and autumn. J Hydrometeorol 13(1):172–188

JICA (Japan International Cooperation Agency) (2014) The project for review and update of Nigeria national water resources master plan, vol 2. Japan International Cooperation Agency: Yachiyo Engineering Co., Ltd.: CTI Engineering International Co., Ltd.: Sanyu Consultants Inc

Kahsay KD, Pingale SM, Hatiye SD (2018) Impact of climate change on groundwater recharge and base flow in the subcatchment of Tekeze basin, Ethiopia. Groundwater Sustain Develop 6:121–133. https://doi.org/10.1016/j.gsd.2017.12.002

Lee TM, Markowitz EM, Howe PD, Ko C-Y, Leiserowitz A. (2015) Predictors of public climate change awareness and risk perception around the world. Nat Clim Change 10. https://doi.org/10.1038/nclimate2728

Macdonald AM, Cobbing J, Davies J (2005) Developing groundwater for rural water supply in Nigeria: a report of the May 2005 training course and summary of the groundwater issues in the eight focus states. In: British geological survey commissioned report, CR/05/219N, 32 pp

McNally A, Arsenault K, Kumar S, Shukla S, Peterson P, Wang S, Funk C, Peters-Lidard CD, Verdin JP (2017) Sci Data 4:170012:19. https://doi.org/10.1038/sdata.2017.12

Meddi M, Boucefine A (2013) Climate change impact on groundwater in Cheliff-Zahrez basin (Algeria). Asia-Pacific Chem Biol Environ Eng Soc (APCBEE) Proc 5:446–450. https://doi.org/10.1016/j.apcbee.2013.05.077

Merietu TS, Olarewaju IO (2009) Resource conflict among farmers and Fulani herdsmen: implications for resource sustainability. African J Polit Sci Int Rel 3(9):360–364. Available online at http://www.academicjournals.org/ajpsir. Accessed on 05 Oct 2019

Morice CP, Kennedy JJ, Rayner NA, Jones PD (2012) Quantifying uncertainties in global and regional temperature change using an ensemble of observational estimates: the HadCRUT4 data set. J Geophys Res 117:D08101. https://doi.org/10.1029/2011JD017187

Naumann G, Alfieri L, Wyser K, Mentaschi L, Betts RA, Carrao H, Spinoni J, Vogt J, Feyen L (2018) Global changes in drought conditions under different levels of warming. Geophys Res Lett 45:3285–3296. https://doi.org/10.1002/2017GL076521

Oguntunde PG, Lischeid G, Abiodun BJ, Dietrich O (2016) Analysis of long-term dry and wet conditions over Nigeria. Int J Clim 37(9):3577–3586. https://doi.org/10.1002/joc.4938

Ojiako GU (1985) Nigerian water resources and their management. Water Int 10:2:64–72. https://doi.org/10.1080/02508068508686310

Okoli AIC, Atelhe AG (2014) Nomads against natives: a political ecology of herder/farmer conflicts in Nasarawa State, Nigeria. Am Int J Contemp Res 4(2):76–88. Available http://www.aijcrnet.com/journals/Vol_4_No_2_February_2014/11.pdf. Accessed 15 Nov 2019

Olaniran OJ, Sumner GN (1989) a study of climatic variability in Nigeria based on the onset, retreat, and length of the rainy season. Int J Climatol 9:253–269

Oloruntade AJ, Mohammad TA, Ghazali AH, Wayayok A (2017) Analysis of meteorological and hydrological droughts in the Niger-South Basin, Nigeria. Global Planet Change 155:225–233. https://doi.org/10.1016/j.gloplacha.2017.05.002

Omondi PA, Awange JL, Forootan E, Ogallo LA, Barakiza R, Girmaw GB, Fesseha I, Kululetera V, Kilembe C, Mbati MM, Kilavi M, King'uyu SM, Omeny PA, Njogu A, Badr EM, Musa TA, Muchiri P, Bamanya D, Komutunga E (2014) Changes in temperature and precipitation extremes over the Greater Horn of Africa region from 1961 to 2010. Int J Climatol 34:1262–1277. https://doi.org/10.1002/joc.3763

Oteze GE (1981) Water resources in Nigeria. Environ Geol 3:177–184

Piao S, Ciais P, Huang Y, Shen Z, Peng S, Li J, Zhou L, Liu H, Ma Y, Ding Y, Friedlingstein P, Liu C, Tan K, Yu Y, Zhang T, Fang J (2010) The impacts of climate change on water resources and agriculture in China. Nature 467(2):43–51

Perez-Valdivia C, Sauchyn D, Vanstone J (2012) Groundwater levels and teleconnection patterns in the Canadian Prairies. Water Resour Res 48:W07516. https://doi.org/10.1029/2011WR010930

Qutbudin I, Shiru MS, Sharafati A, Ahmed K, Al-Ansari N, Yassen ZM, Shahid S, Wang X (2019) Seasonal drought pattern changes due to climate variability: case study in Afghanistan. Water 11:1096. https://doi.org/10.3390/w11051096

Ranjan P, Kazama S, Sawamoto M (2006) Effects of climate change on coastal fresh groundwater resources. Glob Environ Change 16:388–399. https://doi.org/10.1016/j.gloenvcha.2006.03.006

Rashid MM, Beecham S, Chowdhury RK (2015) Statistical downscaling of CMIP5 outputs for projecting future changes in rainfall in the Onkaparinga catchment. Sci Total Environ 530–531:171–182

Sa'adi Z, Shiru MS, Shahid S, Ismail T (2019) Selection of general circulation models for the projections of spatio-temporal changes in temperature of Borneo Island based on CMIP5. Theoret Appl Climatol. https://doi.org/10.1007/s00704-019-02948-z

Salem GSA, Kazama S, Komori D, Shahid S, Dey NC (2017) Optimum abstraction of groundwater for sustaining groundwater level and reducing irrigation cost. Water Resour Manage 31:1947–1959. https://doi.org/10.1007/s11269-017-1623-8

Salem GSA, Kazama S, Shahid S, Dey NC (2018) Impacts of climate change on groundwater level and irrigation cost in a groundwater dependent irrigated region. Agric Water Manag 208:33–42. https://doi.org/10.1016/j.agwat.2018.06.011

Salman SA, Shahid S, Afan HA, Shiru MS, Al-Ansari N, Yaseen ZM (2020) Changes in climatic water availability and crop water demand for Iraq region. Sustain 12:3437. https://doi.org/10.3390/su12083437

Schneider U, Becker A, Meyer-Christoffer A, Ziese M, Rudolf B (2011) Global precipitation analysis products of the GPCC. In: Global precipitation climatology centre (GPCC) Deutscher Wetterdienst, 13 pp. Available online: ftp://www.dwd.de/pub/data/gpcc/PDF/GPCC_intro_products_v2011.pdf

Schneider U, Becker A, Finger P, Meyer-Christoffer A, Ziese M, Rudolf B (2014) GPCC's new land surface precipitation climatology based on quality-controlled in situ data and its role in quantifying the global water cycle. Theoret Appl Climatol 115(1–2):15–40

Sediqi MN, Shiru MS, Nashwan MS, Ali R, Abubaker S, Wang X, Ahmed K, Shahid S, Asaduzzaman M, Manawi MA (2019) Spatio-temporal pattern in the changes in availability and sustainability of water resources in Afghanistan. Sustainability 11:5836

Shahid S, Alamgir M, Wang X, Eslamian S (2017) Climate change impacts on and adaptation to groundwater. In: Eslamian S, Eslamian F (eds) Handbook of drought and water scarcity. CRC Press, Taylor & Francis Group, 6000 Broken Sound Parkway NW, Suite 300, Boca Raton, FL 33487 3742, pp 107–123

Shiru MS, Shahid S, Alias N, Chung E-S (2018) Trend Analysis of droughts during crop growing seasons of Nigeria. Sustainability 10(871):13. https://doi.org/10.3390/su10030871

Shiru MS, Shahid S, Chung E-S, Alias N (2019a) Changing characteristics of meteorological droughts in Nigeria during 1901–2010. Atmos Res 223:60–73. https://doi.org/10.1016/j.atmosres.2019.03.010

Shiru MS, Shahid S, Chung E-S, Alias N, Scherer L (2019b) A MCDM-based framework for selection of general circulation models and projection of spatio-temporal rainfall changes: a case study of Nigeria. Atmos Res 225:1–16. https://doi.org/10.1016/j.atmosres.2019.03.033

Shiru MS, Shahid S, Shiru S, Chung E-S, Alias N, Ahmed K, Dioha CE, Sa'adi Z, Salman S, Noor M, Nashwan MS, Idlan MK, Khan N, Momade MH, Houmsi MR, Iqbal Z, Qutbudin I, Sediqi MN (2019c) Challenges in water resources of Lagos mega city of Nigeria in the context of climate change. J Water Clim Change. https://doi.org/10.2166/wcc.2019.047

Simmons AJ, Jones PD, Bechtold V-C, Beljaars ACM, Kallberg PW, Saarinen S, Uppala SM, Viterbo P, Wedi N (2004) Comparison of trends and low-frequency variability in CRU, ERA-40, and NCEP/NCAR analyses of surface air temperature. J Geophys Res 109:D24115. https://doi.org/10.1029/2004JD005306

Spinoni J, Naumann G, Carrao H, Barbosa P, Vogt J (2014) World drought frequency, duration, and severity for 1951–2010. Int J Clim 34(8):2792–2804. https://doi.org/10.1002/joc.3875

Thorncroft CD, Nguyen H, Zhang C, Peyrille P (2011) Annual cycle of the West African monsoon: regional circulations and associated water vapour transport. Q J R Meteorol Soc 137:129–147

Tremblay A, Larocque M, Anctil F, Rivard C (2011) Teleconnections and interannual variability in Canadian groundwater levels. J Hydrol 410:178–188. https://doi.org/10.1016/j.jhydrol.2011.09.013

Ubelejit NT (2016) Fulani Herdsmen and communal conflicts: climate change as precipitator. J Polit Sci Leadership Res 2(1):26–32

Wang T, Hamann A, Spittlehouse D, Carroll C (2016) Locally downscaled and spatially customizable climate data for historical and future periods for North America. PLoS ONE 11(6):e0156720. https://doi.org/10.1371/journal.pone.0156720

Ward FA (2014) Economic impacts on irrigated agriculture of water conservation programs in drought. J Hydrol 508:114–127

Watto MA (2015) The economics of groundwater irrigation in the Indus Basin. Tube-well Adoption, Technical and Irrigation Water Efficiency and Optimal Allocation. The University of Western Australia, Pakistan

Weezel S-V (2017) Drought severity and communal conflict in Nigeria. Available online https://www.researchgate.net/publication/320374674_Drought_severity_and_communal_conflict_in_Nigeria. Accessed on 06 Oct 2019

Wilhelmi OV, Wilhite DA (2002) Assessing vulnerability to agricultural drought: a Nebraska case study. Drought Mitigation Center Faculty Publications, vol 9. Available http://digitalcommons.unl.edu/droughtfacpub/9/. Accessed 05 Dec 2019

World Bank Group (2019) Agriculture, value added (% of GDP). Available online: http://data.worldbank.org/indicator/NV.AGR.TOTL.ZS (Accessed on 18/03/2019)

Yang Y, Wang G, Wang L, Yu J, Xu Z (2014) Evaluation of gridded precipitation data for driving SWAT model in area upstream of three gorges reservoir. PLoS ONE 9(11):e112725. https://doi.org/10.1371/journal.pone.0112725

Yu Y, Disse M, Yu R, Yu G, Sun L, Huttner P, Rumbaur C (2015) Large-scale hydrological modeling and decision-making for agricultural water consumption and allocation in the main stem Tarim River, China. Water 7:2821–2839. https://doi.org/10.3390/w7062821

# Chapter 20
# Prediction of River Water Quality Parameters Using Soft Computing Techniques

**Kulwinder Singh Parmar, Kirti Soni, and Sarbjit Singh**

## 20.1 Introduction

Water is essential for every living being on earth. Besides drinking purposes, most of the household requirements are met with water. The water cycle is sustained by various hydrological natural processes such as groundwater recharge, a river system and rainfall-runoff processes. River plays an important role to satisfy these requirements of water. Rivers are not only the most vital source for maintaining ground level of water but also have a significant impact on the climate of surroundings also. Most of old well-known civilizations were evolved on the banks of rivers to meet the daily needs of water. At that time, the river water was very pure. But now most of rivers are polluted due to modern civilizations and the future of river water is in a dangerous phase. To diagnose the pollution in the river water, we have to check first the quality of water with the future trends of the river water. So, to avoid the occurrence of any natural crisis, the need of the hour is to model these processes for estimating the future trends by using the past trends of such events.

To calculate the future quality of river water, there is a need for accurate models with least error. In the mathematical modeling, lots of models are popular, which are used for the prediction. Time series models like AR, ARMA, ARIMA and regression models (linear, multiple) are very popular and widely used in the area of forecasting. In spite of having number of benefits, one major drawback of statistical modeling

K. S. Parmar (✉)
Department of Mathematics, IKG Punjab Technical University, Jalandhar, Kapurthala, India
e-mail: kulmaths@gmail.com

K. Soni
CSIR-National Physical Laboratory, New Delhi, India

S. Singh
Guru Nanak Dev University College, Narot Jaimal Singh, Pathankot, Punjab, India

Guru Nanak Dev University, Amritsar, Punjab, India

© Springer Nature Singapore Pte Ltd. 2021

R. C. Deo et al. (eds.), *Intelligent Data Analytics for Decision-Support Systems in Hazard Mitigation*, Springer Transactions in Civil and Environmental Engineering,
https://doi.org/10.1007/978-981-15-5772-9_20

approach is its inability to handle nonlinear data. Linear correlation given by correlation coefficient forms the basis of the statistical models, so statistical modeling approach cannot explore nonlinear characteristics of data. Neural networks (NN), artificial neural network (ANN) and adaptive neuro-fuzzy interface system (ANFIS), etc., models are the soft computing models, which consists of strong tools to model the nonlinear situations with least error (Parmar et al. 2009).

The capacity of a forecasting model can be enhanced with the use of wavelet transformation as a pre-processing approach by seizing essential information at different levels of resolution. This chapter deals with to understand soft computing models and its applications that will lead to find the best prediction model to predict the biochemical oxygen demand (BOD) by using the averaged data of past month at Nizamuddin (Delhi) sample site of river Yamuna, India. The statistical analysis of data recommends considering the constitutive series in wavelet domain. A comparison of the results obtained by neuro-fuzzy-wavelet joint model, neuro-fuzzy, artificial neural network (ANN) and regression models reveal significant forecast with least errors for neuro-fuzzy-wavelet joint model.

ANN is a method of comparing the pattern, which are having mainly three major layers input layer, hidden layer and output layer. The final output layer contains nodes based on learning algorithm. The application of ANN method helps in forecasting streamflow, drought and precipitation (Furundzic 1998; Bodri and Cermak 2000; Sajikumar and Thandaveswara 1999; Luk et al. 2001; Soni et al. 2014a, b; Moustris et al. 2011). Like artificial neural networks (ANN), fuzzy interface system (FIS) is data-driven technique and based on the if-then fuzzy rules. This model is used when it is difficult to formulate the model with crisp parameters. Nowadays, Mamdani method with coupled with FIS model is used widely and in differently many areas of research. Data set is divided into two parts, first for training of model and second for testing of model. The estimation of future trends using fuzzy-based models lead to more accurate results (French et al. 1992; Aksoy et al. 2004; Kahya and Kalayci 2004; Torprak et al. 2004, 2009; Torprak 2009; Soni et al. 2015).

To analyze a nonlinear data, ANN-based models are found to be extremely effective. The training and testing datasets for the development of ANN model are created using analytic element method (AEM). ANN models have the potential to reduce the computational task and their outcomes reveal the significant performance of these models (Nayak et al. 2004; Kisi 2005; Jeong et al. 2012; Pinto et al. 2012). Wavelet analysis, being a significant mathematical tool for analyzing rapidly changing signals, has ability to effectively diagnose signal mainly in frequency component. Multi-resolution of signal is a technique used to extract local information at different scales, resolution of good time and good frequency received at same time. Wavelet analysis fulfills the drawbacks of Fourier series as it provides a better time scale picture of the signal. The trends and non-stationarities of the time series are revealed by carefully examining the relationships among these wavelet functions.

Wavelet transform carries the decomposition of given time series data at several resolution levels and thus collecting the necessary information to enhance the accuracy of the model leading to reduced forecasting errors (Daubechies 1992; Adamowski 2007; Adamowski and Sun 2010; Soni et al. 2017). As compared to the Fourier series and short-time Fourier transform (STFT), wavelets are capable of capturing a signal in both time and frequency domain effectively and hence produce better results (Mallat 2001; Lafrenière and Sharp 2003; Can et al. 2005; Loboda et al. 2006; Pasquini and Depetris 2007; Partal and Kisi 2007; Moosavi et al. 2013).

## 20.2 Methodology

### *20.2.1 Artificial Neural Network (ANN)*

Artificial neural networks (ANNs) modeling techniques are well known in this era of research and usually deal with nonlinear systems which are based on the functioning of human brain. As discussed above, statistical modeling techniques are unable to handle the nonlinear data, ANN can investigate nonlinearity efficiently. The framework of neural networks consists of simple processing nodes or neurons which are interconnected to each other in a specific order. These structured neurons have ability to perform simple mathematical calculations (See and Openshaw 1999).

The three-layered structure of ANNs is comprised of input, hidden and the output layers each having a number of nodes interconnected with each other (Haykin 1994). The research problems related to forecasting are logically handled by ANN techniques and feed forward method in the ANN structure is a widely used method. We can generalize the prevalent least-mean-square (LMS) process by clubbing ANN architecture with the corresponding learning algorithm (Haykin 1994).

Unlike the feed forward neural network, the transmission of information in a fully recurrent network is no longer in one direction only but also back to the hidden layer after being fed the output layer. Partially recurrent network begin with a fully recurrent network topology and then add a feed forward connection that sidesteps the recurrence, efficiently treating the recurrent part as a state memory. The best estimates of a nonlinear time series are procured by recurrent network methodology.

### *20.2.2 The Adaptive Neuro-Fuzzy Inference System (ANFIS)*

When ANN structure is combined with the fuzzy interface system (FIS), it forms the ANFIS which is used for estimating and forecasting the time series. More accurate and reliable prediction results are obtained by ANFIS (Jang et al. 1997). The mapping of input characteristics functions to input membership functions is linked with the output properties. The dependence of the output membership functions upon the output properties results into a unique output or a decision related with the output. This output membership provides the output (decision) of the problem (Jang et al. 1997). This type of fuzzy modeling is divided into three important phases namely fuzzifier, fuzzy database and de-fuzzifier. The fuzzy database is further subdivided into two stages namely fuzzy rule base and inference structure (Karmakar and Mujumdar 2006; Seyed et al. 2013; Sahay and Srivastava 2014).

Fuzzy interface system (FIS), being a data-driven and soft computing model, has dynamics parallel with ANN. It obeys fuzzy if-then rules which a crisp parameter model finds crucial to develop (Jeong et al. 2012). The implementation of Mamdani technique involves the division of whole data into two parts, one is training and the other is testing dataset. Training is used to train the data set and testing stage to test the model's output with the actual raw data set. The consistency of the model and accuracy of the model's output is achieved by making use of fuzziness (Aksoy et al. 2004; Toprak et al. 2004; Hung et al. 2009; Toprak et al. 2009; Chen and Chang 2010).

*ANFIS Architecture*

As per the Sugeno and Takagi type, fuzzy inference system contains two inputs and a single output as shown in Fig. 20.1.

$$\text{Rule 1: } f_1 = p_1 x + q_1 y + r_1 \tag{20.1}$$

as $x$ is $A_1$ and $y$ is $B_1$

$$\text{Rule 2: } f_2 = p_2 x + q_2 y + r_2 \tag{20.2}$$

as $x$ is $A_2$ and $y$ is $B_2$

Layer 1: In this layer, here, with a node function each node $i$ is a square node

$$O_i^1 = \mu A_i(x) \tag{20.3}$$

In Eq. (20.3), $x$ denotes input to node $i$, where $A_i$ is the input to node $i$ and $A_i$ the syntactical brand with the node function. With these observations, $O_i^1$ become membership mapping to $A_i$. The membership function is represented by $\mu A_i(x)$ have the highest value 1 and lowest to 0, as in the general bell mapping or Gaussian mapping as discuss below

$$\mu A_i(x) = \frac{1}{1 + \left[\left(\frac{x - c_i}{a_i}\right)^2\right]^{b_i}} \tag{20.4}$$

$$\mu A_i(x) = \mathrm{e}^{\left[-\left(\frac{x - c_i}{a_i}\right)^2\right]} \tag{20.5}$$

Here, the parameter set is defined as $\{a_i, b_i, c_i\}$ (Hus et al. 1995).

Layer 2: Every node in this layer is a circle node which multiplies the incoming signals and sends the product out. For instance,

$$O_i^2 = w_i = \mu A_i(x) \times \mu B_i(x), \quad i = 1, 2, 3 \ldots \tag{20.6}$$

Each node output represents the firing strength of a rule.

Layer 3: In this layer, each node is a circle node labeled as N. Here, *i*th node calculates the ratio of the *i*th rule's firing strength to sum of all rule's firing strengths

$$O_i^3 = w_i = \frac{w_i}{w_1 + w_2}, \quad i = 1, 2, 3 \ldots \tag{20.7}$$

For convenience, outputs of this layer will be called normalized firing strength.

Layer 4: Every node $i$ in this layer is a square node with a node function

$$O_i^4 = \bar{w}_i f_i = \bar{w}_i = \bar{w}_i(p_i x + q_i y + r_i) \tag{20.8}$$

where $\bar{w}_i$ is the output of layer 3 and ($p_i$, $q_i$, $n_i$) is the parameter set. Parameters in this layer will be referred to as consequent parameters.

Layer 5: The single node in this layer is a circle node labeled that computes the overall output as the summation of all incoming signals, i.e.,

$$O_i^5 = \sum \bar{w}_i f_i = \sum_i w_i f_i / \sum_i w_i \tag{20.9}$$

In this way, an adaptive network is constructed which is functionally corresponding to a type three fuzzy inference systems.

### *20.2.3 Wavelet Transforms*

A small wave decaying rapidly can be termed as a wavelet. Wavelets are magnificent signal processing tool compact support and finite energy. Unlike the Fourier transform, which is suitable for sinusoids in frequency domain only, wavelet transform can analyze a signal both in time and frequency domains. The inability in applying Fourier transforms for studying a signal is to describe frequencies-time resolution. Wavelet transforms overwhelm many limitations of Fourier transforms by preserving an important conciliation between frequency information and time location. Wavelet transform is an important tool to find information about a specific hydrological event and comparing the trends of the hydrological data. Mother wavelet $h(t)$ works as a prototype mapping in the wavelet analysis. Wavelet analysis works with a prototype function termed as the mother wavelet $h(t)$. Mother wavelet can be used for scaling and translating a data set. Continuous wavelet transforms (CWT) of data series $y(t)$

with mother wavelet is defined as in Eq. (20.10).

$$CWT(c,d) = \frac{1}{\sqrt{c}} \int_{-\infty}^{+\infty} y(t) h\left(\frac{t-d}{c}\right) dt \tag{20.10}$$

Here, $c$ is scaling and $d$ is the translation factor (Adamowski 2007). The wavelet depiction of $y(t)$ related to $h(t)$ is provided by CWT $(c, d)$. Meanwhile, CWT furnishes substantial redundant information by continuously scaling and translating the mother wavelet. Alternatively, the discretization of CWT can be accomplished by scaling and translating the mother wavelet using definite scales and position based on power of two. This discreet structure is more effective and as precise as CWT and known as discrete wavelet transforms (DWT).

$$DWT(m,t) = \frac{1}{\sqrt{c^m}} \sum_n y(n)\, g\left(\frac{t - ndc^m}{c^m}\right) \tag{20.11}$$

Here, again $c$ and d are scaling and translation function, respectively, of the integer variable $m$; $t$ is an integer variable referring to the input signal of a point; $n$ is the discrete time index; $y(t)$ is a given signal and $g(t)$ is the mother wavelet (Adamowski 2007).

Many properties of a signal like periodicity, seasonality and points of abrupt changes can be disposed by wavelet transform. The plane formed by variables $(c, d)$ is called time-frequency plane or scale space. Lipschitz exponent is regularly used to measure the local regularity of mapping. The asymptomatic decay of wavelet transformation at small scales is used to characterize the local and the global Lipschitz regularity. At level 1, the decomposition of $g$ is given by $g = a_1 + d_1$ where $a_1$ and $d_1$ denote, respectively, the low frequency part and high frequency part. Following the same decomposition procedure performed on $a_1$ results into next level finer scales given by $a_1 = a_2 + d_2$. At $n$th level, the decomposition details of $g$ are given by the low frequencies as $a_1, a_2, a_3, \ldots, a_n$ and the high frequencies as $d_1, d_2, d_3, \ldots, d_n$. It is also classified as follows:

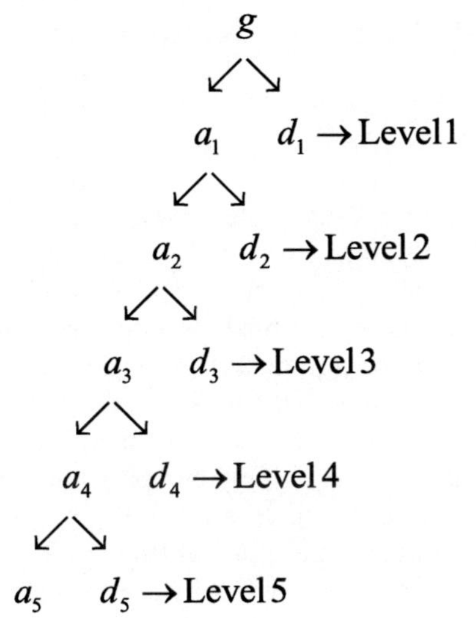


Let $g = (g_1, g_2, g_3, \ldots, g_N)$, $a_m = \frac{g_{2m-1}+g_{2m}}{\sqrt{2}}$, $m = 1, 2, 3, \ldots, N/2$ for $N$ is even integer, $a^1 = (a_1, a_2, \ldots, a_{N/2})$. The approximation sub-composition can be established by the below given formula

$$d_m = \frac{g_{2m-1} - g_{2m}}{\sqrt{2}}, \quad \text{for } m = 1, 2, 3, \ldots, N/2$$

First-level Transform

$$g \longrightarrow H_1(a^1/d^1)$$

These wavelets are defined as

$$\begin{aligned}
W_1^1 &= \left(\tfrac{1}{\sqrt{2}}, -\tfrac{1}{\sqrt{2}}, 0, 0, \ldots, 0\right) \\
W_2^1 &= \left(0, 0, \tfrac{1}{\sqrt{2}}, -\tfrac{1}{\sqrt{2}}, 0, 0, \ldots, 0\right) \\
&\ldots \\
W_{N/2}^1 &= \left(0, 0, 0, \ldots, 0, \tfrac{1}{\sqrt{2}}, -\tfrac{1}{\sqrt{2}}\right)
\end{aligned}$$

$$\begin{aligned}
d_1 &= g \cdot W_1^1 \\
d_2 &= g \cdot W_2^1 \\
&\ldots \\
d_m &= g \cdot W_m^1 \quad \text{for } m = 1, 2, 3, \ldots, N/2
\end{aligned}$$

## 20.3 Application to Numerical Solution

### *20.3.1 Time Series Analysis*

The time series processes are mainly of the two types; stationary and non-stationary processes. In stationary processes, mean and variance does not change with time. Most of the time series analysis procedures demand stationarity, quiet often non-stationary time series data is transformed to stationary data. In addition to some natural events like wind speed, earthquakes, large forest fires and landslides, etc., that can bring about quick changes in the time series, the man-made changes mainly contribute to non-stationarity in our time series data problem. Some other man-made changes like dam construction, tree cutting, pumping out of ground water, etc., also affect the hydrological process in a time series.

The first step in time series analysis is to check our time series data for stationarity. In case of non-stationary series, it will be firstly transformed to the stationary series. Stationarity of time series can also be checked by autocorrelation function (ACF) and partial autocorrelation function (PACF) plots. A rapidly decaying ACF plot reveals stationarity, whereas a comparative slow decay points out the non-stationarity (Wiee 1990).

### *20.3.2 Neural Network Model*

In this section, the data set of the BOD level at the Nizamuddin (Delhi) sample site is preprocessed and artificial neural network (ANN) model is applied to it. Feed forward neural network (FFNN) together with the backpropagation conditions is preferred for accurate forecasting results. The first three values are used as input to initialize, hence, three input nodes used along with the sigmoid transfer mapping to forecast the time series. In the part to train, the model 2000 epochs are used to get the best results with keeping error tolerance at 0.1 levels.

### *20.3.3 Artificial Neural Network and Fuzzy Coupled Model*

To get the best forecasting results, search of better forecasting model are always high. To forecast the BOD level of river Yamuna at Nizamuddin (Delhi), ANN model herewith coupled with the fuzzy model as described in Fig. 20.1. In the first stage, model is trained with backpropagation method with properties of Sugeno. 2000 epochs are used for training model with tolerance level 0.001.

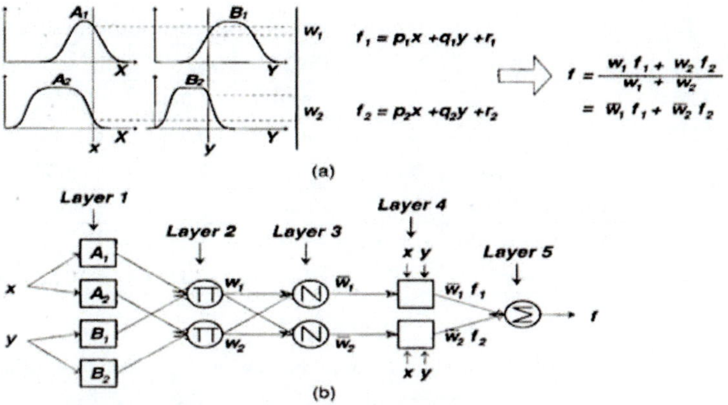


Fig. 20.1 Graphical representation of neuro-fuzzy model

## 20.3.4 Artificial Neural Network-Fuzzy-Wavelet Model

Artificial neural network-fuzzy coupled model is established and functional to the current problem for forecasting the BOD level of the river Yamuna as in the previous section. This model yields better results than the traditional multiple regression models. In this section, the data set is decomposed into component series by wavelet decomposition and again neuro-fuzzy model is applied on each decomposed series. The choice, order and the level of the mother wavelet are important factors which are responsible for wavelet decomposition. Daubechies wavelet is one of the most important types of wavelets and is most suitable for rapidly varying data. To check the appropriateness of each order, the data set is decomposed using Daubechies mother wavelet with orders ranging from 2 to 8. The values given in Table 20.1 indicate that Daubechies wavelet of order 2 can produce accurate results.

The accuracy of combined forecasting approach depends upon the level of decomposition of input data signal. In Fig. 20.1, A1 to A6 explains the approximations of the BOD level signal and D1 to D6 depicts the detailed parts of the same. Approximations

**Table 20.1** Skewness of approximation coefficient using Daubechies wavelet (actual skewness = 0.178)

| | Db1 | Db2 | Db3 | Db4 | Db5 | Db6 | Db7 | Db8 |
|---|---|---|---|---|---|---|---|---|
| A1 | 0.0925 | 0.1886 | 0.2078 | 0.1909 | 0.1570 | 0.1406 | 0.1487 | 0.1674 |
| A2 | 0.7470 | 0.3348 | −0.2060 | 0.2982 | 0.5952 | 0.0251 | 0.1082 | 0.5099 |
| A3 | **−0.4231** | **−0.3282** | **−0.5505** | **−0.4073** | **−0.4806** | **−0.5553** | **−0.5075** | **−0.7342** |
| A4 | −1.2792 | −0.7797 | −1.6852 | −1.0051 | −1.4447 | −1.5709 | −1.1607 | −1.5563 |
| A5 | −0.7866 | −1.4623 | −1.4796 | −0.8731 | −1.2227 | −1.2687 | −0.8703 | −1.0421 |
| A6 | 0.1353 | −1.2334 | −0.6817 | −0.1484 | −0.9704 | −0.8893 | −0.4580 | −0.5777 |

**Table 20.2** Kurtosis of detailed coefficient using Daubechies wavelet (actual kurtosis = −0.5334)

| | Db1 | Db2 | Db3 | Db4 | Db5 | Db6 | Db7 | Db8 |
|---|---|---|---|---|---|---|---|---|
| D1 | **0.2258** | **0.4587** | **0.4704** | **0.4641** | **0.8677** | **1.0461** | **0.5968** | **0.3528** |
| D2 | **−0.7115** | **2.0243** | **0.7236** | **−0.6528** | **0.5240** | **0.3384** | **−0.3406** | **0.3655** |
| D3 | **0.4736** | **2.2410** | **0.5335** | **0.5988** | **−0.0699** | **0.1694** | **−0.4777** | **0.5898** |
| D4 | −1.0426 | −0.1322 | 1.2886 | −0.1182 | 0.8445 | −0.3216 | 1.1246 | −0.8106 |
| D5 | −0.3973 | 2.4869 | −0.6120 | −1.1349 | 1.7942 | −1.3350 | −1.3246 | −0.1674 |
| D6 | −1.2407 | −0.7972 | −0.6077 | −0.9044 | −1.1731 | −1.3253 | −0.8020 | −0.5263 |

**Table 20.3** Training and testing results of each model for monthly BOD level forecasting

| | Training results | | Testing results | |
|---|---|---|---|---|
| | MAE (mm) | ε (%) | MAE (mm) | ε (%) |
| Neuro-fuzzy-wavelet | **0.5130** | **0.23** | **0.4281** | **2.00** |
| Neuro-fuzzy | 0.1729 | 0.79 | 1.4864 | **6.93** |
| Artificial neural network | 0.9949 | 4.20 | 2.456 | **13.67** |
| Multiple linear regression | **2.878** | **7.29** | **6.7768** | **20.54** |

show the signal's trend are the low frequency components, whereas the detailed parts represent the local or short phase variation are mapping to high frequency. Skewness is a measure of symmetry and its value is zero for a perfect normal curve. Kurtosis is a measure of dispersion with a value 3 for ideal normal curve. In Table 20.1, only A1, A2 and A3 are toning with the original signal. Table 20.2 shows that D1 to D3 are more harmonizing with the original signal than the other details part D4 to D7. So, for the further investigation, A3 for approximation and D1 to D3 for the detailed parts are to be considered (Table 20.3).

## 20.4 Conclusion

The potential of neural-fuzzy-wavelet model in comparison with the clubbed neuro-fuzzy, simple ANN and multiple linear regression models for 9 months prediction of BOD level of river Yamuna at the sample site Nizamuddin (Delhi) has been investigated. Here, the time series data is highly nonlinear, non-stationary and shows seasonality in its behavior. In order to develop the new forecasting model to predict BOD level, a number of approaches such as wavelet transformation, artificial neural network and fuzzy approaches have been combined. Firstly, multiple linear regression model was used to forecast, but this model generated a high forecasting error of 20.54%; after that ANN model produced a forecasting error of 13.67% in prediction results; then, neuro-fuzzy coupled model was used and an error of 6.93% seen in prediction values of the BOD level; and finally the same data of BOD level of Yamuna river modeled with neuro-fuzzy-wavelet coupled model yielded very sharp prediction with 2% error. Neuro-fuzzy-wavelet hybrid combined model considerably lowered forecasting errors and proved the best forecasting model.

## References


Adamowski J, Sun K (2010) Development of a coupled wavelet transform and neural network method for flow forecasting of non-perennial rivers in semi-arid watersheds. J Hydrol 390:85–91

Aksoy H, Toprak ZF, Aytek A, Ünal NE (2004) Stochastic generation of hourly mean wind speed data. Renew Energy 29:2111–2131

Bodri L, Cermak V (2000) Prediction of extreme precipitation using a neural network: application to summer flood occurrence in Moravia. Adv Eng Softw 31:311–321

Can Z, Aslan Z, Oguz O, Siddiqi AH (2005) Wavelet transform of metrological parameter and gravity waves. Ann Geophys 23:659–663

Chen HW, Chang NB (2010) Using fuzzy operators to address the complexity in decision making of water resources redistribution in two neighboring river basins. Adv Water Resour 33:652–666

Daubechies I (1992) Ten lectures on wavelets. SIAM, Philadelphia, PA

French MN, Krajewski WF, Cuykendall RR (1992) Rainfall forecasting in space and time using neural networks. J Hydrol 137:1–31

Furundzic D (1998) Application example of neural networks for time series analysis: rainfall-runoff modeling. Signal Process 64:383–396

Haykin S (1994) Neural networks, a comprehensive foundation. Macmillan College Publishing Company, New York

Hsu K, Gupta HV, Sorooshian S (1995) Artificial neural network modeling of the rainfall runoff process. Water Resour Res 31:2517–2530

Hung NQ, Babel MS, Weesakul S, Tripathi NK (2009) An artificial neural network model for rainfall forecasting in Bangkok Thailand. Hydrol Earth Syst Sci 13:1413–1425

Jeong C, Shin JY, Kim T, Heo JH (2012) Monthly precipitation forecasting with a neuro-fuzzy model. Water Resour Manage 26:4467–4483

Kahya E, Kalayci S (2004) Trend analysis of streamflow in Turkey. J Hydrol 289:128–144

Karmakar S, Mujumdar PP (2006) Grey fuzzy optimization model for water quality management of a river system. Adv Water Resour 29(7):1088–1105

Kisi O (2005) Suspended sediment estimation using neuro fuzzy and neural network approaches. Hydrol Science Journal 50:683–696

Lafrenière M, Sharp M (2003) Wavelet analysis of inter-annual variability in the runoff regimes of glacial and nival stream catchments, Bow Lake, Alberta. Hydrolog Process 17:1093–1118

Loboda NS, Glushkov AV, Knokhlov VN, Lovett L (2006) Using non decimated wavelet decomposition to analyse time variations of North Atlantic Oscillation, eddy kinetic energy, and Ukrainian precipitation. J Hydrol 322:14–24

Luk W, Fleischmann M, Beullens P, Bloemhof-Ruwaard JM (2001) The impact of product recovery on logistics network design. Prod Oper Manage 10:156–173

Mallat S (2001) A wavelet tour of signal processing, 2nd edn. Academic Press, San Diego

Moosavi V, Vafakhah M, Shirmohammadi B, Behnia N (2013) A wavelet-ANFIS hybrid model for groundwater level forecasting for different prediction periods. Water Resour Manage 27:1301–1321

Moustris KP, Larissi IK, Nastos PT, Paliatsos AG (2011) Precipitation forecast using artificial neural networks in specific regions of Greece. Water Resour Manag 25:1979–1993

Nayak PC, Sudheer KP, Ranjan DM, Ramasastri KS (2004) A neuro fuzzy computing technique for modeling hydrological time series. J Hydrol 291:52–66

Partal T, Kisi O (2007) Wavelet and neuro fuzzy conjunction model for precipitation forecasting. J Hydrol 342:199–212

Parmar KS, Chugh P, Minhas P, Sahota HS (2009) Alarming pollution levels in rivers of Punjab. Indian J Env Prot 29:953–959

Pinto SC, Adamowski J, Oron G (2012) Forecasting urbanwater demand viawavelet-denoising and neural network models. Case study: city of Syracuse, Italy. Water Resour Manage 26:3539–3558

Sahay RR, Srivastava A (2014) Predicting monsoon floods in rivers embedding wavelet transform, geneticalgorithm and neural network. Water Resour Manag 28(2):301–317

Sajikumar N, Thandaveswara BS (1999) A non-linear rainfall-runoff model using an artificial neural network. J Hydrol 216:32–55

See L, Openshaw S (1999) Applying soft computing approaches to river level forecasting. Hydrolog Sci J 44:763–777

Seyed AA, Ahmed E, Jaafar O (2013) Improving rainfall forecasting efficiency using modified adaptive neurofuzzy inference system (MANFIS). Water Resour Manag 27(9):3507–3523

Soni K, Kapoor S, Parmar KS (2014a) Long-term aerosol characteristics over eastern, southeastern, and south coalfield regions in India. Water Air Soil Pollut 225:1832

Soni K, Kapoor S, Parmar KS, Kaskaoutis DG (2014b) Statistical analysis of aerosols over the Gangetic-Himalayan region using ARIMA model based on long-term MODIS observations. Atmos Res 149:174–192

Soni K, Parmar KS, Kapoor S (2015) Time series model prediction and trend variability of aerosol optical depth over coal mines in India. Environ Sci Pollut Res 22:3652–3671

Soni K, Parmar KS, Agarwal S (2017) Modeling of air pollution in residential and industrial sites by integrating statistical and Daubechies Wavelet (Level 5) analysis. Model Earth Syst Environ 3:1187–1198

Toprak ZF (2009) Flow discharge modeling in open canals using a new fuzzy modeling technique (SMRGT). CLEAN-Soil Air Water 37:742–752

Toprak ZF, Sen Z, Savci ME (2004) Comment on Longitudinal dispersion coefficients in natural channels. Water Res 38:3139–3143

Toprak ZF, Eris E, Agiralioglu N, Cigizoglu HK, Yilmaz L, Aksoy H, Coskun G, Andic G, Alganci U (2009) Modeling monthly mean flow in a poorly gauged basin by fuzzy logic, CLEAN-soil, air. Water 37:555–564

Wiee WWS (1990) Time series analysis. Addision Wesley Publishing Company, New York, 478p

# Chapter 21
# Soft Computing Applications in Air Pollution Meteorology

**Kirti Soni and Kulwinder Singh Parmar**

## 21.1 Introductory Note

Air pollution is a big environmental issue affecting both emerging and established countries around the globe. Air pollutants contain gaseous pollutants like $SO_2$, $NO_2$, CO, odors, and SPM in the form of dust, fumes, mist, and smoke. The concentration of these in and nearby the urban areas causes severe pollution to the surroundings. The major part of air pollution created by humans is from construction, transportation, heating, energy generation, industries, etc. Delhi is among the top 10 for cities with the awful air quality. In the past two decades, Delhi was expanding rapidly by both means the population and pollution. Delhi's air pollution is a combination of factors including industries, power plants, domestic combustion of coal, biomass, and transport (direct vehicle exhaust and indirect road dust) that contribute to air pollution. The major air pollutant, which could cause potential harm to human health has been included are $SO_2$, $NO_2$, particulate matter (SPM), etc.

## 21.2 Soft Computing Techniques

### *21.2.1 Auto-regressive Integrated Moving Average (ARIMA)*

ARIMA model is used to simulate the different types of data variation and to forecast future trends. ARIMA method is an exploratory and iterative method intending

K. Soni
CSIR-National Physical Laboratory, New Delhi, India

K. S. Parmar (✉)
Department of Mathematics, IKG Punjab Technical University, Jalandhar, Kapurthala, India
e-mail: kulmaths@gmail.com

© Springer Nature Singapore Pte Ltd. 2021

R. C. Deo et al. (eds.), *Intelligent Data Analytics for Decision-Support Systems in Hazard Mitigation*, Springer Transactions in Civil and Environmental Engineering,
https://doi.org/10.1007/978-981-15-5772-9_21

to best fit long-term observations by using three stages—identification, estimation, and diagnostic checking in the development of a suitable model for a time series. ACF (autocorrelation function) PACF (partial autocorrelation function) are used to calculate the time series variable dependency.

Parmar and Bhardwaj (2013a) suggest that the ARIMA model is based on (*p*, *d*, *q*) operators. The auto-regressive (AR) model in the time series AR (*p*), is signified as

$$Y_t = \alpha_0 + \alpha_1 Y_{t-1} + \alpha_2 Y_{t-2} + \cdots + \alpha_p Y_{t-p} + \varepsilon_t \tag{21.1}$$

where

$\varepsilon_t$ is a source of randomness
$\alpha_i$ are constants. It is supposed to have the below-mentioned characteristics:

$$E[\varepsilon_t] = 0,$$
$$E\left[\varepsilon_t^2\right] = \sigma^2,$$
$$E[\varepsilon_t\,\varepsilon_s] = 0 \quad \text{for all } t \neq s$$

Aerosol optical depth series may have auto-regressive and moving average constituents, therefore both types of correlations are essential to model the patterns. If both components are existing only at lag 1, so as to recognize this let the linear equation at time *t*.

$$y_t = x_t\beta + \varepsilon_t \tag{21.2}$$

$$\varepsilon_t = \rho\varepsilon_{t-1} + \nu_t \tag{21.3}$$

where $-1 < \rho < 1$ and $\nu_t$ is *iid* (independent and identically distributed) and from the above-mentioned expectation values

$$E(\nu_t, \nu_{t-1}) = 0 \tag{21.4}$$

The error in this model is said to follow a first-order auto-regressive (AR1) procedure. Therefore, the present error is part of the earlier error plus some shock. Thus, Eq. (21.2) can be represented as

$$\begin{aligned} y_t &= x_t\beta + \rho\varepsilon_{t-1} + \nu_t \\ y_{t-1} &= x_{t-1}\beta + \varepsilon_{t-1} \end{aligned} \tag{21.5}$$

Similarly, we know that

$$\Rightarrow \varepsilon_{t-1} = y_{t-1} - x_{t-1}\beta$$

From Eq. (21.5)

$$y_t = x_t\beta + \rho(y_{t-1} - x_{t-1}\beta) + v_t$$
$$y_t = x_t\beta + \rho y_{t-1} - \rho x_{t-1}\beta + v_t \quad (21.6)$$

ARIMA model is developed by clubbing AR model and MA model (Parmar and Bhardwaj 2013a, c, 2014; Soni et al. 2014; Yang and Jin 2010). The prediction results depend on both past and present values of response $y$ and also taken the current and past values of the residuals.

### 21.2.2 The Adaptive Neuro-Fuzzy Inference System (ANFIS)

The ANFIS is a universal estimator, which helps to approximate any real continuous function on a compact set (Jang 1993). In the fuzzy inference system, input characteristics are transformed to input membership functions and then relate the input membership functions to rules. Then, the rules are transformed into a set of output characteristics with output membership functions, which are finally transformed into a single output. There are three main steps in the fuzzy system, namely the fuzzifier, the fuzzy database, and the defuzzifier. The fuzzy database includes two main parts, the fuzzy rule base, and the inference engine.

**ANFIS Structural Design**

In the artificial neural-fuzzy inference system, the Takagi and Sugeno method have been used; the ANFIS has two inputs $x$ and $y$, while it provides one output $f$ as shown in Fig. 21.1.

$$\text{Rule 1: If } x \text{ is } A_1 \ \& \ y \text{ is } B_1 \text{ then } f_1 = r_1 + p_1 x + q_1 y \quad (21.7)$$

$$\text{Rule 2: If } x \text{ is } A_2 \ \& \ y \text{ is } B_2 \text{ then } f_2 = r_2 + p_2 x + q_2 y \quad (21.8)$$

Layer 1: Every node $i$ is a square node in this layer with a node function

$$O_i^1 = \mu A_i(x) \quad (21.9)$$

where $x$ is the input to the node $i$, and $A_i$ is the linguistic label associated with this node function. $O_i^1$ is the membership function of $A_i$ and it specifies the degree to which the given $x$ satisfies the quantifier $A_i$. Let $\mu A_i(x)$ assume to be bell-shaped with a maximum equal to 1 and minimum equal to 0, such as the generalization bell function

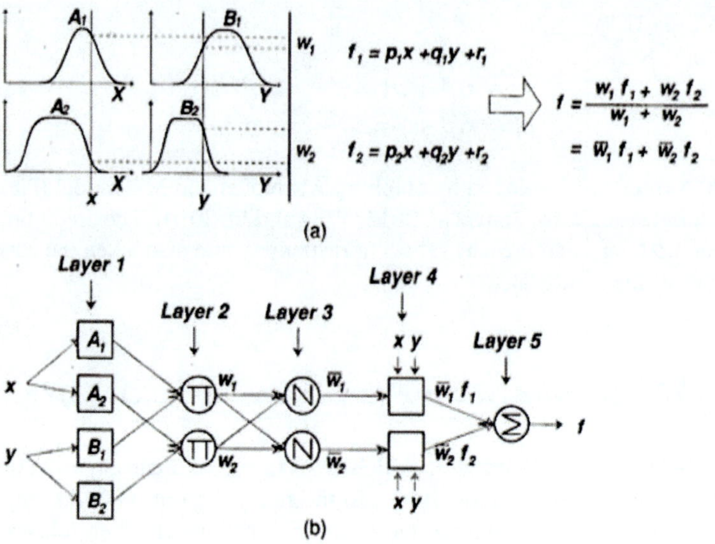


**Fig. 21.1** Graphical representation of a neuro-fuzzy system

$$\mu A_i(x) = \frac{1}{1 + \left[\left(\frac{x-c_i}{a_i}\right)^2\right]^{b_i}} \tag{21.10}$$

or the Gaussian function

$$\mu A_i(x) = \mathrm{e}^{\left[-\left(\frac{x-c_i}{a_i}\right)^2\right]} \tag{21.11}$$

where $\{a_i, b_i, c_i\}$ is the parameter set. Bell-shaped functions vary accordingly with the change of the parameter set (Hsu et al. 1995).

Layer 2: Every node is a circle node. In this layer, multiplication of the incoming signals and product out is done. Each node output depicts the firing strength of a rule. For example,

$$O_i^2 = w_i = \mu A_i(x) \times \mu B_i(x), \quad i = 1, 2, 3 \ldots \tag{21.12}$$

Layer 3: Every node is a circle node in this layer labeled by N. The $i$th node observes the ratio of the $i$th rule's firing strength to the sum of all rule's firing strengths.

$$O_i^3 = w_i = \frac{w_i}{w_1 + w_2}, \quad i = 1, 2, 3 \ldots \tag{21.13}$$

This layer's output will be called normalized firing strength.

Layer 4: In this layer, every node $i$ is a square node with a node function.

$$O_i^4 = \bar{w}_i f_i = \bar{w}_i = \bar{w}_i (p_i x + q_i y + r_i) \tag{21.14}$$

where $\bar{w}_i$ is the output of the previous layer and ($p_i$, $q_i$, $n_i$) is the parameter set.

Layer 5: In this layer, the single node is a circle node labeled that calculates the overall output as the summation of all incoming signals.

$$O_i^5 = \sum \bar{w}_i f_i = \frac{\sum_i w_i f_i}{\sum_i w_i} \tag{21.15}$$

Thus, it constructed an adaptive network, which is functionally equivalent to a type 3 fuzzy inference system.

### 21.2.3 Wavelet

Wavelet analysis is a refinement of Fourier analysis (Daubechies 1992; Georgiou and Kumar 1994; Hu and Nitta 1996; Jang et al. 1997; Kumar and Foufoula 1997; Mallat 1998; Siddiqi 2004; Siddiqi et al. 2003 ; Can et al. 2004, 2005; Manchanda et al. 2007) which has been used for prediction of time series of air pollutants parameter, meteorological parameter such as wind speed, temperature, humidity, etc. (Yousefi et al. 2005; Stanislaw and Garanty 2007).

Wavelet Transform decomposes a signal into a number of groups (vectors) of coefficients and different coefficient vectors contain information about features of the sequence at different scales. Wavelet analysis employs a prototype function called mother wavelet $f(t)$, which has a null mean, sharply drops in oscillatory way and data are denoted via superposition of scaled and translated versions of the pre-specified mother wavelet. It can be calculated by using the following equation (Parmar and Bhardwaj 2013a, b, c)

$$\mathrm{DWT}(m, t) = \frac{1}{\sqrt{a^m}} \sum_n x(n)\, f\left(\frac{t - nba^m}{a^m}\right) \tag{21.16}$$

where

| | |
|---|---|
| $a$ and $b$ | are scaling and translation function of integer variable $m$; |
| $t$ | integer variable that refers to a point of the input signal; |
| $n$ | discrete-time index; |
| $x(t)$ | given signal |
| $f(t)$ | mother wavelet. |

The wavelet transform is a prism that shows properties of a signal such as points of abrupt changes, seasonality, or periodicity. The plane defined by variables ($a$, $b$) is called a scale-space or time-frequency plane. The wavelet transform measures a variation of $f$ in the neighborhood of $b$. For a compactly supported wavelet or for a wavelet vanishing outside a closed and bounded interval, the value of the wavelet transform depends upon the value of $f$ in a neighborhood of $b$ of size proportional to the scale $a$. The local regularity of a function (or signal) is frequently measured with Lipschitz exponent. The global and local Lipschitz regularity can be characterized by the asymptomatic decay of wavelet transformation at small scales.

The first step of DWT corresponds to the mapping signal $f$ to its wavelet coefficients and form these coefficients two components are received, i.e., a smooth version called approximation and a second component that corresponds to the deviations or details of the signal.

A decomposition of $f$ into a low-frequency part $a_1$ and a high-frequency part $d_1$ at level 1 is represented by $f = a_1 + d_1$. The same method is performed on $a_1$ in order to find decomposition in finer scales: $a_1 = a_2 + d_2$. Recursive decomposition for the low-frequency parts follows the directions. The resulting low-frequency parts $a_1, a_2, \ldots, a_n$ are approximations of $f$ and the high-frequency parts $d_1, d_2, \ldots, d_n$ contain the details of $f$. It also classifies as follows:

$$
\begin{array}{l}
f \\
\swarrow \searrow \\
a_1 \quad d_1 \rightarrow \text{Level1} \\
\swarrow \searrow \\
a_2 \quad d_2 \rightarrow \text{Level2} \\
\swarrow \searrow \\
a_3 \quad d_3 \rightarrow \text{Level3} \\
\swarrow \searrow \\
a_4 \quad d_4 \rightarrow \text{Level4} \\
\swarrow \searrow \\
a_5 \quad d_5 \rightarrow \text{Level5}
\end{array}
$$

Let $f = (f_1, f_2, f_3, \ldots, f_N)$, N is an even integer.

$$a_m = \frac{f_{2m-1} + f_{2m}}{\sqrt{2}}, m = 1, 2, 3, \ldots, N/2$$

$$a^1 = (a_1, a_2, \ldots, a_{N/2})$$

The first trend sub-signal (approximation)

$$d_m = \frac{f_{2m-1} - f_{2m}}{\sqrt{2}}, \quad m = 1, 2, 3, \ldots, N/2$$

One-level Transform

$$f \xrightarrow{H_1} (a^1/d^1)$$

These wavelets are defined as

$$W_1^1 = \left(\frac{1}{\sqrt{2}}, -\frac{1}{\sqrt{2}}, 0, 0, \ldots, 0\right)$$
$$W_2^1 = \left(0, 0, \frac{1}{\sqrt{2}}, -\frac{1}{\sqrt{2}}, 0, 0, \ldots, 0\right)$$
$$\ldots$$
$$W_{N/2}^1 = \left(0, 0, 0, \ldots, 0, \frac{1}{\sqrt{2}}, -\frac{1}{\sqrt{2}}\right)$$

$$d_1 = f \cdot W_1^1$$
$$d_2 = f \cdot W_2^1$$
$$\ldots$$
$$d_m = f \cdot W_m^1 \quad \text{for } m = 1, 2, 3, \ldots, N/2$$

The main point in the prediction system is to select an appropriate number of decomposition levels based on the nature of the signal.

### *21.2.4 Artificial Neural Network (ANN)*

The ANN model consists of powerful non-linear approaches based on the function of the human brain. It identifies and learns the correlated patterns between input data sets and target values. The neural network firstly creates a network of simple processing nodes or neurons, which interconnected to each other in a specific order, performing simple numerical manipulations (Kisi et al. 2017).

The networks are made up of an input layer consisting of nodes that represent different input variables, the hidden layer consisted of many hidden nodes and an output layer consisted of output variables. In the current research, the ANN tool is widely applied in applications of AOD forecasting.

## 21.3 Implementation of Wavelet Analysis on Air Pollutants Data

Wavelet analysis is becoming an important tool for analyzing the localized variations of power in a time series. By decomposing a time series into time-frequency space, one is able to determine both the dominant modes of variability and how those modes vary in time. Wavelet transforms provide useful decompositions of the original time series; so that wavelet transformed data improve the ability of a forecasting model by capturing useful information on various resolution levels (Quiroz et al. 2011; Grossmann and Morlet 1984; Hong and Zhang 2008; Pellegrini et al. 2012).

Wavelet analysis is a more effective tool than the Fourier transforms in analyzing non-stationary time series (Parmar and Bhardwaj 2013a, b, c). Yousefi et al. (2005)

introduced a wavelet-based prediction procedure for the investigation of market efficiency in the future market for oil prices (Yousefi et al. 2005). Osowski and Garanty (2007) presented the method of daily air pollution forecasting by using support vector machine (SVM) and wavelet decomposition. Fourier transform is a powerful tool for analyzing the components of a stationary signal. However, it is unsuccessful for analyzing the non-stationary signal while wavelet transform allows the components of a non-stationary signal to be analyzed (Sifuzzaman et al. 2009). This chapter deals with wavelet analysis using Daubechies wavelet at level 5 (Db 5) of air pollutants such as $SO_2$, $NO_2$, and SPM at three different sites two residential and one industrial for the more than 20 years period from 1987 to 2010 over Delhi. Central Pollution Control Board (CPCB) is monitoring different criteria pollutants such as $SO_2$, $NO_2$, and SPM at eight different sites from which the data of three sites, two residential, Janakpuri and Nizamuddin, and one industrial, Shahazada Bagh, is used for the present study.

Statistical analysis involves mean, mode, median, minimum, maximum, standard deviation, range, Skewness, coefficient of variation, and kurtosis. Mean describes the average value. The median provides a middle value of an ordered sequence, moreover, it is named as positional average. The mode is defined as a value which occurs the maximum number of time that is having the maximum frequency. Minimum and maximum describe a minimum and maximum value of sample data. Standard deviation (SD) gives a measure of "spread" or "variability" of a sample. Skewness discusses the symmetry of data. The range is used for studying a variation of data. The coefficient of variation gives a relative measure of the sample. Kurtosis mentions to a degree of flatness in a region about the mode of a frequency curve. Skewness talks over the symmetry of data (Parmar and Bhardwaj 2013a, b, c). Table 21.1 shows a statistical analysis of air pollutants at three different sites.

**Table 21.1** Statistical analysis of air pollutants at three sites in Delhi

| | Observation site | Mean | Median | Mode | SD | Skewness | Kurtosis | Coefficient of variation |
|---|---|---|---|---|---|---|---|---|
| $SO_2$ | Janakpuri | 12.18 | 12.85 | 4.20 | 5.42 | −0.15 | −1.18 | 0.45 |
| | Shahzada Bagh | 14.23 | 11.90 | 4.00 | 9.34 | 1.19 | 2.20 | 0.66 |
| | Nizamuddin | 12.61 | 13.40 | 4.00 | 5.60 | 0.18 | 0.50 | 0.44 |
| $NO_2$ | Janakpuri | 38.85 | 38.50 | 34.50 | 11.03 | −0.37 | 0.89 | 0.28 |
| | Shahzada Bagh | 37.99 | 37.10 | 33.10 | 11.31 | 0.77 | 2.25 | 0.30 |
| | Nizamuddin | 36.97 | 37.40 | 39.30 | 10.18 | −0.26 | 0.80 | 0.28 |
| SPM | Janakpuri | 369.88 | 359.00 | 337.00 | 112.40 | 0.48 | 0.31 | 0.30 |
| | Shahzada Bagh | 399.98 | 380.00 | 311.00 | 130.46 | 0.78 | 0.87 | 0.33 |
| | Nizamuddin | 353.44 | 345.00 | 312.00 | 114.63 | 0.52 | 0.26 | 0.32 |

Observation shows the mean concentration of $SO_2$ decreased for both residential (Janakpuri, Nizamuddin) as well as industrial (Shahzada Bagh) area whereas $NO_2$ increased but under the NAAQS prescribed limits.

In practical applications, an estimated Daubechies wavelet level should be used to obtain the finer approximation and decomposition (Parmar and Bhardwaj 2012, 2013a, b, c). In the present paper, the Daubechies wavelet at level 5 ($Db_5$) is used to get the finer approximation and decomposition.

One dimensional discrete wavelet analysis of air pollutant concentrations namely $SO_2$, $NO_2$, and SPM at three locations in Delhi (India) has been discussed. Db5 wavelet decomposition of each data presented in seven parts namely s, $a_5$, $d_1$, $d_2$, $d_3$, $d_4$, and $d_5$ where "s" denotes signal or raw data; low-frequency part "$a_5$" gives an approximate of signal at level 5; high-frequency parts $d_1$, $d_2$, $d_3$, $d_4$, and $d_5$ contains the detail of "s" at different levels correspondingly(Table 21.2).

Statistical and wavelet analysis of air pollutants $SO_2$, $NO_2$, and SPM at three different sites two residential, i.e., Janakpuri and Nizamuddin, and one industrial, i.e., Shahazada Bagh over Delhi for the more than 20 year period from 1987 to 2010 in India have been discussed. It has been observed that Janakpuri and Nizamuddin represent almost equal but lower mean values of $SO_2$ concentration than Shahzada Bagh. For $NO_2$, the mean and median values are approximately equal but the mode is different for all three sites. For SPM Mean and median values are approximately equal for Janakpuri and Nizamuddin but Shahzada Bagh shows slightly higher values. Skewness value of SO2, NO2, and SPM suggests that for all sites curve is symmetrical and platykurtic excluding Shahzada Bagh it is leptokurtic for $SO_2$. In this work, we have decomposed the air pollution parameter $SO_2$, $NO_2$, and SPM concentration data which gives satisfactory results on employing wavelet analysis.

The wavelet transform proofed as the strength of generalization to neural network and specialization to inference fuzzy logic for training the non-stationary data and predicting the output. The average normalized error for testing can be reduced by using wavelet decomposition. The variability of data was unable to be trained by using only neuro-fuzzy. Consequently, the wavelet decomposition plays a very important role in the prediction analysis of air pollution data.

From this study, it has been observed that the values of signal data at five different levels for air pollutants vary between −3 to 3 (Janakpuri), -8 to 8(Nizamuddin) and -5 to 5 (Shahzada Bagh) for $SO_2$, −12 to 12 (Janakpuri and Nizamuddin), and −20 to 20 (Shahzada Bagh) for NO2 and −180 to 180 (Janakpuri), −225 to 225 (Nizamuddin) and −140 to 140 (Shahzada Bagh) for SPM, respectively. These wavelet decomposition outputs can be used in the Adaptive Network-based fuzzy inference system for prediction.

**Table 21.2** 1D Daubechies wavelet level (5) air pollutants values

| Air pollutants | Range of "s" | Range of "$a_5$" | Range of "$d_5$" | Range of "$d_4$" | Range of "$d_3$" | Range of "$d_2$" | Range of "$d_1$" |
|---|---|---|---|---|---|---|---|
| $SO_2$ at Janakpuri | 0–25 | 7.5–17.5 | −2 to 3 | −3 to 3 | −2 to 2 | −5 to 5 | −3 to 3 |
| $SO_2$ at Nizamuddin | 0–60 | 11–21 | −3 to 3 | −3 to 3 | −5 to 5 | −5 to 5 | −8 to 8 |
| $SO_2$ at Shahzada Bagh | 0–35 | 10–17 | −2.5 to 2.5 | −1.1 to 1.1 | −5.1 to 5.1 | −15 to 15 | −5 to 5 |
| $NO_2$ at Janakpuri | 0–62 | 30–46 | −3 to 3 | −4 to 4 | −6 to 6 | −12 to 12 | −12 to 12 |
| $NO_2$ at Nizamuddin | 0–62 | 30–46 | −3 to 3 | −4 to 4 | −6 to 6 | −12 to 12 | −12 to 12 |
| $NO_2$ at Shahzada Bagh | 0–70 | 30–46 | −3 to 3 | 7 to −7 | −12 to 12 | −10 to 10 | −20 to 20 |
| SPM at Janakpuri | 0–700 | 300–350 | −50 to 50 | −100 to 100 | −100 to 100 | −100 to 100 | −180 to 180 |
| SPM at Nizamuddin | 0–900 | 325–410 | −30 to 30 | −50 to 50 | −120 to 120 | −120 to 120 | 225 to 225 |
| SPM at Shahzada Bagh | 0–700 | 298–360 | −30 to 30 | −50 to 50 | −75 to 75 | −140 to 140 | 140 to 140 |

## 21.4 Implementation of ARIMA Technique

The objective of this chapter is to construct a forecast model for future pollution prediction based on long-term (1993–2012) pollutant concentration data for air pollutant management. In this study, the ARIMA model was used to evaluate the contribution of different pollutants in Delhi. ARIMA is a popular linear model in time series forecasting during the past years (Table 21.3).

This study is important because Delhi exhibits high air pollution and first time such a long-term (about twenty years) data have been studied. Also, in order to develop environmental policies and alternatives of producing agents of these pollutant sources and their implementation, such type of study is very important. The ambient air quality, long-term data used in the present study covers the period 1993–2012 and was obtained from the Central Pollution Control Board (CPCB). CPCB is monitoring different criterion pollutants such as $SO_2$, $NO_2$, and SPM at eight different sites from which the data at three sites, two residential, i.e., Janakpuri and Nizamuddin, and one industrial, i.e., Shahazada Bagh, is used for the present study.

The average $SO_{2,}$ over Janakpuri, Shahzada Bagh, and Nizamiddin are, respectively, (11.6 ± 5.7), (12.98 ± 8.6), and (11.52 ± 5.54). From Fig. 21.2, it is found that Janakpuri and Nizamuddin represent almost equal, but lower mean values of $SO_2$ concentration than Shahzada Bagh. Similarly, the average value of $NO_2$ for Janakpuri, Shahzada Bagh, and Nizamuddin are observed (41.46 ± 9.4), (41.36 ± 11), and (39.8 ± 8.5), respectively. In addition, the average value of SPM for all

**Table 21.3** ARIMA model fitting statistics

| Location | Stationary *R*-squared | *R*-squared | RMSE | MAPE | MAE | Normalized BIC | Model type |
|---|---|---|---|---|---|---|---|
| $SO_2$ Nizamuddin | 0.666 | 0.893 | 1.818 | 12.716 | 1.244 | 1.241 | Simple seasonal |
| $NO_2$ Nizamuddin | 0.648 | 0.504 | 6.031 | 26.561 | 4.231 | 3.639 | Simple seasonal |
| SPM Nizamuddin | 0.620 | 0.524 | 84.846 | 21.906 | 66.414 | 8.927 | Simple seasonal |
| $SO_2$ Shahzabagh | 0.614 | 0.882 | 2.955 | 14.593 | 1.917 | 2.236 | Winters' multiplicative |
| $NO_2$ Shahzabagh | 0.552 | 0.515 | 7.694 | 13.780 | 5.467 | 4.127 | Simple measonal |
| SPM Shahzabagh | 0.530 | 0.437 | 100.198 | 22.153 | 74.399 | 9.260 | Simple measonal |
| $SO_2$ Janakpuri | 0.585 | 0.898 | 1.825 | 11.727 | 1.318 | 1.272 | Winters' multiplicative |
| $NO_2$ Janakpuri | 0.622 | 0.641 | 5.670 | 14.326 | 4.200 | 3.516 | Simple seasonal |
| SPM Janakpuri | 0.655 | 0.453 | 87.175 | 24.598 | 68.859 | 8.982 | Simple seasonal |

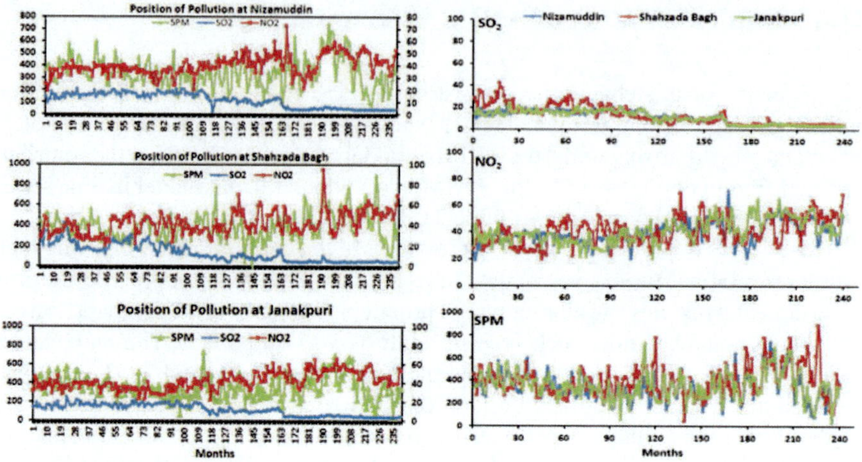


**Fig. 21.2** Variability of air pollutants ($SO_2$, $NO_2$, and SPM) for Nizamuddin, Shahzada Bagh and Janakpuri over Delhi

three sites is found to be 355 ± 117.5 (Janakpuri), 388.3 ± 133.3 (Shahzada Bagh), and 344.4 ± 122.7 (Nizamuddin). From this study, it has been observed in the mean concentration of $SO_2$ decreased for both residential (Janakpuri, Nizamuddin) as well as industrial (Shahzada Bagh) area, whereas $NO_2$ increased but under the National Ambient Air Quality Standards (NAAQS) prescribed limits.

The trend and prediction of air pollutants series are computed by the ARIMA model with SPSS 17 software. First, for the identification of a suitable model, it is necessary to determine whether the time series is stationary or not. The time series is stationary if the probable values of the series and the autocorrelation function (ACF) are independent of time. Auto-correlate is the correlation between time series and the same time series lag, while partial autocorrelations are the correlation coefficients between the basic time series and the same time series lag, but with the elimination of the influence of the members in between. The autocorrelation function (ACF) shows the degree of correlation with past values of the series as a function of the number of periods in the past (i.e., the lag) at which the correlation is computed.

It is observed from autocorrelation analysis that for $SO_2$, $NO_2$, and SPM over all three sites most of the autocorrelation coefficients of the residuals are within the confidence limits, indicating that the model is a good fit. However, over Shahzada Bagh for $SO_2$, $NO_2$, and SPM and at Janakpuri for $SO_2$, it can be noted that all of the 24 ACF are not statistically significant, suggesting that residuals are not autocorrelated with each other, hence the selected model is acceptable for the future prediction of air pollutants. From the original signal, ARIMA makes a new signal using the previous one, and RMSE gives us (Table 21.1) the error between the original and ARIMA signal. This produces a new time series based on the original (actual) series and can predict the unknown future values using the predicted ARIMA model series signal.

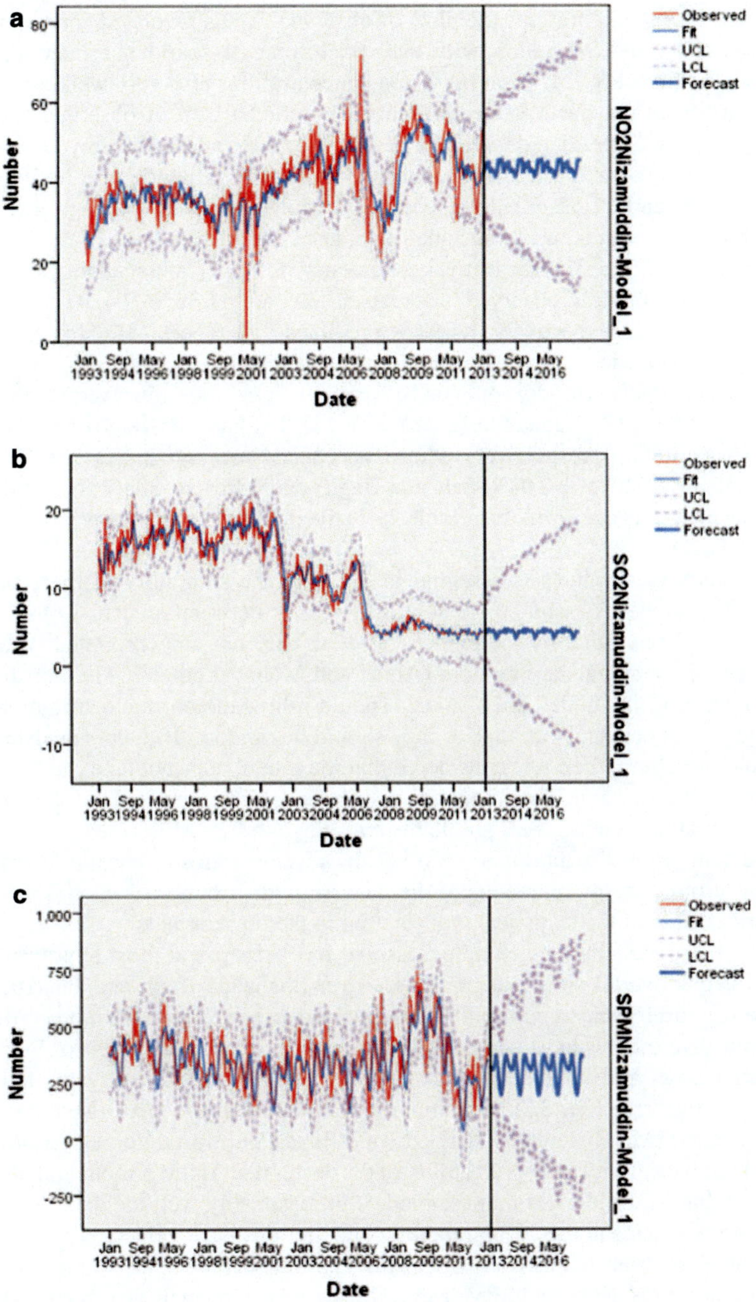


**Fig. 21.3** Time series for observed (red line) and ARIMA model forecast (blue line) over Nizamuddin, Shahzada Bagh and Janakpuri over Delhi

Figure 21.3a–c, represents the plots of observed versus predicted concentrations for $SO_2$, $NO_2$, and SPM alongwith their 95% forecast confidence intervals. It is clearly seen from Fig. 21.3 that predicted concentrations also very well capture the variability from the mean value. Moreover, by comparison of observed "average ± SD" to that of predicted "average ± SD", it is found that observed "average ± SD" for $SO_2$ over Janakpuri, Shahzada Bagh and Nizamuddin are 11.59 ± 5.7, 12.98 ± 8.6, and 11.52 ± 5.5, respectively, while predicted "average ± SD" in case of $SO_2$ over Janakpuri, Shahzada Bagh, and Nizamuddin is 11.57 ± 5.5, 12.97 ± 8.4, and 11.60 ± 5.4, respectively. Similarly, for $NO_2$ at Janakpuri, Shahzada Bagh, and Nizamuddin observed "average ± SD" are 41.46 ± 9.4, 41.36 ± 11.0, and 39.79 ± 8.5, respectively, whereas predicted "average ± SD" for $NO_2$ over Janakpuri, Shahzada Bagh, and Nizamuddin are 41.34 ± 8.3, 41.1 ± 10.1 and 39.55 ± 7.1, respectively. In addition, for SPM at all three sites observed "average ± SD" are 355 ± 117.5 (Janakpuri), 388.3 ± 133.3 (Shahzada Bagh) and 344.4 ± 122.7 (Nizamuddin), respectively, whereas predicted "average ± SD" are 356.03 ± 96 (Janakpuri), 389.5 ± 108 (Shahzada Bagh) and 345.1 ± 103.7 (Nizamuddin). Consequently, it is observed that (Table 21.1) the model fits reasonably well with the observed data.

The study of residuals is essential in deciding the suitability of the statistical model. The model is assumed suitable if the errors between an actual observation and the predicted value by the model (residuals) are not autocorrelated with each other, i.e., the residuals are random (Abish and Mohanakumar 2011). For the best performance of the model, the residuals should follow the normal distribution with zero mean and constant variance or they should be random. If the residuals display any kind of pattern, then it is considered that the model does not take care of all the systematic information. The frequency distributions of the residuals of the ARIMA models for all sites and for all pollutants are shown in Fig. 21.4.

The histogram distribution shows that the residuals are, in general, distributed equally around zero approaching the Gaussian distribution, which shows the appropriateness of the statistical models used in the present paper.

ARIMA models namely Simple Seasonal has been found most appropriate for all pollutants overall sites excepting $SO_2$ over Shahzada Bagh and Janakpuri for simulating and forecasting air pollutants over the three selected locations. Winters' Multiplicative models have been developed as the most suitable only for $SO_2$, over Shahzada Bagh and Janakpuri. The statistical parameters stationary $R^2$, $R^2$, Root Mean Square Error (RMSE), Maximum absolute percentage error (MAPE), Mean absolute error (MAE), and Normalized BIC (Bayesian Information Criterion) were used to test the validity and applicability of the developed ARIMA models at different stages, Table 21.1 indicates that the models fit reasonably well the data series with some discrepancies in the air pollutants concentrations peak values.

Long-term observations from 1993 to 2012 (nearly two decades) and forecast of air pollutants $SO_2$, $NO_2$, and SPM over Delhi are presented in this study and ARIMA models are developed to forecast the air pollutant ($SO_2$, $NO_2$, and SPM) concentrations. The performance evaluations of the adopted models are carried out on the basis

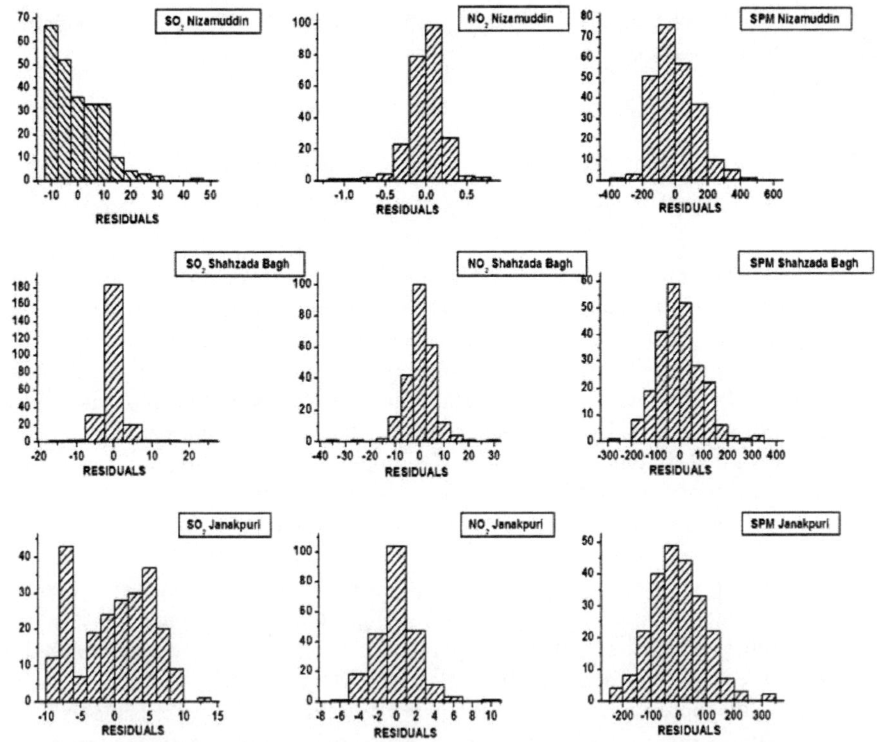


**Fig. 21.4** Frequency distribution of residuals of ARIMA model application Nizamuddin, Shahzada Bagh and Janakpuri in Delhi concerning air pollutant data ($SO_2$, $NO_2$, and SPM)

of correlation coefficient ($R^2$), Root Mean Square Error (RMSE), Maximum absolute percentage error (MAPE), Mean absolute error (MAE), and Normalized BIC (Normalized Bayesian Information Criterion). The results indicate that the seasonal ARIMA model provides reliable and satisfactory predictions for air quality parameters. Comparing the observed values to the predicted values in this study, it is obvious that the forecasting performances of the present study are adequate. The comparison between forecasted and observed values by models suggests that this model can be reliably used for air pollution predictions.

Additionally, the present study has successfully applied the ARIMA modeling procedure on the three of the air pollutants ($SO_2$, $NO_2$, and SPM) and found satisfactory results. The average values of $SO_2$, $NO_2$, and SPM concentration for the next five years (2013–2017) at all three sites are found to be 1.99 ± 1.25(Janakpuri), 1.11 ± 2.0 (Shahzada Bagh), 4.25 (Nizamuddin); 48.31(Janakpuri), 61.6(Shahzada Bagh), 43.7(Nizamuddin) and 262.02(Janakpuri), 260.02(Shahzada Bagh), 304.30(Nizamuddin), respectively. The developed model can also be applied to predict other pollutant concentrations. ARIMA modeling technique in the present study has worked well in forecasting air pollutants and can be efficiently used for air quality forewarning purposes.

## 21.5 Implementation of Time Series Model on Satellite and Ground Aerosol Optical Depth Data

Aerosols play an important role in Earth-atmosphere, ocean system by means of their direct and indirect impact on climate. This is a key factor in atmospheric management. Keeping in view of the importance and sensitivity of aerosol properties over the IGP region, the study has been carried out using both ground-based AERONET data and the MODIS satellite-derived data. Also, statistical variation and time series prediction of satellite and ground data sets are analyzed by using past data of the period 2001–2012. In this study, an ARIMA model is built and fitted to time series data of ground-based AERONET and the MODIS satellite-derived AODs, to allow a deeper understanding of that data and to predict the future level of AODs. Figure 21.5 represents the monthly variation of the ground and MODIS AOD.

As discussed before, the development of ARIMA includes four steps: Identification, Estimation, Verification, and Forecasting. Figure 21.6 depicts the residual autocorrelation and partial autocorrelation functions of the monthly-average MODIS and ground $AOD_{550}$ time series in the IGP region. The two vertical lines in the ACF and PACF plots represent the 95% confidence intervals for the assessed autocorrelation and partial autocorrelation coefficients. Both positive and negative correlations have been detected.

On the basis of the above discussion, the model is performing satisfactorily for both data sets, however, the low RMSE for Ground AODs data indicates a good fit (Fig. 21.7) for the model and a perfect prediction over the mean.

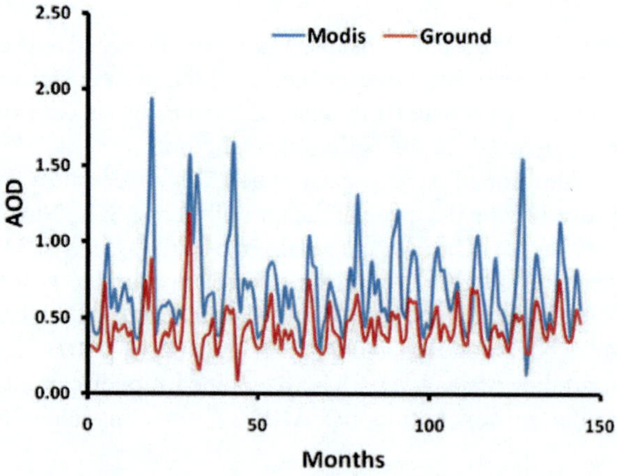


**Fig. 21.5** Monthly variation of MODIS and Ground AODs over Kanpur

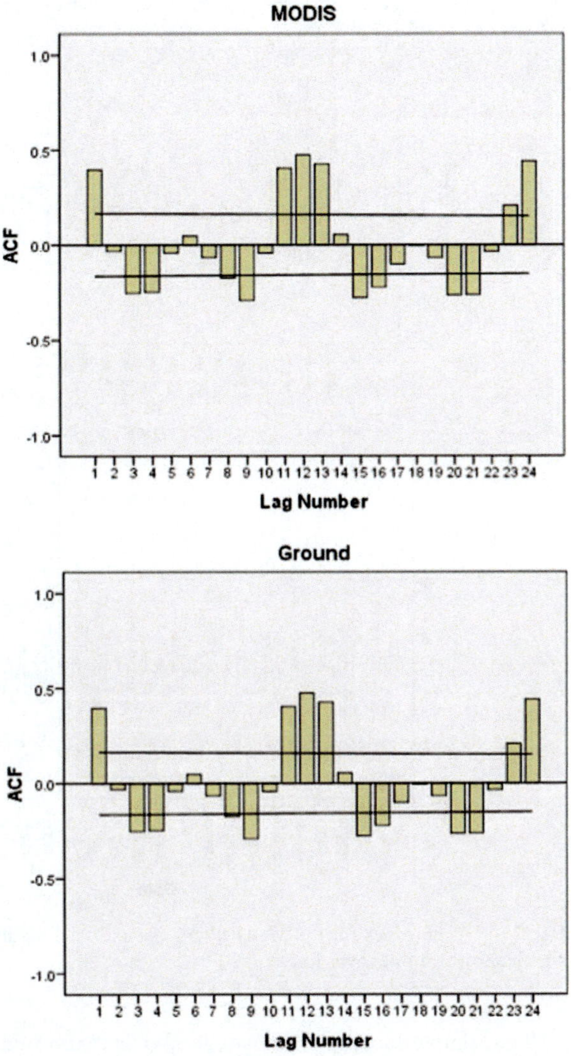


**Fig. 21.6** ACF and PACF residuals for the MODIS and Ground AODs. The vertical lines indicate the upper and lower limits

Figure 21.8 shows the P-P plot for the MODIS and ground AODs. The P-P PLOT statement creates a probability-probability plot, it compares the empirical cumulative distribution function (ecdf) of a variable with a specified theoretical cumulative distribution function such as the normal. If the two distributions match, the points on the plot form a linear pattern that passes through the origin and has a unit slope. Thus, the P-P plot can be used to determine how well a theoretical distribution model a set of measurements. In the present study for both MODIS and AERONET AODs data are consistent with a sample from a normal distribution.

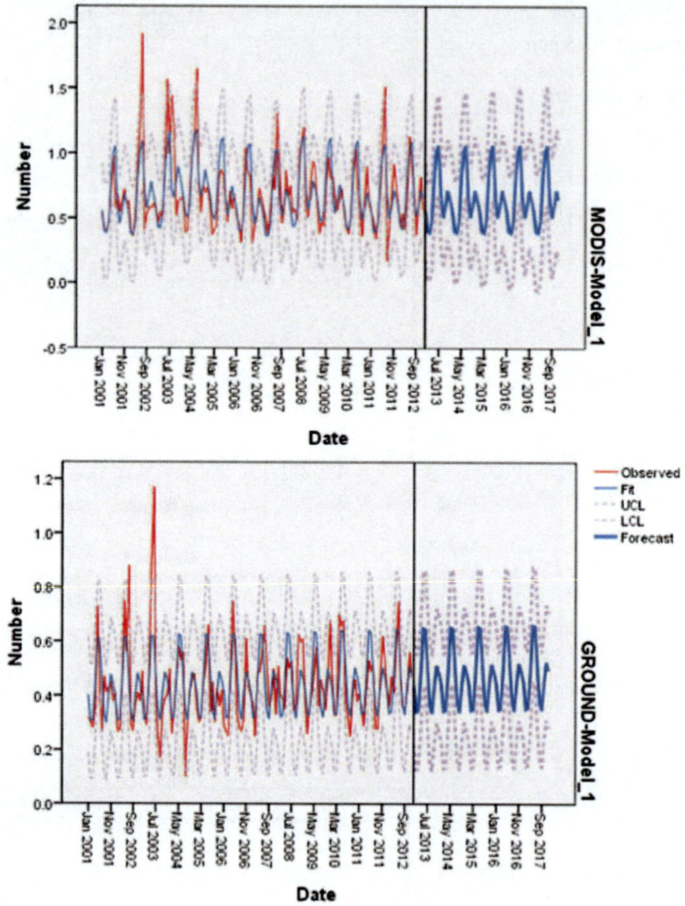


**Fig. 21.7** Time series for the MODIS and Ground AODs values (red line) and ARIMA model simulations(blue line) over Kanpur

The relative success of statistical models in reproducing the measured time series can also be estimated in terms of the residuals of error. For ARIMA models the frequency distributions of the residuals at each site are presented in Fig. 21.9.

The histogram distribution trend indicates that the residuals are, in general, dispersed in the same way around zero approaching the Gaussian distribution, which also points out the suitability of the statistical models used in the present work. The Ljung-Box statistic, also known as the modified Box-Pierce statistic, provides an indication of whether the model is correctly specified. A significance value of less than 0.05 implies that there is a structure in the observed series, which is not accounted for by the model. On the basis of this statistical parameter, the comparison shows that ARIMA models reveal better performance with ground data series in comparison to MODIS data.

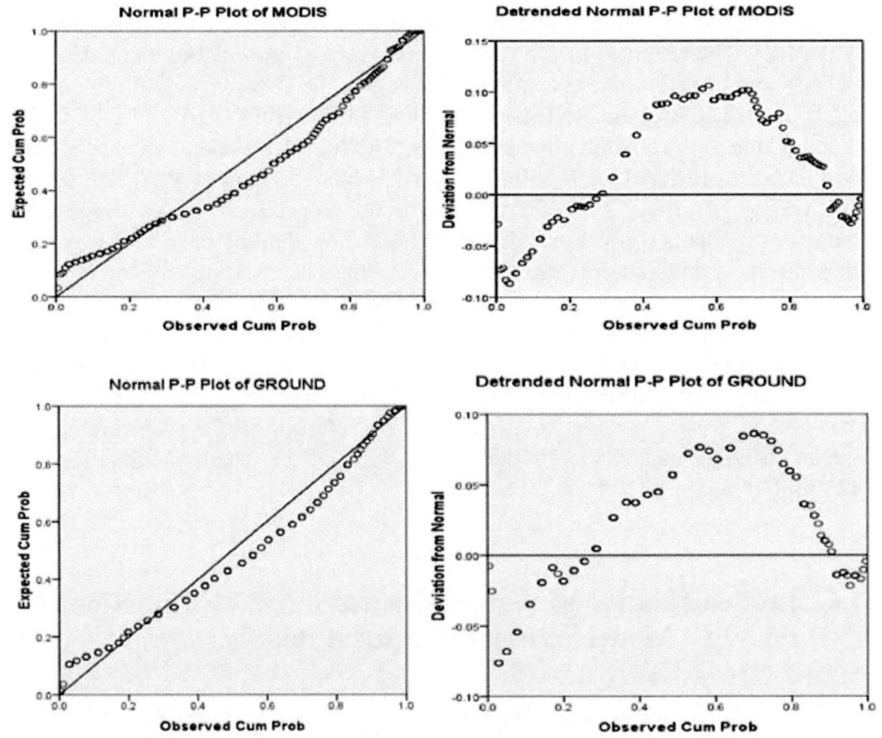


**Fig. 21.8** Normal P-P and detrended normal P-P plot for the MODIS and ground AODs values over Kanpur

**Fig. 21.9** Frequency distributions of the residuals of the ARIMA model

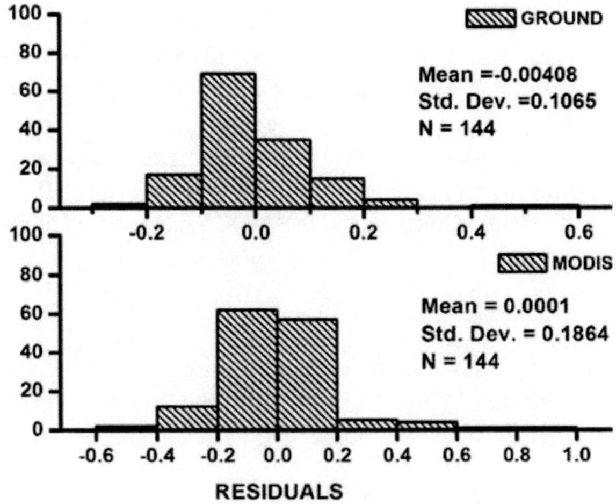

Time series analysis and forecasting are an active research area over the last few decades. The accuracy of time series forecasting is essential to many decision processes and therefore, the research for improving the effectiveness of forecasting models has never stopped. ARIMA has gained abundant popularity in time series prediction due to its reliability and simplicity. The ARIMA model is used to predict the AOD level of MODIS and AERONET. ARIMA model has MAE 0.127, 0.076, and RMSE 0.187, 0.107 for MODIS and ground, respectively, which suggest the accuracy of model performance. ACF and PACF show that the satellite and ground follow the six-month patterns. MODIS and AERONET are positively correlated at the 0.01 significance level. Using two independent sample t-test and one way another, it is analyzed that the population means of MODIS and AERONET are significantly different. From the P-P plot, it reveals that the data in both cases are consistent with a sample from a normal distribution. Time series and statistical variability of MODIS and AREONET using past data of 2001–2012, concluded that the pattern of both is same and MODIS data can be produced from AREONET data by adding $0.251 \pm 0.133$ in IGP region.

## 21.6 Implementation of Wavelet Neural-Fuzzy Conjunction (WNFC) Model on Aerosol Optical Depths

A new wavelet neural-fuzzy conjunction (WNFC) model is introduced and developed for long-term predictions of the aerosol characteristics over Kanpur in the Indo-Gangetic plains (IGP), India, this area is extremely polluted over the world with anthropogenic absorbing aerosols and natural particles. The WNFC model is a conjunction of three popular soft computing models, namely wavelet, artificial neural networks, and fuzzy models. The aerosol optical depth (AOD) derived from the AERONET (level 2) during the time period 2001–2012 is used as model input. On the ground of mean absolute error (MAE), the introduced WNFC model is compared with the outputs of multiple regression, artificial neural network, and neural-fuzzy coupled models. The WNFC model predicts the AOD with the least error, which is helpful for the management of air quality over IGP.

The methodology used in AOD forecasting using the WNFC model follows these steps:

(a) *Data Collection*

The first step is data collection (primary or secondary). In our application, the monthly-averaged values of AERONET data level 2 were used during the period of January 2001–December 2012 over Kanpur.

(b) *Pre-processing of Data*

The data collected in the earlier step is pre-processed in the second step. The dataset is filtered using the de-noising method, followed by normalization.

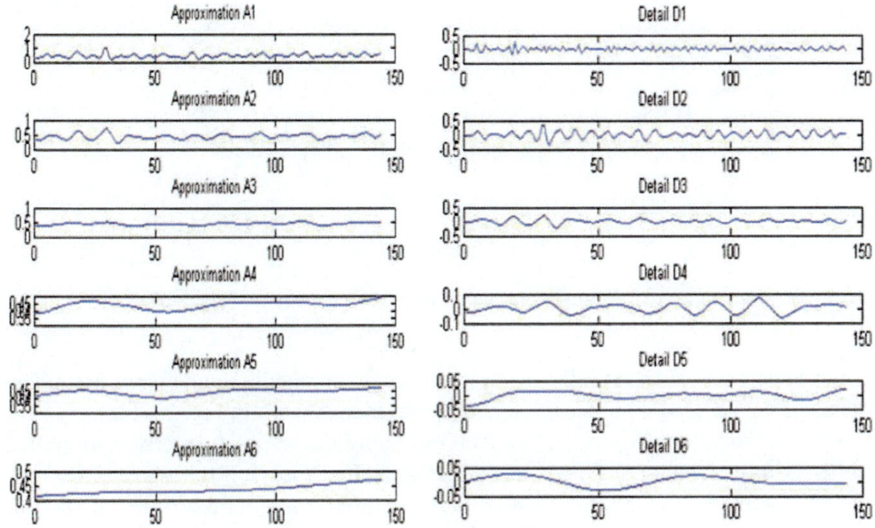


**Fig. 21.10** Six-level detailed and approximation coefficient using the Daubechies 8 ($Db_8$) wavelet

(c) *Wavelet Decomposition*

The wavelet transform was used to decompose the normalized AOD pattern into several wavelet components as shown in Fig. 21.10. The original normalized signal of AOD is decomposed to high and low-frequency components by using $Db_8$, mother wavelet for calculating the coefficient of the details ($d$) and approximate ($a$) components.

(d) *Selection of training pattern*

The first step in training is obtained accurate and sufficient historical data after pre-processing. The input data is chosen as the relevance to the model because it directly affects the performance of the model.

The wavelet decomposed components are used for training. The training pattern consists of decomposed wavelet components of giving load pattern at time $t$, $t-1$, $t-2$ (past three points) as input, and the forecasted wavelet component at $t+1$ as output. Training patterns are expressed as the pair of a set of input and output.

$$\text{Training pattern} = \left[\text{Input vector}\right]\left[\text{Output vector}\right]$$

85% of total data are used for training and the remaining 15% are used for testing the model.

For the execution of Wavelet-Neuro-Fuzzy, Neuro-Fuzzy, artificial neural network models, programs are developed using Matlab 2009 software. The monthly-averaged values of $AOD_{550}$ over Kanpur from 2001 to 2012 have been used. Out of the 144 monthly values, 117 have been used to train a model and for the model testing, 24

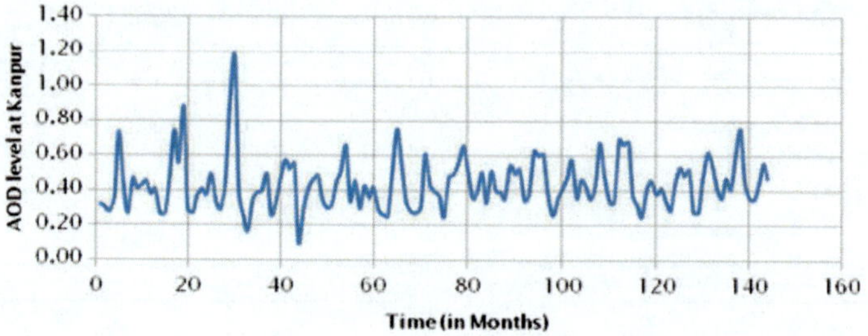


**Fig. 21.11** Time series of $AOD_{550}$ over Kanpur, India during the period 2001–2012

values have been used. Initially, as an input value, three data points were used. The values of the $AOD_{550}$ in time series over Kanpur are shown in Fig. 21.11.

The statistical description of the $AOD_{550}$ using the wavelet decomposition as a multi-scale analysis is analyzed in this section. For applying the wavelet decomposition method, it is necessary to select a mother wavelet, the order of mother wavelet along with the number of level decomposition. Daubechies are highly suitable for treating random and spike series because in these families of wavelet the smoothness increases and, thus, suitable higher-order wavelet must be taken into account. In the present work, Daubechies wavelets of order 2–8 have been considered for the analysis. Comparison of all Daubechies wavelet levels using the WNFC model. Least forecasting error has been achieved by applying the decomposed series using wavelet of order eight.

Figure 21.10 shows $AOD_{550}$ signal approximation, $a_1$ to $a_7$ corresponds to low-frequency details, while the detailed parts $d_1$ to $d_7$ correspond to the high-frequency band, which contains the local short period variation in the $AOD_{550}$ signals. Table 21.4 shows that $a_1$, $a_2$, and $a_3$ series are in similar shapes out of other approximation levels to the actual signal.

The statistical characteristics of $a_3$ taking into account the skewness show the best matching, whereas $a_4$ to $a_6$ does not give any significant information. Thus, the third level illustrates better the normal behavior of the $AOD_{550}$ time series. On

**Table 21.4** Skewness of approximation coefficient using Daubechies wavelet (Actual skewness = 1.387)

| | Db2 | Db3 | Db4 | Db5 | Db6 | Db7 | Db8 |
|---|---|---|---|---|---|---|---|
| A1 | 1.730714 | 0.879948 | 0.491798 | 0.604409 | 1.009702 | 1.340546 | 1.38227 |
| A2 | 1.130094 | 0.699018 | 0.503047 | 0.690856 | 0.602819 | 0.319444 | 0.555065 |
| **A3** | **−0.01082** | **0.43781** | **0.045571** | **0.217759** | **0.033235** | **0.080581** | **0.124083** |
| A4 | −0.70375 | −0.34761 | 0.038208 | −0.27032 | −0.24985 | 0.110933 | −0.63452 |
| A5 | −1.24834 | −0.41823 | -0.75971 | −0.29417 | −0.22453 | 0.401722 | −0.81777 |
| A6 | −1.48468 | −0.36925 | 0.091794 | 0.298949 | 0.067991 | 0.884141 | 0.813783 |
| A7 | −0.5673 | 0.142622 | −0.04124 | 0.029649 | −0.17785 | 0.092772 | 0.178305 |

**Table 21.5** Kurtosis of detailed coefficient using Daubechies wavelet (Actual kurtosis = 3.967)

| | Db2 | Db3 | Db4 | Db5 | Db6 | Db7 | Db8 |
|---|---|---|---|---|---|---|---|
| **D1** | **1.948139** | **2.067158** | **3.426446** | **4.098181** | **2.346865** | **0.872187** | **0.866418** |
| **D2** | **1.840861** | **2.922505** | **0.435124** | **-0.47003** | **0.465486** | **1.477135** | **0.859114** |
| **D3** | **5.857645** | **1.015409** | **1.290579** | **0.709717** | **1.846732** | **1.083561** | **1.343721** |
| D4 | 0.441802 | 0.71691 | 0.085832 | 0.463258 | 0.754677 | −0.9673 | −0.49347 |
| D5 | −0.19328 | 0.559339 | −0.78137 | 0.25651 | −0.69479 | −0.71274 | 0.459116 |
| D6 | 0.060007 | −0.2195 | −0.03947 | −1.07704 | −1.30181 | −0.16938 | −0.91752 |
| D7 | −1.28299 | −0.42151 | −0.91393 | −1.40241 | 0.221023 | −1.14821 | −1.40936 |

observing the detail series $d_1$ to $d_7$in Fig. 21.10, it can be found that the range of detail parts is lesser than the approximation parts. Furthermore, the variation of $d_4$ to $d_7$ is lower than that from $d_1$ to $d_3$, suggested that the detailed part $d_4$ to $d_7$ is preserved random noise, as was shown in the kurtosis characteristics (Table 21.5). Localized variation detects in $d_1$ to $d_3$ and provides better results. On statistical and visual checks only $d_1$ and $d_2$ have valuable higher frequency information and show the irregularities representing random variations. Furthermore, $d_3$indicates peaks that allow time localization of the peak.

In the WNFC model, the general spirit is to decompose the $AOD_{550}$ series using Daubechies 8 wavelet ($Db_8$) and then model it by separately fitting of each level of resolution. Each component $a_3$, $d_3$, $d_{2,}$ and $d_1$ is modeled separately, and the final result is obtained by reconstruction of the signal by using these four forecasts. Figure 21.12 represents the mechanism for the forecasting procedure, i.e., using three previous time lags for creating a new forecasted time lag.

The total predicted AOD at any instant ($i$) is given by

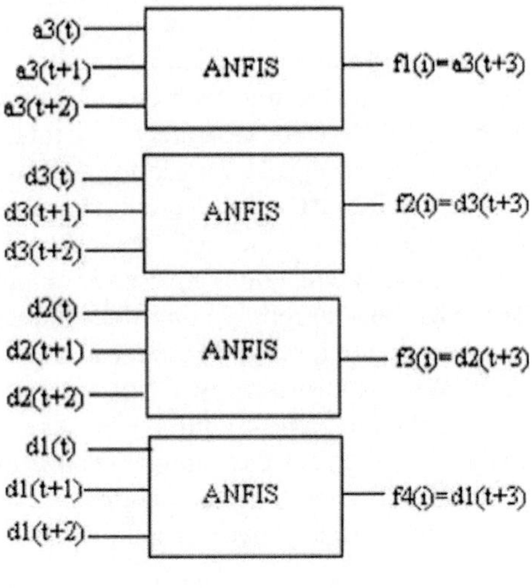


**Fig. 21.12** Flow chart of the mechanism for the forecasting procedure

**Table 21.6** Training and testing results from the WNFC model for the prediction of two years (24 months) ahead $AOD_{550}$ in Kanpur using $db_2$ to $db_8$ wavelet

| Wavelet order | Training results | | Testing results | |
|---|---|---|---|---|
| | MAE | ε (%) | MAE | ε (%) |
| db2 | 0.0119 | 2.72 | 0.0072 | 1.60 |
| db3 | 0.0148 | 3.38 | 0.0034 | 7.58 |
| db4 | 0.0079 | 1.81 | 0.0019 | 0.43 |
| db5 | 0.0102 | 2.33 | 0.0153 | 3.40 |
| db6 | 0.0017 | 0.39 | 0.0059 | 1.30 |
| db7 | 0.0159 | 3.63 | 0.0100 | 2.33 |
| **db8** | **0.0021** | **0.48** | **0.0009** | **0.21** |

$$f(i) = f_1(i) + f_2(i) + f_3(i) + f_4(i) \tag{21.17}$$

For attaining the best results, the WNFC model is estimated using wavelet subseries. The past three inputs were used to predict the AOD for orders 2–8. Table 21.6 shows the results of training and testing data for different decomposition levels of the Daubechies wavelet. The presented numerical experiment confirms the accuracy of prediction is obtained in the past three months. The testing results show a 0.21% relative error for 24-months ahead predictions.

The WNFC model is used to predict the wavelet coefficients $a_3$, $d_1$, $d_2$, and $d_3$. Every coefficient is added up to predict the next future values. The outcome of predicted coefficients for training data is shown in the Figs. 21.13a–d.

The trained predicted output is obtained from the decomposed wavelet coefficients by simple summation represented by $S(n)$.

$$S(n) = d_1 + d_2 + d_3 + a_3 \tag{21.18}$$

The actual and predicted trained signal is shown in Fig. 21.14 for the $AOD_{550}$ data series. The actual testing coefficients acquired from wavelet decomposition are given to the ANFIS system separately. The total predicted and the actual testing outputs are presented in Fig. 21.7 using the $AOD_{550}$ data during the period 2001–2012.

The monthly $AOD_{550}$ prediction has been also performed by the classical ANN, which was trained by the segregate Levemberg–Marquardt method. The linear activation and sigmoid functions are used as outputs and hidden layer neurons, respectively. The ANN (3,4,1) representing three inputs, four hidden neurons, and one output node was found to result in the most accurate $AOD_{550}$ estimates. The performances of the WNFC, Neuro-fuzzy, ANN, and regression models in predicting the $AOD_{550}$ time series for both training and testing datasets are summarized in Table 21.7. Figure 21.15 depicts the training, testing, and validation of the model.

Table 21.7 shows the comparative results for all different models applied for the prediction of $AOD_{550}$ over Kanpur. WNFC model exhibits the smallest relative errors

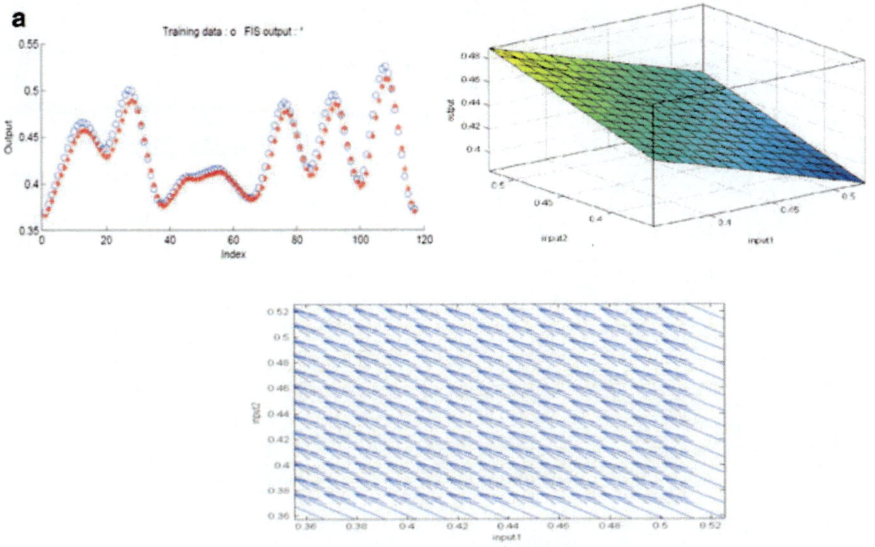


**Fig. 21.13** **a** $a_3$ Training results and model output with surface and quiver view. **b** $d_1$ Training results and model output with surface and quiver view. **c** $d_2$ Training results and model output with surface and quiver view. **d** $d_3$ Training results and model output with surface and quiver view

of 0.21% for the testing data and 0.48% for the training data, indicating the best correspondence to the measured $AOD_{550}$. The relative errors (testing data) for the neuro-fuzzy, ANN, and regression method are 0.62%, 3.34%, and 54.03%, respectively, resulting in inaccurate fits compared to the WNFC model. The mean absolute error for the WNFC model is extremely low (0.000963) for the testing dataset suggesting an accurate representation of the $AOD_{550}$ over Kanpur (Fig. 21.16).

## 21.7 Summary

In this chapter, soft computing models are compared and applied to the prediction of AOD level in Kanpur, IGP region. Also, a new proposed model, the Wavelet Neural-Fuzzy Conjunction (WNFC) is introduced, which is developed by combing the properties of wavelet, ANN, and fuzzy approaches. Output results of the WNFC model are compared with other popular models neuro-fuzzy, neuro, and regression for AOD level forecasting of IGP at Kanpur, India. It is observed that multiple linear regression model has a 54.03% rate of forecasting error; ANN model has a 3.34% error in forecasting future; Neuro-fuzzy coupled model results in 0.62% error; WNFC model predicts future with 0.21% error. Thus, it is concluded that Wavelet Neural-Fuzzy Conjunction (WNFC) model predicts the future accurately and the model is helpful in the management of atmospheric modeling.

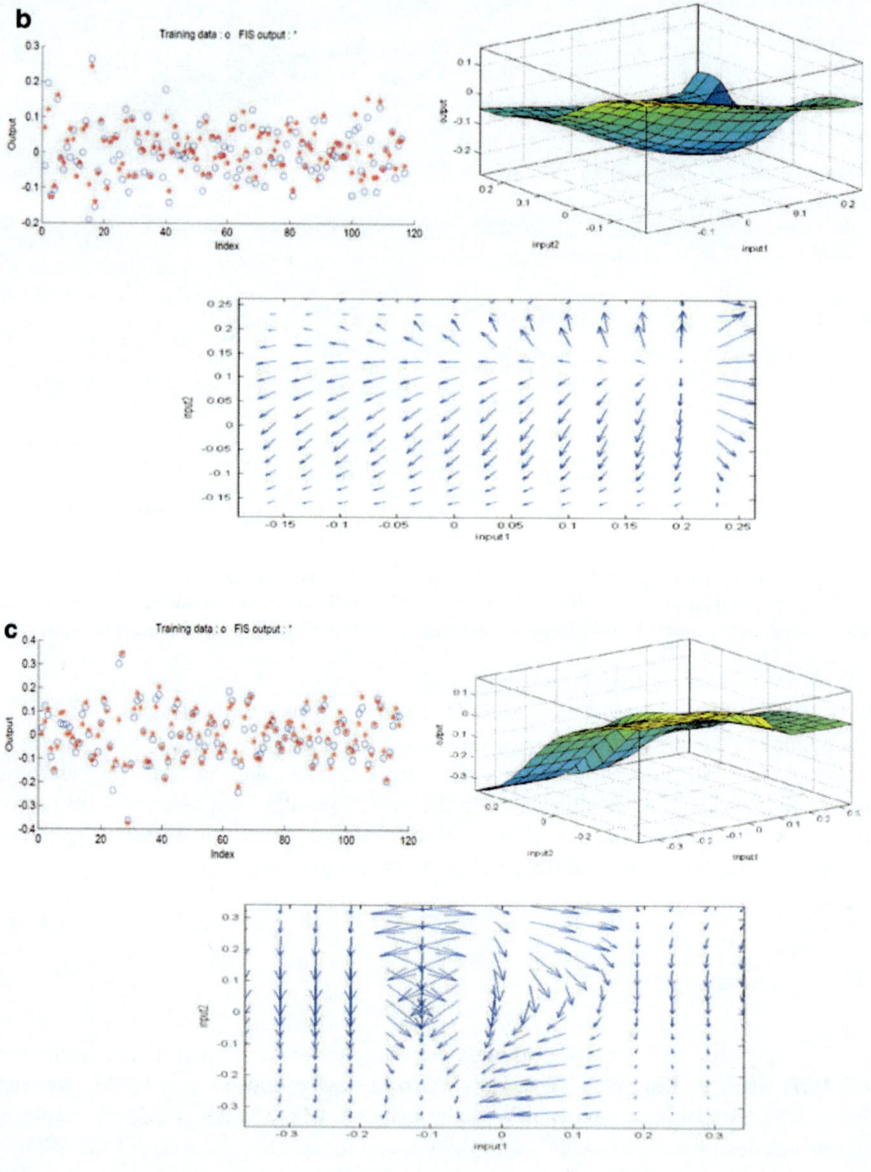


Fig. 21.13 (continued)

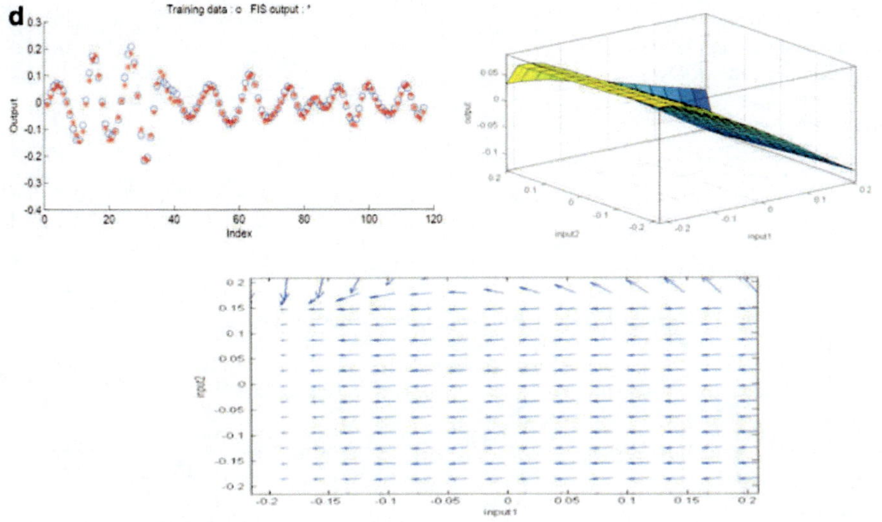


**Fig. 21.13** (continued)

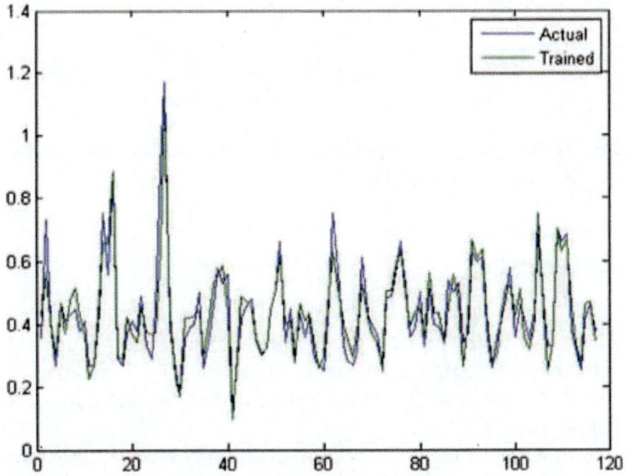


**Fig. 21.14** Actual and predicted output by the WNFC model for $AOD_{550}$ over Kanpur using the training dataset

**Table 21.7** Training and testing error results of each model for monthly AOD level forecasting

| | Training results | | Testing results | |
|---|---|---|---|---|
| | MAE | ε (%) | MAE | ε (%) |
| Wavelet-neuro-fuzzy conjunction (WNFC) | **0.021** | **0.48** | **0.000963** | **0.21** |
| Neuro-fuzzy | 0.055 | 1.25 | 0.0266 | 0.62 |
| Artificial neural network | 0.0514 | 2.34 | 0.0345 | 3.34 |
| Multiple linear regression | 19.934 | 38.34 | 30.450 | 54.03 |

**Fig. 21.15** Actual and predicted output by the WNFC model for $AOD_{550}$ over Kanpur using the testing dataset

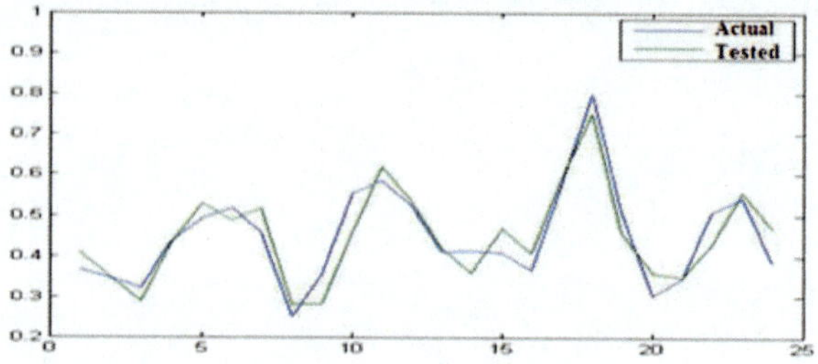


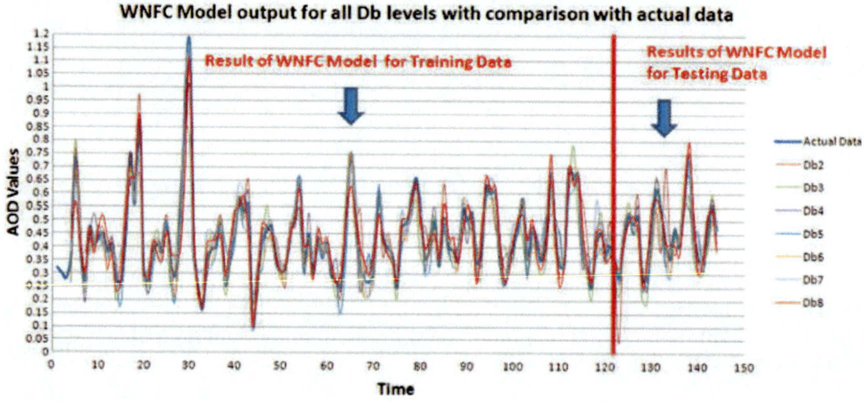


**Fig. 21.16** Comparison between measured $AOD_{550}$ and WNFC Model predictions for the Training and Testing data for all Db levels

## References


Abish B, Mohanakumar K (2011) Biennial variability in aerosol optical depth associated with QBO modulated tropical tropopause. Atmos Sci Lett 13:61–66

Can Z, Aslan Z and Oguz O (2004) One dimensional wavelet Real analysis of gravity waves. Arab J Sci Eng 2c:33–42

Can Z, Aslan Z, Oguz O, Siddiqi AH (2005) Wavelet transform of meteorological parameters and gravity waves. Ann. Geophysicca 23:650–663

Daubechies I (1992) Ten lectures on wavelets. SIAM, Philadelphia

Georgiou F, Kumar P (1994) Wavelet in geophysics. Academic, San Diago

Grossmann A, Morlet J (1984) Decomposition of Hardy functions into square integrable wavelets of constant shape. SIAM J Math Anal 15(4):723–736

Hong G, Zhang Y (2008) Wavelet-based image registration technique for high-resolution remote sensing images. Comput Geosci 34(12):1708–1720

Hsu K, Gupta HV, Sorooshian S (1995) Artificial neural network modeling of the rainfall runoff process. Water Resour Res 31:2517–2530

Hu ZZ, Nitta T (1996) Wavelet analysis of summer rainfall over North China & India and SOI Using 1891–1992 data. J Meteorol Soc Japan 74(6):833–844

Jang JSR (1993) ANFIS: adaptive-network-based fuzzy inference System. IEEE Trans Syst Man Cybern 23(3):665–685

Jang JSR, Sun CT, Mizutani E (1997) Neuro fuzzy and soft computing. Prentice Hall, Upper Saddle River

Kisi O, Parmar KS, Soni K, Vahdettin D (2017) Modeling of air pollutants using least square support vector regression, multivariate adaptive regression spline, and M5 model tree models. Air Quality Atmos Health. https://doi.org/10.1007/s11869-017-0477-9

Kumar P, Foufoula E (1997) Wavelet analysis for geophysical applications. Rev Geo Phys 33:385–412

Mallat S (1998) A wavelet tour of signal processing. Academic, New York

Manchanda P, Kumar J, Siddiqi AH (2007) Mathematical methods for modeling price fluctuations of financial time series. J Franklin Inst 344:613–636

Osowski S, Garanty K (2007) Forecasting of the daily meteorological pollution using wavelets and supportvectormachine. Eng Appl Art Intell 6:745–755

Parmar KS, Bhardwaj R (2012) Analysis of water parameters using Haar Wavelet (Level 3). Int J Curr Eng Tech 2(1):166–171

Parmar KS, Bhardwaj R (2013a) Wavelet and statistical analysis of river water quality Parameters. Appl Math Comput 219:10172–10182

Parmar KS, Bhardwaj (2013b) Analysis of water parameters using Daubechies Wavelet (Level 5) (Db5.). Am J Math Stat 2(3):57–63

Parmar KS, Bhardwaj R (2013c) Water quality index and fractal dimension analysis of water parameters. Int J Environ Sci Technol 10:151–164. https://doi.org/10.1007/s13762-012-0086-y

Parmar KS, Bhardwaj R (2014) Water quality management using statistical analysis and time-series prediction model. Appl Water Sci. https://doi.org/10.1007/s13201-014-0159-9

Parmar KS, Bhardwaj R (2015) Statistical, time series, and fractal analysis of full stretch of river Yamuna (India) for water quality management. Environ Sci Pollut Res 22:397–414

Pellegrini M, Sini F, Taramasso AC (2012) Wavelet-based automated localization and classification of peaks in streamflow data series. Comp Geosci 40:200–204

Quiroz R et al (2011) Improving daily rainfall estimation from NDVI using a wavelet transform. Environ Model Softw 26(2):201–209

Siddiqi AH (2004) Applied functional analysis. Maral Dekkar, New York

Siddiqi AH, Kovin G, Freeden W, Mosco U, Stephan S (2003) Theme issue on wavelet fractal in science & engineering. Arab J Sci Eng 28(1C):Part 1

Sifuzzaman M, Islam MR, Ali MZ (2009) Application of wavelet transform and its advantages compared to fourier transform. J Phys Sci 13:121–134

Soni K, Kapoor S, Parmar KS, Kaskaoutis DG (2014) Statistical analysis of aerosols over the Gangetic-Himalayan region using ARIMA model based on long-term MODIS observations. Atmos Res 149:174–192. https://doi.org/10.1016/j.atmosres.2014.05.025

Stanislaw O, Garanty K (2007) Forecasting of the daily meteorological pollution using wavelet and support vector machine. Eng Appl Artif Intell 20:745–755

Yang X, Jin W (2010) GIS-based spatial regression and prediction of water quality in river networks: a case study in Iowa. J Environ Manage 91:1943–1951

Yousefi S, Weinrich I, Reinarz D (2005) Wavelet based prediction of oil prices. Chaos Solitons Fractals 25(2):265–275